NEW FRONTIERS IN TECTONIC RESEARCH - AT THE MIDST OF PLATE CONVERGENCE

Edited by **Uri Schattner**

New Frontiers in Tectonic Research - At the Midst of Plate Convergence
Edited by Uri Schattner

Published by InTech
Janeza Trdine 9, 51000 Rijeka, Croatia

Publishing Process Manager Ana Pantar

Technical Editor Teodora Smiljanic

Cover Designer InTech Design Team

Additional hard copies can be obtained from orders@intechopen.com

New Frontiers in Tectonic Research - At the Midst of Plate Convergence,
Edited by Uri Schattner
p. cm.
ISBN 978-953-307-594-5

For EU product safety concerns:
IN TECH d.o.o., Prolaz Marije Krucifikse Kozulić 3, 51000 Rijeka, Croatia,
info@intechopen.com or visit our website at intechopen.com.

Meet the editor

Uri Schattner is a lecturer, co-founder and now Chair of the Dept. of Marine Geosciences, at the Charney School of Marine Sciences, University of Haifa, Israel. He has a BSc in both Geology and Biology from the Hebrew University (Jerusalem, Israel), MSc (applied geodesy) and PhD (tectonophysics) in Geophysics from Tel Aviv University (Israel), and two post-docs at the Geological Survey of Israel and the University of Leeds (UK). His research focuses on the tectonic development of the eastern Mediterranean and surrounding plates through geological history. Dr. Schattner specializes in marine geology and geophysics, mainly through seismic reflection perspective. Alongside, he teaches and tutors MSc and PhD students. He planned and participated in eight scientific cruises. He strongly believes that the joy of scientific discovery belongs to everyone, and to implement this view he spent two years as a desert field guide, making science accessible and joyful to hundreds of children.

Contents

Preface XI

Part 1 Margins of the Western Mediterranean 1

Chapter 1 **Tectonic Evolution of the Active "Chelif" Basin (Northern Algeria) from Paleomagnetic and Magnetic Fabric Investigations 3**
Derder M.E.M., Henry B., Amenna M., Bayou B., Maouche S., Besse J., Abtout A., Boukerbout H., Bessedik M., Bourouis S. and Ayache M.

Chapter 2 **Tunisian Transtensive Basins in Tethyan Geodynamic Context and Their Post-Tortonian Inversion 27**
Adel Rigane and Claude Gourmelen

Chapter 3 **The Chotts Fold Belt of Southern Tunisia, North African Margin: Structural Pattern, Evolution, and Regional Geodynamic Implications 49**
Taher Zouaghi, Rihab Guellala, Marzouk Lazzez, Mourad Bédir, Mohamed Ben Youssef, Mohamed Hédi Inoubli and Fouad Zargouni

Chapter 4 **Interplay between Tectonics and Mount Etna's Volcanism: Insights into the Geometry of the Plumbing System 73**
Domenico Patanè, Marco Aliotta, Andrea Cannata, Carmelo Cassisi, Mauro Coltelli, Giuseppe Di Grazia, Placido Montalto and Luciano Zuccarello

Chapter 5 **Transtensional Tectonics Played a Key Role During the Variscan Cycle in the Sardinia-Corsica Massif 105**
Nello Minzoni

Chapter 6 **Relationships between Lithospheric Flexure, Thrust Tectonics and Stratigraphic Sequences in Foreland Setting: the Southern Apennines Foreland Basin System, Italy 121**
Salvatore Critelli, Francesco Muto, Vincenzo Tripodi and Francesco Perri

Part 2 Around the Adriatic 171

Chapter 7 **Two Tectonic Geomorphology Studies on the Landscape and Drainage Network of Chain and Piedmont Areas of the Abruzzi Region (Central Apennines, Italy) 173**
Miccadei Enrico and Piacentini Tommaso

Chapter 8 **Neogene Tectonics in Croatian Part of the Pannonian Basin and Reflectance in Hydrocarbon Accumulations 215**
Tomislav Malvić and Josipa Velić

Part 3 Dynamics of the Eastern Mediterranean 239

Chapter 9 **Architecture of Kinematics and Deformation History of the Tertiary Supradetachment Thrace Basin: Rhodope Province (NE Greece) 241**
Adamantios Kilias, George Falalakis, Aristides Sfeikos, Eleftheria Papadimitriou, Agni Vamvaka and Chara Gkarlaouni

Chapter 10 **The North Anatolian Fault Zone: an Evaluation of Earthquake Hazard Parameters 269**
Yusuf Bayrak, Hakan Çınar and Erdem Bayrak

Chapter 11 **Tectonic Escape Mechanism in the Crustal Evolution of Eastern Anatolian Region (Turkey) 289**
Ömer Elitok and M. Nuri Dolmaz

Chapter 12 **Early-to-mid Pleistocene Tectonic Transition Across the Eastern Mediterranean Influences the Course of Human History 303**
Uri Schattner

Part 4 The Persian Gulf 323

Chapter 13 **The Qatar–South Fars Arch Development (Arabian Platform, Persian Gulf): Insights from Seismic Interpretation and Analogue Modelling 325**
C.R. Perotti, S. Carruba, M. Rinaldi, G. Bertozzi, L. Feltre and M. Rahimi

Preface

Nestled in the midst of Africa-Arabia plate convergence with Eurasia, the fate of the Mediterranean basin and the Persian Gulf is seemingly a known fact. Over millions of years through the Cenozoic era they progressively shrunk, leaving only a glimpse of the Tethys Ocean. Eventually, the basins will adhere to the Alpine-Himalaya orogen and dissipate. Tectonic processes transform this world scale ocean closure into an orogen similar to older belts such as the Caledonian, Appalachian and Hercynian. This book focuses on a unique stage in the ocean closure process, when significant convergence already induced major deformations, yet the inter-plate basins and margins still record the geological history.

Ocean closure involves a variety of parallel and converging tectonic processes that reshape the shrinking basins, their adjacent margins and the entire earth underneath. Following continental breakup, margin formation and sediment accumulation, tectonics normally relaxes and the margins become passive for millions of years. However when final convergence is at the gate, the passive days of the Mediterranean and Persian Gulf will soon be over. Studies presented in this book delve into the tectonics "At the Midst of Plate Convergence" to simplify the obscurity involved in progressive ocean closure.

Basins of the Mediterranean region and their associated arcs exhibit different evolutionary phases. Formation of the western Mediterranean and the Aegean as back arc basins are strongly associated with slab roll-back since the mid-Cenozoic. East of Sicily, the eastern Mediterranean, as well as the Adriatic and Black seas are older relicts of the Tethys Ocean. Along the entire convergence front, slab roll-back dominates the upper plate deformation by creating extended basins, tectonically thickened and in places volcanic arcs. Differences in age, crustal structure and tectonic history of the basins nested in the midst of convergence dictate a wide range of structural deformation styles. Therefore, this region serves as an exceptional natural laboratory for studying an assortment of evolutionary stages of complex convergence dynamics, including subduction, ophiolite emplacement, orogeny, tectonic faulting and tectonic geomorphology, back-arc and foreland basin evolution, and prevalent seismicity.

Parts of this book are presented from west to east, similar to the present growth in convergence rates between Africa-Arabia and Eurasia. In the **first part**, *Margins of the Western Mediterranean, El-Messaoud Derder et al.* (Ch. 1 on Fig. 1) present a model for Arabia-Eurasia convergence through the paleomagnetic record of the active "Chelif" basin of northern Algeria. Further east along the northern African margin, *Rigane and Gourmelen* (Ch. 2) explore the post-Tortonian transtensive inversion of the Tunisian basins, in Tethyan geodynamic context. The work of *Zouaghi et al.* (Ch. 3) focuses on the Chotts fold belt of southern Tunisia, unraveling its structural patterns based on the interpretation of seismic reflection land data. Crossing the Strait of Sicily, the work of *Patanè et al.* (Ch. 4) dives into the volcanic plumbing system geometry of Mount Etna to infer on regional tectonics. The following work of *Minzoni* (Ch. 5) shows that transtensional tectonics played a key role during the development of the Sardinia-Corsica massif, during the Variscan geodynamic cycle from the middle Ordovician to the Silurian. And, concluding this part, *Critelli et al.* (Ch. 6) highlight the southern Apennines foreland basin with new insights on the relationships between lithospheric flexure.

Part two, *Around the Adriatic,* focuses on the margins of the Apulian plate. *Miccadei and Piacentini* (Ch. 7) apply tectonic geomorphological methodologies to study the Abruzzi region of the central Apennines. Whereas, on the eastern Adriatic, *Malvić and Velić* (Ch. 8) investigate the Neogene tectonics of the western Pannonian Basin in view of hydrocarbon system development.

Part three, moves further east to explore the *Dynamics of the Eastern Mediterranean* through four studies. *Kilias et al.* (Ch. 9) looks into the architecture and kinematics of the Thrace basin at the northern Aegean region. Further east, *Bayrak et al.* (Ch. 10) analyze contemporary seismicity patterns along the North Anatolian Fault Zone, along the northern margin of the Anatolian plate. *Elitok and Dolmaz* (Ch. 11) examine the mechanism of tectonic escape through the crustal evolution of eastern Anatolian plate. *Schattner* (Ch. 12) ends this section with a synthesis on the pan-eastern Mediterranean early-to-mid Pleistocene tectonic transition, and suggests its influence on the course of human history.

In **part four**, which conclude this book, *The Persian Gulf* is at the center of the stage. *Perotti et al.* (Ch. 13) use seismic interpretation and analogue modeling to study the development of Qatar–south Fars arch.

Editor,
Uri Schattner
Chair and co-founder, Department of Marine Geosciences,
Charney School of Marine Sciences,
University of Haifa,
Israel

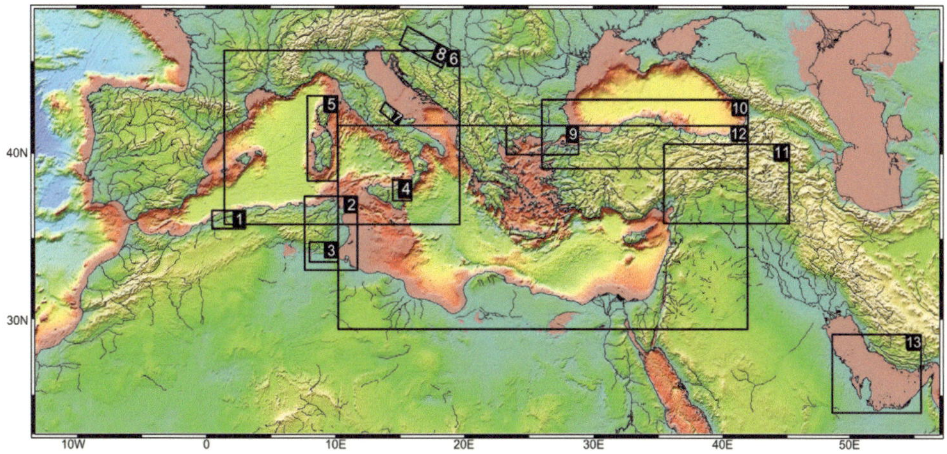

Fig. 1. Location of the studies presented in this volume marked across a relief map of Africa-Arabia plate convergence with Eurasia (base a map from Smith et al., 1997). Numbers correspond to the text and chapter order

References

Smith, W. H. F., and D. T. Sandwell, Global seafloor topography from satellite altimetry and ship depth soundings, *Science*, v. 277, p. 1957-1962, 26 Sept., 1997.

Part 1

Margins of the Western Mediterranean

1

Tectonic Evolution of the Active "Chelif" Basin (Northern Algeria) from Paleomagnetic and Magnetic Fabric Investigations

Derder M.E.M.[1], Henry B.[2], Amenna M.[1], Bayou B.[1], Maouche S.[1], Besse J.[2], Abtout A.[1], Boukerbout H.[1], Bessedik M.[3], Bourouis S.[1] and Ayache M.[1]

[1]C.R.A.A.G., BP 63, 16340 Bouzaréah, Alger,
[2]Paléomagnétisme, IPGP, PRES Sorbonne Paris Cité and CNRS,
[3]Laboratoire de Paléontologie Stratigraphie et Paléoenvironnement,
Université d'Oran & Université Hassiba Benbouali Chlef,
[1,3]Algeria
[2]France

1. Introduction

The present-day crustal deformation in the North African area is mainly driven by the NW-SE to NNW-SSE convergence (4–5 mm/yr) between the African and Eurasian plates (Mckenzie, 1972; Nocquet & Calais, 2004; Serpelloni et al., 2007). This convergence is accommodated over a wide deformation zone (the Tellian Atlas) implying the existence of significant seismic activity. This domain represents the southern part of the Alpine ranges at the boundary between the African and Eurasian plates. It is characterized by E-W to NE-SW trending folds and thrust belt and composed by an "external zone" with Mesozoic to Tertiary rocks and an "internal zone" consisting of a thrust stack of metamorphic complexes and Mesozoic calcareous and flysch. Overlying these previous series, Neogene to Quaternary sediments fill the intramountainous basins, limited by E-W and NE–SW faults as evidenced in the "Chelif" basin (Chiarabba et al, 1996; Meghraoui, 1982; Meghraoui, 1988; Thomas, 1985). Few studies were dedicated to these faults (Aoudia & Meghraoui, 1995; Chiabbara et al., 1996; Meghraoui, 1988; Phillip & Meghraoui, 1983), consequently the timing and amount of displacement of these faults, as well as their relationship with the Neogene basin development, are still a matter of debate (Thomas, 1985; Neurdin-Trescartes, 1992; Belkebir et al., 1996; Bessedik et al., 2002). The seismic activity is mainly characterized by moderate to destructive magnitude events as it was shown by the recent 21 May 2003 Zemmouri earthquake (Mw=6.8) (Meghraoui et al., 2004) and the 10 October 1980 El Asnam (presently Chlef) earthquake (Mw=7.3) (Ouyed et al.,1981). These two earthquakes are the strongest recent ones that occurred in this part of Western Mediterranean area. According to the Algerian earthquake catalogue (Benouar, 1994) the earthquake activity is mainly concentrated in the intramountainous seismogenic basins which belongs to the Tellian Atlas. The convergence movement between Africa and Eurasia plates involves a tectonic transpression with N-S to NNW-SSE direction, which initiate faulting activity along "en echelon" NE-SW trending folds and associated reverse faults. This "en echelon" system should be likely due to the presence of deep E–W strike-slip faults (Meghraoui et al., 1996 and references therein). In the Chellif Mio-Plio-Quaternary intramountainous seismogenic basin, the active tectonic witnesses are highlighted by recent Quaternary tectonic structures activated during the earthquakes: e.g. "El-Asnam" reverse fault (2 m coseismic average deformation during the 1980 Mw 7.3 El-Asnam earthquake) and "Tenes Abou El Hassan" fault (1 m average coseismic deformation during the 1922 Ms 6 earthquake - Philippe & Meghraoui, 1983; Aoudia & Meghraoui, 1995).These active NE-SW faults are associated with asymmetric folds. These different tectonic structures define several NE-SW oriented tectonics blocks (Morel & Meghraoui, 1996). As suggested by Meghraoui et al., (1996), the active deformation could be explained by a simple kinematic model of block rotation related to the transpression with NNW-SSE direction of plates convergence.

This new study aims to quantify the tectonic activity of the Chelif basin. This work will particularly focus on rotations in this active basin (as suggested by preliminary paleomagnetic results - Aïfa et al., 1992) in order to improve a block rotation model based on the oblique Africa-Europe convergence (Meghraoui et al., 1996). Paleomagnetic and Anisotropy of Magnetic Susceptibility (AMS) analyses were thus performed (using larger sampling area 66 sites) in the Miocene and Pliocene sedimentary rocks cropping out in this basin

2. Geological setting

The intramountainous Chelif basin is located in Northern Algeria within the Tellian atlas (Fig. 1). It is a "post-thrust" basin elongated in an E-W direction. It is bordered to the north by the "Dahra" and "Bou-Maad" hills, mainly constituted of allochtonous Cretaceous and Upper Oligocene allochtonous formations. To the south, it is limited by the "Ouarsenis" mountains which are formed by Lower Jurassic and Cretaceous allochtonous formations. The Chelif basin constitutes one of the Neogene "post-thrust" basins located in the northern Algeria (e.g. the "Constantine", "Soumam" and "Mitidja" basins), resulting from the distention, which affected this area during the Lower-Middle Miocene (Thomas, 1985; Neurdin-Trescartes, 1992; Belkebir et al., 1996).

From stratigraphical point of view, it is constituted by Mio-Plio-Quaternary deposits overlaying a basement formed by thrust sheets with Mesozoic series (Fig. 1) (Anderson, 1936; Perrodon, 1957; Belkebir et al., 1996). The bottom of the basin series is represented by the blue marls and gray calcareous sandstones of Uppermost Burdigalian and Langhian - Serravallian ages (Belkebir et al., 1996; Bessedik et al., 2002). Overlying the latter, the Tortonian formation is constituted by limestones, marls and sandstones. The Messinian ones are formed by blues marls alternating with diatomites, crowned by (*Lithothamnium* and/or coral) limestones on the platforms, and finally massive gypsum or interbedded stratifications within marls and sandstones intercalations. The Pliocene formation is mainly represented by marls, sandstone, calcareous sandstone, calcarenite, and conglomerate. Friable red sandstones and sandy to silty deposits constitute the Quaternary formation. The history of the Chelif basin began by a subsidence during the Late Burdigalian (Meghraoui, 1982; Bessedik et al., 2002) and was stopped at the end of the Langhian detritic sedimentation located in the northern part of the Ouarsenis relief which reveals the absence of the Serravallian sedimentation and testifies the end of the Ouarsenis thrust establishment. The sedimentation continued during the Lower Serravallian in the Northern part of the Chelif basin (Bessedik et al., 2002), followed by a Lower Tortonian extensional phase with

formation of large grabens. After the deposit of the Lower Pliocene, an important NNE-SSW compressive phase deformed the Tortonian and Messinian formations (Meghraoui et al., 1986). A second important compression phase occurred during the Quaternary, with a NNW-SSE to NW-SE shortening direction. It affected the Quaternary deposits and accentuated the deformation of the older series (Meghraoui, 1982).
The neotectonic studies (Philippe & Meghraoui, 1983, Meghraoui et al., 1986) showed that the present deformation in the Chelif basin is mainly related to a transpression with N-S to NNW-SSE shortening direction, which is expressed by active tectonic responsible of the earthquake activity. Indeed, NE-SW trending folds and NE-SW active sinistral transpressive faults were activated during the 1954 and 1980 destructive earthquakes (Bezzeghoud et al., 1995; Ouyed et al., 1981). These reverse faults and related folding are disposed on right lateral echelon and should be coupled with NW-SE to E-W trending strike-slip deep active faults.

3. Sampling and analysis procedure

Different formations outcropping in this basin have been extensively (879 oriented cores distributed over 66 sites) sampled (Fig. 1). The sampling was made in the gray calcareous sandstone of Lower Miocene age (65 samples from 4 sites), the Tortonian gray limestone (147 samples from 10 sites), the dark marls, the fine red limestone and gray marls of Messinian age (191 samples from 12 sites), and the Pliocene marls, sandstone, calcareous sandstone and calcarenite (478 samples from 40 sites).
One to three specimens were cut from each core in order to have additional specimens for paleomagnetic pilot studies, rock magnetic analysis and magnetic fabric measurements. Prior to any demagnetization analysis, the specimens were stored in a zero field for at least one month, in order to reduce a possible viscous magnetization component. The Natural Remanent Magnetization (NRM) of the specimens was measured using a JR5 spinner magnetometer (AGICO, Brno). Several pilot specimens from each site were subjected to a stepwise alternating field and thermal demagnetizations in order to characterize their magnetic behavior.
The results of the pilot study evidenced that significant part of the samples have very weak magnetization, showing moreover very unstable remanent magnetization behavior during the thermal demagnetization process. The corresponding sites have been then discarded.
For the other samples, the obtained results showed that the AF procedure did not allow complete demagnetization. Thermal demagnetization was then applied to the remaining specimens. It was performed using a cylindrical furnace with a fast heating level while forced air enhanced the cooling. In order to correctly isolate and identify the magnetization components, numerous steps were performed, with increments ranging from 70°C at the lowest temperatures to 10°C at the highest one, up to 670°C maximum temperature.
Finally, 350 samples were successfully demagnetized in this study. The demagnetization data analysis was carried out using classical methods: the direction of the magnetization components was computed using principal component analysis (Kirschvink, 1980) and plotted on orthogonal vector plots (As & Zijderveld 1958; Zijderveld 1967); the remaining vectors and vectorial differences of the magnetization were plotted on equal-area projections. The mean direction of the different components was computed using (Fisher, 1953) statistics. Fold tests of McElhinny (1964) and reversal test of McFadden & McElhinny (1990) were performed when applicable.

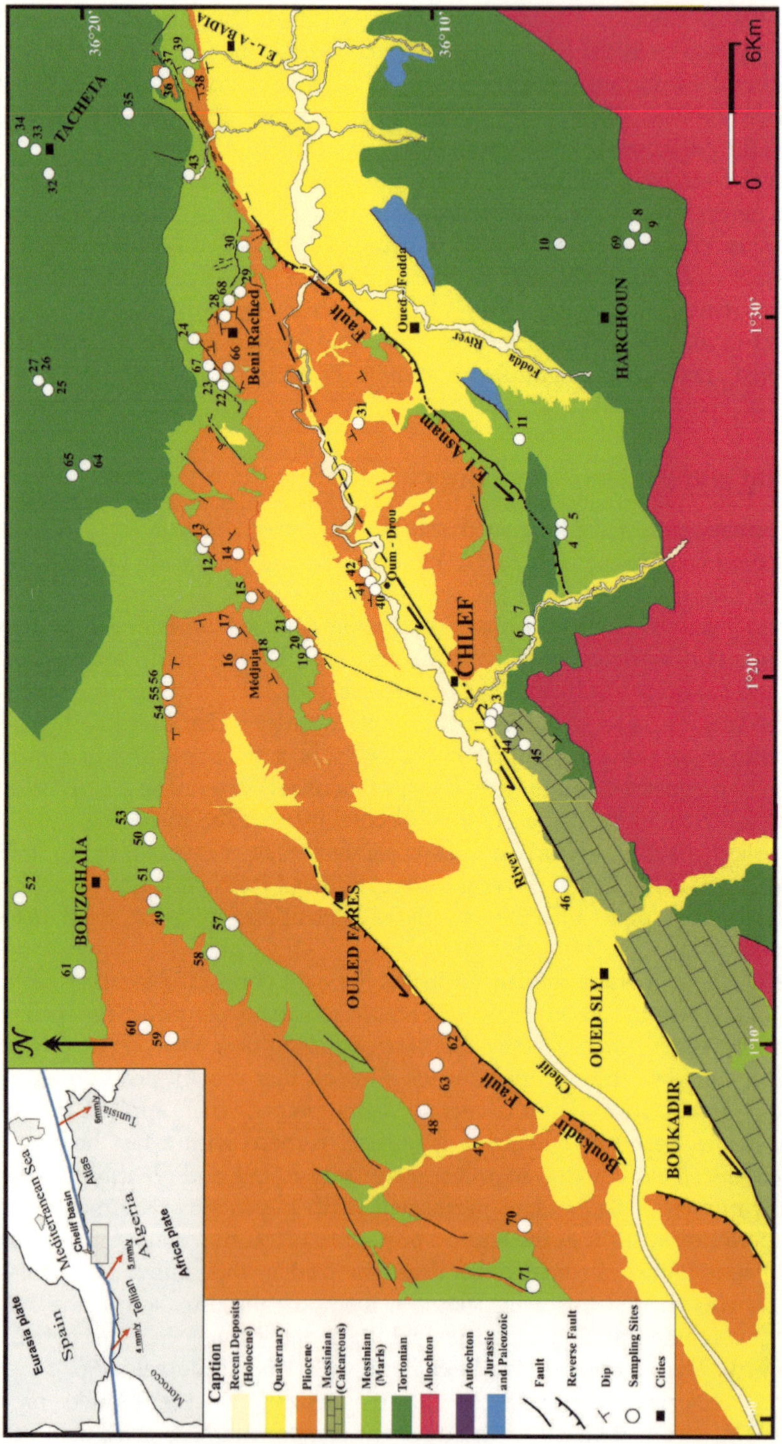

Fig. 1. Geological map of the "Chelif" basin area (after Anderson, 1936; Meghraoui et al., 1986; Meghraoui, 1988) modified, and location of the sampling sites

4. Rock magnetism

The NRM intensity measured on the samples from the different geological formations is often very weak. It varies from 0.01 to 9 10^{-6} A/m. The maximum unblocking temperature (Tunb) spectra seems very large, from 670 to 250°C. The maximum Tunb changes within each formation; indeed, for the Lower Miocene samples, it varies from 350 to 670°C, (mostly between 480 and 600°C). For the Tortonian formation, it is between 260 and 630°C, mainly below 440°C, while for the Messinian samples, the window is from 350 to 600°C (mostly between 430 and 510°C). Ending, for the Pliocene formation, the maximum Tunb spectra is between 250 and 635°C, but mainly with values below 480°C. Notice that even, if some maximum Tunb values greater than 580°C have been obtained (evidencing the presence of hematite), the majority of the Tunb values are lower than 580°C.

Thermomagnetic curves (low-field susceptibility as a function of temperature) were determined using CS3 equipment - Kappabridge KLY3 (AGICO, Brno, Czech Republic) on crushed representative samples. To point possible mineralogical alteration due to heating, heating and cooling curves have been compared, even during partial cooling loops during the initial heating. Few samples (all from the Messinian formation) yield reversible curves (Fig. 2 – sample 93 of Messinian age) showing the absence of chemical alteration of the ferromagnetic minerals during heating. Some others samples show weak alteration (Fig. 2- sample 315 of Pliocene age) indicated by slightly different heating and cooling curves. But most samples present important mineralogical alteration (shown by a strong magnetic susceptibility irreversible increase which occurs during the heating at about 400-500°C (e.g. Fig. 2 - sample 469 from Tortonian formation). The main reversible variation of the susceptibility up to 580°C corresponds to the magnetite Curie temperature (Fig. 2; samples 93, 315 and 469). However, part of the evidenced magnetite was clearly formed during mineralogical alteration in samples with no reversible heating and cooling curves. In some samples (on Fig. 2 - e.g. 315 of Pliocene age), small variation of the magnetic susceptibility at temperatures higher than 600°C could correspond to the presence of hematite.

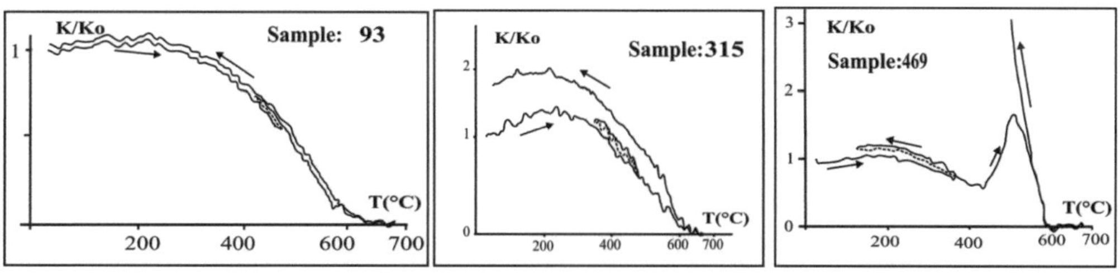

Fig. 2. Thermomagnetic curves: Normalized low field susceptibility (K/Ko) as a function of the temperature T (in °C) for samples 93, 469 and 315

Hysteresis loops were measured on small samples (about 3 cm^3) using a translation inductometer within an electromagnet. The loops are almost linear, pointing out a largely dominant effect of the paramagnetic minerals. Because of the weak magnetization, only loops corrected for the paramagnetism yield usable indications about the magnetic mineralogy (Fig. 3). The total saturation of the magnetization is not reached despite application of magnetic field of 0.8T intensity, confirming the presence of hematite (Fig. 3 - samples 469, 315, 93). A partial saturation, related to a low coercive mineral, explains that the coercive force varies mainly from 114 mT (Tortonian samples) to 44 mT for (samples from the Messinian formation); few weaker values of 27 mT have been also obtained for the Pliocene samples.

As a conclusion, the magnetic mineralogy includes mainly magnetite (relatively low coercivity), but hematite (high coercivity) is also often present, mostly in weak amount.

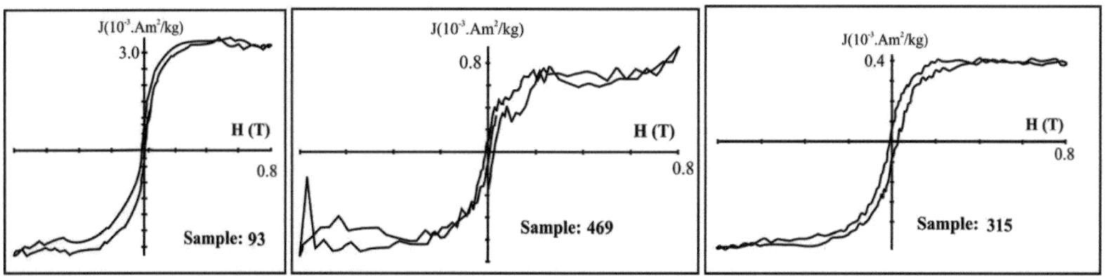

Fig. 3. Hysteresis loops after correction for paramagnetism for samples samples 93, 469 and 315 - J in 10^{-3} $A.m^2/kg$ and H in T

5. Paleomagnetic results

The weakness of the magnetization intensity and the frequent magnetization instability observed during the thermal treatment due to mineral alteration did not allow obtaining a reliable paleomagnetic datum for part of the samples. A dramatic decrease of the magnetization intensity is observed at about 80-100°C, during the thermal demagnetization process of most Pliocene and Messinian samples. In some other samples, mineralogical alteration, shown by a strong increase of the magnetization intensity, occurs at 200-300°C. The corresponding Zijderveld diagrams present then an erratic behavior and no magnetic component can be isolated. However, few Pliocene and Messinian samples show, after elimination of a viscous component, a linear segment in these diagrams defining a Characteristic Remanent Magnetization (ChRM) carried by magnetite (illustrated by maximum Tunb about 500°C - Fig. 4a, samples 688 and 387 of Pliocene age and Fig. 4b, samples 11 and 25 of Messinian age). Similarly, in some samples of Messinian age, after an evolution of the magnetization direction along great circles (remagnetization circles) for the lowest temperatures, a stable direction is reached at the highest temperatures defining a ChRM carried by hematite (maximum Tunb reaches ~ 600°C - e.g. Fig. 4c, sample 117). For most Tortonian samples, after an evolution along remagnetization circles at the lowest temperatures, either the magnetization intensity became too weak or mineralogical alteration occurs, and no magnetic component can be isolated. However, few Tortonian samples show, after elimination of a viscous component, a linear segment defining a ChRM carried by magnetite (maximum Tunb ~ 500°C - Fig. 4d, sample 073). In the Lower Miocene samples, an evolution along remagnetization circles occurs at the lowest temperatures (until ~ 250°C). At highest temperatures, two cases are observed: either a mineralogical alteration occurs shown by erratic behavior on the Zijderveld diagrams, or a ChRM, carried by magnetite, can be defined (e.g. sample 154, Fig. 4e), or hematite (e.g. sample 125, Fig. 4f - maximum Tunb ~ 650°C). ChRM, with normal or reversed polarity, has been isolated in the different sampled formations and are distributed in several sites (Fig. 5). Its direction varies, depending on the sites location and on the studied geological formations (Fig. 5a, b, c, and d).

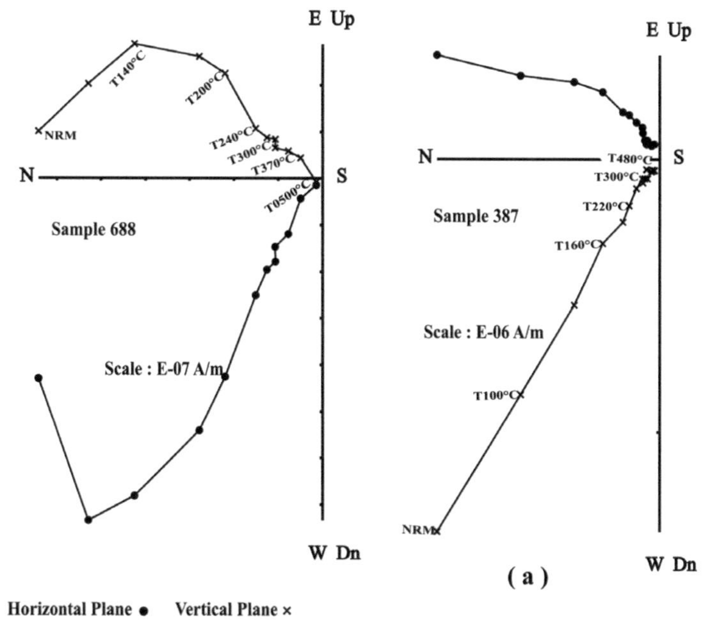

Fig. 4a. Orthogonal vector plots (filled circles: horizontal plane, crosses: vertical plane), in geographic coordinates for Pliocene samples 387 and 688

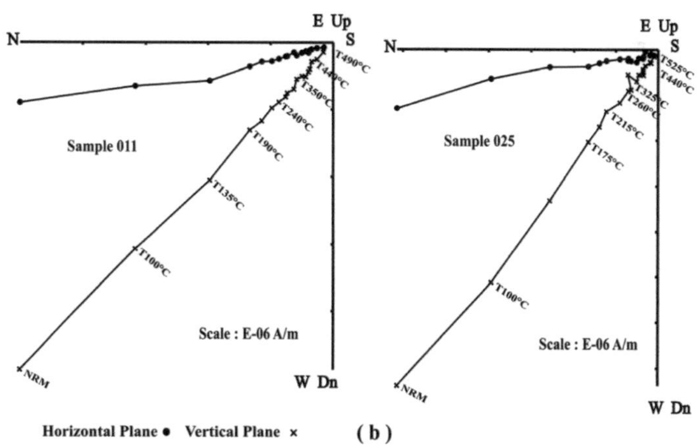

Fig. 4b. Messinian samples 11 and 25. See caption of fig. 4a

In order to determine the acquisition age of this magnetization, paleomagnetic tests were considered when possible for each site. The conglomerate test was possible for two sites: site 8 (Lower Miocene), and site 32 (Tortonian). Within the conglomerate, the ChRM directions from matrix samples are relatively grouped (Fig 5a - site 8 and Fig 5b - site 32, matrix) while ChRMs with random orientation have been obtained for pebbles samples. This conglomerate test provides the evidence that the matrix ChRM (with reversed polarity - site

8 - and with normal polarity - site 32) is the primary magnetization. The fold test can be applied also only in few sites. Though this fold test never gives statistically significant (McElhinny, 1964) result here, the fact that the bedding correction increases the scattering in some sites (Fig. 5d – station k, sites 47-48-62-63) argues rather in favor of a secondary character of the magnetization for these sites. This secondary character is reinforced by the results obtained by an attempt of magnetostratigraphy study performed in the "Djebel Menni" area (Western Chelif basin). This study has been carried out on 47 samples regularly distributed along a continuous section of well known stratigraphical ages (Belkebir et al., 2008) corresponding to a period including magnetic Earth field reversals (Langhian, Serravallian and Tortonian). The only normal polarity obtained for the ChRM (Fig. 6) indicates that these levels were affected by a remagnetization.

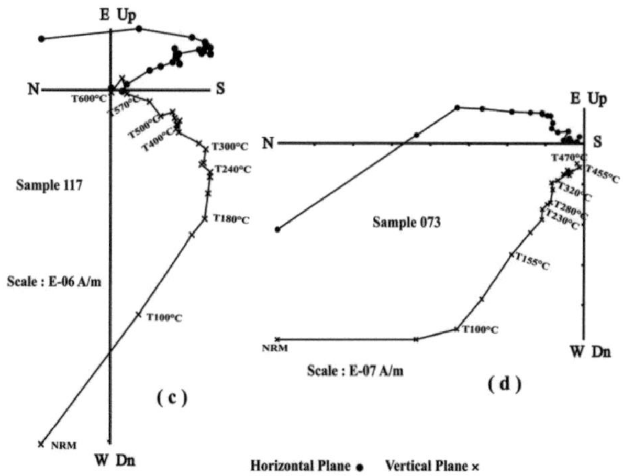

Fig. 4c and 4d. c) Messinian sample 117; d) Tortonian sample 73. See caption of fig. 4a

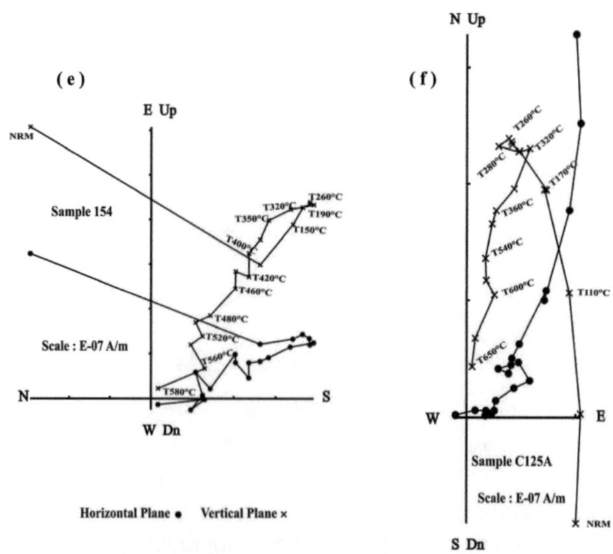

Fig. 4e and 4f. Lower Miocene sample 154, and 125. See caption of fig. 4a

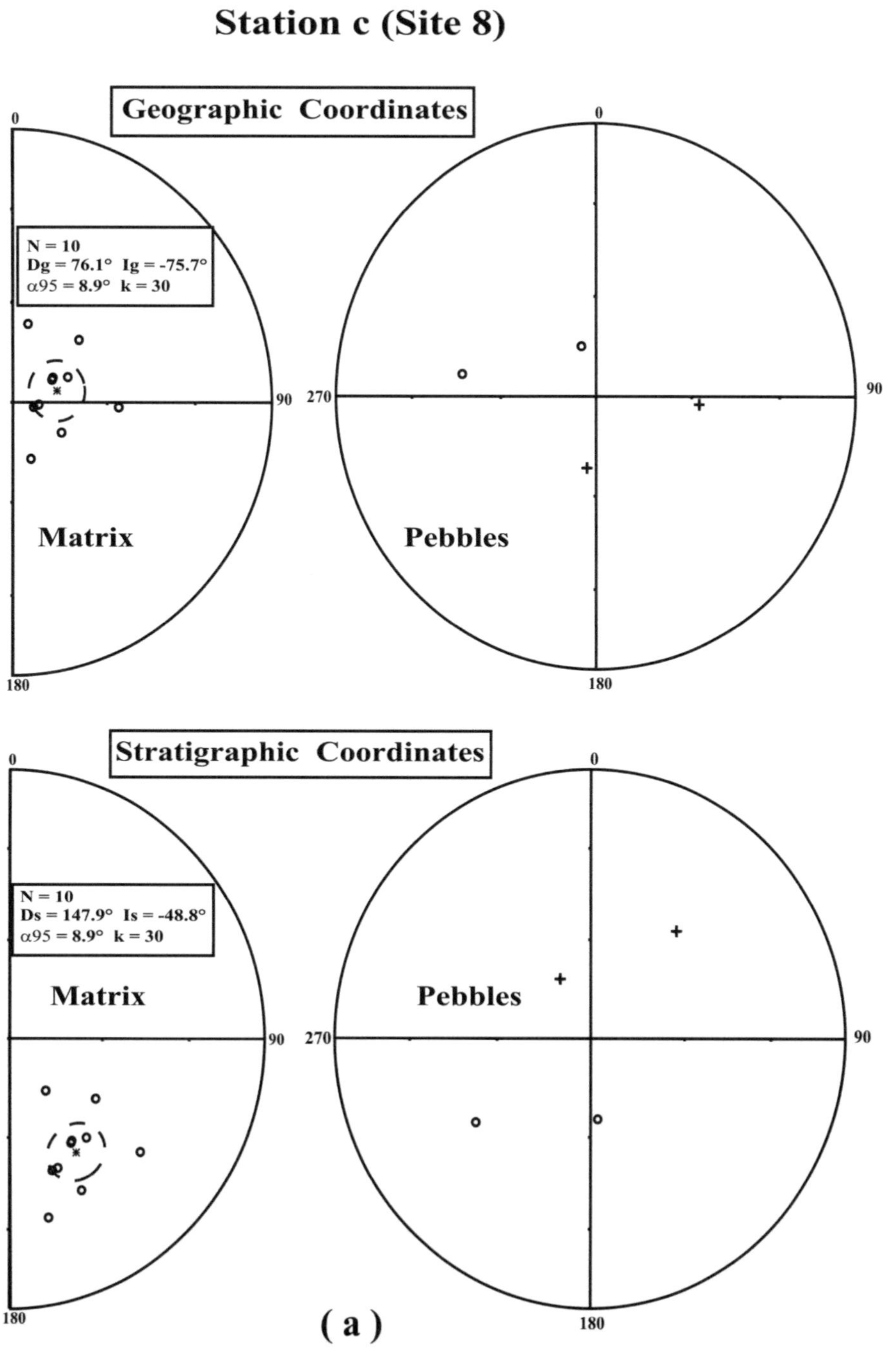

Fig. 5a. Mean paleomagnetic directions and their associated 95% confidence cone, before and after bedding correction for station c (site 8), (stereographic projection - open symbols and dotted line (confidence cone) for the upper hemisphere; full symbols and continuous line (confidence cone) for the lower hemisphere)

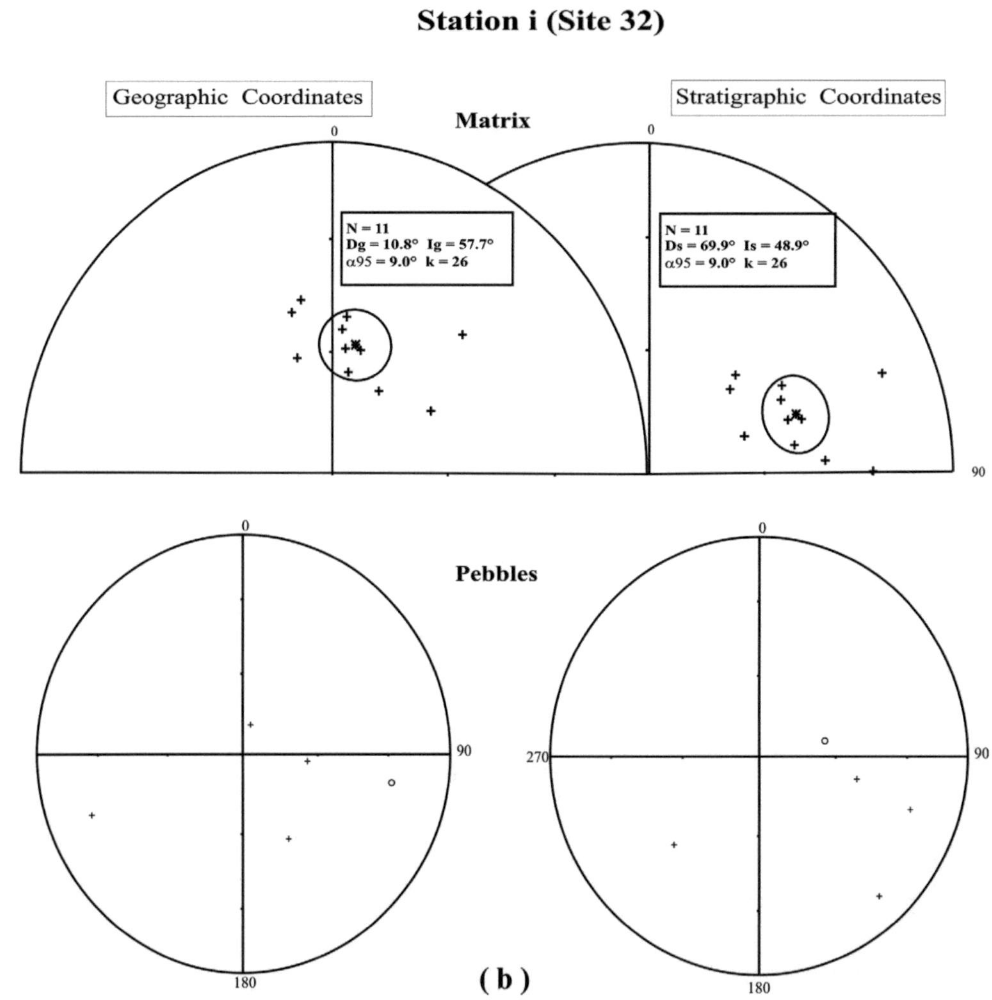

Fig. 5b. Station i (sites 32). See caption of fig. 5a

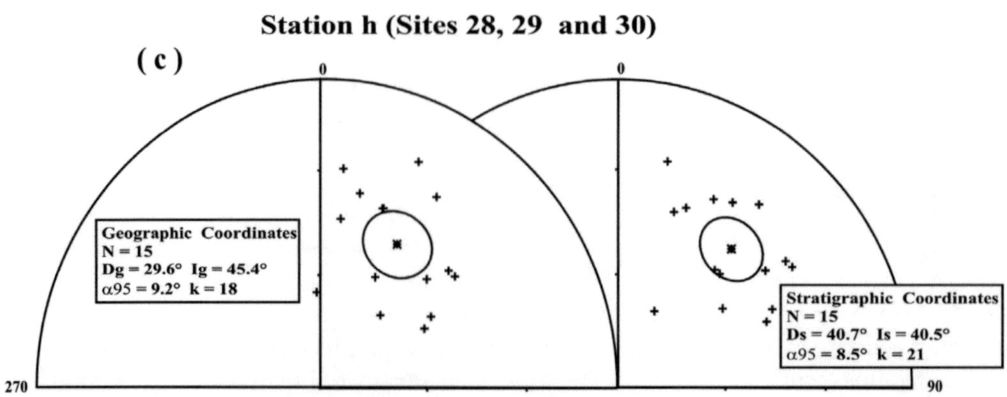

Fig. 5c. Station h (sites 28-29-30). See caption of fig. 5a

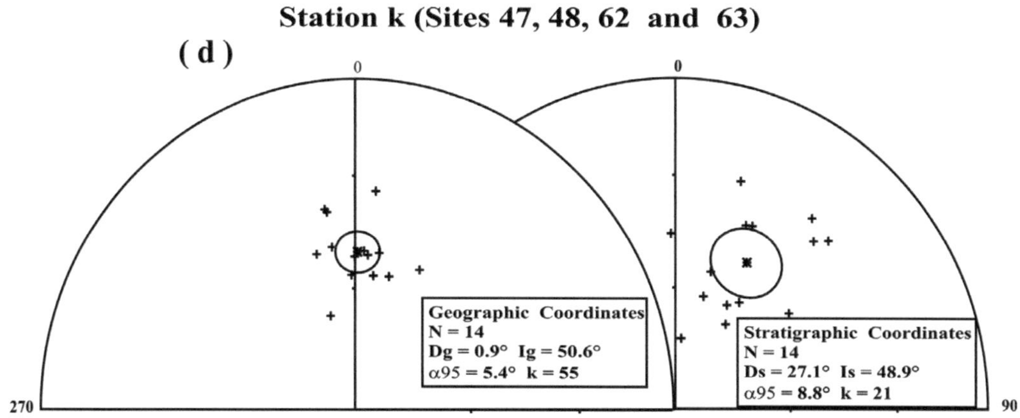

Fig. 5d. Station k (sites 47, 48, 62, 63). See caption of fig. 5a

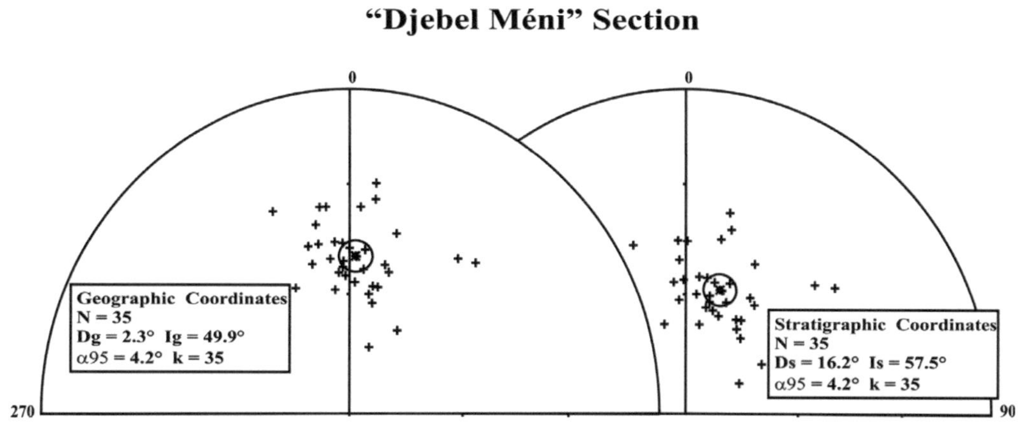

Fig. 6. ChRM directions determined in the "Djebel Meni" section. See caption of fig. 5a

6. Magnetic fabric results

Previous studies of depositional AMS in sediments have shown that the shape and orientation of the magnetic fabric can be related to different sedimentological processes during the deposition of particles from suspension in water (Ellwood, 1980; Rees, 1965; Rees et al., 1968; Rees & Woodall, 1975; Weiler, 2007). The initial sedimentary fabrics are generally determined by gravity and currents. Typical primary depositional fabric tends to be characterized by vertical or subvertical minimum axis. In the second case, the deviation of this axis (after bedding correction) from the vertical is related to the effects of currents (imbrication); the magnetic foliation generally slightly dips toward the current sense. The maximum axis can be parallel or perpendicular to the current direction, according to the shape of the particles and the velocity of the flow. The best approach to determine the paleocurrents direction is therefore to look for the deviation sense of the minimum axis relatively to the vertical in stratigraphical coordinates.

This initial sedimentary fabric may have been modified if the rock was affected by a deformation. During a progressive deformation, first the maximum axis becomes perpendicular to the shortening direction within the stratification plane (intersection lineation) and later the minimum axis moves to a direction parallel to this shortening direction (e.g. Graham, 1978; Weiler, 2007). For a stronger deformation, the orientation of all the principal susceptibility axes becomes similar to that of the corresponding strain axes.

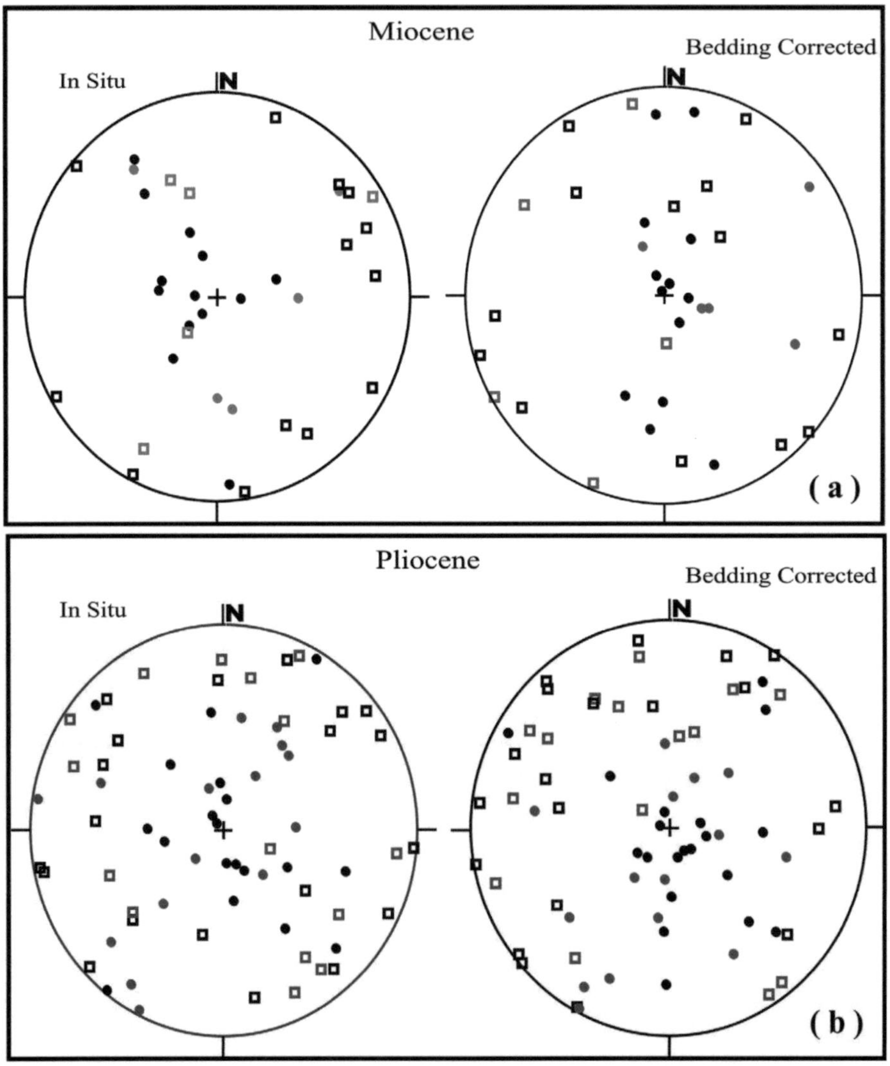

Fig. 7. Principal mean maximum (squares) and minimum (circles) susceptibility axes before and after bedding correction obtained in (a) the Lower Miocene, Tortonian and Messinian and (b) the Pliocene sites (stereographic projection in the lower hemisphere). Gray symbols (associated confidence zone at 95% of 20° or larger than 20°), black symbols (associated confidence zone at 95% lower than 20°)

The AMS in low field, measured using a KLY3 Kappabridge (AGICO, Brno), yields the principal magnetic susceptibility axes: maximum K_1 (magnetic lineation), intermediate K_2 and minimum K_3 (pole of the magnetic foliation). The Jelinek (1981) intensity P' and shape T

(possibly varying from +1 for uniaxial oblate fabric to -1 for uniaxial prolate fabric) parameters were used to describe the magnetic fabric. The data for a group of samples were analyzed using normalized tensor variability (Hext, 1963; Jelinek, 1978). AMS was measured on 344 specimens. For the Lower Miocene, Tortonian and Messinian sites, the mean susceptibility varies between 29 and 641 in 10^{-6} SI (mean 203.10^{-6} SI). The uncertainty (confidence zone at 95%) related to measurement is moderate (radius of less than 20° for the axes orientation) for about 75% of the samples. The results from the different samples of a same site are coherent in most sites and the principal axes mean direction per site is significantly defined (confidence zone at 95% of the mean axes lower than 20°) for 14 of the 18 sites. The minimum axes are scattered around the stratification pole and the maximum axes around the stratification plane (Fig. 7a). For the mean tensor fabric per site (Fig. 8), the anisotropy degree is weak (P' parameter between 1.001 and 1.036 with a mean value of 1.012) and the shape of the susceptibility ellipsoid is dominantly oblate (T between -0.47 and 0.91 with a mean value of 0.40). The two sites with a prolate mean fabric correspond to sites with incoherent fabric of their different samples and this prolate shape could be an artifact.
For the Pliocene sites, the mean susceptibility is mostly low (between 0.06 and 1481 in 10^{-6} SI with a mean of 128.10^{-6} SI). The uncertainty (confidence zone at 95%) related to measurement is moderate (radius of less than 20° for the axes orientation) for about 70% of the samples, but the results in the different samples of a same site are less coherent than for the Lower Miocene, Tortonian and Messinian sites, and the principal axes mean direction per site is precisely defined for only 13 of the 35 Pliocene sites. For the 22 other Pliocene sites, the confidence zone at 95% of the axes are larger than 20° and the mean results have to be considered with care. In the Pliocene sites, the minimum axes are also scattered around the stratification pole and the maximum axes around the stratification plane (Fig. 7b). For the mean tensor fabric per site (Fig. 8), the anisotropy degree is weak (P' parameter between 1.002 and 1.048 with a mean value of 1.013) and the shape of the susceptibility ellipsoid is dominantly oblate (T between -0.53 and 0.98 with a mean value of 0.20).
Though the weak anisotropy and the low susceptibility of the studied rocks, reliable magnetic fabric has been obtained for half of the sites. For the other sites, AMS results have to be considered with more care. However these sites give mostly relatively similar axes orientation as their neighboring reliable sites. Their results were therefore not discarded but will be presented with a gray symbol on the figures of the discussion to underline the higher uncertainty of these data.

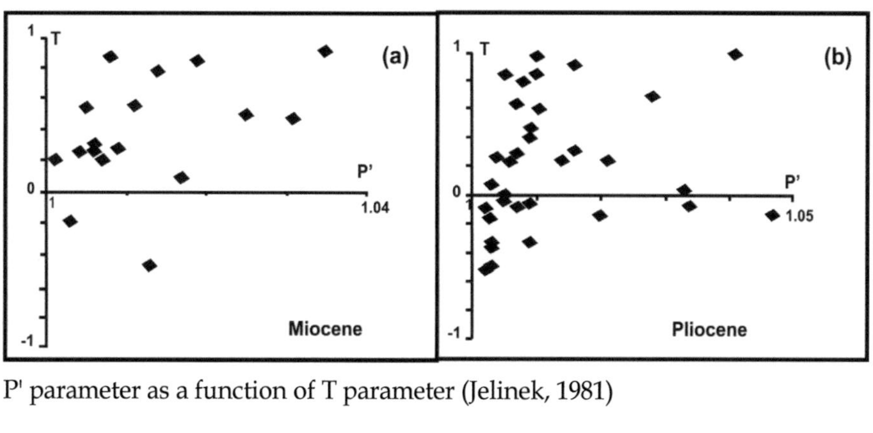

Fig. 8. P' parameter as a function of T parameter (Jelinek, 1981)

7. Discussion

7.1 Paleomagnetic data

The lack of well-established age of magnetization acquisition for some sites makes the interpretation of the paleomagnetic data very difficult. Aïfa et al. (1992) obtained negative fold test in the area of Beni Rached and El Abadia and therefore concluded for a postfolding remagnetization. In the present work, we came to the same conclusion with the section of well-known Langhian to Tortonian age (Belkebir et al., 2008; Bessedik et al., 2002) in the "Djebel Meni" area (in the western part of the Chelif basin) because of the obtained uniform normal polarity. However, in the sites 44 to 46, both normal and reversed polarities have been obtained. Moreover, in two similar conglomeratic facies sites (8 and 32), the conglomerate test indicates a primary magnetization. We therefore tried to obtain more information about the magnetization age in the different studied sites.

Because of the low proportion of reliable paleomagnetic results obtained in most studied sites, when possible we grouped as "stations" the results from several neighboring sites corresponding to formations of the same age. At the total, 14 mean paleomagnetic data (stations) were retained (Table 1).

In three neighboring sites, normal and reversed polarities have been obtained (in equal proportion in the site 44, mainly normal in the site 45 and mainly reversed in the site 46). An indeterminate (γ_c= 24.8° and 23.0° after and before bedding correction, respectively) reversal test (McFadden & McElhinny, 1990) has been obtained for this group of sites (station ***j***, Fig. 10) though neighboring mean normal and reversed directions (γ= 6.3° and 7.6° after and before bedding correction, respectively). This is due to the very high scattering of the reversed directions and only the normal polarity directions will be thus considered for this station *j* (Table 1). On the other hand, the obtained polarity in the sites 44 and 46 is not homogeneous in a same bed. This unhomogeneity shows that at least one of the polarities is related to a remagnetization. This can be due only to unhomogeneity of the rock characteristics that is likely of sedimentological (e.g. local variation of permeability) or brittle tectonic origin. The remagnetization process in such a basin cannot be thermal and the obtained Tunb during thermal demagnetization also exclude a viscous magnetic overprint. It is therefore chemical, probably related to fluids circulation (Henry et al., 2004a and references therein).

The station ***k*** is the only one giving a statistically significant fold test. The progressive unfolding (Fig. 9) clearly indicates a postfolding remagnetization. This fold test is significant at 95% (ratio of the Fisher's – 1953 - precision parameter k after/before bedding correction of 2.56 for a critical value of 1.84 – McElhinny, 1964). The conglomerate test shows that the magnetization is primary in the stations ***c*** and ***i*** (Fig. 5a and 5b). 3 stations (***e***, ***k*** and ***m***) show before bedding correction a magnetization direction coinciding with the recent magnetic field (Table 1), and their result was therefore interpreted as a postfolding remagnetization. For the station ***k***, the secondary character of the magnetization was already shown by the fold test. Finally, because of the aberrant value of magnetization inclination in some sites during part of a progressive unfolding, we compared (Fig. 11) the obtained inclination of the paleomagnetic direction before and after bedding correction with the expected inclination of the Earth magnetic field for the Upper Neogene in the Chelif basin (of the order of 50° - Besse and Courtillot, 2002). That indicates a postfolding remagnetization for the paleomagnetic directions of the stations ***a*** (inclination of 45° and of 29° before - *Ig* - and after - *Is*- dip correction, respectively), ***d*** (*Ig*=53° and *Is*=30°) and ***h*** (*Ig*=45° and *Is*=40°), but a prefolding magnetization for the stations ***b*** (*Ig*=68° and *Is*=55°), ***g*** (*Ig*=62° and *Is*=56°), ***j***

(*Ig*=61° and *Is*=56°) and ***l*** (*Ig*=60° and *Is*=50°). Finally only the two stations ***f*** and ***n*** have paleomagnetic direction of undetermined age. The retained paleomagnetic declination for each station is presented on a map (Fig. 12), which shows different dominant orientations.

- The orientation **A** corresponds to 6 declinations indicating strong clockwise rotation. It is well-represented in a large northeastern part of the studied area. Four of these declinations (stations ***b***, ***g***, ***i*** and ***j***) represent a prefolding magnetization (i.e. very likely a primary magnetization because of the synsedimentary age of the basin structuration – see magnetic fabric data section) and one is postfolding (***h***). It is important to notice that, for the other station (***f***) of this group **A**, with undetermined age of magnetization, the rotation is also clockwise, whatever the assumed magnetization age. The only difference is the magnitude of the rotation, which is a little bit lower for a postfolding remagnetization hypothesis. A seventh declination (station ***l***), again corresponding to a primary magnetization, indicates clockwise rotation of smaller magnitude and could be associated with the orientation **A**. The results obtained by (Aïfa et al., 1992) in 3 sites at Beni Rached area and 2 sites at El Abadia area also belong to the orientation **A**. They correspond to magnetic overprint and evidenced a clockwise rotation of 10 to 25° (Fig. 12).
- The orientation **B** on the contrary indicates counterclockwise rotation. It has been found in 3 stations (***a***, ***c*** and ***d***), all corresponding to area close to the main southeastern border-faults of the basin pointed out by the gravimetric data (Abtout et al., 2009; Idres et al., 1996).
- No significant rotation is associated with the declination **C** obtained in 3 stations (***e***, ***k*** and ***m***). However, this orientation also corresponds to postfolding remagnetization and assumption of a pre-remagnetization rotation cannot be excluded. These data therefore do not yield to usable information about the structural evolution of the basin.
- The declination **D** corresponds to a single station (***n***), located close to a secondary fault (Fig. 1). It highlights a very strong rotation. The fact that this datum is well-defined (Table 1) and obtained from 11 different specimens distributed on a large area (more than 100 m^2) shows that this declination is not an artifact. This rotation is attributed to a very local deformation associated to this secondary fault, and thus cannot be used as information for improving the structural evolution model of the basin.

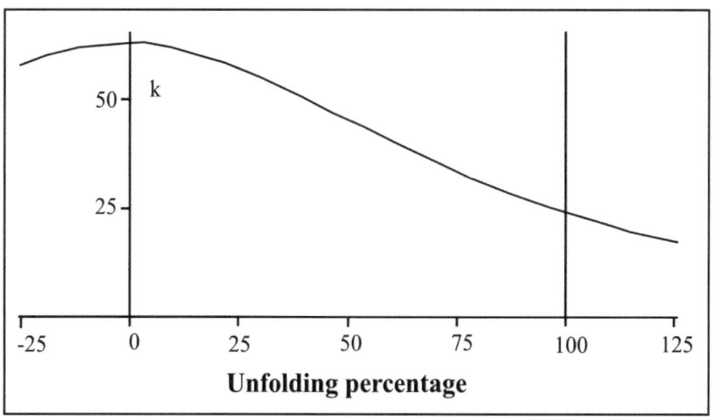

Fig. 9. Variation of the Fisher's (1953) precision parameter k during progressive unfolding for the station ***k*** (sites 47-48-62-63)

S	Sites	N	Geographic coordinates				Stratigraphic coordinates			
			D (°)	I (°)	k	α_{95} (°)	D (°)	I (°)	k	α_{95} (°)
a	1-2-6-7	15	341.7	45.2	24	8.0	342.6	29.1	21	8.6
b	4-5	4	18.2	67.9	39	14.8	46.8	54.6	40	14.7
c	8	10	76.1	-75.7	30	8.9	147.9	-48.8	30	8.9
d	10	7	155.2	-52.7	47	8.9	175.8	-29.6	47	8.9
e	16-17	6	5.4	56.2	27	13.2	343.2	47.3	17	16.7
f	22-23	6	224.1	-46.8	11	20.8	251.1	-50.1	10	21.8
g	25	5	11.2	62.4	22	16.7	54.2	56.4	22	16.7
h	28-29-30	15	29.6	45.4	18	9.2	40.7	40.5	21	8.5
i	32	11	10.8	57.6	26	9.0	69.8	48.9	26	9.0
j	44-45-46	9	0.1	61.2	35	8.8	17.0	56.5	20	11.8
k	47-48-62-63	16	1.2	50.5	62	4.7	26.7	49.7	24	7.7
l	60-61	6	14.9	60.5	26	13.3	5.9	50.5	13	19.2
m	64-65	4	1.2	49.4	53	12.7	78.0	59.8	41	14.5
n	71	11	176.4	51.2	91	4.8	236.5	55.2	46	6.8

Table 1. Stations, corresponding sites, number of samples N, magnetic declination D, magnetic Inclination I and corresponding Fisher's (1953) parameters k and α_{95} in geographic and stratigraphic coordinates (D, I and α_{95} in degrees)

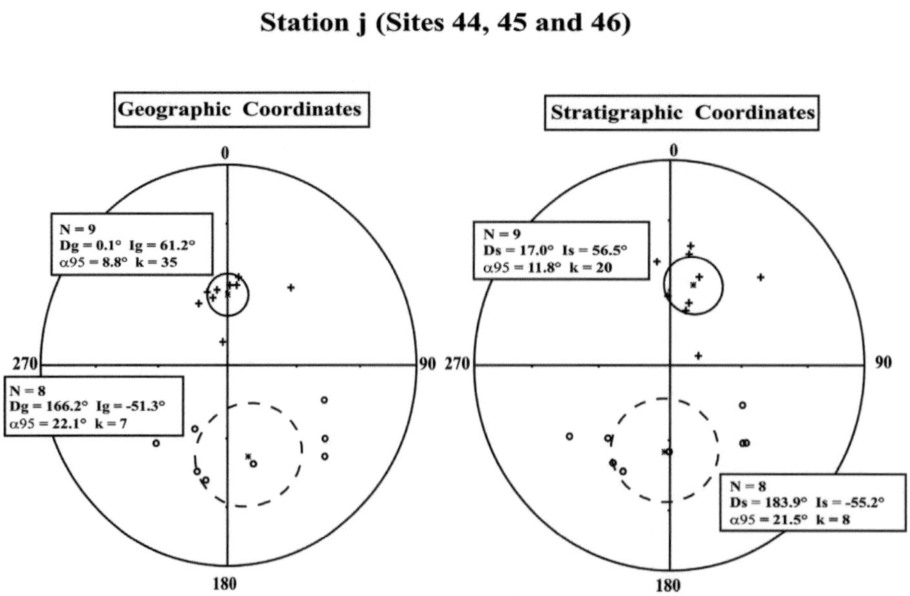

Fig. 10. Mean paleomagnetic directions and their associated 95% confidence cone (for the normal and reversed directions), before and after bedding correction for station j (sites 44-45-46), (See caption of fig. 5a)

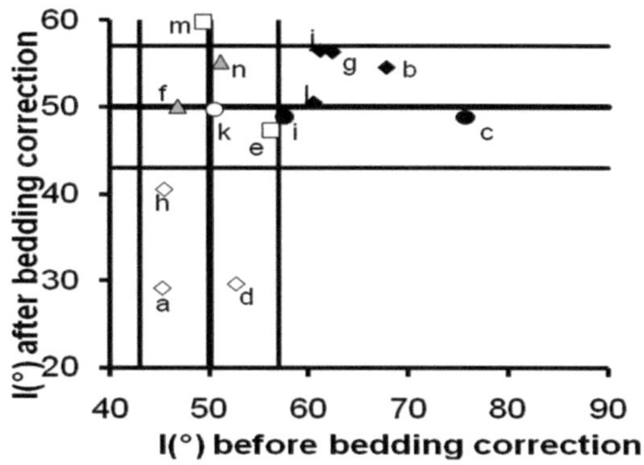

Fig. 11. Inclination after bedding correction versus inclination before bedding correction (in absolute values). Expected inclination around 50° (calculated after Besse & Courtillot, 2002) with an uncertainty window of 7° (corresponding to the uncertainty of 2° on the paleomagnetic pole - calculated after Besse & Courtillot, 2002 - and an arbitrary additional uncertainty of 5° associated with the paleomagnetic data). Prefolding (full symbols), undated (gray triangles) and postfolding (open symbols) magnetizations. Magnetization acquisition age according to paleomagnetic tests (circles), coincidence with the recent field (squares) and inclination value (diamonds)

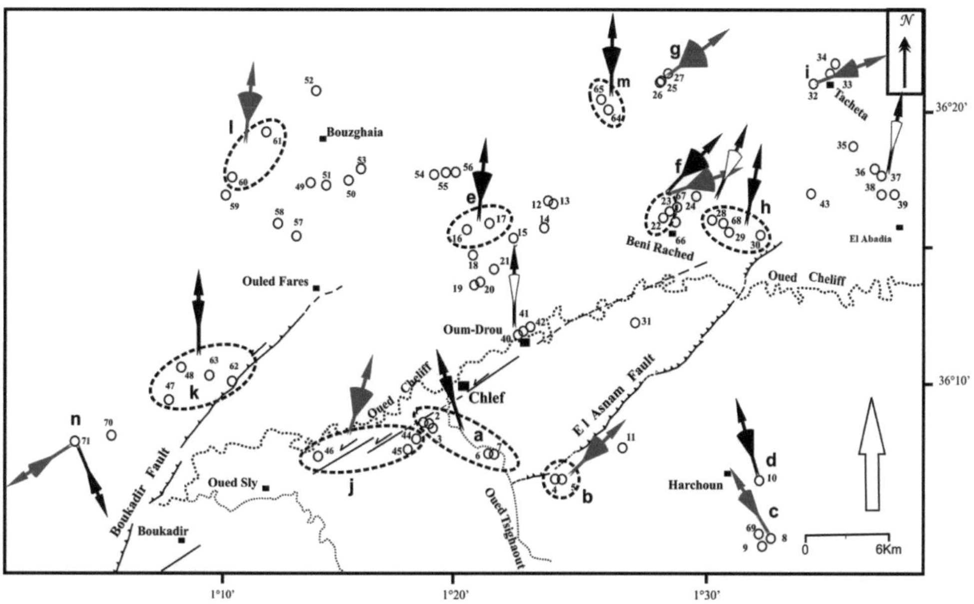

Fig. 12. Retained paleomagnetic declination for the different stations with uncertainty associated with the rotation amplitude (Desmarest, 1983). Gray and black small arrows correspond to prefolding and postfolding magnetizations respectively (gray and black arrows in the stations f and n with undetermined magnetization age). The partly open small arrows represent the Aïfa et al. (1992) data. The large open arrow on the bottom right indicates the expected magnetization direction after Besse and Courtillot (2002)

For the structural interpretation at the basin scale, only the declinations **A** and **B** can thus be considered. The declination **A** has been obtained in widespread sites and is interpreted as resulting from the clockwise rotation of large blocks, as in a "bookshelf" model (Biggs et al., 2006 and reference therein). The uncertainty about the rotation amplitude has been determined according to the Demarest (1983) procedure. The stations *g* and *i* with prefolding magnetization of Tortonian age, recorded clockwise rotations of ~ 54 ± 33.5°and ~ 70 ± 16° respectively, giving a mean of ~ 62° for the north-eastern part of the basin. That corresponds to a rotation rate between ~ 5°/My and 8.6°/My according to the age, not precisely known, of the studied level within the Tortonian (between 11.6 and 7.2 Ma – Gradstein et al., 2004). Whereas for the stations *b* and *j* with prefolding magnetization of Messinian age, the obtained clockwise rotations of ~ 47 ± 20.5° and ~ 17 ± 18.6° respectively, gives a mean of ~ 32° as rotation for the central part of the basin, corresponding to a rotation rate of ~ 4.4°/My to 6 °/My according to the age, not precisely known, of the studied level within the Messinian (between 7.2 and 5.3 Ma – Gradstein et al., 2004).

To summarize, this area has thus undertaken clockwise rotation since the deposition of the Tortonian formation. The obtained rotation rate should be between 4.4°/My and 8.6°/My according to our data. Taking into account the large uncertainty associated with our data, these window of rotation rates is compatible with that (2.5 to 3.9 °/My) proposed by (Meghraoui et al., 1996) using the convergence velocity of the Africa and Eurasia plates.

Local deformation due to shearing close to the main sinistral faults gave on the contrary the counterclockwise rotation associated with the orientation **B** (Henry et al., 2004b).

7.2 Magnetic fabric data

The magnetic foliation is close to the bedding plane, mostly with a moderate imbrication angle relative to this plane. Most magnetic lineations trend NW-SE and do not have a relation with the fold axes and with the perpendicular to the regional shortening direction (Fig. 13a). In this case, it is clear that the magnetic fabric is only of sedimentary origin. Some other magnetic lineations are NE-SW oriented. We cannot exclude a tectonic effect on the fabric for these samples. However, this orientation is quite perpendicular to the NW-SE lineations. In sedimentary fabric, lineation can be parallel or perpendicular to the deposit current direction, and it is then more probable that the NE-SW lineations are also related to paleocurrents.

To determine the direction of the paleocurrents, the most reliable approach is based on the imbrication. This direction is in the plane including the pole of the magnetic foliation (minimum susceptibility axis) and the pole of the stratification (Fig. 13a). On a projection in the lower hemisphere, the sense of the current is assumed to be opposite to the sense of the deviation of the minimum axis relative to the stratification pole. About ¼ of the sites do not reliably yield this direction because of too weak imbrication angle (i.e too weak angular difference between minimum axis and stratification pole). For part of the remaining sites, this direction also corresponds to the magnetic lineation, confirming the obtained paleocurrent orientation. However, the paleomagnetic data showed that at least part of the sites was affected by a rotation. To infer the initial paleocurrent direction, a correction has to be applied for this rotation and only very few sites (including two sites with large uncertainty on AMS) finally yield at the same time significant rotation and paleocurrent data. We can however remark that the obtained paleocurrent direction (Fig. 13b) for these sites is rather perpendicular to the present fold axes within the basin, except around the main fold axis where it is rather parallel to this axis. That suggests deposit currents already in relation with the basin structure, toward the basin synclines axes on the border and parallel to basin main syncline axis in the middle, and therefore an early occurrence of this structure.

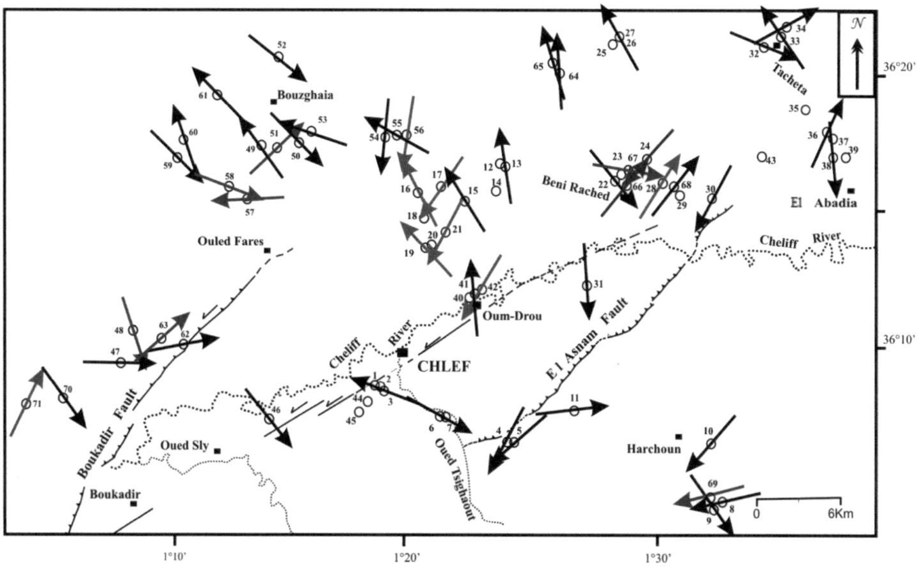

Fig. 13a. Paleocurrents directions in the different sites without correction of the rotation inferred from paleomagnetic data. The gray arrows correspond to sites with large uncertainty on the mean magnetic fabric per site

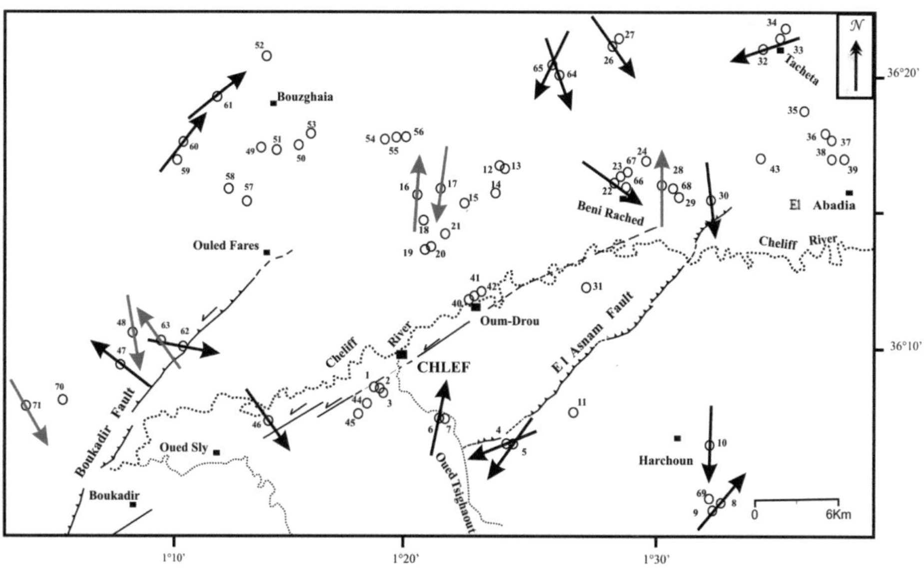

Fig. 13b. Paleocurrents directions in the different sites after correction of the rotation inferred from paleomagnetic data. The gray arrows correspond to sites with large uncertainty on the mean magnetic fabric per site

7.3 Implication for the structural evolution and modeling

On the one hand, several studies (Thomas 1985, Meghraoui et al. 1986, Neurdin-Trescartes, 1992; Meghraoui et al., 1996, Belabbes 2008) have already shown that the deformation in northern Algeria (especially in its western and central parts) is characterized by NE-SW trending folds associated with NE-SW sinistral transpressive faults. As these active structures are distributed in an E-W lengthened zone and have an "en echelon" disposition, these authors suggested the existence of hidden major deep dextral E-W faults in the northern and southern borders of this deformed zone. The existence of such E-W dextral major faults has been also proposed by (Mauffret et al., (1987), along the off-shore domain. Meghraoui et al., (1996) completed this structural model (Fig. 14) suggesting an evolution with clockwise block rotation within the Chelif basin.

On the other hand, the detailed interpretation of the aeromagnetic data (Boukerbout et al., 2008) pointed out the presence of two major deep E-W faults located north and south of the Chelif basin, and a dextral movement on these deep strike-slip faults shown by these geophysical investigations implies a general shearing in all this area. In a similar structural context, block rotation related to strike-slip faulting has been already shown (e.g. Ron et al., 1984). That exactly corresponds to our results pointing out a "bookshelf" evolution.

The new detailed interpretation of the aeromagnetic data and the new results presented here evidence these key elements of the Meghraoui et al., (1996) model of the tectonic evolution of the Chelif basin, confirming the validity of this theoretical model. The tectonic activity is presently still highlighted by movements along the present tectonic structures at the origin of several recent damage earthquakes: Tenes 1922, Ms=6; Orleansville 1954, Ms= 7; and El Asnam 1980, Mw= 7.3 (Aoudia & Meghraoui 1995, Dewey, 1990; Philip & Meghraoui 1983).

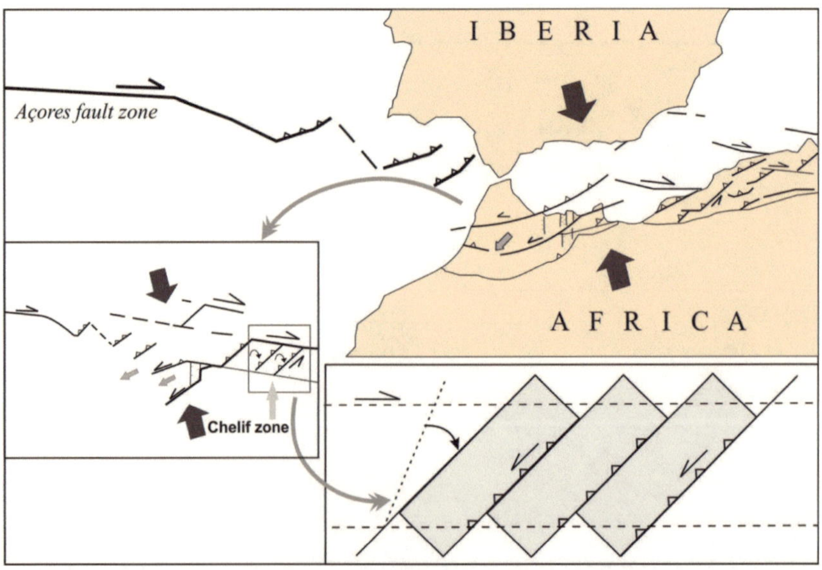

Fig. 14. Main tectonic structures of the "Gorringe-Alboran-Tell" zone of déformation. Model of deformation showing the compatibility between a NNW-SSE shortening and block movements with clockwise rotation within this zone. Kinematic model of a deformed zone with block rotation and without lateral extension, after Meghraoui et al. (1996)

8. Conclusion

The relative convergence motion between the Africa and Eurasia plates is decisive for the area tectonic regime at a regional scale. The active plate boundary is situated along the coastal part of the Tellian Atlas mountains. It is highlighted by the main seismic activity related to the active tectonics, highly concentrated in the intramountainous basins (e.g. the Chelif basin). It is clear that, in the Tellian Atlas domain, the actual deformation is associated with the existence of thrust and strike- slip faults related to the convergent context. The kinematic model of block rotation highlighted here using the paleomagnetic method gives a good image of the tectonic evolution of the Chelif basin. At the scale of the plate boundary, this model can also explain the coexistence of the two kinds of structure (strike slip faulting and fault related folding) in a convergence context. Using the present results and those evidenced recently by Derder et al. (2009) in the "Mitidja" basin (central Algeria), and Derder et al. (2011) in the "Beni Haoua" area (northern Chelif basin), it seems to be clear that all the Tellian Atlas domain is organized as tectonic blocks delimited by strike- slip faults. The relative movements of these blocks, by thrust and strike-slip faulting, are explained by clockwise rotations as in a "bookshelf" evolution. The present work evidenced the existence of such block rotations in the Chelif basin and thus confirms the validity of the Meghraoui et al. (1996) model for all the Gorringe-Alboran-Tellian Atlas area (Fig. 14).

9. Acknowledgments

This work was supported by the Algerian–French cooperation CMEP- PHC "Tassili" program 08MDU752 and we are very grateful to the Algerian DPGRF of the MESRS, and to the French Foreign Office. Special thanks to the "Chlef" local team of the CRAAG and to the civil and military authorities of the Chlef area for help on the field. Dr Toufik Abdelatif is thanked for his strong help with the manuscript.

10. References

Abtout, A., Boukerbout, H., Bouyahiaoui, B., Gibert, D. & Derder, M.E.M. (2009). Gravity anomalies and structure of Chelif seismogenic basin (Algeria). *International Earthquake Symposium Kocaeli,* Turkey, 17-19 August 2009.

Aïfa, T., Feinberg, H., Derder, M.E.M. & Merabet, N. (1992). Rotations paléomagnétiques récentes dans le bassin du Chéliff (Algérie). *Comptes Rendus de l'Académie des Sciences Paris,* 314, SII, 915-922.

Anderson, R.V. (1936). Geology in the coastal atlas of western Algeria. *Memoir Geological Society of America,* 4, 450 p.

Aoudia, A. & Meghraoui, M. (1995). Seismotectonics in theTell Atlas of Algeria: the Cavaignac (Abou El Hassan) earthquake of 25.08.1922 (Ms = 5.9). *Tectonophysics 248,* 263-276.

As, J.A. & Zijderveld, J.D.A. (1958). Magnetic cleaning of rocks in paleomagnetic research. *Geophysical Journal. 1,* 308–319.

Belabbes, S. (2008). Caractérisation de la déformation active par l'Interferométrie Radar (InSar) : Failles sismiques et aveugles de l'Atlas Tellien (Algérie) et du Rif (Maroc) le long de la limite des plaques Afrique -Eurasie. *Doct. Thesis, Univ. Louis Pasteur Strasbourg-I,* 226pp.

Belkebir, L., Bessedik, M., Ameur-Chehbeur, A. & Anglada, R. (1996). Le Miocène des bassins nord-occidentaux d'Algérie : biostratigraphie et eustatisme. *Elf Aquitaine Editions*, Pau, 16, pp. 553-561.

Belkebir, L., Labdi, A., Mansour, B., Bessedik, M. & Saint Martin, J.P. (2008). Biostratigraphie et lithologie des séries serravallo-tortoniennes du massif du Dahra et du bassin du Chélif (Algérie). Implications sur la position de la limite serravallo-tortonienne. *Geodiversitas*, 30 (1).

Bessedik, M., Belkebir, L. & Mansour, B. (2002). Révision de l'âge Miocène inférieur (au sens des anciens auteurs) des dépôts du bassin du Bas Chelif (Oran, Algérie) : conséquences biostratigraphique et géodynamique. *Mémoires des Services Géologiques d'Algérie*, 11, pp. 167-156.

Benouar, D. (1994). Materials for the investigation of The seismicity of Algeria and Adjacent Regions During the Twentieth Century. *Annali di Geofisica*, volume XXXVII, N, 4.

Besse, J. & Courtillot, V. (2002). Apparent and true polar wander and the geometry of the geomagnetic field over the last 200 Myr. *Journal of Geophysical Research*, 107, 2300, doi:10.1029/2000JB000050.

Bezzeghoud, M., Dimitrov, D., Ruegg, J.C. & Lammali, K. (1995). Faulting mechanism of the El Asnam (Algeria) 1954 and 1980 earthquakes from modelling of vertical movements. *Tectonophysics 249*, 249- 266.

Biggs, J., Bergman, E., Emmerson, B., Funning, G.J., Jackson, J., Parsons, B. & Wright, T.J. (2006). Fault identification for buried strike-slip earthquakes using InSAR: The 1994 and 2004 Al Hoceima, Morocco earthquakes. *Geophysical Journal International*, 166, 3, 1347–1362.

Boukerbout, H., Abtout, A. & Gibert, D. (2008). Interpretation of aeromagnetic data in the Chlef region (Algeria) using the wavelet transform in the case 3-D. *Third International scientific and practical Conference and exhibition EAGE (The European Association of Geoscientists and Engineers)* Saint Petersburg, Russia 2008.

Chiarabba, C., Amato, A. & Meghraoui, M. (1997). Tomographic images of the El Asnam fault zone and the evolution of a seismogenic thrust-related fold. *Journal of Geophysical Research.*, 102(B11), 24,485–24,498, doi:10.1029/97JB01778

Derder, M.E.M., Henry, B., Djellit, H., Dorbath, C., Ymel, H., Gharbi, S., Guemache, M. & Abtout, A. (2009). Bloc rotation tectonics in northern Algeria revealed by paleomagnetic investigations in the "Mitidja" basin (Algiers area, Algeria), *International Earthquake Symposium Kocaeli 2009*, Turkey, 17-19 August 2009.

Derder, M.E.M., Henry, B., Amenna M., Bayou B., Maouche S., Besse J., Ayache M., (2011). Bloc rotation tectonics recorded in the Miocene magmatic rocks of "Beni Haoua" area (northern Algeria): preliminary paleomagnetic results. *European Geosciences Union, General Assembly 2010*, Vienna, Austria, 03 - 08 April 2011.

Demarest, H. H. (1983). Error analysis for the determination of tectonic rotation from paleomagnetic data. *Journal of Geophysical Research*, 88, 4321–4328.

Dewey, J.W. (1991). The 1954 and 1980 Algerian earthquake implications for the characteristic displacement model of fault behaviour. *Bulletin of Seismological Society of America, 81*, 446-467.

Ellwood, B. B. (1980). Induced and remanent properties of marine sediments as indicators in depositional processes. *Marine Geology*, 38: 233-244.

Fisher, R. A. (1953). Dispersion on a sphere. *Proceedings of the Royal Society* of *London*, 217, 295-305.

Gradstein, F.M., Cooper, R.A., Sadler, P.M., Hinnov, L.A., Smith, A.G., Ogg, J.G., Villeneuve, M., McArthur, J.M., Howarth, R.J., Agterberg, F.P., Robb, L.J., Knoll, A.H., Plumb, K.A., Shields, G.A., Strauss, H., Veizer, J., Bleeker, W., Shergold, J.H., Melchin, M.J., House, M.R., Davydov, V., Wardlaw, B.R., Luterbacher, H.P., Ali, J.R., Brinkhuis, H., Hooker, J.J., Monechi, S., Powell, J., Röhl, U., Sanfilippo, A., Schmitz, B., Lourens, L., Hilgen, F., Shackleton, N.J., Laskar, J., Wilson, D., Gibbard, P. & Van Kolfschoten, T. (2004). A Geologic Time Scale 2004. *Cambridge University Press*, 589 p.

Graham, R. H. (1978). Quantitative deformation studies in the Permian rocks of the Alpes-Maritimes. *Mémoire du Bureau de Recherche Géologique et Minière*, 91, 219-238.

Henry, B., Merabet, N., Derder, M.E.M. & Bayou, B. (2004a). Chemical remagnetizations in the Illizi basin (Saharan craton, Algeria) and their acquisition process. *Geophysical Journal International*, 156, 200-212.

Henry, B., Rouvier, H. & Le Goff, M. (2004b). Using syntectonic remagnetizations for fold geometry and vertical axis rotation: example of the Cévennes border (France). *Geophysical Journal International*, 157, 1061-1070.

Hext, G. (1963). The estimation of second-order tensors, with related tests and designs. *Biometrika 50*, 353.

Idres, M., Ydri, A. & Lefort, J.P. (1996). Proposition d'un schéma structural du bassin du Chelif (Algérie) à partir de données gravimétriques. *Comptes Rendus de l'Académie des Sciences Paris*, 322, IIa, 85-91.

Jelinek, V. (1978). Statistical processing of magnetic susceptibility measured in groups of specimens. *Studia Geophysica et Geodaetica, 22, 50–62.*

Jelinek, V. (1981). Characterization of the magnetic fabric of rocks. *Tectonophysics 79*, 63–67.

Kirschvink, J.L. (1980). The least-squares line and plane and the analysis of palaeomagnetic data. *Geophysical Journal of the Royal Astronomical Society, 62*, 699–718.

Mauffret, A., El Robrini, M. & Gennesseaux, M. (1987). Indice de la compression récente en mer Méditerranée : un bassin losangique sur la marge algérienne. *Bulletin de la Société Géologique de France (8)*, III, 6, 1195-1206.

McElhinny, M.W. (1964). Statistical significance of the fold test in paleomagnetism. *Geophysical Journal of the Royal Astronomical Society* 8, 338–340.

McFadden P.L. & McElhinny M.W. (1990). Classification of the reversal test in palaeomagnetism. *Geophysical Journal International,* 103, 725-729.

McKenzie, D.P. (1972). Active tectonics of the Mediterranean region. *Geophysical Journal of the Royal Astronomical Society, 30*, 109-185.

Meghraoui, M. (1982). Etude néotectonique de la région nord-est d'El-Asnam: relation avec le séisme du 10 octobre 1980. *3th cycle thesis, Paris7 Univ.*, pp 210.

Meghraoui, M., Cisternas, A. & Philip, H. (1986). Seismotectonics of the lower Chelif basin : structural background of the El-Asnam (Algeria) earthkaque , *Tectonics*, 5, 6, 809-836.

Meghraoui, M. (1988). Géologie des zones sismiques du nord de l'Algérie: Paléosismologie, tectonique active et synthèse sismotectonique. *Doct. Sci. Thesis, Univ. Paris XI*, 356 pp.

Meghraoui, M., Morel J.L., Andrieux J. & M. Dahmani., 1996.Tectonique plio-quaternaire de la chaîne tello-rifaine et de la mer d'Alboran. Une zone complexe de convergence continent-continent. *Bulletin de la Société Géologique de France*, 167, 1, 141-157.

Meghraoui, M., Maouche, S., Chemaa, B., Cakir, Z., Aoudia, A., Harbi, A., Alasset, P.J., Ayadi, A., Bouhadad, Y. & Benhamouda, F. (2004). Coastal uplift and thrust faulting associated with the Mw = 6.8 Zemmouri (Algeria) earthquake of 21 May, 2003. *Geophysical Research Letters*, 31, L19605, doi:10.1029/2004GL020466.

Morel, J.L. & Meghraoui, M. (1996). Goringe-Alboran-Tell tectonic zone; a transpression system along the Africa-Eurasia plate boundary. *Geology*, 1996, 24, 8, 755-758.

Neurdin-Trescartes J. (1992). Le remplissage sédimentaire du bassin néogène du Chelif, modèle de référence de bassins intramontagneux. *Doct. Sci. Thesis, Univ. de Pau et des Pays de l'Adour*, France, 605 pp.

Nocquet, J. M. & Calais, E. (2004). Geodetic measurements of crustal deformation in the Western Mediterranean and Europe. *Pure and Applied Geophysics*, 161, 661–681.

Rees, A.I. (1965). The use of anisotropy of magnetic susceptibility in the estimation of sedimentary fabric. *Sedimentology*, 4, 257271.

Rees, A.I. & Woodall, W.A. (1975). The magnetic fabric of some laboratory-deposited sediments. *Earth and Planetary Science Letters*, 25, 121130

Rees, A.I., Von Rad, U. & Shepard, F.P. (1968). Magnetic fabric of sediments from the La Jolla submarine canyon and fan, California. *Marine Geology*, 6, 145178.

Ron, H., Freund, R., Garfunkel, Z. & Nur, A. (1984). Block rotation by strike-slip faulting: structural and paleomagnetic evidence. *Journal of Geophysical Research*, 89, 6256-6270.

Ouyed, M., Meghraoui, M., Cisternas, A., Deschamp, A., Dorel, J., Frechet, F., Gaulon, R., Hatzfeld, D. & Philip, H. (1981). *Nature*, 292, 5818, 26-31.

Perrodon, A. (1957). Etude géologique des bassins néogènes sublittoraux de l'Algérie du Nord occidental. *Publications du service de la carte géologique de l'Algérie*, 12, 343

Philip, H. & Meghraoui, M. (1983). Structural analysis and interpretation of the surface deformations of the El-Asnam earthquake of October 10, 1980. *Tectonics*, 2, 1, 17-49.

Serpelloni, E., Vannucci, G., Pondrelli, S., Argnani, A., Casula, G., Anzidei, M., Baldi, P. & Gasperini, P. (2007). Kinematics of theWestern Africa-Eurasia plate boundary from focal mechanisms and GPS data. *Geophysical Journal International* 169, 1180–1200

Thomas, G. (1985). Géodynamique d'un bassin intramontagneux: le bassin du bas Chéliff occidental (Algérie) durant le Mio-Plio-Quaternaire. *Doct. Sci. Thesis, Univ. de Pau*, France.

Weiler, P. D. (2007). Magnetic anisotropy, sedimentary rocks and strain alteration. *In* Gubbins D. and Herrero-Bervera E. (Eds), *Encyclopedia of Geomagnetism and Paleomagnetism*, Springer, 475-477.

Zijderveld, J.D.A. (1967). AC demagnetization of rocks: analysis of results. In: Collinson, D.W., Creer, K.M., Runcorn, S.K. (Eds.), *Method in Paleomagnetism. Elsevier, Amsterdam*, 254–286.

2

Tunisian Transtensive Basins in Tethyan Geodynamic Context and Their Post-Tortonian Inversion

Adel Rigane[1] and Claude Gourmelen[2]
[1]*University of Sfax-Tunisia, Faculty of Sciences, Department of Geology, RU: GEOGLOB (03/UR/10-02)*
[2]*University of Maine, Faculty of Sciences, Geosciences Laboratory*
[1]*Tunisia*
[2]*France*

1. Introduction

Generally, the geological and structural context of Tunisia can be described, from the North to the South, in terms of the structural domains illustrated on Fig. 1:

1. The Alpine zone, composed of the Tellian Atlas, characterized by the stacking of allochthonous units (Jauzein et al., 1965; Biely et al., 1974; Rouvier, 1977; Ben Ferjani et al., 2006; Ould Bagga et al., 2006) and a series of imbricate thrust slices, whose front corresponds to the major Teboursouk overthrust (Zargouni, 1975; Perthuisot, 1978).
2. The diapir zone (DZ), represented by the N050-trending Triassic outcrops of northern Tunisia (Perthuisot, 1978).
3. The Atlasic zone includes the Northern Atlas, the central Atlas, the southernmost Atlas and the "North-South Axis". The central Atlas and the Northern Atlas are characterized by a bundle of N040-050 anticlinal folds that are cut orthogonally by a graben system. The southernmost Atlas is made up of E-W folds, inflected towards N060 at their terminations and bounded by the major faults of Gafsa and Négrine-Tozeur striking N120-130 (Zargouni, 1985). The "North-South Axis" comprises a folded and faulted zone delimiting the central Atlas (sensu lato) and the Eastern Atlas. It corresponds to a zone of major heterogeneity giving rise to N-S trending faults (Burollet, 1956; Turki, 1984, 1985, Gourmelen, 1984; Abbès, 2004).
4. The Eastern platform is characterized by a weakly folded thick Neogene succession that extends eastward to the Malta scarp.
5. The Saharan platform is characterized by folded Paleozoic rocks of the African craton, and makes up the foreland of the Atlasic chains (Fig. 1).

The present architecture of the Saharan Atlas in Tunisia is defined by two principal models: The first one (Fig. 1) emphasizes a general SW-NE geological structure in successive and parallel bands represented by the tellian zone and the diapirs zone in the North, the central Atlas then the southernmost Atlas with oblique orientation and installed within a NW-SE corridor. These fields are bordered in the East by the “North-South Axis” (Castany, 1951; Burollet, 1956, 1991; Rouvier, 1977; Zargouni, 1984, 1985).

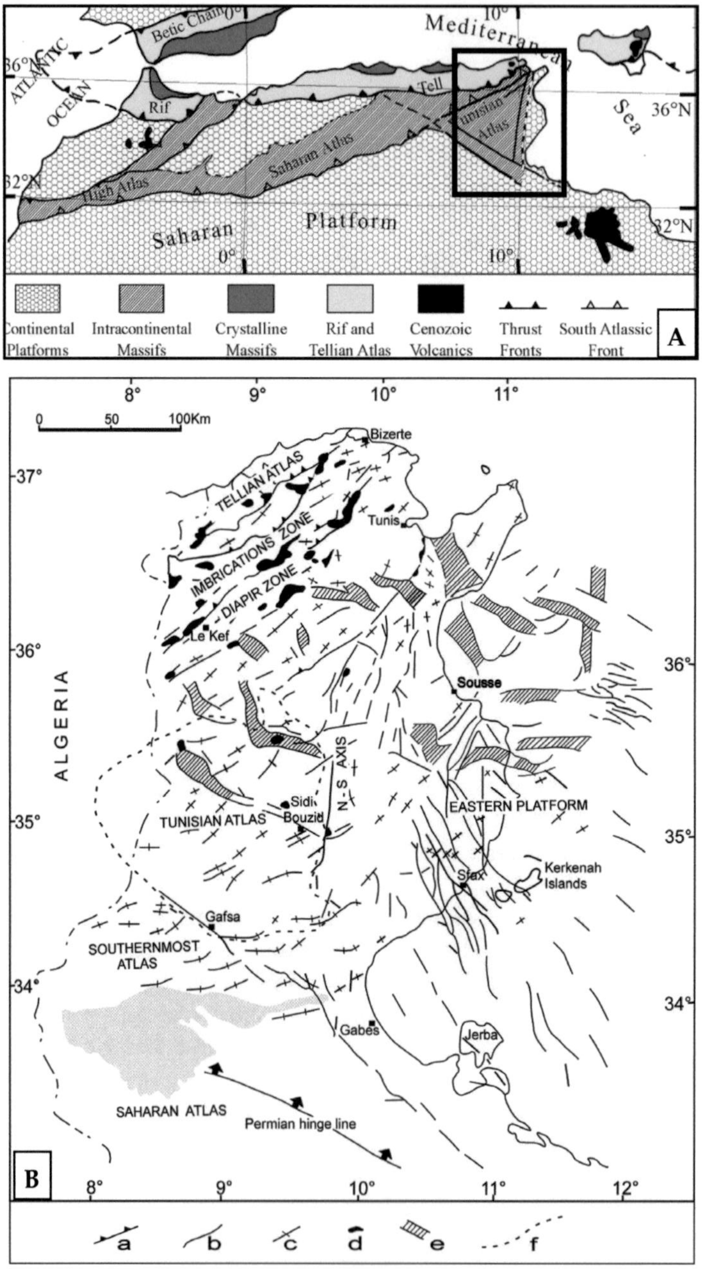

Fig. 1. Location of Tunisia in the eastern Maghreb: A: general context; B- schematic tectonic map of Tunisia. In B : a- major thrust, b- major faults, c- major anticlines, d- Trias, e- grben, f- limit of Kasserine Island

The second model (Fig. 2) frames atlasic Tunisia in a big E-W strike-slip corridor which controls at first the facies distribution during the Meso-Cenozoïc sedimentation and generates later long echelon folds in the sedimentary cover by dextral shear zone (Ben Ayed 1980; Ben Ayed and Viguier, 1981; Ben Ayed, 1986; El Ghali et al., 2003).

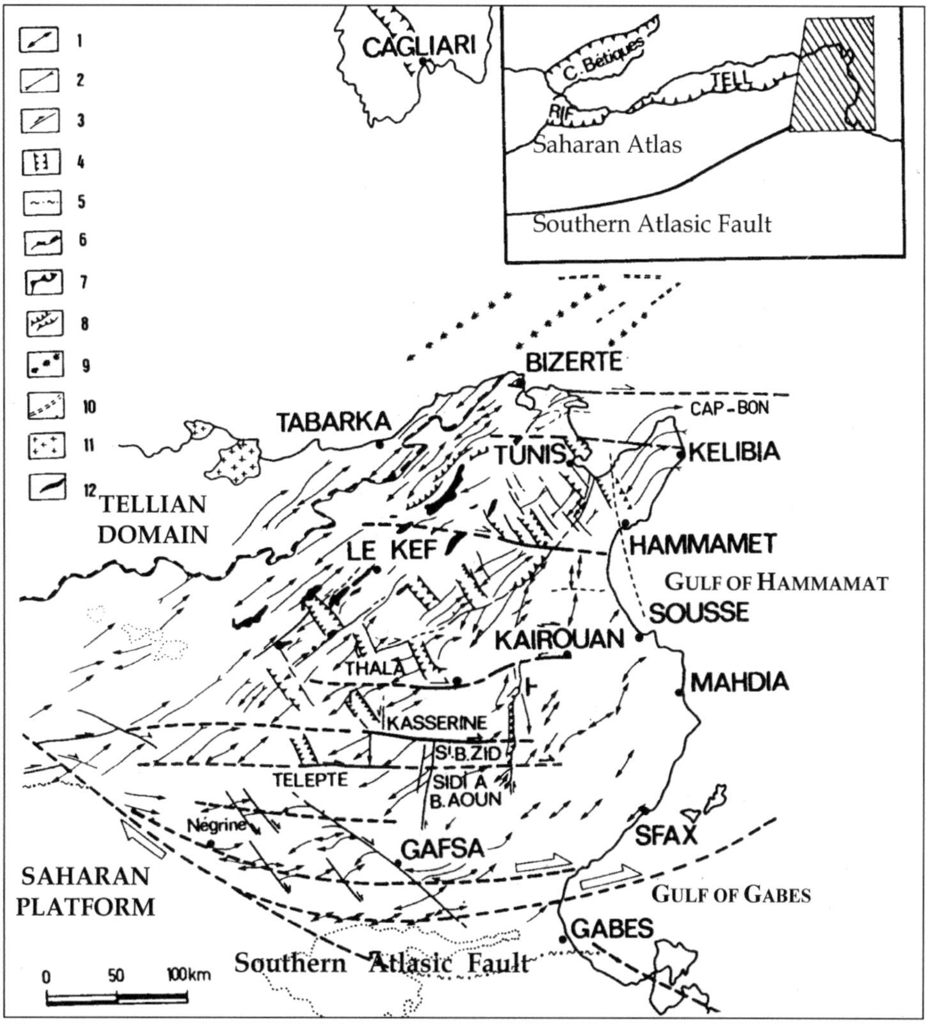

Fig. 2. Simplfied tectonic map of Tunisian Atlasic domain (Ben Ayed and Viguier, 1981). 1: anticline; 2: syncline; 3: strike-slip fault; 4: rift; 5: flexure; 6: slipping thrust front; 7: thrust; 8: imbrications; 9: anticlinal alignment; 10: eastern extension of this alignment from the bathymetrical data; 11: pre-alpine; 12: Triassic outcrops

2. Present-day architecture of Tunisian Atlas

Independently of these two historical models, the present Tunisian Atlas appears as a megabloc that we call "Tunisian Block (TB)".These feature is bordered by three major faults belonging to the Northern Saharan "reghmatic" network (Fig.3): (1) the eastern boundary appears as a complex faulted and folded corridor (NSC), defined by Rigane and Gourmelen (submitted paper) delimiting the folded zone of the Central and the Northern Atlas in the West and the Sahel subsident zone in the East: it corresponds to the "North-South Axis (sensu stricto)"described in the literature (Burollet, 1956; Gourmelen, 1984; Gourmelen and Tricart, 1985; Khessibi, 1978; Rigane, 1991, Piqué et al., 1998; Abbès, 2004; Rigane and

Gourmelen, submitted). (2) The Southern boundary also corresponds to a faulted domain, known as the Gafsa-Negrine-Tozeur (GNTC) (Zargouni, 1986; Zouaghi et al., 2005). The GNTC interrupts the extension of the NSC towards the south in the Gabès area, and corresponds to the Southern Saharan Atlas limited by the Gafsa fault in the North and the Negrine-Tozeur fault in the South (Zargouni, 1985; Boukadi, 1998; Zouari, 1999). (3) The Northern boundary is characterized by a SW-NE orientation. This zone appears rather as a sheaf of reverse faults, facing South-East or South (with oblique convergence), whose major structural feature corresponds to the El Alia-Teboursouk fault (Jauzein, 1967; Martinez and Truillet, 1987). This tectonic bundle, which also coincides with the "diapir zone (DZ)" corresponds to the Southernmost limit of the TB in Algeria. This Northern boundary cuts across and limits the NSC in the North, so this feature is bounded at both extremities. Finally, within the TB corresponds to a mosaic of second-order blocks (Gourmelen et al., 2000; Zouaghi et al., 2005; Rigane and Gourmelen, submitted). In this domain, we observe widely spaced SW-NE anticlinal folds separated by vast plains very often occupied by sebkhas. The paleogeographic and structural evolution of the studied area shows that the TB became individualized very early in Mesozoic and Paleogene times and characterized, thus, by an extensive (and/or trantensive) tectonic regime.

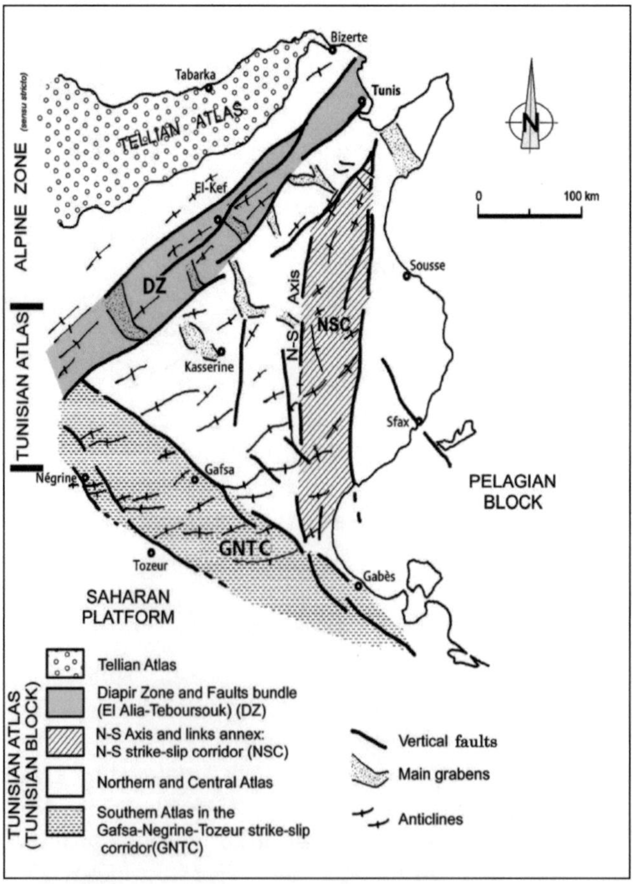

Fig. 3. The present-day architecture of Tunisian Atlas: the faulted block and its components

3. Pre-orogenic history of the "Tunisian Block"

3.1 Paleogeographic and structural evolution of the basin

The studied area, as illustrated in Fig. 4, shows that this TB became individualized very early in Mesozoic times along with its determined limits.

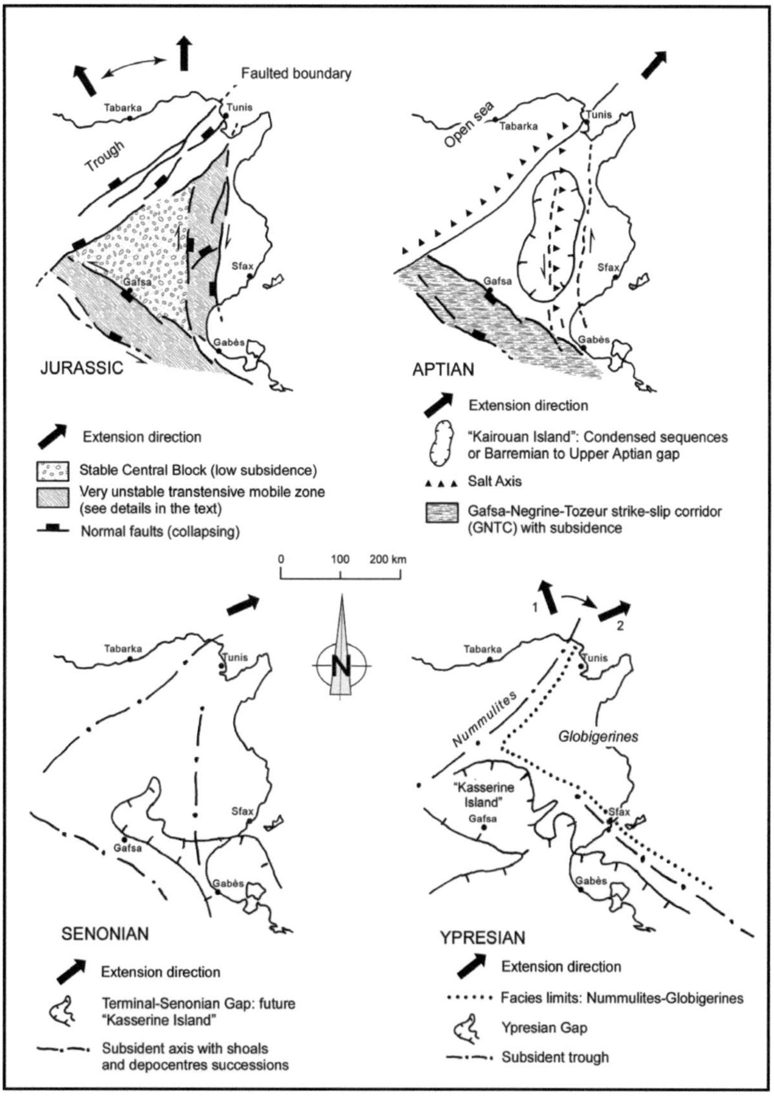

Fig. 4. Tunisian Basin and continuously mobile boundaries

Whereas the Saharan Atlas in Algeria is composed of a folded and faulted chain elongated in the SW-NE direction, it forms an almost isosceles triangle in Tunisia. This latter terrain corresponds to a true continental block, whose origin is precisely linked to its paleogeographic history, which is dominated by its three boundaries, undergoing continuous tectonic remobilization, controlling the geometry, morphology and filling of the basin (Fig.4).

More precisely, these boundaries correspond to mobile zones from the early Mesozoic onwards, with the formation of sedimentary wedges, gaps and unconformities (Burollet, 1956, 1991 and Burollet et al., 2000; Rouvier, 1977; Hlaïem, 1999; Jauzein, 1967; Soyer and Tricart, 1987). In this way, the Tunisian Atlas appears to inherit its present morphology from the early and continual compartmentalization of the basin, which is tectonically inverted from Tortonian times onward to form the folded Atlas chain.

3.1.1 During the Jurassic

The three boundaries of the block form mobile zones of variable width associated at depth with vertical faults of the pre-Triassic basement (Midassi, 1982; Morelli and Nicolih, 1990; European Geotraverse, 1992; Piqué et al., 2002; Gabtni et al., 2005).

- The southern boundary, which, moreover, is prolonged in the extreme south-east of Tunisia, corresponds to a coastal depositional environment, sometimes of evaporitic type but in a subsident zone, (Kamoun et al., 1999; Soussi, 2002), with a series of depocentres and tectonic highs succeeding each other from NW to SE: the Chotts basin, the Tébaga high, the Tatatouine basin and the Néfuza highs in Libya (Aissaoui, 1984; Bouaziz, I995). All these morphostructural features result from transtensive movements of the GNTC in an overall N-S extensional regime (en echelon E-W normal faults). At this epoch, sedimentation was occurring in a typical intracratonic basin.
- The northern boundary corresponds to the faulted edge of the "deep trough" (Burollet, 1956) i.e. an open marine basin. This basin contains pelagic deposits displaying a margin-type character with SW-NE-trending active faults (El Alia-Téboursouk, Boujabeur-Lorbeus-Gafour, for the principal ones). These faults are probably inherited from the basement, and are picked out by diapirs. They could be integrated into a transform margin on a much larger scale associated with a sinistral strike-slip fault zone in relation to the Açores-Gibraltar-Sicily system (Dercourt et al., 1985; Dewey et al., 1989).
- The Eastern boundary is generally expressed as a N-S-trending axis of dome-type diapiric highs with condensed series, gaps, wedges and discordances (Gourmelen et al., 1989), intersected by local depocentres associated with grabens or hemi-graben structures linked to the regional extension and arranged according to transverse faults following the axial trend.

Finally, in the Jurassic, in spite of the scarcity of outcrops and subsurface data, a synthesis of sedimentological and paleogeographic studies (Kamoun et al., 1999; Soussi, 2002) combined with the existing tectonics allows us to propose a simple geometrical model of the Tunisian basin at that time: three unstable zones are arranged around a resistant central block with little or no subsidence, and are subject to the remobilization of pre-Triassic basement fractures. As a result, this basin displays clearly intracontinental characteristics. Subsequently, these three unstable zones exert a more or less important influence on the geometry and development of the basin at different times.

3.1.2 During Early Cretaceous

A positive zone appears corresponding to the "Kairouan Island" (M'Rabet, 1981), clearly delimited in the East by the CNS and the GNTC in the South undergoing more subsidence. The tectonic regime is thus extensional with a maximum lengthening towards the NE (Rigane et al., 2010). At this time, the Eastern boundary becomes increasingly positive, while, throughout the region, we observe the development of condensed series, gaps and wedges of Aptian age called the Gafsa group by M' Rabet (1981); sometimes, the entire Gafsa group is even reduced to a metre-thick quartz-bearing dolomitic level named the Bouzer dolomite by Gourmelen (1984). Among other factors, this mobile and positive zone is the result of diapiric phenomena which probably begins at the end of the Triassic and which ceases, at least temporarily, at the end of the Aptian, since, along the entire length of this boundary, the Cenomanian is discordant on various units ranging from the Triassic to the Aptian.

At the end of the Aptian, the tectonic regime is shown to consist of a SW-NE extension (Bismuth et al., 1982; Boltenhagen, 1985; Soyer and Tricart, 1987; Feki et al., 2005, Rigane et al., 2010), with a strong strike-slip component (transtension) on the megablock boundaries.

3.1.3 During Cenomanian

The Tunisian block temporarily loses its individual character as a consequence of the generalized transgression expressed in Tunisia by the Zebbag formation sensu lato (Burollet, 1956, Lüming et al., 2004). In detail, however, the N, S and E boundaries remain mobile with highs zones in the East, a deep basin towards the NW (formation Bahloul) and a subsident southern corridor. During the Cenomanian, the tectonic regime is predominantly extensional (Bouaziz et al., 2002).

3.1.4 At End-Senonian

The distribution of depocentres is inverted along the three boundaries of the block. The CNS becomes subsident, with the development of a succession of depocentres and highs, frequently of halokinetic origin. The Western extremity of the block shows a tendency towards emergence reflecting the appearance of the "Kasserine Island".

We may conclude that, since the Jurassic period, the eastern part of Tunisia, mainly to the east of the CNS, remains a subsident zone, with, however, the existence of some more marked basins (Ellouz, 1984).

3.1.5 The Paleogene

Throughout this era, some major changes occur in the basin geometry:

3.1.5.1 Ypresian

The "Kasserine Island" is superposed on the southernmost part of the Tunisian block. Indeed, isopach and facies maps (Bishop, 1975; Rigane, 1991; Rigane et al., 1994; Zaier et al., 1998; Ben Ferjani et al., 2006) clearly show the presence of shallow basins, in the NW and the SW (nummulitic and phosphatic facies), which become deeper in the East (troughs of the Gulfs of Tunis and Gabes). During this epoch, a NW-SE extensive and/or transtensive tectonic regime controls the sedimentation (thickness and facies) in the Ypresian basin (Rigane et al., 1994).

3.1.5.2 Lutetian-Priabonian

The deposits are arranged in isofacial bands almost parallel to the CNGT trend. The "Kasserine Island" gradually loses its individual character to become connected finally to the Saharan continent. The predominant tectonic regime is also transtensive, but we observe a change in orientation towards a SW-NE trend (Gourmelen et al., 2000). This radical change in basin cartography is probably linked to the modification of Africa/Europe kinematics at this time (Dewey et al., 1989).

3.1.5.3 Oligocene

Although the area of emerged land tends to migrate northwards, the shoreline is always constrained by the NW-SE trend. The tectonic regime is characterized by a SW-NE extension (Blondel, 1991; Yaïch, 1997).

3.2 Major crises

Since the Jurassic to the Oligocene period, a succession of basins and shoals is observed and controlled by continuous mobility of the three faulted corridors (DZ, NSC, GNTC) delimiting the TB associated to diapiric movements. The corridors mobility attains his paroxysm during three major tectonic crises: at the end of the Aptian, the Ypresian and the Oligocene. Each crisis is marked by a momentary stopping (Aptian, Ypresian) or definitive (Oligocene) of halocinetic ascension.

3.2.1 The Aptian crisis

At that time, Tunisia can be regarded as an intracontinental basin at the northern margin of the African craton, characterized by depocentres directly related to the activity (reactivation) of pre-existing faults. In this case, mobility is concentrated within an N-S trending tectonic bundle in a transtensive regime. For example, the structural and tectono-sedimentary study of Jebel El Hamra to the West of Kasserine (central Tunisia) underlines the predominant role of paleostructures acquired during the evolution of the South-Tethyan basin, and their influence of the present-day geometry of the fold belt. Indeed, the existing pattern of folding and faulting results directly from tectonic blocks formed during the Aptian. Hence, the depocentres appear as two pull-apart basins bounded by a lower-shaped structure (Fig. 5). shown by small arrows. Black arrows: direction of Extension (Rigane et al., 2010).

Lastly, the existence of tectonic nodes supports the hypothesis that diapirism can induce local tectonic extension. The tilted blocks of Jebel El Hamra can be, consequently, the final result of this transtensive tectonics, locally associated to halocinetic vertical movements of Triassic salt. At a large scale, in the African geodynamic context of this period (Guiraud and Maurin, 1991; Guiraud and Maurin, 1992; Guiraud et al., 2005; Basile et al., 2005), this late Aptian phase, very exactly dated in the studied area to the end of Clansaysian age, is integrated in the rifting phase shown by Guiraud and Maurin (1991) in all the intracontinental basins of the Northern part of the African craton, in particular in Libya (gulf of Syrte). In addition, as noticed by these authors (Guiraud and Maurin, 1991, 1992), we conclude also that the opening was done on a mobile zone of the crust in Tunisia, more precisely on a submeridian fault, inherited from the African craton. Finally, the SW–NE detected extension originates from the rifting and oceanic activity in Northern basin (tethyan western branch) and the Eastern basin (Tethyan southernmost branch: Mesogea) (Dercourt et al., 1985; Martinez and Truillet, 1987; Ricou, 1994; Stampfli et al., 2002; Schettino and Scotese, 2002).

In this context, the northern margin behaves like a strike-slip margin (Gibraltar-Messine transform fault) from the upper Liassic onwards, while the Eastern margin represents a classical passive margin that takes over from or develops alongside the transcurrent shear zone from the Early/Late Cretaceous boundary (Fig. 6).

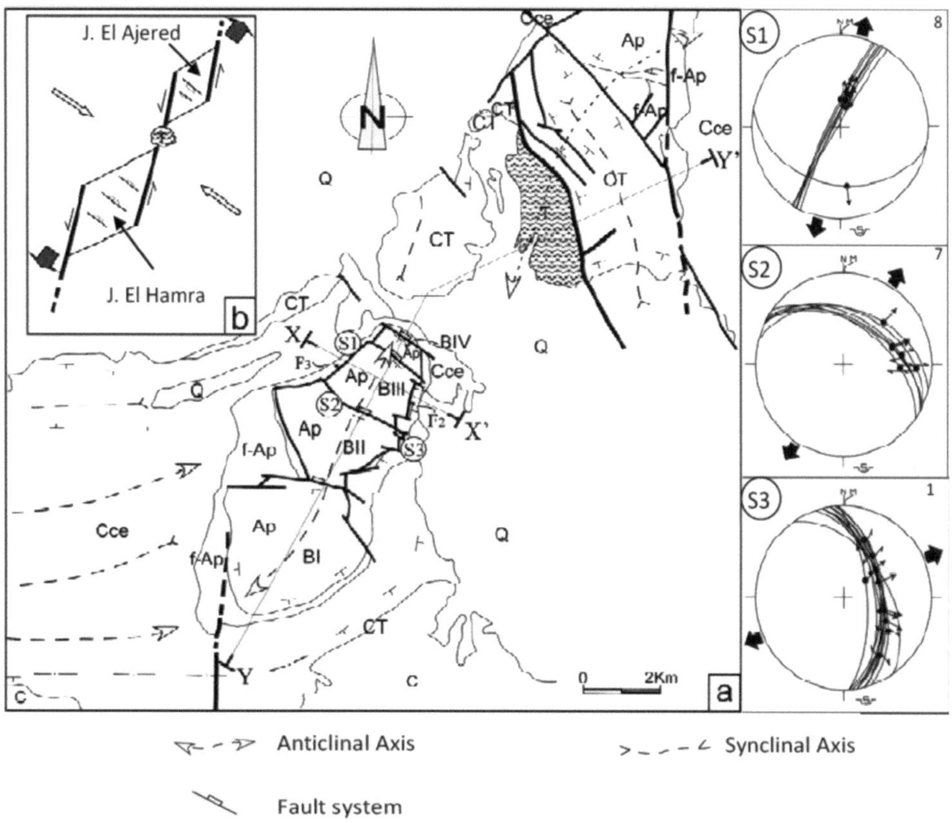

T: Trias; Ap: Aptian; f-Ap: Uppermost Aptian; Cce: Cenomanian; C: Cenozoïc; Q: Quaternary; CT: Turonian

Fig. 5. a) Simplified geological and structural map of El Hamra-El Ajered area and stereographic plots of faults (equal-area, lower hemisphere). Slickenside lineations are shown by small arrows. Black arrows: direction of Extension (Rigane et al., 2010).
b) Interpretation of the study area: two negative lozenge-shaped depocentre bounded by a N-S shear zone (Rigane et al., 2010).

3.2.2 The Ypresian crisis

The Ypresian limestones, in Central area or Tunisia, are cut up by normal dip-slip and strike-slip faults of late Ypresian age. All these fractures are arranged mainly in submeridian N060, E-W and N140 directional systems. The first, the commonest and the most pronounced associates locally according to sectors with the other systems to form Riedel-type, pull-apart or horst and graben structures. The detailed structural analysis of these Ypresian limestones clearly shows that this period is characterized by normal fault and extensive strike-slip fault movements. The normal and/or strike fault throw vary from a few centimeters to a few meters and evolve laterally into a flexure.

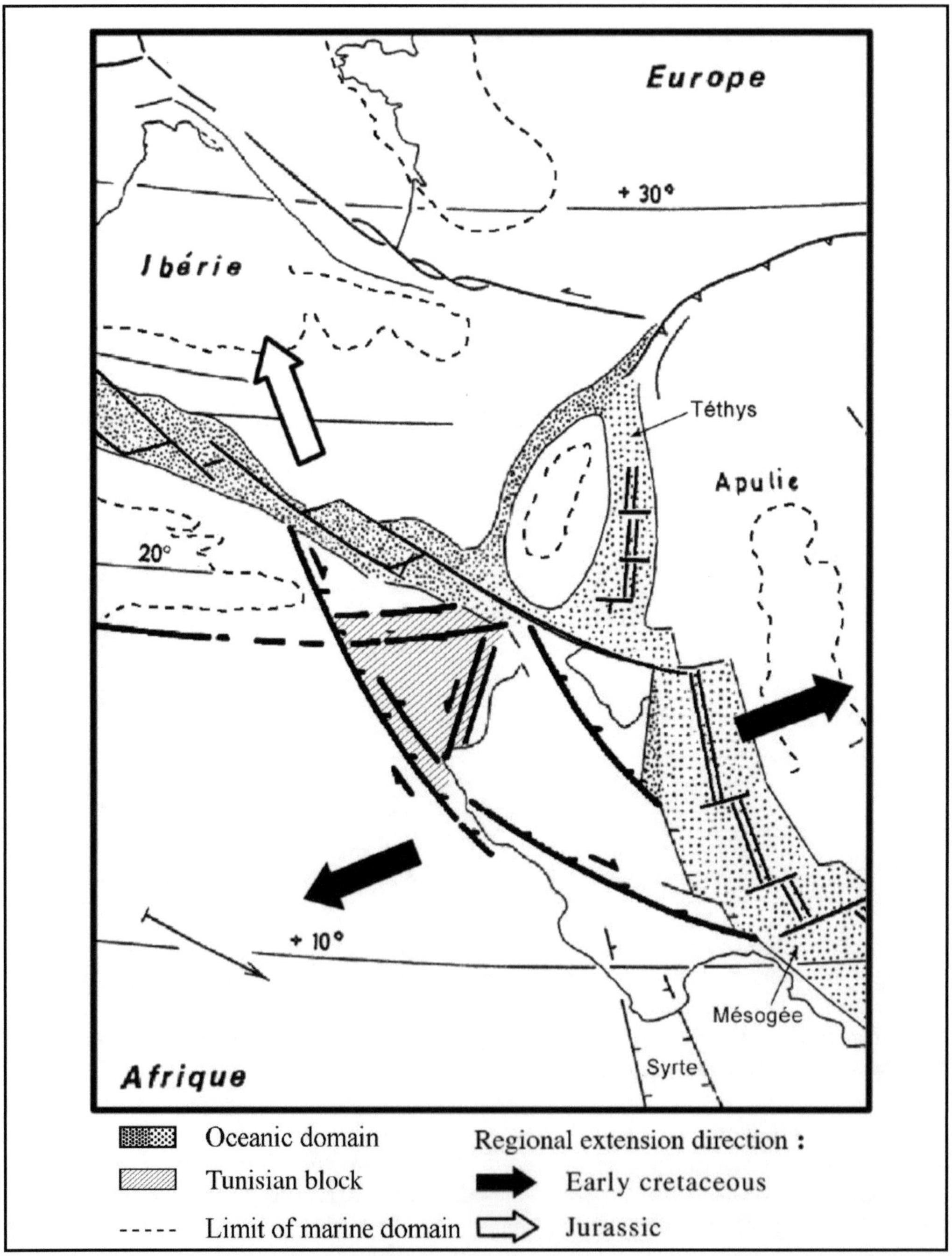

Fig. 6. Geodynamic context of Tunisian Basin from Jurassic to the end of Lower Cretaceous (after Martinez et al. 1991, modified)

The results obtained from the different measures carried out for each system and their processing show that the finite Ypresian faulting event is the result of two successive

episodes, very close in time (Fig. 7). They are characterized first by a NW-SE extension (episode 1) followed by a SW-NE extension (episode 2).
The geometric and kinematic study of this area clearly indicates that, during late Ypresian, there was a remobilization of ancient fractures, mainly submeridian and NW-SE, in an extensive strike-slip context.

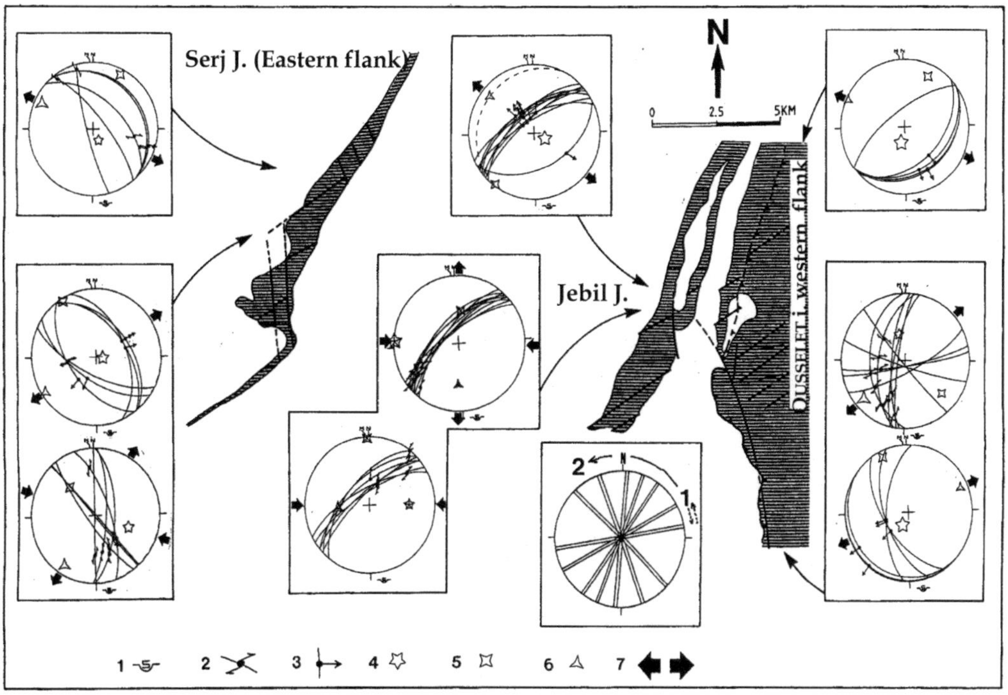

Fig. 7. Quantitative analysis of terminal Ypresian striated slickensides revealing the two strain sequences and their directions of extension in Central Tunisia. 1) Schmidt pattern, lower hemisphere; 2) strike-slip fault; 3) normal fault; 4, 5, 6) major axes of the stress tensor; 7) direction of extension

Extensive structures (normal faults, normal strike-slip-faults and flexures) appeared and evolved in this context within the Ousseltia block (Fig. 8) which, therefore, presented, at that time, a mainly-vertical mobility the origin of which stays in the basement. This mobility results in extensive structures in the basin. The corresponding extension is manifested in the sedimentary pile by horizontal shearing in the ductile level and notably in the evaporites of the Trias. The strong rotation of the stress field observed in this block corresponds exactly to the period when important variation in the drifting of Africa and Europe or recorded.
In fact, according to Dewey et al. (1989), after a noticeable slowing down and "erratic movements" of the two converging plates during Paleocene, Africa move again towards the NE precisely during late Ypresian around 51 Ma (anomaly 21). We are tempted to believe that this brutal change of behavior of the African plate was recorded in the late Ypresian faulting episode in this part of the Saharan Platform and corresponds to the rotation of the detected stress fields.

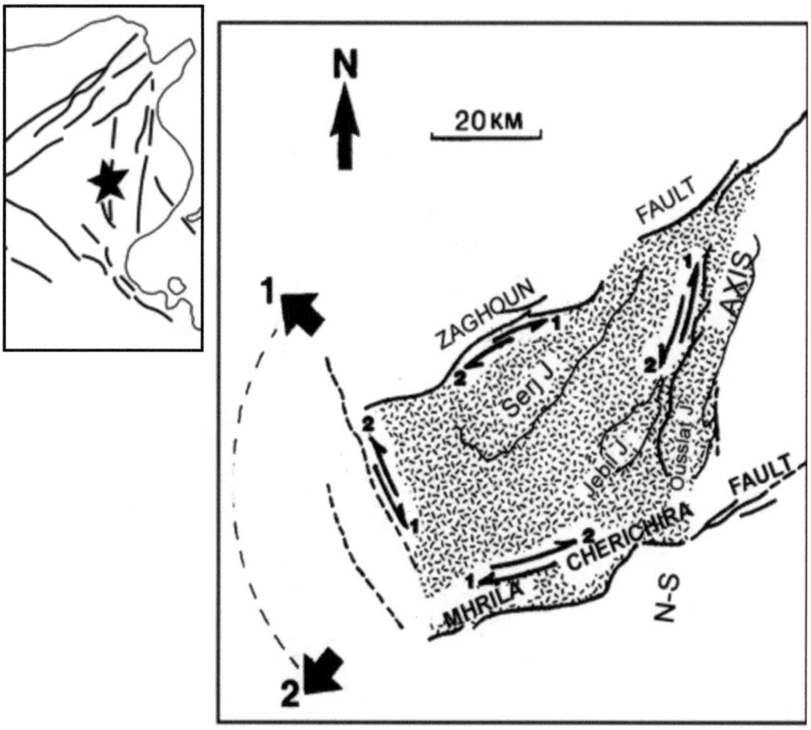

Fig. 8. Terminal Ypresian movements on the boders of the Ousseltia Block associated with NW-SE then NE-SW transtension (1-2: successive extension direction at the end of Ypresian)

At the end of the Ypresian, the Tunisian platform looked like a jigsaw puzzle made up of huge blocks limited by ancient fractures. The Ousseltia block is an example of this blocks. The remobilization of the block boundaries occurred in an extensive context. The stress field developed a strain within the blocks themselves that corresponds to the stress pattern. In this context, the material affected during lithification, reacted in an original way by developing a ductile/fragile strain according to the rheological characteristic of the levels.

This intra-block deformation appears superficial and space limited compared to the one that can be seen on the block boundaries. It is immediately registered and memorized in the deposits and then sealed by successive sedimentation periods.

The later stress fields which strained these blocks again caused a remobilization at the boundaries, more and less masking the earlier movements, and provoked new intra-block deformation which, on the contrary, will be fossilized. Thus, that is how a great number of palaeo-structures have been yielded to us nearly intact for observation since Jurassic. The example of the Ousseltia block evolution at late Ypresian may represent a model of the tectonic behavior of the Africa plate in Tunisia during its pre-Alpine history.

Finally, this geological structure (blocks) could correspond to a similar geodynamic context observed during Aptian crisis: northern tethyan transtensive and eastern mesogean margins (Fig. 9).

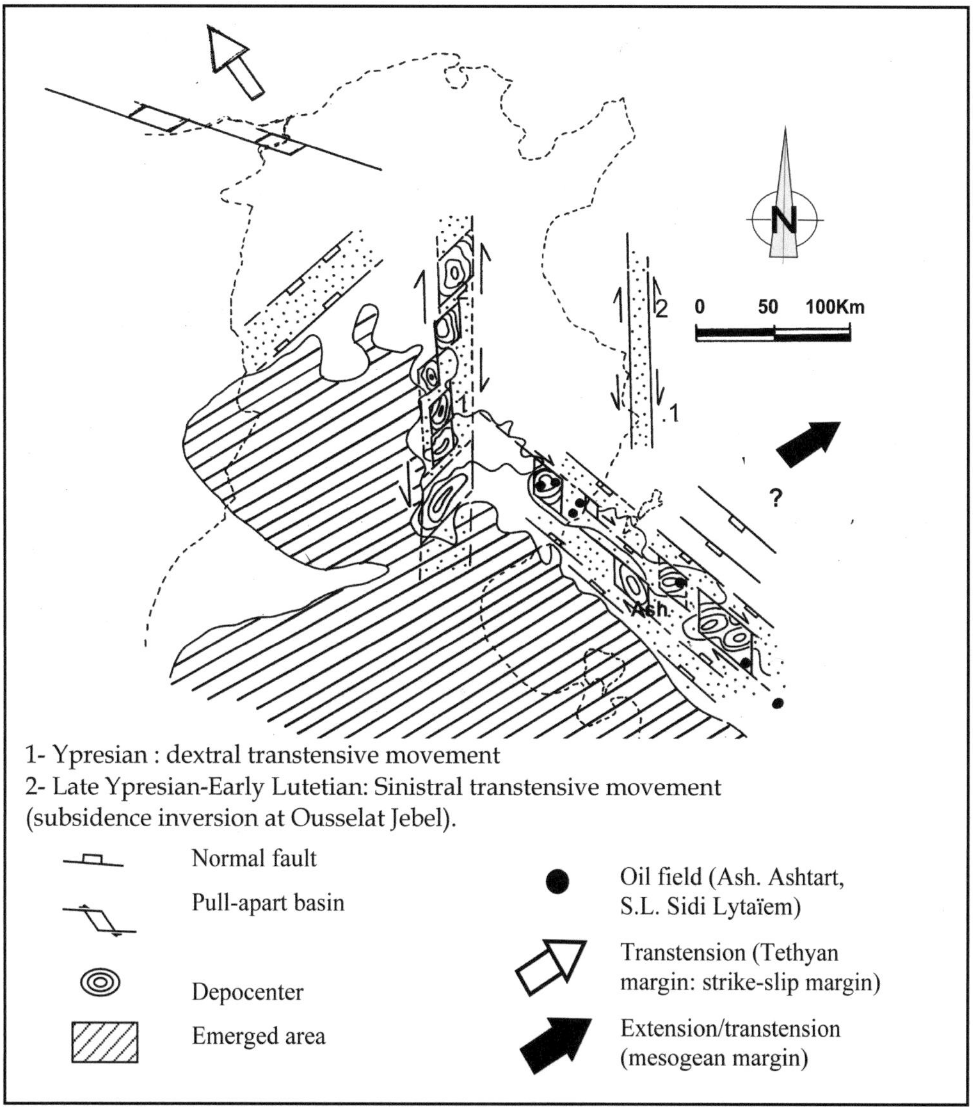

Fig. 9. Geodynamic context of Ypresian Basin in Tunisia

3.2.3 Pre-Burdigalian crisis

This crisis is identified in Central Tunisia by Yaïch (1984, 1997) and confirmed by Blondel (1991). It is characterized by extension related to the Western Mediterranean, Red Sea and Suez Gulf opening and Western Europe Rifts genesis.

4. Structural inversion and its consequences

4.1 Compressive phases in Tunisia

For a long time, the scientific community (Castany, 1951; Richert, 1971; Rouvier, 1977) accepted that the Alpine chain in Tunisia was built up by two major compressive phases: the first is dated as middle Miocene (Tortonian) and the other as Quaternary (post-Villafranchian). However, recent work has revealed the existence of an early tectonic event known as the "Atlasic phase" of Middle Eocene age (Khomsi et al., 2006; Frizon de Lamotte et al., 2006). These authors base their argument on the presence of angular unconformities located on the flanks of anticlines observed on seismic sections (Fig. 10). In agreement with the present study, other authors (Brahim and Mercier, 2007) have shown that these discordances are actually located on the flanks of diapiric domes and result from a slow and progressive diapirism. Zouaghi et al. (2005) also propose a compressive phase during the Late Cretaceous and even at the end of the Aptian (Ben Ayed, 1986; Rabhi and Ben Ayed, 1990; Ben Ayed et al., 1997; Bouaziz et al., 2002). In these two latter cases, we consider that the discordances observed by these authors are either of the same type as previously described (on diapir flanks) or relate to disharmonic structures within the Miocene and Quaternary folds or are located on top of drape fold zones (Gourmelen et al., 2000; Rigane et al., 2010).

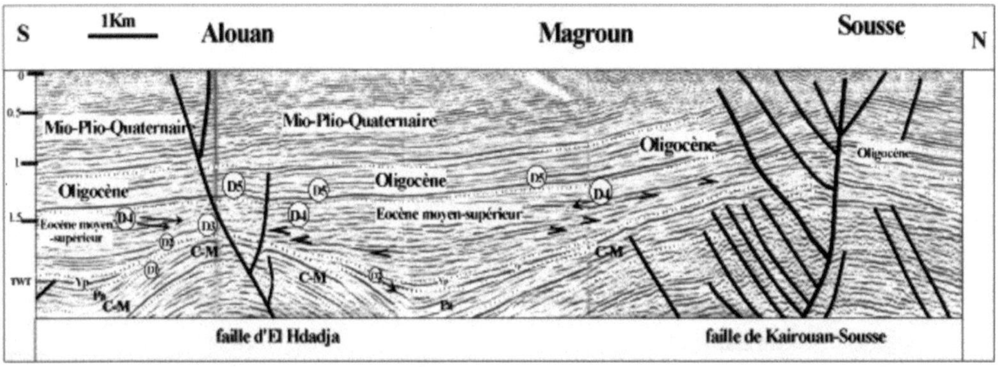

Fig. 10. Interpreted seismic line L1, showing the angular unconformities D1 to D5, the compressional Eocene tectonics and the extensional tectonic affecting the Oligocene (Khomsi et al., 2006). CM: Late Cretaceous, Pa: Palaeocene, Yp: Ypresian

However, we should stress that these discordances are always local and/or linear. Moreover, at these times, typical compressive structures (folds, reverse faults, etc.) are not observed or described in the Tunisian Atlas. On the contrary, as discussed above, the geometrical, kinematic and dynamic characteristics of these structures indicate that the studied area corresponds to an extensional (or transtensive) setting.

4.2 Inherited structures, deformation style and compressive tectonics

The deformation style resulting from these two major tectonic phases (i.e. Tortonian and Villafranchian) are different from one region to another (Fig. 11).

We find a geographical distribution of deformation that is similar to the pattern that prevailed during the basin history: central block and mobile borders. In addition, the variety of structural styles draws its origin from the structures gradually acquired since the Jurassic and accumulated until the Oligocene. The compressive deformation phases appear strongly controlled by the pre-existing geometries: folds, faults and strike-slip faults are directly induced by the paleostructures orientation of the TB.

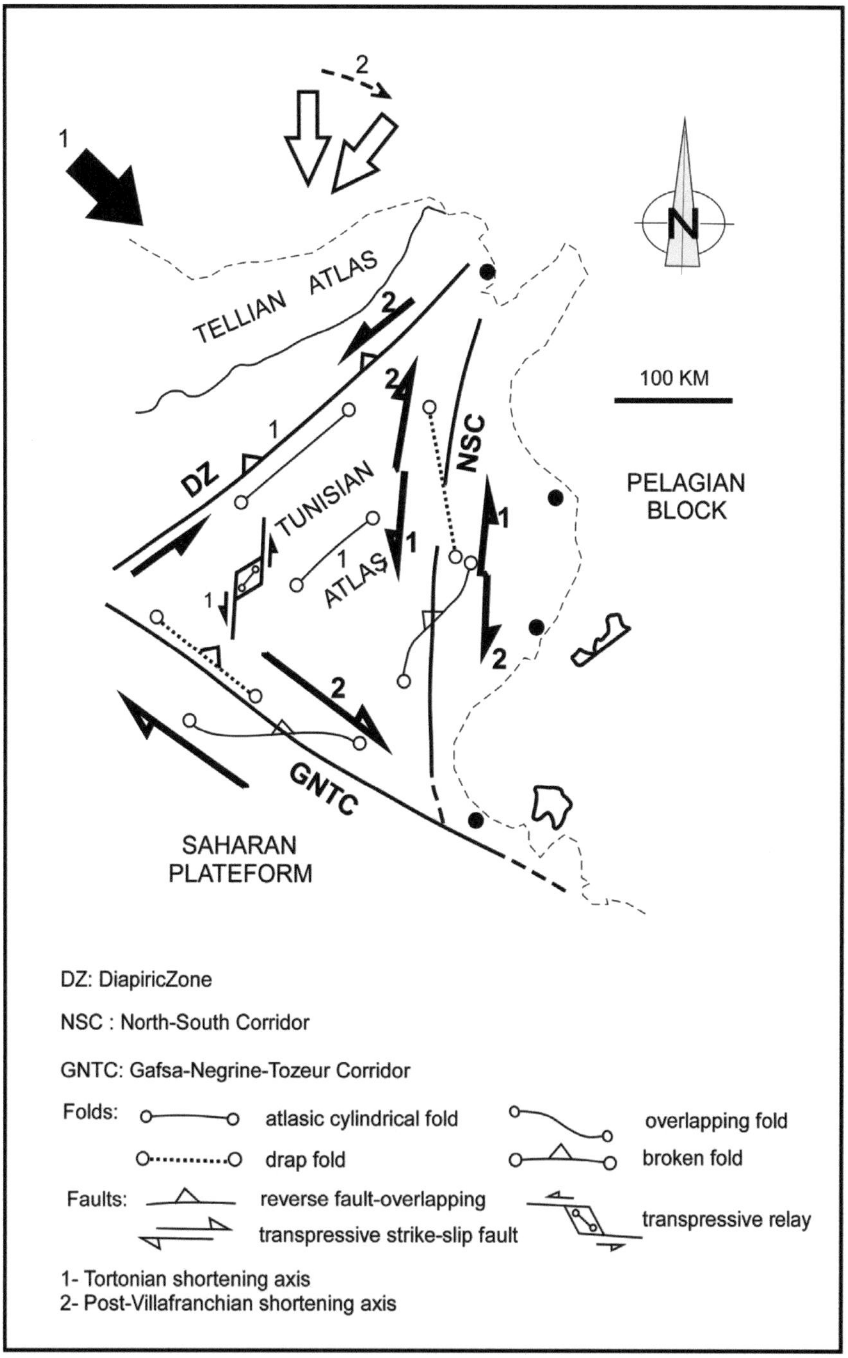

Fig. 11. Tectonic control of compressive structures by inherited structure

On the boundaries, the tectonic reactivation affects essentially vertical accidents (localized at the block limit) inducing with ancient sedimentary wedges a very superficial overthrust (Fig. 12).

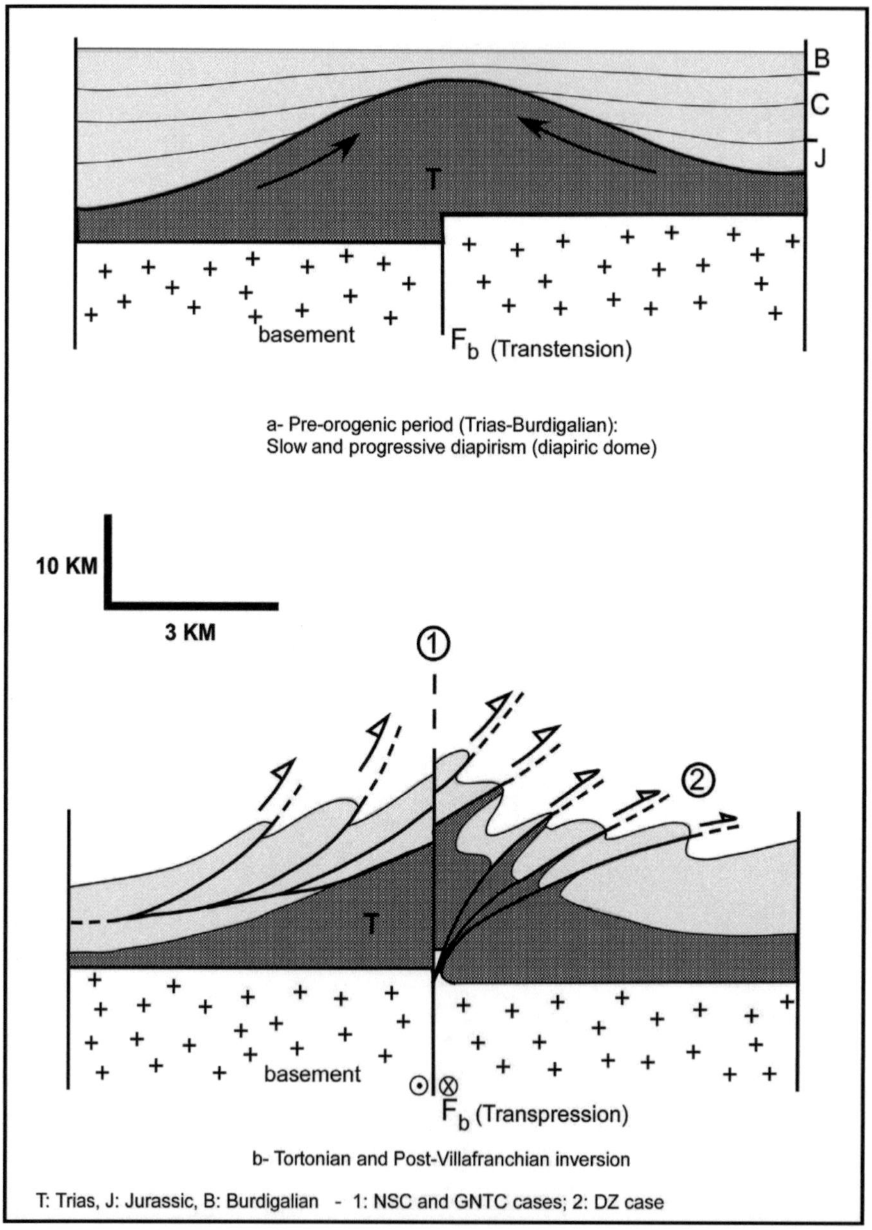

Fig. 12. Basement fault, diapiric dome, sedimentary wedges and superficial structures associated to structural inversion

Within the TB, the folds or broken-fold are, still, the superficial expressions of deep and vertical faults observed in tow major direction:

- The atlasic direction (SW-NE) corresponding to major folds very spaced with vast plain characterized by tabular structure;
- The N-S direction associated to a transpressive relay corresponding to ancient depocenters.

5. Conclusions

From the Triassic to the Middle Miocene, the Tunisian basin represents a typical intracontinental basin for the reason that there is no detectable evidence of ocean-floor accretion or a classical passive margin setting. Throughout its history, this basin shows a structure composed of blocks, all of which are incorporated into a main block (the Tunisian block: BT) of triangular form and bounded by continuously mobile shear corridors which thus result from polyphase deformation. The mobility of these corridors corresponds to the reactivation of inherited faults in an extensive or transtensive setting.

The changes in stress regime (or rotations), and hence the extension directions, are related to the existence of two sedimentary basins: the first in the North corresponds to a strike-slip margin (Western Tethys with numerous pull-apart basins), while the second, in the East, corresponds to the present-day Tunisian basin, which is developed in an oceanic setting on a passive-type margin (Mesogea), with predominance of the activity of one or the other of the basins at different times. Three major crises can be recognized, the first at the end of Jurassic, the second at the end of the Aptian (Soyer and Tricart, 1987; Rigane et al., 2010) and the last at the end of the Ypresian (Rigane et al., 1994; Gourmelen et al., 2000). This vertical tectonics is often characterized by abundant drape folds giving rise to extensional-fault related folding and strike-slip/dip-slip fault geometry, creating frequent discordances that are always localized and linear.

The salt tectonic phenomena related to the Triassic are linear (diapir zone axis) or localized (tectonic node), form high zones during the basin evolution that accentuate the vertical tectonics associated with the extension and/or transtension of blocks. The diapiric movements are slow and progressive, taking place from the Late Triassic until the Langhian, and inducing the formation of many sedimentary wedges on the dome flanks.

The folded chain resulting from the structural inversion of the Tunisian paleoblock and its boundaries show mainly SW-NE-trending folds since the Tortonian, in accord with the NW-SE major shortening direction. However, other folds are characterized by oblique axes compared to this shortening direction, which appear completely aberrant in relation to the two main folding phases: Tortonian (NW-SE) and post-Villafranchian (N-S). These axial trends can be interpreted in the context of an intracontinental chain where the inherited major vertical faults can only develop at the surface as drape folds in a transpressive regime.

Finally, the Tunisian block has an eastern front that has been subsiding since Middle Miocene times, with SSW-NNE oriented depocentres on the edge of the NSC, but which follow an almost orthogonal NW-SE trend close to the Malta scarp. A renewed vertical inversion of wide extent thus affects the Tunisian block, in contrast to the Pelagian block, as a result of the present-day N-S compression.

6. References

Abbès, Ch. (2004). Structurations et évolutions tectono-sédimentaires mésozoïques et cénozoïques, associés aux accidents reghmatiques, à la jonction des marges téthysienne et Nord-africaine (chaîne Nord-Sud, Tunisie centrale). Thèse Doctorat ès-Sciences. Univ. Tunis El Manar, 437p.

Aissaoui, D. (1984). Les structures liées à l'accident sud-atlasique entre Biskra et le Jebel Manndra, Algérie, évolution géométrique et cinématique. Thèse 3ème cycle, Univ. L. Pasteur, Strasbourg, 145p.

Ben Ayed, N. (1980). Le rôle des décrochements E-W dans l'évolution structurale de l'Atlas tunisien. C. R. somm. Soc. Géol. France, Vol.1, pp.29-32

Ben Ayed, N. (1986). Evolution tectonique de l'avant-pays de la chaîne alpine de Tunisie du début du Mésozoïque à l'Actuel. Thèse Doct. Ès-Sciences Naturelles, Univ. Paris Sud-Orsay, 286p.

Ben Ayed, N.; Chekhma, H. & Gueddiche, M. (1997). La zonation des structures compressives et distensives de la Tunisie centrale et ses relations avec les décrochements. *C. R. Acad. Sci. Paris*, Vol.324, pp.485-490.

Ben Ayed, N. & Viguier, C. (1981). Interprétation structurale de la Tunisie atlasique. *C. R. Acad. Sc.*, Paris, Vol.292, pp.1445-1448.

Ben Ferjani A.; Burollet P.F. & Mejri F. (2006). Petroleum Geology of Tunisia. A Renwed Synthesis. *Entreprise Tunisienne d'Activités Pétrolières*, Mém. N°22.

Biely, A.; Burollet, P.F. & Lajmi, T. (1974). Etude géodynamique de la Tunisie et des secteurs voisins de la méditerranée. *Notes Serv. Géol. Tunisie*, Vol.41, pp.23-38.

Bishop, W.F. (1975). Geology of Tunisia and adjacent parts of Algeria and Libya. *AAPG Bulletin*, Vol.59, pp.413-450.

Bismuth, H.; Boltenhagen, C.; Donze, P.; Le Fevre, J. & Saint-Marc, P. (1982). Etude sédimentologique et biostratigraphique du Crétacé moyen et supérieur du Jebel Semmama (Tunisie Centre Nord). *Cretaceous Research*, Vol.3, pp.171-185.

Blondel, T.J.A. (1991). Les séries à tendance transgressive marine du Miocène inférieur à moyen de Tunisie centrale. Thèse Univ. Genève, n° 2469, Swiss, 409p.

Boltenhagen, C. (1985). Paléogéographie du Crétacé moyen de la Tunisie centrale. *Proceedings of 1er Congrès Nat. Sc. Terre*, Tunis, 1981, 1, 97-114.

Bouaziz, S. (1995). Etude de la tectonique cassante dans la plate-forme et l'Atlas sahariens (Tunisie méridionale) : évolution des pléochamps de contraintes et implications géodynamiques. Thèse Doc. Etat, Univ.Tunis II, 485p.

Bouaziz, S.; Barrier, E.; Soussi M.; Turki, M.M. & Zouari, H. (2002). Tectonic evolution of the northern African margin in Tunisia from paleostress data sedimentary record. *Tectonophysics*, Vol.357, pp. 227-253.

Boukadi, N.; Bédir, M. & Zargouni, F. (1998). Geometric and kinematic analyses of pull-apart basins produced by en echelon strike-slip of Gafsa fault systems (Southern Tunisia). *African Geoscience Review*, Vol.5, pp. 327-338.

Brahim, N. & Mercier, E., (2007). Commentaire à la note intitulée « Mise en évidence en subsurface d'événements compressifs Eocène moyen-supérieur en Tunisie orientale (sahel) : généralité de la phase atlasique en Afrique du Nord de Sami Khomsi, Mourad Bédir, Mohamed Soussi, Mohamed Ghazi Ben Jemia et Kmar Ben Ismail-Lattrache [C. R. Geoscience 338 (2006) 41-49]. *C. R. Geoscience*, Vol.338, pp. 41-49.

Burollet, P.F. (1956). Contribution à l'étude stratigraphique de la Tunisie centrale. *Ann. Mines Géol. Tunis*, 18, 350p.

Burollet, P.F. (1991). Structures and tectonics of Tunisia. *Tectonophysics*, Vol.195, pp. 359-369.

Burollet, P.F.; Mejri, F. & Ben Kilani, H. (2000). What about the deep structure of Tunisia? *Proceedings of 7th Tunisian Petroleum Exploration and Production Conference*, Tunis, April, 167-178.

Castany, G. (1951). Etude géologique de l'Atlas tunisien oriental. Ann. Min. et Géol., Tunisie, 8. Thèse Doct. Ès-Sc., Paris

Dercourt, J.; Zonenshain, L.P. & Ricou, L.E. (1985). Présentation de 9 cartes paléogéographiques au 1/20 000 000 s'étendant de l'Atlantique au Pamir pour la période du Lias à l'Actuel. *Bull. Soc. Géol. France*, Vol.8, pp. 635-652.

Dewey, J.F.; Helman, M.L.; Turko, E.; Hutton, D.H.W. & Knott, S.D. (1989). Kinematics of the western Mediterranean. *Geol. Soc. Spec. Publ.*, Vol.45, pp.265-83.

El Ghali, A.; Bobier, C. & Ben Ayed, N. (2003). Rôle du système de failles E-W dans l'évolution géodynamique de l'avant-pays de la chaîne alpine de Tunisie. Exemple de l'accident de Sbiba-Chérichira en Tunisie centrale. *Bull. Soc. Géol. France*, Vol.174, pp.373-381.

European Geotraverse, (1992). The EGT'85 seismic experiment: a reconnaissance of the deep structure. *Tectonophysics*, Vol.207, pp. 245-267.

Feki, M.; Rigane, A. & Gourmelen, C. (2005). Tectonique distensive fini-aptienne au Jebel El Hamra de Kasserine (Tunisie centro-occidentale). *Notes Serv. Géol. Tunisie*, Vol.73, pp.77-89.

Frizon de Lamotte, D.; Michard, A. & Saddiqi, O. (2006). Quelques développements récents sur la géodynamique du Maghreb. *C. R. Geoscience*, Vol.336, pp.1-10.

Gabtni, H.; Jallouli, C.; Mickus, K.; Zouari, H. & Turki, M.M. (2005). Geophysical constraints on the location and nature of the North Saharan Flexure in Southern Tunisia. *Pure appl. Geophys*. Vol.162, pp.2051-2069.

Gourmelen, C. (1984). Serrage polyphasé de paléostructures distensives dans l'Axe Nord-Sud tunisien : le segment Bou Zer-Rhéouis. Thèse 3ème cycle, Grenoble, 216p.

Gourmelen, C. & Tricart, P. (1985). Nouvelle interprétation de l'Axe Nord-Sud en Tunisie centrale : le chaînon du Bou Zer. *Terra Cognita*, EUG, Strasbourg, Vol.5, pp.258-259.

Gourmelen, C.; Ouali, J. & Tricart. P. (1989). Les blocs basculés mésozoïques dans l'Axe nord-sud de Tunisie centrale: importance et signification. *Bull. Soc. Géol. France*, Vol.8, pp.117-122.

Gourmelen, C.; Rigane, A.; Broquet, P.; Truillet, R. & Aïte, M.R. (2000). Caractères structuraux et dynamiques d'un bassin en transtension : la plate-forme tunisienne à l'Yprésien terminal. *Bull. Soc. Géol. France*, Vol.171, pp.559-568.

Guiraud, R. & Maurin, J.C. (1991). Le rifting au Crétacé inférieur: synthèse structurale, mise en évidence de deux étapes dans la genèse des bassins, relations avec les couvertures océaniques péri-africaines. *Bull. Soc. Géol. France*, Vol.162, pp.811- 823.

Guiraud, R. and Maurin, J.C. (1992). Early Cretaceous rifts of western and central Africa: an overview. *Tectonophysics*, Vol.213, pp.153–168.

Guiraud, R.; Bosworth, W.; Thierry, J. & Delplanque, A. (2005). Phanerozoic geological evolution of Northern and Central Africa: an overview. *J. African Earth Sci*. Vol.43, pp.83–143.

Hlaiem, A., (1999). Halokinesis and structural evolution of the major features in eastern and southern Atlas. Tectonophysics, Vol.306, pp79-95.

Jauzein, A. & Rouvier, H. (1965). Sur les formations allochtones de Kroumirie (Tunisie septentrionale). C. R. somm. Soc. Géol., Vol.2, pp.36-38.

Jauzein, A. (1967). Contribution à l'étude géologique des confins de la bordure tunisienne (Tunisie septentrionale). *Ann. Mines et géol.*, Tunisie, Vol.22, 475p.

Kamoun, F.; Peybernès, B. & Fauré P., (1999). Evolution paléogéographique de la Tunisie saharienne et atlasique au cours du Jurassique. C. R. Acad. Sciences, Paris, Vol.328, pp.547-552.

Khessibi, M., 1978. Etude géologique du secteur Maknassy-Mezzouna et du Jebel Kebar (Tunisie centrale). - Thèse Univ. Claude Bernard, Lyon, 175p.

Khomsi, S.; Bedir, M.; Soussi, M.; Ben Jemia, M.G. & Ben Ismail-Lattrache, K. (2006). Mise en évidence en subsurface d'événements compressifs Eocène moyen-supérieur en Tunisie orientale (Sahel) : généralités de la phase atlasique en Afrique du Nord. *C. R. Géoscience*, Vol.338, pp. 41-49.

Lüning, S.; Kolonic, S.; Belhaj, E.M.; Belhaj, Z.; Cota, L.; Baric, G. & Wagner, T. (2004). Integrated depositional model for the Cenomanian-Turonian organic-rich strata in North Africa. *Earth-Science Reviews*, Vol.64, pp.51-117.

Martinez, C., & Truillet, R. (1987). Evolution paléogéographique et structurale de la Tunisie.- *Mém. Soc. Géol. It.*, Vol.38, pp.35-45.

Midassi, M., (1982). Regional gravity of Tunisia. Masters Thesis, Univ. South Carolina. USA, 125p.

Morelli, C. & Nicolih, R. (1990). A cross section of the lithosphere along the European Geotraverse Southern Segment (from the Alps to Tunisia). *Tectonophysics*, Vol.176, pp.229-243.

M'Rabet, A. (1981). Stratigraphie, sédimentation et diagenèse carbonatée des séries du Crétacé inférieur de Tunisie centrale. Thèse Doct. Ès Sciences, Univ. Paris Sud Orsay, 540p.

Ould Bagga, M.A.; Saadi, A. & Mercier, E. (2006). La "zone des nappes" de Tunisie: une marge méso-cénozoïque en blocs basculés modérément inversé (région de Tabarka/Jendouba ; Tunisie nord-occidentale). *Bull. Soc. Géol. France*, Vol.177, pp.145-154.

Perthuisot, V. (1978). Dynamique et pétrogenèse des extrusions triasiques en Tunisie septentrionale. Thèse Doct. Ès-Sc. Nat., Ecole Norm. Sup., Paris, 312p.

Pervinquière, L. (1903). Etude géologique de la Tunisie centrale. In: F.R. DE Rudeval. (Ed.), Thèse doct. es-sciences, Paris. 360p.

Piqué, A.; Ait Brahim, L.; Ait Ouali, R.; Amrhar, M.; Charroud, F.; Gourmelen, C.; Laville, E.; Rekhiss, F. & Tricart, P. (1998). Evolution structurale des domaines atlasiques du Maghreb au Méso-Cénozoïque ; le rôle des structures héritées dans la déformation du domaine atlasique de l'Afrique du Nord. *Bull. Soc. Géol. France*, Vol.169, pp.797-810.

Piqué, A.; Tricart, P.; Guiraud, R.; Laville, E.; Bouaziz, S.; Amrhar, M. & Ait Ouali, R. (2002). The Mesozoic-Cenozoic Atlas belt (North Africa): an overview. *Geodinamica Acta*, Vol.15, pp.185-208.

Rabhi, M. & Ben Ayed, N. (1990). Mise en évidence d'une tectonique compressive d'âge Aptien supérieur en Tunisie centrale. *Notes Serv. Géol. Tunisie*, Vol.56, pp.91-98.

Richert, J.P. (1971). Mise en évidence de quatre phases tectoniques successives en Tunisie. *Notes Serv. Géol. Tunisie*, Vol.34, pp.115-125.

Ricou, L.E. (1994). Tethys reconstructed: plates Continental fragments and their boundaries since 260 Ma from Central America to South-Eastern Asia. *Geodinamica Acta*, Vol.7, pp.160–218.

Rigane, A. (1991). Les calcaires de l'Yprésien en Tunisie centro-septentrionale : cartographie, cinématique et dynamique des structures. Thèse d'Univ. Franche-Comté, 236, 214p.

Rigane, A.; Gourmelen, C.; Broquet, P. & Truillet, R., (1994). Originalité des phénomènes tectoniques synsédimentaires fini-yprésiens en Tunisie centro-septentrionale (région de Kairouan). *Bull. Soc. Géol. France*, Vol.156, pp.27-35.

Rigane, A.; Feki, M.; Gourmelen, C. & Montacer, M. (2010). The « Aptian Crisis » of the South-Tethyan margin: New tectonic data in Tunisia. *Journal of African Earth Sciences*, Vol.57, pp.360-366.

Rigane, A. & Gourmelen, C. (submitted, to Journal of African Earth Sciences, under revision). Inverted intracontinental Basin and Vertical Tectonics: The Saharan Atlas in Tunisia.

Rouvier, H. (1977). Géologie de l'extrême nord tunisien : tectoniques et paléogéographie superposées à l'extrémité orientale de la chaîne nord maghrébine. Thèse d'Etat, Université Paris-VI, 898 p.

Schettino, A. & Scotese, C. (2002). Global kinematic constraints to the tectonic history of the Mediterranean region and surrounding areas during the Jurassic and Cretaceous. In: In: Rosenbaum, G. and Lister, G. S. 2002. Reconstruction of the evolution of the Alpine-Himalayan Orogen. *Journal Virtual Explorer*, Vol.8, pp.148-168

Soussi, M. (2002). Le Jurassique de la Tunisie atlasique : stratigraphie, dynamique sédimentaire, paléogéographie et intérêt pétrolier. Ed. Villeurbanne : Univ. Claude Bernard-UFR des Sciences de la Terre, 363p.

Soyer C. & Tricart, P. (1987). La crise aptienne en Tunisie centrale : approche paléostructurale aux confins de l'Atlas et de l' « axe Nord - Sud », C. R. Acad. Sc., Paris, Vol.305, pp.301-305.

Stampfli, G.M.; Borel, G.; R., Marchant, R. & Mosar J. (2002). Western Alps geological constraints on western Tethyan reconstructions. In: Rosenbaum, G. and Lister, G. S. 2002. Reconstruction of the evolution of the Alpine-Himalayan Orogen. *Journal Virtual Explorer*, Vol.8, pp.77-106.

Turki, M.M. (1984). L'Ousselat-Boudabbous : un exemple de bassin subsident sénonien supérieur-éocène inférieur. Implication sur la notion d' « axe Nord-Sud » (Atlas tunisien). 10ème RAST Bordeaux, p. 529.

Turki, M.M. (1985). Polycinématique et contrôle sédimentaire associé sur la cicatrice Zaghouan-Nebhana. Thèse ès-Sc., Univ. Tunis, Revue Sc. Terre, édit. INRST, Tunis, 1988, 262p.

Yaich, C. (1984). Etude géologique des chainons du Chérahil et du Khechem El Artsouma (Tunisie centrale). Liaison avec les structures profondes des plaines adjacentes. Thèse 3ème cycle. Univ. Franche-Comté. Besançon, 165p.

Yaich, C. (1997). Dynamique sédimentaire, eustatisme et tectonique Durant l'Oligo-Miocène en Tunisie centro-septentrionale. Thèse d'Etat, Univ. de Tunis, 479p.

Zaier, A.; Béji-Sassi, A. & Moody, R.T.J. (1998). Basin evolution and deposition during the early Paleogene in Tunisia. *Geological Society*, London, Special Publication, Vol.132, pp.375-393.

Zargouni, F. (1975). Etude géologique de la chaîne de Lansarine (région de Tébourba, Atlas tunisien). Thèse de 3ème cycle. Univ. Pierre et Marie-Curie (Paris 6).

Zargouni, F. (1985). Tectonique de l'Atlas méridional de Tunisie, évolution géométrique et cinématique des structures en zone de cisaillement. Thèse Doct. ès-Sc., Univ. Louis Pasteur (Strasbourg). Ed. Rev. Sc. Terre, INRST, Tunisie.

Zargouni, F. (1986). Style tectonique de l'Atlas méridional de Tunisie. Evolution géométrique et cinématique des structures en zone de cisaillement. Thèse es Sciences, Univ. Louis Pasteur Strasbourg. Revue Sciences Terre. Ed. Centre Sciences, Institut National Recherches Scientifiques Techniques, Tunisie (1985), 304p.

Zouaghi, T.; Bedir, M. & Inoubli, M.H. (2005). 2D Seismic interpretation of strike-slip faulting, salt tectonics, and Cretaceous unconformities, Atlas Mountains, central Tunisia. *Journal of African Earth Sciences*, Vol.43, pp.464-486.

Zouari, H.; Turki, M.M.; Delteil, J. & Stephan, J.F. (1999). Tectonique transtensive de la paléomarge tunisienne au cours de l'Aptien-Campanien. *Bull. Soc. Géol.* France, Vol.170, pp.295-301.

3

The Chotts Fold Belt of Southern Tunisia, North African Margin: Structural Pattern, Evolution, and Regional Geodynamic Implications

Taher Zouaghi[1], Rihab Guellala[1], Marzouk Lazzez[2], Mourad Bédir[1], Mohamed Ben Youssef[1], Mohamed Hédi Inoubli[3] and Fouad Zargouni[3]
[1]*Laboratoire de Géoressources, CERTE, Pôle Technologique de Borj Cédria, Soliman,*
[2]*Entreprise Tunisienne d'Activités Pétrolières, Monplaisir, Tunis,*
[3]*Département des Sciences de la Terre, FST, Université Tunis El Manar, Tunis*
Tunisia

1. Introduction

At the North of the old African continent, craton and shields having more than two billion years, Tunisia, Algeria and northern Morocco underwent a complex geodynamic and structural evolution during the Mesozoic and Cenozoic times (Dercourt et al., 1985; Bouillin, 1986; Frizon de Lamotte et al., 2009). This evolution resulted in the development of varied paleogeographic fields, in relation with the Tethyan and Atlantic movements. Its end led to the genesis of the North-African alpine orogen (Dercourt et al., 1985; Martinez et al., 1990) formed by the Maghrebid and Atlassic domains (Fig. 1).

Tunisia occupies the eastern part of this orogen, located at the north of a large Saharan platform, developed on the stable African craton, not deformed during the alpine cycle and bounded by a major structural lineament « South Atlassic fault » composed of complex overlapping folds trending NE-SW, E-W and NW-SE (Caire,1971; Zargouni, 1985; Turki, 1988; Zouari et al., 1990; Ben Ayed, 1993; Boukadi, 1994; Bédir, 1995; Bouaziz, 1995; Zouari, 1995; Bouaziz et al., 1999, 2002; Abbès, 2004; Zouaghi et al., 2005a, b, 2011; Ouali, 2007; Melki et al, 2010).

Structures of the North African margin were usually subject of discussion. This domain could be considered as a passive margin, close to the oceanic opening, characterized by a strong subsidence marked by accumulations of prograding deposits (Dercourt et al., 1985; Biju-Duval et al., 1976). For others, it is a transform margin related to displacements of the African plate compared to the Eurasian plate. These movements generated opening of the Paleo-Tethys (Arthaud and Thomas, 1977). The Africa-Europe relative motions would be at the origin of the recent ocean floor spreading of the Mediterranean (Taponnier, 1977; Reading, 1980; Olivet et al., 1982; Alvarez et al., 1984; Ricou, 1994).

The study area belongs to the North African margin and the northern edge of the Saharan platform. Studies undertaken on Paleo-Tethys show the development of deformed and subsiding zones between the cratonic blocks and the basins (Caire, 1974; Arthaud and Thomas, 1977; Aubouin and Debelmas, 1980; Bernoulli and Lemoine, 1980; Durand-Delga and Fonrbote, 1980; Bousquet and Philip, 1981; Dercourt et al., 1992). The geodynamic aspects at the Mediterranean scale are the origin of the tectonic mechanisms responsible for the structural evolution of the study area during the Mesozoic and Cenozoic periods. These aspects correspond to: (1) Permian-Triassic Tethyan rifting. Mesozoic divergence between the blocks continues and results in opening of the central Atlantic and the Tethys ligure and development of the Mesogea following the fragmentation of the Pangea super-continent, which generates the Gondwana to the South and the Laurasia to the North. An extensional tectonic event was consequently generated during the beginning of Mesozoic times, recorded in the African and European margins. (Biju-Duval, 1980; Dercourt et al., 1985 and 1992). (2) During the Cenozoic times, blocks located on both sides of the Mediterranean Basin converge, involving compressional phases, which induced formation of the European alpine chains and the Maghrebides (Bouillin, 1977, 1986).

The tectonic polyphasage in North Africa domains presents one of the most discussed subjects from an area to another. Some interpretations concerning the role of inherited features , halokinesis and later inversion, showed by outcrop studies, remain still not well argued even if the majority of authors agree with the influence of the ante-Triassic basement on the sedimentary lapout. In this work we try to study the geometry and structural evolution of the various morphostructural units during extensional and contractional periods.

2. Paleogeographic summarize

The studies carried out on outcropping strata and on well data of southern Tunisia allow to identify the lithostratigraphic series from Paleozoic to the Quaternary one. However, there are some divergences between the authors concerning the age of various geological Formations (Burollet, 1956; Fournié, 1978; M'Rabet, 1987). The sedimentary series show lateral variations of thickness and facies and local gaps. In this section we try to describe and discuss briefly the sedimentary history of the study area.

The Paleozoic outcrops in the Tebaga mountain is represented only by Permian deposits. In addition the Paleozoic has been crossed by many petroleum wells in Saharan platform of Tunisia where various units have been identified. (Busson, 1969, 1970a, b; Ben Ismail, 1982, 1991). During the Paleozoic periods, the Saharan field is characterized by clayey detrital and sandy facies indicating a continental to margino-continental deposition (Fig. 2) with a general tendency to marine platform towards the North in the Djeffara domain (Bellini and Massa, 1980). For the southern Atlas, no information exists yet concerning the Paleozoic. But based on its lithostratigraphy, the Early Paleozoic could be marked by a progradational series followed by a transgressive interval corresponding to the clays of Silurian-Devonian known in the Saharan field (Busson, 1969, 1970a, b; Ben Ismail, 1982).

Except the Triassic, which keeps a relative homogeneity of facies from the South to the North (Kamoun et al., 2001), the Mesozoic paleogeographic scheme is characterized by a marginal platform environment with continental influence marked by detrital, carbonate-evaporite and evaporite to the South and by a marine deposition (clayey, more carbonated-clay, less evaporitic and sandy) to the North.

In Tunisia the Jurassic rarely outcrops and is classically represented by three carbonated members of the Nara Formation (Burollet, 1956; Soussi, 2002) (Fig. 2), indicating deposition in a moderate deep marine of external platform in central Tunisia. Towards the South, in the Saharan platform, presence of detrital and evaporite layers indicate fluvio-deltaic and lagunal internal platform under a restricted and confined marine environment. (Ben Ismail, 1982; Chandoul et al., 1993 ; Kamoun et al., 1999).

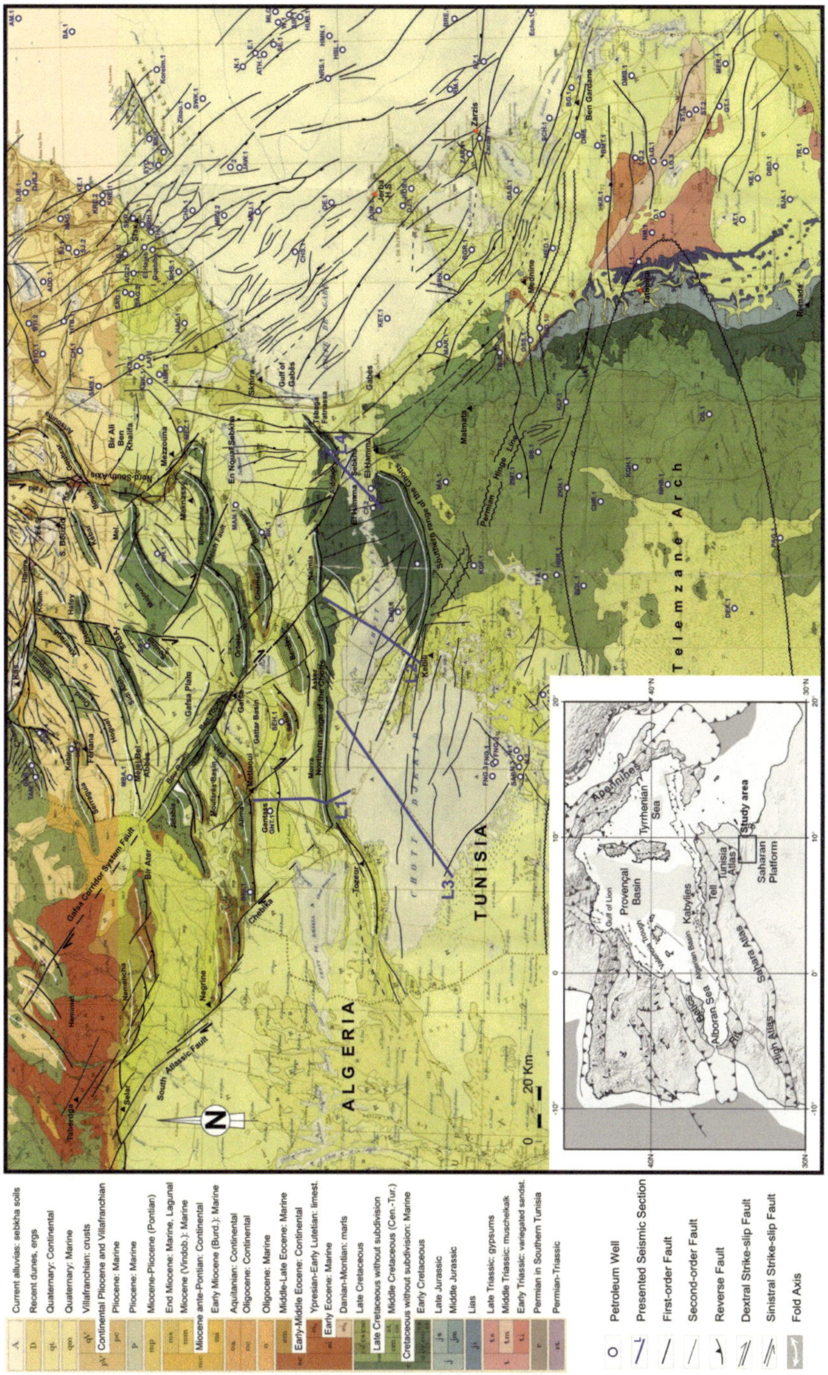

Fig. 1. Geological (Castany, 1951) and tectonic (Zargouni, 1985; Ben Ferjani, 1990; Bédir, 1995; Zouaghi et al., 2009, 2011) setting of the southern Atlas and Saharan platform of Tunisia and eastern Algeria showing distribution of anticline axis trends, location of main outcrops, deep faults and distribution of Mesozoic basins and paleohighs

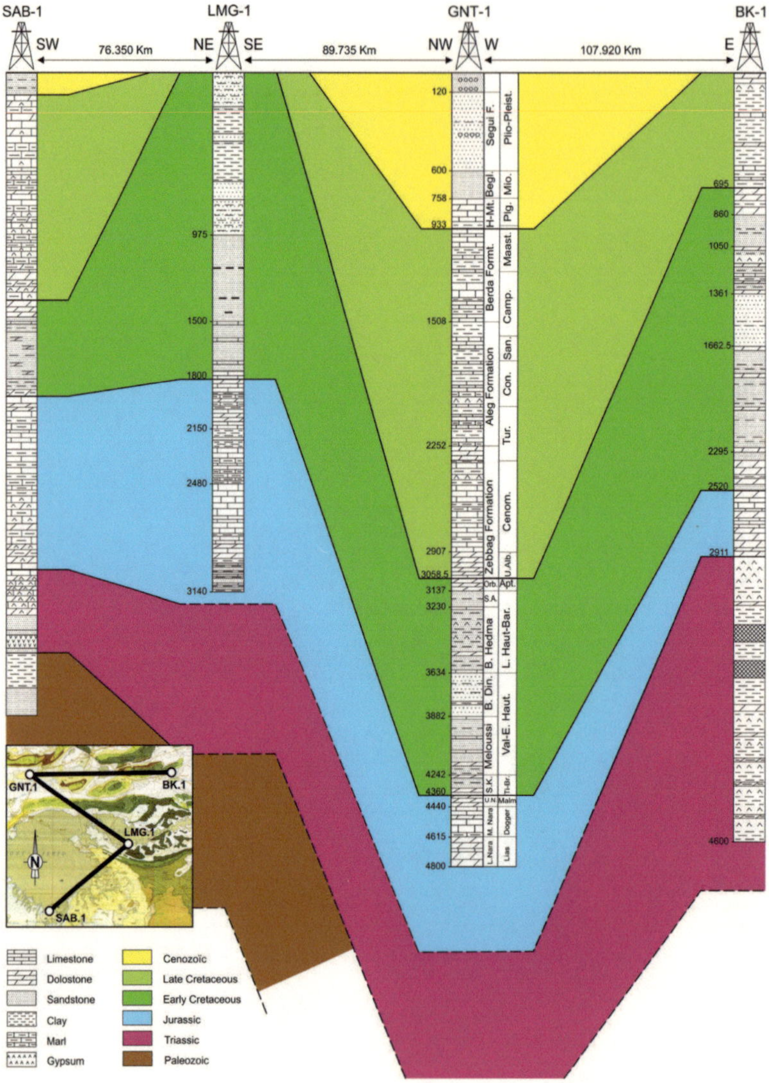

Fig. 2. Lithostratigraphic correlation of Paleozoic to Cenozoic series highlighted in the petroleum wells, showing thickness and facies variation between structures of southern Tunisia

The Cretaceous, which largely outcropped can be subdivided into two great mega-units: the first essentially clastic until Aptian corresponds to the lower Cretaceous fluvial and deltaic to marine environments (Marie et al., 1984; M'Rabet, 1987; Ben Youssef, 1999; Azaïez et al., 2007 ; Lazzez et al., 2008; Guellala, 2010; Zouaghi et al., 2011); the second represented by carbonates, clays and rare evaporitic layers, corresponds to the Late Cretaceous (Marie et al., 1984; Fakraoui, 1990; Abdallah et al., 2000). The southern part of Tunisia belongs to sub-continental field of Saharan platform, where the sedimentation rate is low to null in some localities of the vast stable platform (Burollet, 1956; Bishop, 1975; Ben Ferjani et al., 1990; Negra, 1994; Chaabani, 1995; Zouaghi et al., 2011).

The Paleogene represented by clays, carbonates and evaporites is identified in the Gafsa-Metlaoui phosphate basin (Sassi, 1974; Chaabani, 1995), showing varied thicknesses. The

Saharan platform domain and the Chotts zone, already emerged since the end of the Cretaceous, are deprived of Paleogene sedimentation (Fig. 2). The Cenozoic is represented in the Saharan domain by the Neogene-Quaternary continental sandy and silty deposits (Zargouni, 1985; Fakraoui, 1990; Addoum, 1995). The marine deposition has evolved to continental since the end of the Cretaceous. In the Atlassic domain, the marine environment continued at least until the end of the Eocene. It is marked by the clayey, carbonated and evaporite series of the Paleogene and changed to frankly continental detrital sedimentation during the Neogene and Quaternary periods.

3. Tectonic setting

Placed within the Maghrebin framework, Tunisia occupies a privileged geological position in the African structuring. It belongs to the old African frame by its southern Saharan part, and to the alpine field by its northern area. The boundary between these two domains is marked by the South-Atlassic morphostructural master fault system (Fig. 3).

3.1 Southern Atlas

The southern Atlas, extension of the western Saharan Atlas in Algeria, includes mountains of the Gafsa area trending E-W to NE-SW and NW-SE (Burollet, 1956; Boltenhagen, 1985; Zargouni, 1985; Abdeljaoued and Zargouni, 1985; Boukadi, 1994; Zouari, 1995; Bédir, 1995; Bédir et al.; 2001, Hlaiem, 1999; Bouaziz et al, 2002; Zouaghi et al., 2005a, b, 2009, 2011). These chains consist of overlapping folds poured to the South and separated by synclines filled with Neogene and Quaternary deposits (Fig. 1).

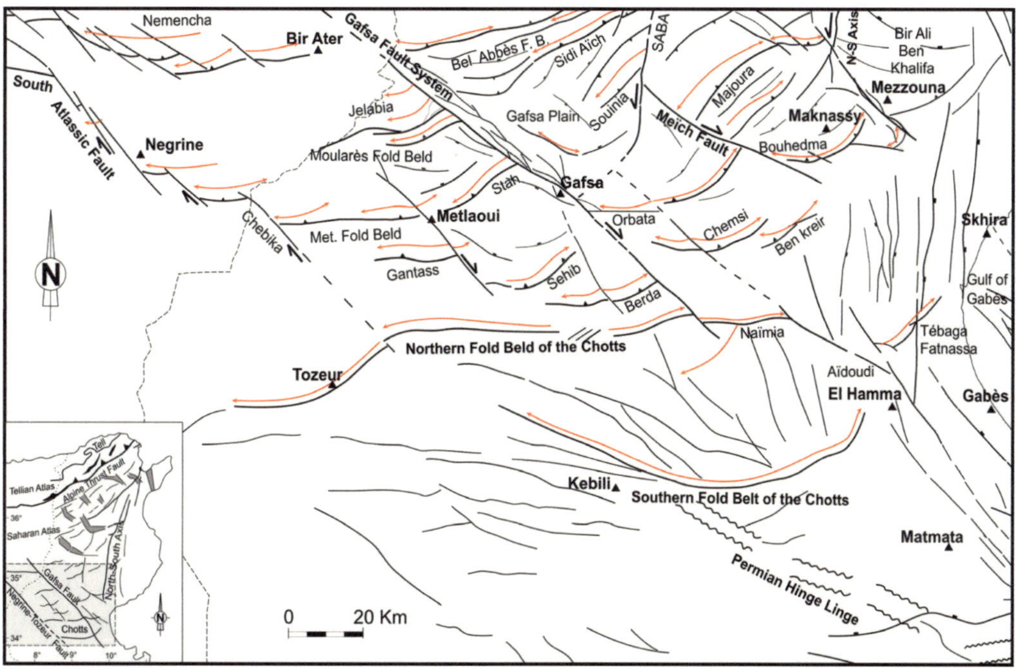

Fig. 3. Structural map, showing location of the master trending faults and folds. Rhombic structures are highlighted between the northwest-southeast and east-west right-steeping, dextral strike-slip faults (Zargouni, 1985; Ben Ferjani, 1990; Bédir, 1995; Zouaghi et al., 2009, 2011)

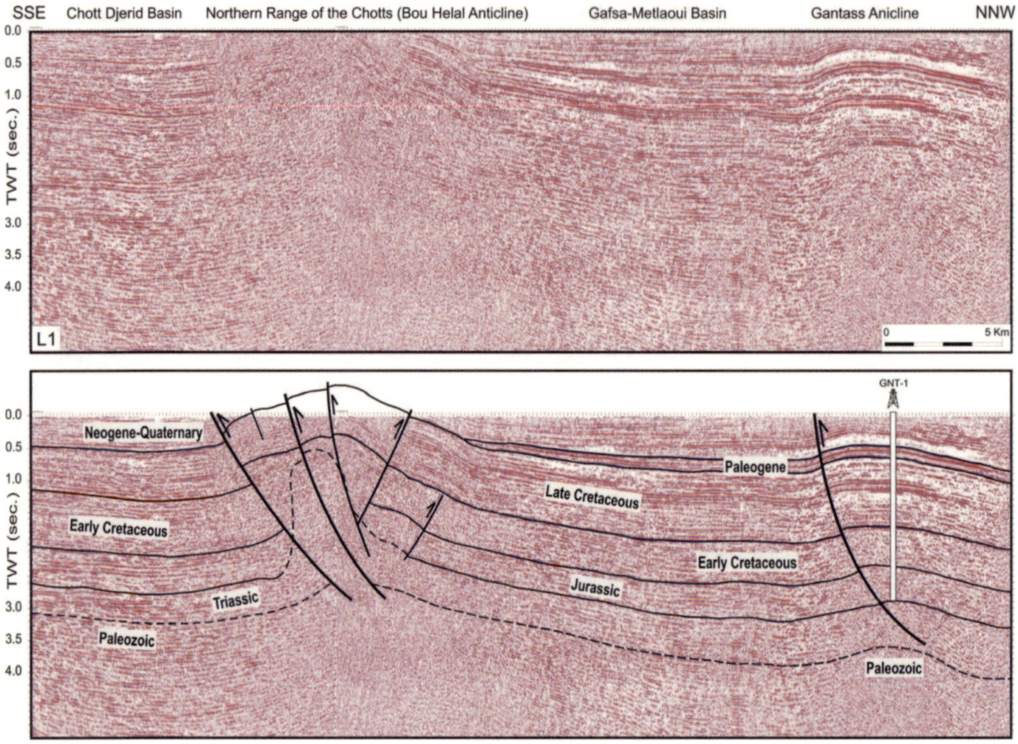

Fig. 4. NW-SE interpreted seismic section L1 crossing the Gantass and northern range of the Chotts. The reduction of thickness and unconformities are related to rejuvenation of master faults associated with Triassic evaporate risings

The Metlaoui tectonic bundle, which is made from West to East by the Bliji, Alima, Oum El Khecheb and Stah anticlines, is extended to the West in Algeria by the Mandra anticlines and is truncated to the East by the Gafsa master strike-slip fault. The Gafsa fold belt generally trending NW-SE is composed by the Moulares, Bou Ramli and Ben Younes anticlines and is crossed by the N120 Gafsa master dextral wrench fault (Zargouni, 1985; Zargouni et al., 1985; Boukadi, 1994; Bédir, 1995; Zouari, 1995; Boutib and Zargouni, 1998; Zouaghi et al., 2005a, b, 2009) (Fig. 1). The Orbata fold belt located at the East of Gafsa town is constituted by the Orbata and Bou Hedma anticlines, which are often asymmetric with sigmoidal shape (Boukadi et al., 1998 ; Bensalem et al., 2009). The Bou Hedma structure is affected to the West by the Mech dextral strike-slip fault. Between the Metlaoui and Orbata chains and northern range of the Chotts are located the Gantass, Sehib, Berda, Chemsi and Ben Kheir separated overlapping folds (Figs. 3 and 4).

The Chotts fold belt, which includes northern and southern chain of Chotts separated by the Chott El Fedjedj depocenter, is the most external structure of the Atlassic domain (Rabia, 1985; Zargouni, 1985; Abdeljaoued and Zargouni, 1985; Fakraoui, 1990; Ben Ayed, 1993; Bouaziz, 1995; Hlaiem, 1999). The eastern end of this fold belt is affected by several faults; the most significant one is that of Bir Oum Ali. The Hadifa diapir appears to be composed by Triassic saliferous located in the eastern end of northern chain of the Chotts shows the Triassic halokinesis in the study area.

These folded structures, located in northern edge of the Saharan platform, are affected and truncated by faults trending NW-SE and E-W (Fig. 3). They are often anchored on these lineaments and are characterized by axial virgations and echelon along a WSW- ENE axial direction (Rabia, 1985; Zargouni, 1985; Fakraoui, 1990; Ben Ayed, 1993; Bouaziz, 1995). The great anticline structures are asymmetric and marked by faulted southern side with steep dip (Fig. 4).

3.2 Saharan platform

Located on the northern edge of the African craton, this domain is composed of a Precambrian substratum surmounted by a thick Paleozoic cover (Figs. 4 and 5). However, Dahar field, high since the Carboniferous times is unconformably overlain by Mesozoic deposits on the Medenine Upper Permian, which is the only Paleozoic outcrop in Tunisia (Busson, 1967; Burollet and Desforges, 1982; Bouaziz, 1986; Ben Ayed, 1993; Bouaziz, 1995; Bouaziz et al., 1999). Except the upper Permian marine deposits of the Tebaga, the Precambrian and Paleozoic are recognized, in southern Tunisia by deep petroleum wells (Fig. 2). Mesozoic series of Dahar slightly tilted towards the West, have not recorded the Alpine and Atlassic shortening phases. To the East, the Djeffara plain, bounded by a NW-SE network of normal faults (Castany, 1954) and marked by Carboniferous and Permian high subsidence related to NE-SW extension (Ben Ayed, 1993; Bouaziz, 1995). The Talemzane Arch is the most significant structure appreciably trending E-W, corresponds to substratum of the Saharan platform and generated following the Hercynian tectonic phase (Busson, 1970a, b). On both sides of the Talemzane Arch, the Mesozoic strata, Triassic in particular, rest with angular unconformity on the eroded Paleozoic (Busson, 1967). This ridge extends from the Algerian Sahara crossing the Dahar structure of Southern Tunisian and constitutes the northern edge of the Ghedames basin and the limit between the Saharan craton and the extensive Mesogean domain to the North.

Study of seismic lines (Figs. 5, 6 and 8) shows that the transitional zone between the little deformed Saharan platform and the Atlassic folded zone, being complex with deep E-W direction in the Djerid Chotts area. This zone corresponded to a major lineament represented by North-Saharan faults and flexures separating the northern subsiding domain with relatively thick sedimentary cover in the Chotts and Gafsa basins from the southern domain with thin sediments.

4. Chotts structuring

4.1 Fold and fault geometry

The Chotts structures contain the Chott Djerid and El Fedjedj depocenters, which are bounded by the North and South chains. The Chotts domain generally trends E-W extending from the Gulf of Gabes in the East to the Nefta-Tozeur zone in the West of Chott Djerid.

On both sides of the chain of Chotts, the thicknesses and facies changed (Figs. 2 and 4). These significant and abrupt variations testify of pronounced subsidence in the Chotts furrow, controlled by normal faults during the Mesozoic times. The synsedimentary activity seems to be accompanied by Triassic halokinetic activity during the Jurassic and Early Cretaceous (Figs. 4, 5 and 8). Triassic saliferous facies moved and caused a local thickness variation of overlying series near the normal faults, inducing therefore an early structuring of the northern chain of Chotts.

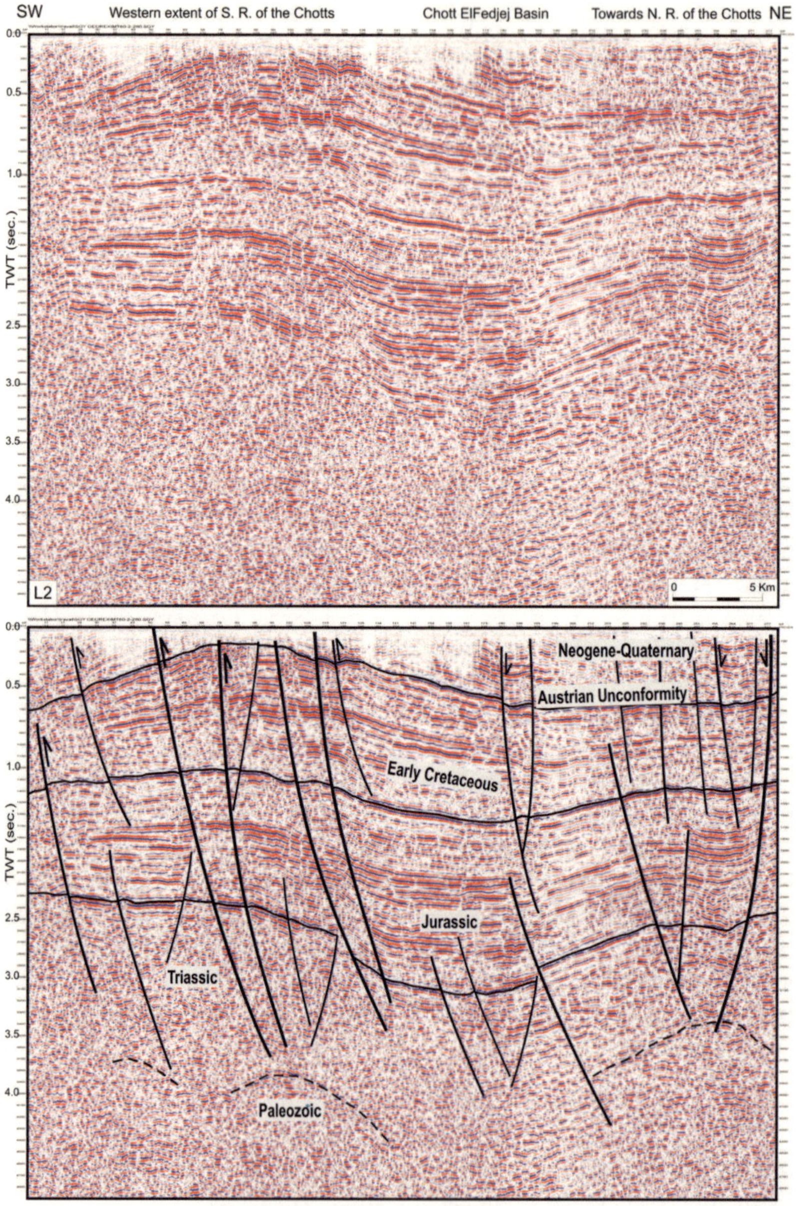

Fig. 5. Seismic section L2 of the Chott El Fedjaj, showing two positive flower fault structures trending east-west and evolution of Mesozoic series under faulting change and halokinesis of northern and southern range of the Chotts

The early halokinesis and its influence on the Chotts structuring was also observed in areas of the southern Atlas of Tunisia (Bédir, 1995; Boukadi and Bédir, 1996; Zitouni, 1997; Boukadi et al., 1998; Hlaiem, 1999; Bédir et al., 2000, 2001; Ben Timzal, 2000; Tanfous-Amri et al., 2005; Zouaghi et al., 2005a, b, 2007 ; Azeiez et al., 2007) and in the Saharan Atlas of Algeria (Vially et al., 1994). Thicknesses of the Mesozoic and Cenozoic series increase gradually from the Saharan platform to the Chotts depocenters (Figs. 5 and 6).

Study of seismic sections shows that the Djerid basin could has a half-graben geometry bounded to the North by a major fault of northern chain of the Chotts (Figs. 5 and 6). Towards the East, the Chott El Fedjej seems to correspond to a graben structure limited to the South and to the North by fault systems. In this area, seismic reflectors emphasize migration of depocenter from North to South since the Triassic until the Early Cretaceous. This inversion would be related to the synsedimentary tectonic and halokinetic activity of the Chotts faults during the extensional intervals. Inversion of structures has continued during Late Cretaceous and Neogene.

4.2 Major unconformities

The Hercynian unconformity defined by toplap structures of the eroded Paleozoic series is related to intense erosion which succeeded the Hercynian orogenesis (Figs. 4-6). The aggradational/retrogradational onlaps mark the transgression of Early Triassic composed by sandy-clay, showing an angular unconformity. The seismic reflector, tilted to the North by slight slope, is characterized by continuous reflections and high amplitude.

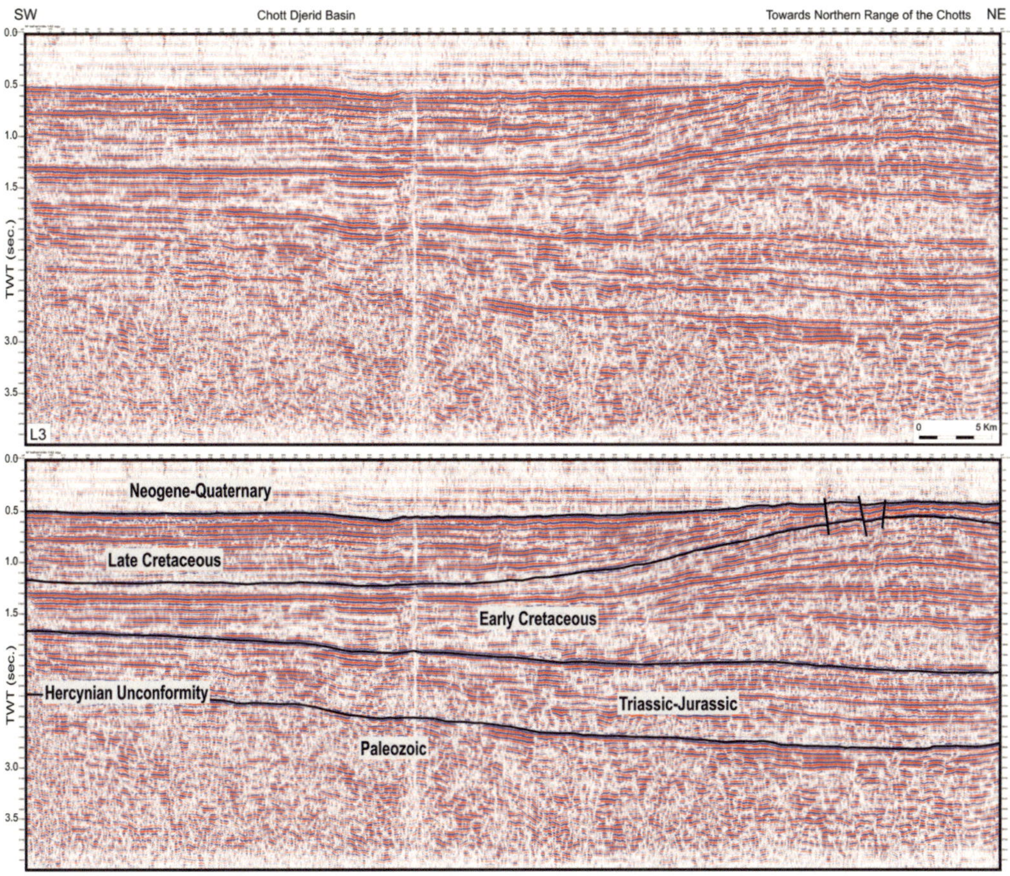

Fig. 6. Seismic section L3 of the Djerid Chott, illustrating distribution of major unconformities and their evolution towards the northern Chotts east-west strike-slip fault and associated Triassic salt intrusion (see Fig. 4)

The top of the Jurassic Nara dolomite formation is marked by moderate continuous and high amplitude seismic doublet (Fig. 6). This reflector is sealed by progradational downlaps of the overlying Neocomian fluvial continental series. This unconformity marked the change from a marine carbonate platform deposition to a fluvial continental one related to a regional marine regression, which reached central Tunisia.
The Late Barremian-Early Aptian erosional surface, which is observed on the entire Saharan platform, is defined by moderate to well continuous and high amplitude reflections forming locally seismic doublets (Figs. 4 and 6). The toplaps correspond to erosional truncation of the Barremian-Neocomian upper detrital series that are onlapped by Aptian to Cenomanian aggradational/retrogradational transgressive carbonates and clays to the South. The occurrence of this regional erosional surface could explained by an uplifting generated by a tectonic deformation and a sea level fall.
The upper Cretaceous unconformity is characterized by basal toplaps, corresponding to upper strata of Senonian truncated by post-Cretaceous erosion. It is marked by moderate continuous and high amplitude reflections followed by the Neogene-Quaternary sandy-clay continental series deposed in angular unconformity by onlap structures (Fig. 6). Absence of all Paleogene strata indicates general elevation of the Saharan platform at the end of Cretaceous related to the Late Cretaceous-Eocene compressional tectonics associated with sea level falls.

5. Evolution and kinematics

Previous synthesis works described the geodynamic evolution of the southern margin of Maghrebid Tethys and established approximate models showing the relation between current structuring and the ancient Hercynian and later-Hercynian events. The outcrops results (Zargouni, 1985; Delteil et al., 1991; Boukadi, 1994; Zouari, 1995) seem agree with those elaborate in subsurface (Ben Ismail, 1991; Bédir, 1995; Zitouni, 1997; Jallouli and Mickus, 2000; Gabtni et al., 2005; Zouaghi et al., 2005a, b, 2009; Gabtni, 2006). These tectonic models, particularly related to the Tertiary and Quaternary deformations were used for deduce the geodynamic evolution of the Mesozoic basins.

5.1 Mesozoic rifting and extensions

Extensional tectonics started with the Carboniferous-Permian times known in the Dahar plateau and the Djeffara plain of South-East Tunisia (Bishop, 1975; Ben Ferjani et al., 1990 Bouaziz, 1995; Bouaziz et al., 1999, 2002). This extensional event continued and accentuated during Mesozoic periods in relation with the sub-meridian Tethyan extensional framework, which affected the whole of the North-African margin. The relatively thick sedimentary strata of Triassic, Jurassic and Early Cretaceous in the Chotts area, the Gulf of Gabes, and the Gafsa area (Figs. 2, 4-6) testify to an active subsidence during these periods.
Triassic structuring seems to be inherited from the Paleozoic. Mechanisms of the opening are accompanied by thick sedimentary sequences and probably volcanic in subsiding grabens with installation of progradational systems tracts of Early to Middle Triassic sandy-clay and carbonate sequences. The evaporites and salts of the upper Triassic sequences seem to be accumulated on down sides of faults bordering grabens. The Triassic NNW-SSE to

NW-SE oriented extension (Bouaziz et al., 2002) is associated with the alkaline magmatism documented from petroleum well data in the Chotts basin (Laaridhi-Ouazaa, 1994). Basin structuring and sedimentary lapout of the Triassic have been described in Algerian outcrops (Bouillin, 1977; Vila, 1980; Obert, 1984, 1986; Kazi Tani, 1986). This kinematic corresponds to the beginning of opening of the Tethys and the central Atlantic and therefore separation of Africa from Eurasia. This event coincides with the beginning of the anti-clockwise rotational migration of African plate towards the East (Olivet et al., 1982, Dercourt et al., 1985).

During the Jurassic and Early Cretaceous, the synsedimentary and halokinetic activity of faults, probably developed since Triassic, persists in relation with N-S extensional stage (Fig. 7). This activity involved formation of the pre-existing structures.

The opening of basins, which began with the Triassic has continued and accentuated during the Jurassic where the grabens and subsiding depocenters in southern Atlas showed a geodynamic mechanism similar to that prevailed with the Triassic. Synchronism between the opening of basins and the Jurassic halokinetic rising has accentuated the paleogeographic differentiation, which characterized by progradational deposits on sides of Rim Synclines (Figs. 4 and 5) and by carbonated reefal platforms appeared since the Jurassic.

The Early Cretaceous is marked by a notable change in the tectonic structuring, induced by the reorientation of tectonic extensional stress and sedimentary evolution. This change is fossilized by general unconformity marked by lower Cretaceous progradational downlaps of Sidi Khalif Formation above the Jurassic carbonates (Figs. 4-6). This discordance has been also highlighted in Algeria by Vila (1980) and Obert (1984, 1986). The opening movements observed during the Triassic and Jurassic, are clearly decreased and even sealed during the Early Cretaceous around the majority of Atlassic blocks. The deep fault network starts to undergone deviations of some directions following the strike-slip movements, the rotations of blocks and the rising of the Triassic intrusions and domes across master fault intersections (Zouaghi, 2008; Zouaghi et al., 2005b, 2011).

The Jurassic and lower Cretaceous kinematics show an influence of extensional stresses trending near NNW-SSE (Fig. 7), which is well integrated in the context of the Tethyan openings (Vila, 1980; Olivet et al., 1982; Obert, 1984, 1986; Dercourt et al., 1985).

During Late Cretaceous several complex deformations have been showed controlled by both extensional and compressional stages. The highlighted structures indicate irregularity of the tectonic and sedimentary mode in Tunisia. These intervals are marked by major Triassic evaporite extrusion indicating the saliferous movements and diapirism well characterized in northern Tunisia (Perthuisot, 1978; Vila, 1980). These deformations result in thickness and facies variations associated with unconformities and gaps recorded in middle and upper Cretaceous strata (Figs. 4-6 and 8) highlighted in many localities of the Tunisian Atlas. The end of the Late Cretaceous corresponds to the beginning of the compressional stresses marked by installation of several anticlines. This extensional stress change to strike-slip movements caused deposition of thick Albian-Cenomanian black shales rocks in the subsiding blocks and reefal carbonated platforms on the high blocks. Regionally, this evolution is related to the bringing together of the African plate with the Iberia to the West and Eurasia to the East, under the effect of its rotation but also related to the East-West relative movement of these plates (Olivet et al., 1982; Dercourt et al., 1985; Guiraud and Bosworth, 1997).

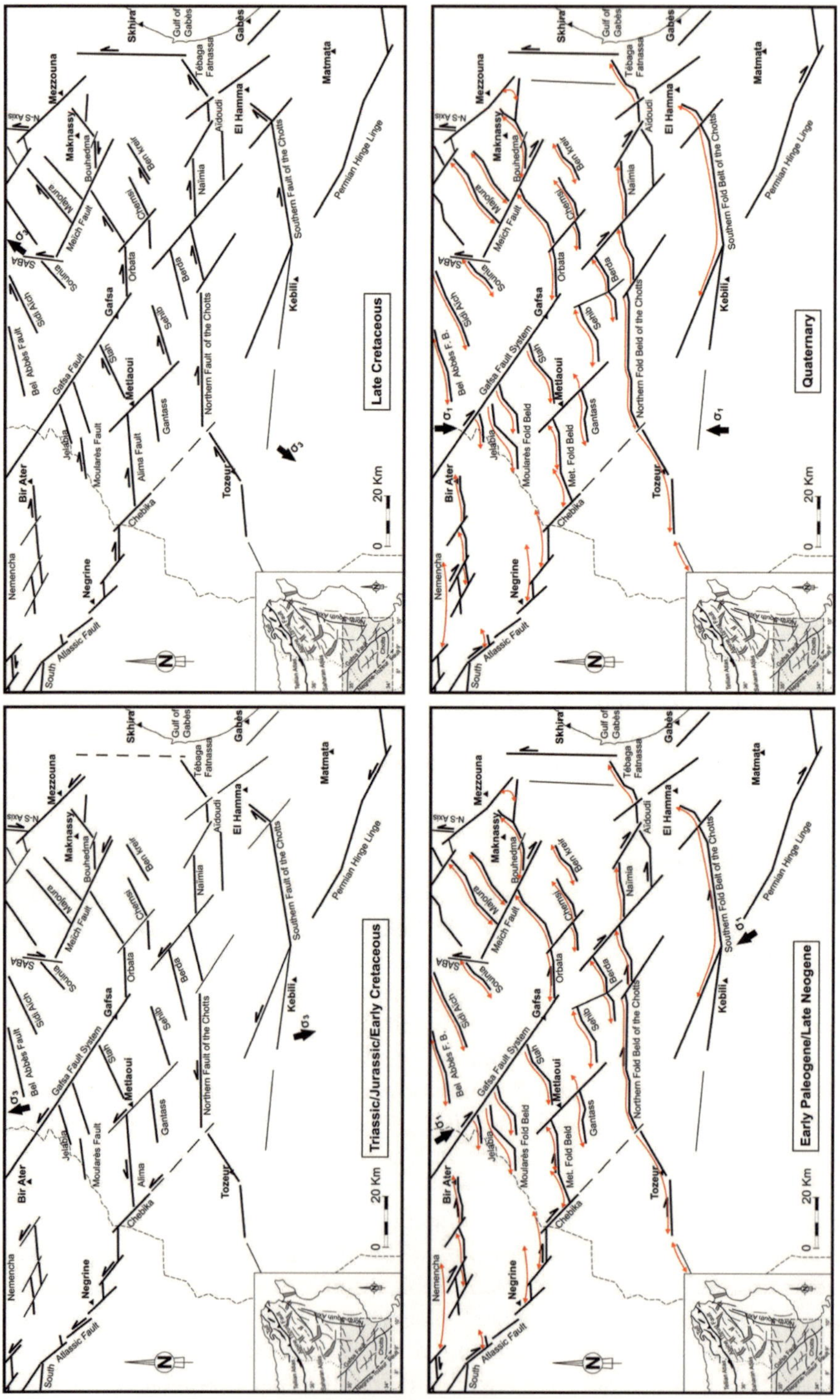

Fig. 7. Effect of the regional tectonic stress field on the geodynamic evolution integrating strike-slip movements, basin geometry, filling and Triassic halokinesis

5.2 Cenozoic inversions

With the beginning of Tertiary times and especially since the Eocene until Quaternary, the compressional events were largely highlighted in the Tunisian areas. These periods are marked by principal tectonic phases of compression trending NW-SE to N-S, which are related to the mechanisms of collision between the African and European plates.

The Eocene corresponds to the final emergence of the Saharan Platform, which started since the Late Cretaceous with the development of the Gafsa-Metlaoui intracontinental folds and synclines. During this period, the Saharan domain that is marked by absence of the Paleogene Sediments, is exposed to erosive action assigning upper strata of the Early Cretaceous (Figs. 4 and 6). Formation of the Eocene folded structures, are locally accompanied by the opening of grabens along the strike slip fault corridors (Ben Ayed, 1993; Melki et al., 2010; Zouaghi et al., 2010). The contractional regional constraint is still NW-SE generating transpressional dextral strike-slip movements on lineaments trending N90 and N120 (Figs. 3 and 7).

The Eocene contractional events detected on the southern Tunisian margin are related to formation of the alpine arc following collision of Europe with the Apulia margin and the accentuation of the Africa-Europe bringing together (Olivet et al., 1982; Dercourt et al., 1985)

At Late Miocene, contractional tectonics become more evident and fossilized by development of folded structures and infilling of the intracontinental basins in southern Atlas (Figs. 4, 5 and 8). The structures previously started by the halokinesis since the Jurassic are reactivated and accentuated during these compressional deformations. The NW-SE upper Miocene contractional events could be correlated to the processes of Africa and Europe bringing together and its collision. The Tortonian major contractional phase, which induces thrust sheets of the North African margin, is well documented (Vila, 1980; Obert, 1984, 1986).

The Quaternary is marked by complex structures resulting from the combined effect of the tectonic polyphasage dominated by the N-S contractional stress (Fig. 7) and saliferous tectonics. We think that it is the result of a combination of a cover tectonic style and deep basement movements.

6. Discussion

6.1 Fold belt of the Tunisia-Algeria Atlassic domain

The Atlassic folds and synclines seem to be evolved under three contractional tectonic events. The first, known as Pyrenean dated end Cretaceous-Eocene has NW-SE (N120 to N140) dominant paleo-stress, the second of Late Miocene named Alpine trending NW-SE, and the third corresponds to post-Villafranchian phase oriented N-S. Some authors have showed another minor and local contractional event dated of Late Pliocene and characterized by N150 to N160 direction of shortening (Zargouni, 1985; Fakraoui, 1990; Addoum, 1995; Bouaziz, 1995).

During the compressional events a local extensional episodes were highlighted. In particularly the Middle Miocene (Langhian-Serravalian) phase trending NE-SW, which induced the opening of the grabens of central Tunisia (Philip et al., 1986; Ben Ayed, 1993; Chihi, 1995; Zouaghi, 2008; Zouaghi et al., 2010, 2011). The N-S post-Villafranchian compressional phase, which largely marked the Atlassic structuring, is associated with normal faults in the Khenchla depression of the eastern Saharan Atlas (Addoum, 1995). In southern Atlas the Eocene-Paleocene phosphate series of the Gafsa-Metlaoui basin shows existence of synsedimentary normal faults (Bouaziz, 1995) that coexist sometimes with other reverse faults (Ben Ayed, 1993; Melki et al., 2010).

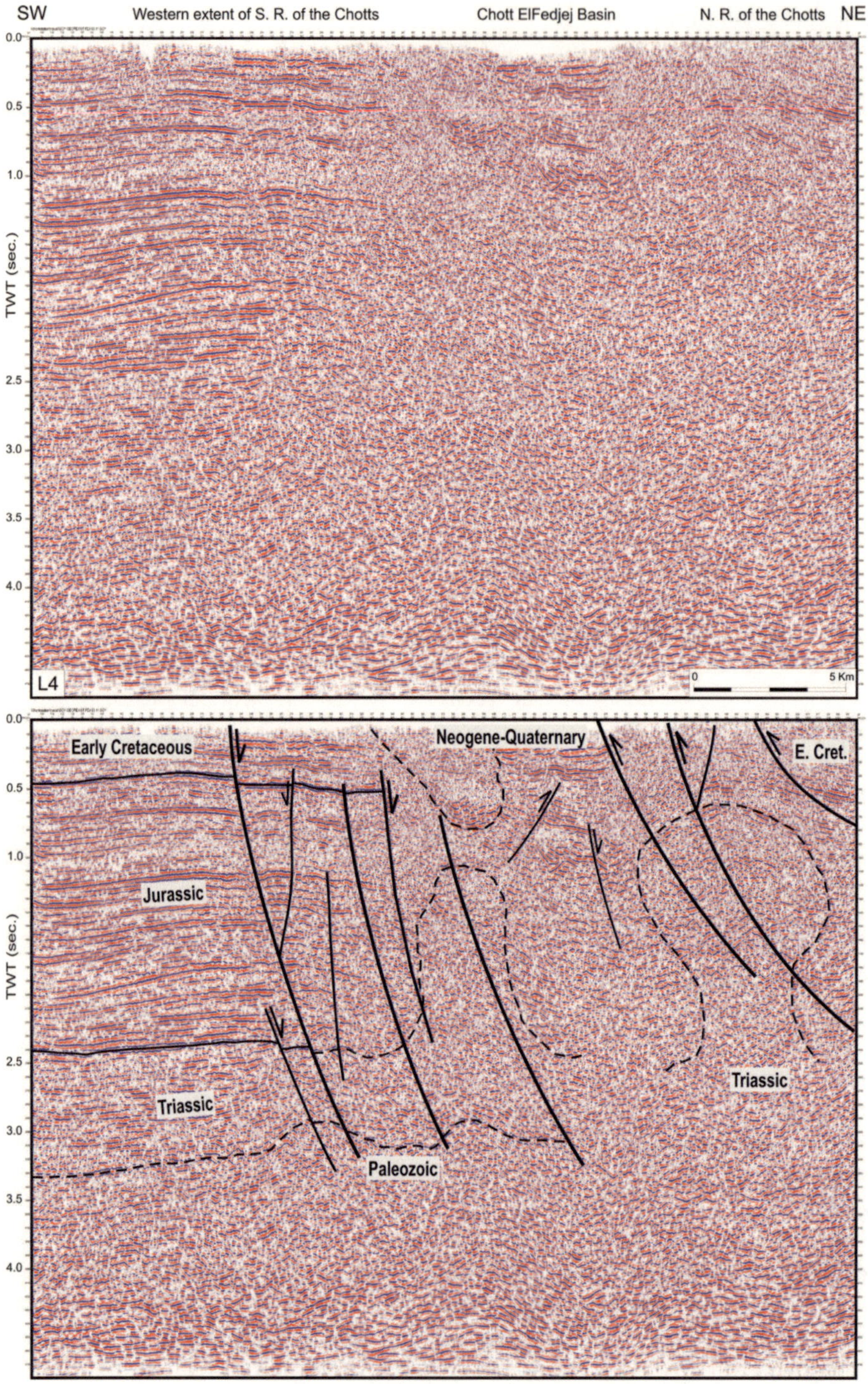

Fig. 8. Seismic section L4 across the El Fedjaj Chott, showing folding and wrench salt-intrusion at the intersection of the northeast-southwest Tebaga-Fatnassa and northwest-southeast Hadifa master lineaments

The main works established on the kinematic of deformation in the Atlassic domain (Creusot et al., 1992 ; Creusot et al., 1993 ; Outtani et al., 1995 ; Addoum, 1995) show the coexistence of two modes of folding; the first corresponds to the folds formed on inherited faults of the infra-Triassic basement, where the ancient extensional structures are reactivated and evolved to reverse faults during the compressional deformations; the second model of propagation folds was also highlighted.
Because the oblique position of the preexistent faults compared to the direction of tectonic stress (Fig. 7), formation of folds becomes more complex because of the strike-slip fault movements (Letouzey, 1990; Ben Ayed, 1993). This kinematic of deformation results in the positive flower-structures organized into overlapping fold belts recognized at the southern Atlas, the most known are the Metlaoui chains, the Chotts chains and the Gafsa chains. In fact, formation of the Atlassic chains was generated by interaction between the major effect related to reactivation in transcurrent and inversion of the old basement faults (thick-skinned style) and a surface effect related to decollement and overfolding of the supra-Triassic cover (thin-skinned style) (Hlaiem, 1999; Zouaghi et al., 2005b; Bensalem et al., 2009). Moreover the early halokinesis during Jurassic and Early Cretaceous associated with the synsedimentary activity of some normal faults contributed to pre-structuring and guiding the genesis of Atlassic fold belts.

6.2 Genesis of the Chotts fold belt

Lithostratigraphic column of the study area reveals the existence of several Formations, which could constitute potential levels of decollement of the overlying cover at the time of folding. Although the main level of decollement remains the Triassic evaporites (Hlaiem, 1999; Zouaghi et al., 2005b; Bensalem et al., 2009), the Cretaceous clays and sandy-clays are locally considered as levels of secondary detachments (Outtani et al., 1995).
The seismic reflection sections suggest a model of broken folds on break-thrusting associated with reverse deep-seated faults in the Triassic. We suggest than these folds are controlled by the basement inherited extensional structures (Figs. 4, 5 and 9).
The kinematic study on seismic lines allows to propose folded structures associated with reverse deep-seated faults reaching the Triassic strata (Figs. 4 and 8). Position and orientation of existing structures would be related to the position of ancient normal faults, which controlled genesis and evolution of the folds structures and associated reverses faults during the compressional events.
Thickness variations on sides of the master faults, which bounded the basins in southern Atlas during the Jurassic and Lower Cretaceous, suggest the existence of normal fault generated during the rifting and extensional phases and caused subsidence increasing (Figs. 5 and 6). The resulted in steps structuring seem to be in agreement with the Knowledge model, which characterizes the North Africa passive margin (Biju-Duval, 1980; Ben Ayed, 1993; Bédir, 1995; Zouaghi, 2008; Zouaghi et al., 2011).
The deep master faults, which are associated with the anticline structures could have a role in the structuring of the cover since the extensional periods by their synsedimentary and halokinetic activity inducing local structural anomalies marked by variations of thicknesses and facies, unconformities, pinching outs and gaps of depositions (Figs. 8 and 9). These inherited faults could influence the localization of the future Atlassic chains before starting of the compressional deformations (Vially et al., 1994).

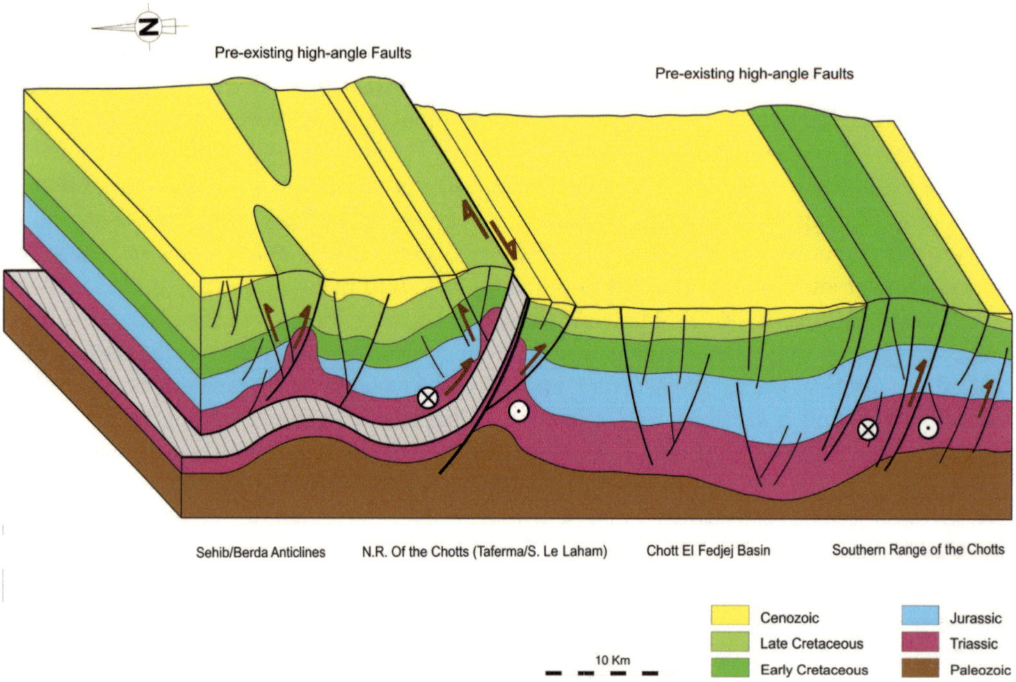

Fig. 9. Block diagram imaging the strain partitioning and resulted folds during the Cenozoic transpressional inversion of Chotts inherited structures

In addition to the role of Triassic halokinesis during the extensional periods that consist in a pre-structuring of the chains, the inherited normal faults would be reactivated and reversed and therefore control the evolution of folds during the contractional times. According to the seismic data we suggest that anticlines structures of southern Atlas are comparable to broken folds on break thrusts (Fig. 9). The slope is generated by the rejuvenation of pre-existing basement faults.

7. Conclusions

Study of major unconformities using the seismic reflections had permitted to identify the principal tectono-sedimentary events, which marked the history of infilling and deformation of the Chotts basins during the Mesozoic and Cenozoic periods. In spite of the local tectonic history, eustatism has marked Saharan craton, in particular during the great falls of sea level. The history of deposition of Chott domain recorded the combined effect of the tectonic deformation and eustatic change, which marked the Saharan platform.

The significant thickening of the Mesozoic series from the South to the North indicates a high subsidence in the Chotts areas controlled by master faults of the Chotts. The southern range of Chotts is characterized by a reduced Mesozoic sedimentation on the platform, which corresponded to a relatively resistant butte rests on the ante-Mesozoic substratum of the northern side of the Talemzane Arch. Increase of thickness in the distal zone to the North, suggests a reduction of deposition space to the South.

Deposition in the Saharan intra-cratonic and marginal basin suggests effect of the Chotts faults related to the regional geodynamic evolution of the peri-tethyan platforms in North Africa.

During the Cenozoic compressive phases trending NW-SE to N-S, rejuvenation of the major faults of Chotts is marked by dextral strike-slip movements, which generate the overlapping fold belts and overthrust folds on the cover of the North chain of Chotts. Structures resulted from deep tectonic deformations as Thick-skinned style and from tectonics of cover as Thin-skinned style are marked by folding and decollement of the sedimentary cover. The southern sides of the majority of chains are vertically straightened sometimes inverted, and transpressive structures are associated with reverse faults. The overlapping folds, which characterize the southern Atlas chains result from dextral strike-slip motion trending near E-W at the level of the basement.

Evolution of the structures around the tectonic blocks of Southern Tunisia from the Jurassic to the Neogene is guided by the rejuvenation of deep crustal lineaments trending E-W and NW-SE and have controlled geometry and evolution of the following sedimentary deposits.

Tectonic deformations have induced halokinesis along master inherited faults. Intersection of these faults during regional extensional and contractional events in the Triassic subsalt basement caused its vertical rising. Interaction of folding and salt diapirism accentuates overthrusting along strike-slip faults.

The strike-slip faults delimited the asymmetric tectonic blocks and differently moved during the geological history. The extensional and transtensional movements during the Triassic, Jurassic and Early Cretaceous, then contractional and transpressional from the end of Late Cretaceous, induced opening of quadratic basins and formation of platforms then closing of the basins and migration of subsiding depocenters, resulting sometimes in blocking stages with compressions following the ancient structural inheritances. The reorientations of the regional stresses along major tectonic discontinuities appear to be induced by movements of the African plate compared to the Eurasia and Iberia (Olivet et al., 1982; Dercourt et al., 1985). Thus we highlight the effects of the principal tectonic events related to the migration of the African plate since the Triassic-Jurassic rifting and the geodynamic answers to these movements at times of shortening on the Tunisian margin.

8. Acknowledgements

The study is based on subsurface data set recorded during exploration and production surveys. It has been partly supported by national and international geological and geophysical projects and research programs on southern Atlas of Tunisia. We thank the ETAP, AGIP, SHELL, CPG and ONM companies for its coordinations and helps.

We thank reviewers and editors of the book for comments and suggestions on an early version of the chapter.

9. References

Abbès, C. (2004). Structurations et évolutions tectono-sédimentaires mésozoïques et cénozoïques associées aux accidents réghmatiques à la jonction des marges téthysienne et nord-africaine (Chaîne Nord-Sud, Tunisie centrale). *Unpublished thesis ès-Sciences, Université Tunis El Manar*, 440 p.

Abdallah, H.; Sassi, S.; Meister, C. & Souissi, R. (2000). Stratigraphie séquentielle et paléogéographie à la limite Cénomanien-Turonien dans la région de Gafsa-Chotts (Tunisie centrale). *Cretaceous Research*, Vol.21, pp. 35-106.

Abdeljaoued, S. & Zargouni, F. (1985). Mise en évidence d'une tectonique intra-crétacée dans l'extrémité orientale de la chaîne des Chotts. *1er Cong. Natio. Sci. Terre, Tunis, Tunisia*, 1, pp. 285-290.

Addoum, B. (1995). L'Atlas saharien sud-oriental : cinématique des Plis-chevauchements et reconstruction du bassin du sud-est constantinois confins algéro-tunsiens). *Unpublished PhD thesis, Université Paris Sud*, Orsay, 200 p.

Alvarez, W.; Virieux, J. & Le Pichon, X. (1984). Thermal consequences of lithosphere extension over continental margins: the initial stretching phase. *Geohys. J. Astr. Soc.*, Vol.124, pp. 255-262.

Arthaud, F. & Thomas, P. (1977). Late Paleozoic strike-slip faulting in southern Europe and northern Africa: results of a right-lateral shear zone between the Appalachians and the Urals. *Geol. Soc. Am. Bull.*, Vol.88, pp. 1305-1320.

Aubouin, J. & Debelmas, J. (1980). L'Europe alpine, les chaines préméditerranéennes. *XXVI Cong. Geol. Int. Coll. C5, Paris, Mém. BRGM*, No.115, pp. 62-66.

Azaïez, H.; Bédir, M.; Tanfous, D. & Soussi, M. (2007). Seismic sequence stratigraphy and platform to basin reservoir structuring of Lower Cretaceous deposits in the Sidi Aïch-Majoura region (Central Tunisia). *Journal of African Earth Sciences*, Vol. 48, pp. 1-18.

Bédir, M. (1995). Mécanismes géodynamiques des bassins associés aux couloirs de coulissements de la marge atlasique de la Tunisie. Seismo-stratigraphie, seismo-tectonique et implications pétrolières. *Unpublished thesis ès-Sciences, Université Tunis II,*, 412 p.

Bédir, M.; Zitouni, L.; Boukadi, N.; Saadi, J.; Alouani, R.; Ben Timzal, F.; Tlig, S. & Bobier, C. (2000). Rifting, halocinèse et structuration des bassins péri-téthysiens jurassiques et crétacé inférieur de subsurface du domaine atlasique central de la Tunisie (région de Gafsa-Sidi Ali Ben Aoun). *Africa Geosciences Review*, Vol.7, pp. 289-306.

Bédir, M.; Boukadi, N.; Tlig, S.; Ben Timzal, F.; Zitouni, L.; Alouani, R.; Slimane, F.; Bobier, C. & Zargouni, F.(2001). Subsurface Mesozoic Basins in the Central Atlas of Tunisia, tectonics, sequence deposit distribution and hydrocarbon potential. *American Association of Petroleum Geologists*, Vol.85, No.5, pp. 885-907.

Bellini, E. & Massa, D. (1980). A stratigraphic contribution to the Paleozoic cf the southern Basins of Libya. *The Geology of Lbya, Tripoli*, Vol.1, pp 3-56.

Ben Ayed, N. (1993). Evolution tectonique de l'avant-pays de la chaine alpine de Tunisie du début du Mésozoïque à l'Actuel. *Annales des Mines et de la Géologie de Tunisie*, No.32, 286 p.

Ben Ferjani, A.; Burollet, P. F. & Mejri, F. (1990). Petroleum Geology of Tunisia. *Mémoire Entreprise Tunisienne d'Activités Pétrolières*, Tunis, No.1, 194 p.

Ben Ismail, M.H. (1982). Le Trias et le Jurassique inférieur et moyen évaporitiques de l'extrême Sud tunisien : étude de sondages profonds et de terrain, synthèses paléogéographiques. *Unpublished thesis, 3ème cycle, Université Paris VI*, 180 p.

Ben Ismail, M.H. (1991). Les bassins mésozoïques (Trias Aptien) du Sud de la Tunisie stratigraphie intégrée, caractéristiques géophysiques et évolution géodynamique. *Unpublished thesis ès-Sciences, Université Tunis*, 446 p.

Bensalem, M.S.; Ghanmi, M. & Zargouni, F. (2009). Modelling Genesis of Intracratonic Chains Related to Tectonics Inheritance. Case Study from Gafsa Basin (Southern Central Tunisia). *Journal of Geography and Geology*, Vol.1, No.2, pp. 58-70.

Ben Timzal, F., 2000. Géodynamique et intérêt pétrolier des bassins du Crétacé inférieur en subsurface (Atlas centro-méridional de Tunisie). *Unpublished PhD thesis, Université Tunis II*, 153 p.

Ben Youssef, M. (1999). Stratigraphie génétique du Crétacé de Tunisie : micropaléontologie, stratigraphie séquentielle et géodynamique des bassins de la marge Sud et péri-téthysienne. *Unpublished thesis ès-Sciences, Université Tunis II*, 402 p.

Bernoulli, D. & Lemoine, M. (1980). Birth and early evolution of the Tethys: the overall situation. *Coll. C5, XXVIeme, Cong. Int. Geol., Mém. BRGM*, Paris.

Biju-Duval, B.; Dercourt, J. & Le Pichon, X. (1976). La genèse de la Méditerranée. *La Recherche*, Vol.7, pp. 811-822.

Biju-Duval, B. (1980). De la Téthys aux mers intra-alpines actuelles : Introduction. *Coll. C5, XXVIeme, Cong. Int. Geol., Mém. BRGM*, Paris.

Bishop, W.F. (1975). Geology of Tunisia and adjacent part of Algeria and Libya. *American Association of Petroleum Geologists*, Vol.59, pp. 413-450.

Boltenhagen, C. (1985). Paléogéographie du Crétacé moyen de la Tunisie centrale. *1er Cong. Natio. Sci. Terre*, Tunis, Tunisia, pp. 97-114.

Bouaziz, S. (1986). La déformation dans la plate-forme du Sud tunisien (Dahar-Jeffara) approche multiscalaire et pluridisciplinaire. *Unpublished thesis, 3ème cycle, Université Tunis*, 180 p.

Bouaziz, S. (1995). Etude de la tectonique cassante dans la plate-forme et l'Atlas Sahariens (Tunisie méridionale): Evolution des paléochamps de contraintes et implications géodynamiques. *Unpublished thesis ès-Sciences, Université Tunis II*, 484 p.

Bouaziz, S.; Barrier, E.; Turki, M.M. & Tricart, P. (1999). La tectonique permo-mésozoïque (anté-Vraconien) dans la marge sud téthysienne en Tunisie méridionale. *Bulletin de la Société Géologique de France*, Vol.170, No.1, pp. 45-56.

Bouaziz, S.; Barrier, E.; Soussi, M.; Turki, M.M. & Zouari, H. (2002). Tectonic evolution of the northern African margin in Tunisia from paleostress data and sedimentary record. *Tectonophysics*, Vol.357, pp. 227-253.

Bouillin, J.P. (1977). Géologie alpine de la petite Kabylie dans les régions de calle et d'El Millia (Algérie). *Unpublished thesis ès-Sciences, Université de Paris*, 509 p.

Bouillin, J.P. (1986). Le bassin maghrébin, une ancienne limite entre l'Europe et l'Afrique à l'Ouest des Alpes. *Bulletin de la Société Géologique de France 8*, pp. 547-558.

Boukadi, N. (1994). Structuration de l'Atlas de Tunisie : signification géométrique et cinématique des nœuds et des zones d'interférences structurales au contact de grands couloirs tectoniques. *Unpublished thesis ès-Sciences, Université de Tunis II*, 249 p.

Boukadi, N. & Bédir, M. (1996). L'halocinèse en Tunisie : contexte tectonique et chronologie des évènements. *Comptes Rendus de l'Académie des Sciences de Paris*, T.322, pp. 587-594.

Boukadi, N.; Bédir, M. & Zargouni, F. (1998). Geometric and kinematic analyses of pull-apart basins produced by en echelon strike-slip of Gafsa fault systems (Southern Tunisia). *Africa Geosciences Review*, Vol.5, pp. 327-338.

Bousquet, J.C. & Philip, H. (1981). Les caractéristiques néotectoniques en Méditerranée occidentale. In: *Sed. Basins. Medit. Marges,* F.C. Wezel, (Ed.), pp. 389-405.

Boutib, L. & Zargouni, F. (1998). Disposition et géométrie des plis de l'Atlas centro-méridional de Tunisie. Découpage et cisaillement en lanières tectoniques. *Comptes Rendus de l'Académie des Sciences de Paris,* T.326, pp. 261-265.

Burollet, P.F. (1956). Contribution à l'étude stratigraphique de la Tunisie centrale. *Annales des Mines et de la Géologie de Tunisie,* No.18, 350 p.

Burollet, P.F. & Desforges, P. (1982). Dynamique des bassins néocrétacés en Tunisie. *Mémoire Géologique, Université Dijon,* Vol.7, pp. 381-389.

Busson, G. (1967). Mesozoic of Southern Tunisia. *Guidebook to the Geology and History of Tunisia, P.E.S.L., 9th Annual Field Gonf.,* Tripoli, pp. 131-152.

Busson, G. (1969). La salifère principale (Trias supérieur et Lias p.p) du Sahara algéro-tunisien. *Comptes Rendus de l'Académie des Sciences de Paris,* T.268, pp. 251-254.

Busson, G. (1970a). Rapports entre terrains mésozoïques et paléozoïques au Sahara Algéro-tunisien: La discordance hercynienne. *Comptes Rendus de l'Académie des Sciences de Paris,* T.269, pp. 685-688.

Busson, G. (1970b). Le Mésozoïque saharien, 2eme partie: essai de synthèse des données des sondages algéro-tunisiens, *2 tomes Edition CNRS, Paris.* France, 811 p.

Caire, A. (1971). Chaînes alpines de la Méditerranée centrale (Algérie et Tunisie septentrionale, Sicile, Calabre et Apennin méridional). *Unesco, Tectonique de l'Afrique, Science de la Terre 6,* pp. 61-90.

Caire, A. (1974). L'Arc Calabro-Sicilien, le promontoire africain et les coulissements des chaines alpines méditerranéennes. *Commission internationale pour l'Exploration de la Mer Méditerranée. Rapports et procès verbaux. Cong. de Monaco,* Vol.23, (décembre 1974), pp. 121-123.

Castany, G. (1951). Carte géologique de la Tunisie, échelle 1/500.000e. *Direction des Travaux Publics, Service des Mines, de l'Industrie et de l'Energie,* Tunisia.

Castany, G. (1954). L'accident Sud-tunisien, son âge et ses relations avec l'accident Sud-atlasique d'Algérie. *Comptes Rendus de l'Académie des Sciences de Paris,* T.238, pp. 916-918.

Chaabani, F. (1995). Dynamique de la partie orientale du bassin de Gafsa au Crétacé et au Paléocène. Etude minéralogique et géochimique de la série phosphatée éocène, Tunisie méridionale. *Unpublished thesis, Doctorat d'Etat, Université Tunis II,* 428 p.

Chandoul, H.; Burollet P.F.; Ben Ferjani A. & Memmi L. (1993). Recueil des coupes types de Tunisie, Trias et Jurassique. *Mémoire Entreprise Tunisienne d'Activités Pétrolières,* Tunis, No.4, 95 p.

Chihi, L. (1995). Les fossés néogènes à quaternaires de la Tunisie et de la mer pélagienne : leur signification dans le cadre géodynamique de la méditerranée centrale. Unpublished *Thèse ès Sciences, Université Tunis II,* Tunisie, 324 p.

Creuzot, G.; Mercier, E.; Ouali, J. & Turki, M.M. (1992). Héritage distensif synsédimentaire et structuration chevauchante : apport de la modélisation du chevauchement alpin de Zaghouan (Atlas tunisien). *Comptes Rendus de l'Académie des Sciences de Paris,* T.314, Série II, pp. 961-965.

Creuzot, G.; Mercier, E.; Ouali, J. & Tricart, P. (1993). La tectogenèse atlasique en Tunisie centrale : apport de la modélisation géométrique. *Eclogae Geologicae Helvetiae,* Vol.36, No.2, pp 609-627

Delteil, J.; Zouari, H.; Chikaoui, M.; Creuzot, G.; Ouali, J.; Turki, M.M.; Yaïch, C. & Zargouni, F. (1991). Relation entre ouvertures téthysienne et mésogéenne en Tunisie. *Bulletin de la Société Géologique de France*, Vol.162, No.6, pp. 1173-1181.

Dercourt, J.; Zonenshain, L.P.; Ricou, L.E.; Kazmin, V.G.; Le Pichon, X.; Knipper, A.L.; Grandjacquet, C.; Sborshchikov, J.M.; Boulin, J.; Sorokhtin, O.; Geyssant J.; Lepvrier, C.; Biju Duval, B.; Sibuet, J.C.; Savostin, L.A.; Westphal, M. & Lauer, J.P. (1985). Présentation de neuf cartes paléogéographiques au 1/20.000.000 s'étendant de l'Atlantique au Pamir pour la période du Lias à l'Actuel. *Bulletin de la Société Géologique de France*, Vol.8, pp. 637-652.

Dercourt, J.; Bassoulet, J.P.; Baud, A.; Butterlin, J.; Camoin, G.; Cavalier, C.; Cacca, F.; Enay, R.; Fourcade, E.; Guiraud, R.; Lorenz, C.; Marcoux, J.; Masse, J.P.; Orzag, F. & Philip, J. (1992). Paleoenvironmental Atlas of the Tethys from Permian to Recent. *28th International Geological Congress, Kyoto, Japan, Proceedings, I-3-24*, Vol.1, pp. 116.

Durand Delga, M. & Fonrbote, L.M. (1980). Le cadre structural de la Méditerranée occidentale. *XXVIeme Cong. Geol. Int., Coll. C5, Paris. Mém. BRGM*, No.115, pp. 67-85.

Fakraoui, M. (1990). Etude stratigraphique et structurale des chaînes des Chotts (Tunisie méridionale) évolution géométrique et cinématique liée l'accident sud-atlasique. *Unpublished thesis, 3ème cycle, Université Tunis II*, 243 p.

Fournié, D. (1978). Nomenclature lithostratigraphique des séries du Crétacé supérieur au Tertiaire en Tunisie. *Bull. Cent. Rech. Exploration-production, Elf-Aquitaine, Pau, 2*, pp. 97-148.

Frizon de Lamotte, D.; Leturmy, P.; Missenard, Y.; Khomsi, S.; Ruiz, G.; Saddiqi, O.; Guillocheau, F. & Michard, A. (2009). Mesozoic and Cenozoic vertical movements in the Atlas system (Algeria, Morocco, Tunisia): An overview. *Tectonophysics*, Vol. 475, pp. 9-28.

Gabtni, H. (2006). Caractérisation profonde et modélisation géophysique des zones de transition entre les différents blocs structuraux de la Tunisie centro-méridionale. *Unpublished PhD thesis, Université Tunis El Manar*, 243 p.

Gabtni, H.; Jallouli, C.; Mickus, K. & Zouari, H. (2005). Geophysical Constraints on the Location and Nature of the North Saharan Flexure in Southern Tunisia. *Pure and Applied Geophysics*, Vol. 162, pp. 2051-2069.

Guellala, R. (2010). Etude géologique et hydrogéologique des séries crétacées inferieures du Djérid (sud-ouest de la Tunisie) apports des méthodes géophysiques. *Unpublished PhD thesis, Université Tunis El Manar*, 165 p.

Guiraud, R. & Bosworth, W. (1997). Senonian basin inversion and rejuvenation of rifting in Africa and Arabia: synthesis and implications to plate-scale tectonics. *Tectonophysics*, Vol.282, pp. 39-82.

Hlaiem, A. (1999). Halokinesis and structural evolution of the major features in eastern and southern Tunisian Atlas. *Tectonophysics*, Vol.306, pp. 79-95.

Jallouli, C. & Mickus, K. (2000). Regional gravity analysis of the crustal structure of Tunisia. *Journal of African Earth Sciences*, Vol.30, pp. 345-357.

Kamourn, F.; Peybernès, B. & Fauré, P. (1999). Palaeogeographic evolution of Sahartin and Atlasic Tunisia during Jurassic times. *C. R. Acad. Sci. Paris*, Vol. 328. pp. 547-552.

Kamoun, F.; Peybernès, B.; Ciszak, S. & Calzada, S. (2001). Triassic palaeogeography of Tunisia. *Palaeogeography, Palaeoclimatology, Palaeoecology*, Vol.172, pp. 223-242.

Kazi-Tani, N. (1986). Evolution géodynamique de la bordure nord africaine, le domaine intraplaque Nord Algérie. Approche mégaséquentielle. *Unpublished thesis ès-Sciences, Université de Pau et des Pays de l'Adour*, 871 p.

Laaridhi-Ouazaa, N. (1994). Etude minéralogique et géochimique des épisodes magmatiques mésozoïques et miocènes de la Tunisie. *Unpublished thesis, Doctorat d'Etat, Université Tunis II*, 466 p.

Lazzez, M.; Zouaghi, T. & Ben Youssef, M. (2008). Austrian phase on the northern African margin inferred from sequence stratigraphy and sedimentary records in southern Tunisia (Chotts and Djeffara areas). *Comptes Rendus Geoscience*, Vol.340, pp. 543-552.

Letouzey, J. (1990). Fault reactivation, inversion and fold-thrust belt. In: *Petroleum and tectonics in mobile belts*, J. Letouzey, (Ed.), 101-128, Edition Technip, Paris.

Marie, J.; Trouvé, P.; Desforges, G. & Dufaure, P. (1984). Nouveaux éléments de paléogéographie du Crétacé de Tunisie. *Notes et Mémoires TOTAL-CFP, Paris, France*, pp. 19, 7-37.

Martinez, C.; Turki, M.M. & Truillet, R. (1990). La signification des plis d'orientation méridienne dans l'Atlas tunisien centro-méridional. *Bulletin de la Société Géologique de France*, Vol.8, pp. 643-852.

Melki, F.; Zouaghi, T.; Ben Chelbi, M.; Bédir, M. & Zargouni, F. (2010). Tectono-sedimentary events and geodynamic evolution of the Mesozoic and Cenozoic basins of the Alpine Margin, Gulf of Tunis, north-eastern Tunisia offshore. *Comptes Rendus Geoscience*, Vol.342, pp. 741-753.

M'Rabet, A. (1987). Stratigraphie, sédimentation et diagenèse carbonatée des séries du Crétacé inférieur de Tunisie centrale. *Annales des Mines et de la Géologie de Tunisie*, No.30, 412 p.

Negra, M.E.H. (1994). Les dépôts de plate-forme à bassin du Crétacé supérieur en Tunisie centro-septentrionale (Formation Abiod et faciès associés), stratigraphie, sédimentation, diagenèse et intérêt pétrolier. *Unpublished thesis, Doctorat d'Etat, Université Tunis II*, 649 p.

Obert, D. (1984). Géologie des Babors (Algérie): importance de la paléotectonique Alpine dans l'orogenèse tellienne. *Revue de Géographie Physique et de Géologie Dynamique*, Vol.23, pp. 99-117.

Obert, D. (1986). La distension liasique dans la zone tellienne d'Afrique du Nord, halocinèse précoce et préfiguration des unités telliennes. *Comptes Rendus de l'Académie des Sciences de Paris*, T.303, Série 2, No.2, pp. 961-965.

Olivet, J.L.; Bonnin, J.; Beuzart, P. & Auzende J.M. (1982). Cinématique des plaques et paléogéographie: une revue. *Bulletin de la Société Géologique de France*, T.24, Série 7, No.5-6, pp. 875-892.

Ouali, J. (2007). Importance du réseau réghmatique dans la tectonogenèse de la Tunisie atlasique à travers l'étude de l'axe Nord-Sud. *Unpublished thesis ès-Sciences, Université Tunis El Manar*, 399 p.

Outtani, F.; Addoum, B.; Mercier, E.; Frizon De Lamotte, D. & Andrieux, J. (1995). Geometry and Kinematics of the South Atlas Front, Algeria and Tunisia. *Tectonophysics*, Vol.249, pp. 233-248.

Perthuisot, V. (1978). Dynamique el pétrogenèse des extrusions triasiques en Tunisie septentrionale. *Thesis, Travaux du Laboratoire de Géologie, Ecole Normale Supérieure*, Paris, 312 p.

Philip, H.; Andrieux, J.; Dlala, M.; Chihi, L. & Ben Ayed, N. (1986). Evolution tectonique mio-plio-quaternaire du Fossé de Kasserine (Tunisie centrale) implications sur l'évolution géodynamique récente de la Tunisie. *Bulletin de la Société Géologique de France*, T.2, Série 8, No.5-6, pp. 559-568.

Rabia, M.C. (1985). Etude géologique de la region des Chotts (Sud tunisien) par Télédetection spatial, détection de la radioactivité naturelle et analyse hydrogéochimique. *Revue des Sciences de la Terre, Institut National de Recherche Scientifique et Technique*, Tunisie, Vol.2, 204 p.

Reading, H.G. (1980). Characteristics and recognition of strike-slip fault systems. *Special Publication of the International Association of Sedimentologists*, No.4, pp. 7-26.

Ricou, L.E. (1994). Tethys reconstructed: plates, continental fragments and their boundaries since 260 Ma from Central America to South-Eastern Asia. *Geodynamica Acta*, Vol.7, No.4, pp. 160-218.

Sassi, S. (1974). La sédimentation phosphatée au Paléocène dans le Sud et le Centre Ouest de la Tunisie. *Unpublished thesis ès-Sciences, Université Paris-Sud*, Orsay, 292 p.

Soussi, M. (2002). Le Jurassique de la Tunisie atlasique: stratigraphie, dynamique sédimentaire, paléogéographie et intérêt pétrolier. *Documents du Laboratoire de Géologie*, Lyon, No.157, 363 p.

Tanfous-Amri, D.; Bédir, M.; Soussi, M.; Azaïez, H.; Zitouni, L.; Inoubli, M.H. & Ben Boubaker, K. (2005). Halocinèse précoce associée au rifting jurassique dans l'Atlas central de Tunisie (région de Majoura-El Hfay). *Comptes Rendus Geoscience*, Vol.337, No.7, pp. 703-711.

Tapponnier, P. (1977). Evolution tectonique du système alpin en Méditerranée: poinçonnement et écrasement rigide-plastique. *Bull. Soc. Géol. Fr., Paris.* Série 7, T. XIX, No.3, pp. 437-460.

Turki, M.M. (1988). Polycinématique et contrôle sédimentaire associé sur la cicatrice Zaghouan-Nebhana. *Revue des Sciences de la Terre, Institut National de Recherche Scientifique et Technique*, Tunisia, 252 p.

Vially, R.; Letouzey, J.; Bernard, F.; Haddadi, N.; Desforges, G.; Askri, H. & Boudjema, A. (1994). Basin inversion along the northern African margin: the Saharan Atlas (Algeria). In: *Peri-Tethyan Platforms*, F. Roure (Ed.), 79-118, Edition Technip, Paris.

Vila, J.M. (1980). La chaîne alpine d'Algérie orientale et des confins algéro-tunisiens. *Thèse es-Sciences, Univ. Paris VI*, 665 p.

Zargouni, F. (1985). Tectonique de l'Atlas méridional de Tunisie : Evolution géométrique et cinématique des structures en zone de cisaillement. *Unpublished thesis ès-Sciences, Université Louis Passeur*, Strasbourg, 292 p.

Zargouni, F.; Rabia, M.C. & Abbès, C. (1985). Rôle des couloirs de cisaillement de Gafsa et de Négrine-Tozeur dans la structuration du faisceau atlasique. *Comptes Rendus de l'Académie des Sciences de Paris*, Vol.306, pp. 831-834.

Zitouni, L.; Bédir, M.; Boukadi, N.; Ibouh, H.; Tlig, S. & Bobier, C. (1997). Géométrie, chronologie et cinématique des accidents N 120-140. Exemple: accident de Majoura-Mech (données de surface et de subsurface). *Africa Geosciences Review*, Vol.4, No.3, pp. 373-380.

Zouaghi, T., 2008. : Distribution des séquences de dépôt du crétacé (Aptien- Maastrichtien) en subsurface : déformation tectonique, Halocinése, évolution géodynamique (Atlas central de Tunisie). *Unpublished PhD thesis, Université Tunis El Manar*, 367 p.

Zouaghi, T.; Bédir, M. & Inoubli, M.H. (2005a). Structuration profonde des dépôts de l'Albien-Maastrichtien en Tunisie centrale: nouvelle limite de l'archipel de Kasserine et implications géodynamiques. *Comptes Rendus Geoscience,* Vol.337, pp. 685-693.

Zouaghi, T.; Bédir, M. & Inoubli, M.H. (2005b). 2D Seismic interpretation of strike-slip faulting, salt tectonics, and Cretaceous unconformities, Atlas Mountains, central Tunisia. *Journal of African Earth Sciences,* Vol.43, pp. 464-486.

Zouaghi, T.; Inoubli, M.H. & Bédir, M. (2007). Contribution of seismic velocities to the structural and the lithostratigraphic studies: Salt-intruded corridor ceiling and Lower Turonian Beida Anhydrite deposits outline in Central-Southern Atlas of Tunisia. *Comptes Rendus Geoscience,* Vol.339, pp. 13-23.

Zouaghi, T.; Bédir, M.; Abdallah, H. & Inoubli, M.H. (2009). Seismic sequence stratigraphy, basin structuring, and hydrocarbon implications of Cretaceous deposits (Albian-Maastrichtian) in central Tunisia. *Cretaceous Research,* Vol.30, pp. 1-21.

Zouaghi, T.; Bédir, M.; Melki, F.; Gabtni, H.; Gharsalli, R.; Bessioud, A. & Zargouni, F. (2010). Neogene sediment deformations and tectonic features of northeastern Tunisia: evidence for paleoseismicity. *Arabian Journal of Geosciences,* DOI 10.1007/s12517-010-0225-z

Zouaghi, T.; Ferhi, I.; Bédir, M.; Ben Youssef, M.; Gasmi, M. & Inoubli, M.H. (2011). Analysis of Cretaceous (Aptian) strata in central Tunisia, using 2D seismic data and well logs. *Journal of African Earth Sciences,* Vol.61, pp. 38-61.

Zouari, H. (1995). Evolution géodynamique de l'Atlas centro-méridional de la Tunisie. Stratigraphie, analyses géométrique, cinématique et tectono-sédimentaire. *Unpublished thesis, Doctorat d'Etat, Université Tunis II,* 278 p.

Zouari, H.; Turki, M.M. & Delteil J. (1990). Nouvelles données sur l'évolution tectonique de la chaîne de Gafsa. *Bulletin de la Société Géologique de France,* Vol.8, pp. 621-628.

4

Interplay between Tectonics and Mount Etna's Volcanism: Insights into the Geometry of the Plumbing System

Domenico Patanè et al.*
Istituto Nazionale di Geofisica e Vulcanologia,
Sezione di Catania – Osservatorio Etneo, Catania
Italy

1. Introduction

Volcanoes are geologic manifestations of highly dynamic and complexly coupled physical and chemical processes in the interior of the Earth. Most volcanism on Earth occurs at plate boundaries in places where tectonic plates move apart (e.g. Iceland) and in places where tectonic plates come together with one plate plunging (subducting) below the other into the mantle (e.g. Pacific ring of fire). Conversely, intraplate volcanism is a type of volcanism occurring far from plate boundaries and whose origins are rather controversial.

To know the working mode of a volcano in a given region it is necessary to understand the interplay between tectonics, deformation processes and magma transport through the lithosphere (e.g. Vigneresse, 1999; Petford et al., 2000). Deformation-induced fault-fracture networks have been regarded as efficient pathways through which magma is transported, stored and eventually erupted at the Earth's surface (e.g. Clemens and Mawer, 1992; Petford et al., 2000). At active volcanoes, magmas rise toward the surface and can stagnate at different levels in the lithosphere, giving rise to magma bodies of different shape and size (Marsh, 2000). Nearly all volcanic eruptions are supplied with magma through dykes and inclined sheets whose initiation and eventual propagation to the surface or, alternatively, arrest at some depth in the volcano, depend on the stress state in the volcano (Gudmundsson, 2006). At the surface of active volcanic edifices, the majority of eruptive fissures have a radial configuration and tangential or oblique fissures are rare. However, within many eroded volcanic edifices, dykes and dyke-fed eruptive fissures commonly have more complex patterns, resulting from regional stresses, magmatic reservoirs, anisotropies or variations in topography (Acocella et al., 2009).

Geophysics can provide information on the geometry of plumbing system and magma chambers, as well as on the mechanisms of emplacement of dykes. Among the different branches of geophysics, seismology is the most powerful tool to obtain information about

---

* Marco Aliotta, Andrea Cannata, Carmelo Cassisi, Mauro Coltelli, Giuseppe Di Grazia, Placido Montalto and Luciano Zuccarello
Istituto Nazionale di Geofisica e Vulcanologia, Sezione di Catania – Osservatorio Etneo, Catania
Italy

the inner structure of volcanoes and the geometry of their plumbing systems. Volcanoes generate seismic energy at frequencies ranging from zero (static displacement) to a few tens of Hz. Generally, two different groups of seismic signals can be distinguished in volcanic areas (Chouet, 1996): the former, involving processes originating in the solid, is associated with shear failures in the volcanic edifice and the related seismic events are called volcano-tectonic (VT) earthquakes; the latter (hereafter referred to as seismo-volcanic signals) involves processes originating in the fluid and includes long-period (LP) events and volcanic tremor, sharing the same spectral components (0.5-5 Hz), and very-long-period (VLP) events characterized by dominant period of 2-100 s (Ohminato et al., 1998).

In the last 30 years, the Earth has been widely investigated through a variety of seismic tomographic methods, leading to many interesting results at both regional and global scale. In volcanic environments, several tomographic high-resolution studies, involving the joint use of local VT earthquakes and artificial explosions have led to important discoveries on volcano plumbing systems both in the shallow and deeper zones (e.g. Achauer et al., 1988; Lees and Crosson, 1990). It is noteworthy that, since the ability to resolve feeding conduits, magma chambers, and zones of solidified magmatic intrusion relies on both the distribution of elastic sources at depth and of receivers at the surface, objects smaller than a few km generally cannot be reliably resolved. In this sense, what we can define with seismic tomography is generally a "large" volcanic structure with minimum dimensions of 1-3 km^3. Smaller structure composing the shallow portion of the plumbing system can be studied by volcanic tremor and LP and VLP events, which, as aforementioned, are driven by fluid processes. The study of these signals can provide information not only on the geometry of the shallow portion of the plumbing system, but also on the variations in time of the magma batches stored inside it (Chouet, 2003).

This chapter deals with the investigation of the plumbing system at Mt. Etna by using seismic signals with the aim of understanding how Etna volcano structure works and its relationship with the geodynamics of eastern Sicily. In particular, **section 2** summarizes the main structural features of Mt. Etna, some theories regarding its origin, as well as some information about the volcano's recent activity. **Section 3** focuses on the investigation of the deep plumbing system by seismic tomography reporting also the previous seismological studies. In **section 4** examples of study of the shallow plumbing system by the analyses of the seismo-volcanic signals are shown. Finally, **section 5** summarizes the main conclusions.

2. Tectonic setting of Mt. Etna volcano

Mt. Etna is one of the most active volcanoes in the world, located on the densely inhabited eastern coast of Sicily (Italy). It is characterized by almost continuous eruptive activity from its summit craters and fairly frequent lava flow eruptions from fissures opened up on its flanks. Mt. Etna is a composite, quaternary, basaltic volcano set in a region of complex geodynamics, where major regional structural lineaments play an important role in the dynamic processes of the volcano (e.g. Gresta et al., 1998; **Fig. 1**). It covers an area of about 1,250 km^2 with a basal circumference of 140 km and reaches a maximum elevation of 3330 m. On the volcano summit four active craters are currently opened: Voragine, Bocca Nuova, South East Crater, and North East Crater (hereafter referred to as VOR, BN, SEC and NEC, respectively; **Fig. 2**).

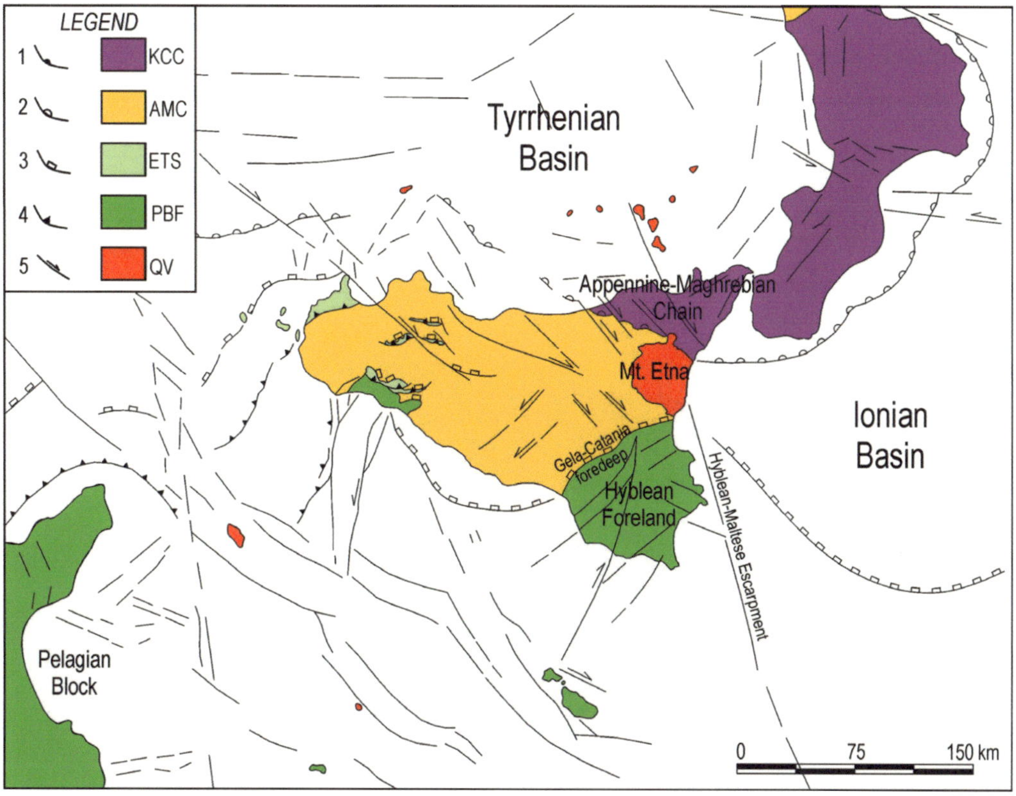

Fig. 1. Structural setting of central Mediterranean Sea (modified from Lentini et al., 2006) and location of Mt. Etna. 1) Regional overthrust of the Sardinia-Corsica block upon Calabride units; 2) Regional overthrust of the Kabilo-Calabride units upon the Apennine-Maghrebian Chain; 3) External front of the Apennine-Maghrebian Chain upon the Foreland units and the External Thrust System; 4) Thrust front of the External Thrust System; 5) Main normal and strike-slip faults. KCC: Kabilo-Calabride Chain Units; AMC: Appennine-Maghrebian Chain Units; ETF: External Thrust System Units; PBF: Pelagian Block Foreland Units; QV: Quaternary Volcanoes. Redrawn from Lentini et al. (2006)

2.1 Structural framework

Mt. Etna lies on the Sicilian continental crust and is located on the external boundary of the Apennine-Maghrebian chain, close to the Gela–Catania Plio-Quaternary foredeep (Bousquet and Lanzafame, 2004). It is bordered by three tectonic domains (**Fig. 1**): the Apennine-Maghrebian Chain northward and westward; the Hyblean Foreland southward belonging to the Pelagian Block, the northernmost part of the African plate (Lentini et al., 2006); the Ionian Basin eastward, an oceanic basin opened during the middle-late Mesozoic and aborted during the Tertiary (Catalano et al., 2001). The thickness of the crust of eastern Sicily has recently been reinterpreted (Cernobori et al., 1996; Continisio et al., 1997; Hirn et al., 1997; Nicolich et al., 2000) allowing the crustal structure in eastern Sicily and the Moho topography beneath the Ionian Sea to be better defined. The Moho has been located at a depth of 30 km beneath the central Hyblean Plateau, rising to 22 km in the Gela-Catania foredeep and to 21–18 km just offshore Catania (Nicolich et al., 2000).

Mt. Etna is sited in an anomalous external position with respect to the arc magmatism and back-arc spreading zones associated with Apennines subduction (Doglioni et al., 2001).

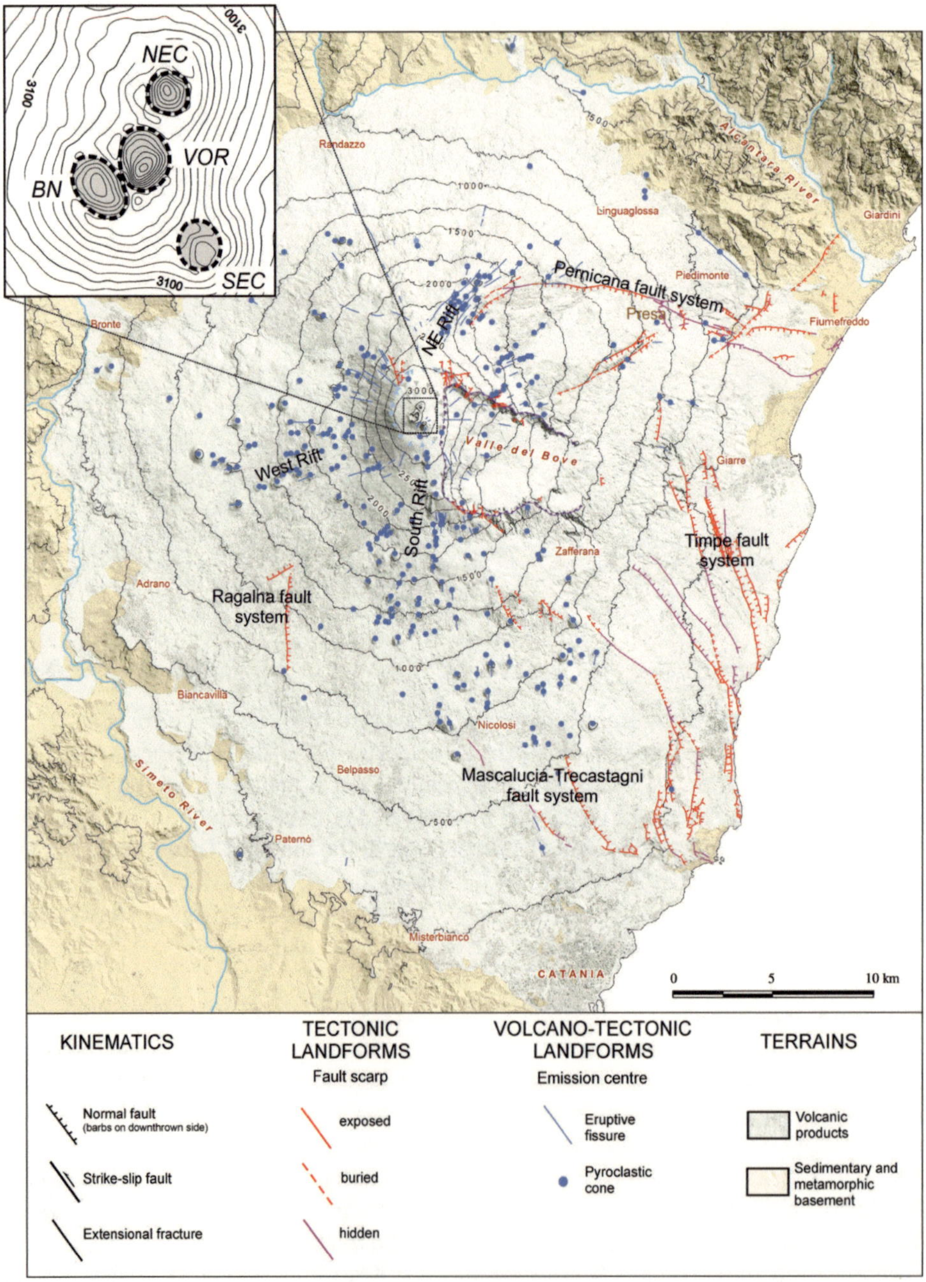

Fig. 2. Structural map of Mt. Etna with the location of the main fault and fissure systems. The location of the summit craters is shown in the inset in the upper left corner (VOR = Voragine, BN = Bocca Nuova, SEC = South-East Crater, NEC = North-East Crater). Redrawn from Azzaro et al (in press)

The structural features of Mt. Etna appear rather complex. On the volcano surface different fault and fissure systems can be recognized (**Fig. 2**). The most outstanding tectonic features at Mt Etna are clearly recognizable on the east and south-east flanks of the volcano, where the clearest morphological evidence of active faulting exists (**Fig. 2**). Here, seismogenic faults can be related to the NNW–SSE Malta Escarpment that is the main lithospheric structure in the eastern Sicily. Other seismogenetic faults (Patanè et al., 2005), though not recognizable on the surface, can be linked to the NE–SW, ENE–WSW fault systems that control the tectonic evolution of the northern margin of the Hyblean Plateau (Torelli et al., 1998). The eastern flank of Mt. Etna is characterized by frequent shallow seismic activity (depth <7 km) and by a seismic creep along some faults. Conversely, the western flank of Mt Etna, normally characterized by a deeper seismicity (depth >5 km), is considered the most stable sector of the volcano. In the western sector, there is only slight morphological evidence of faulting, such as some short segments of faults observable on the south-western flank (e.g. Ragalna fault). However, it must be noted that the faults with morphological evidence may represent only a part of the tectonic structures present in the Etnean area and hidden fault segments could be covered by the huge pile of volcanic products (e.g. Azzaro, 1999).

Following, some of the main tectonic features are discussed:

Timpe fault system. The normal faults belonging to this system dip toward the Ionian Sea and represent the most outstanding structural feature of the volcano. They displace a large part of the eastern flank by a 20 km long and 5 km wide belt of mainly extensional structures, striking from N to NW. Running from the coast to the south toward the inner volcano slope to the north, the fault system consists of a series of parallel seaward-facing step-faults segmented into individual steep fault escarpments up to 8 km long and up to 200 m high, that offset late Pleistocene to Holocene volcanics and historical lava flows. Acireale fault to the south and Moscarello fault to the north, represent the main elements of the Timpe fault system, and the N-S trending S. Alfio fault represents its northernmost apex. Timpe fault system is associated with shallow-depth (<7 km) seismicity including the occurrence of several earthquakes with M equal to 4.5 (Azzaro et al., 2000).

Pernicana fault system. It is located in the north-eastern flank of the volcano, trends E-W and can be considered the most active fault in the Etnean area, as testified by the slip rate estimations and geodetic measurements (up to 2.8 cm/y; Rasà et al., 1996; Neri et al., 2004; Bonforte et al., 2011). This system develops eastward from the NE rift (from 1850 m a.s.l.) to the coastline, over a distance of about 20 km. The westernmost part is mainly characterized by normal dip-slip motion, whereas the easternmost one by left strike-slip motion. The Pernicana fault system is partially characterized by a scarp with a maximum morphological height of 70-80 m between 1000 and 1500 m a.s.l. At lower elevations (starting from 800 to 700 m a.s.l.) this system has a less defined morphological expression (Acocella and Neri, 2005). Despite its continuity, Pernicana can be roughly divided into two main portions (western and eastern), characterized by differential times and amounts of displacement as evidenced during the 2002-2003 eruption (Neri et al., 2004). The western Pernicana, about 11 km long (from the NE Rift to Presa), shows the larger displacement even though the long-term slip rates are similar in both portions. Moreover, the western portion is associated with shallow (< 2-3 km) and moderate seismic activity ($2<M<3.5$; Azzaro et al., 1998). The eastern portion, about 9 km long (from Presa to the coastline), is aseismic and was recognized during the 2002-2003 eruption. These different styles of deformation may be due to the different rheological properties of their substratum (Neri et al., 2004).

Mascalucia-Trecastagni fault system. It is located in the southeastern flank and is composed of NNW-SSE-striking faults displaying prominent linear scarps near the towns of Mascalucia and Trecastagni (Azzaro, 2004). It is mainly characterized by strike-slip motion and by shallow seismicity, with focal depth of 1-2 km (Lo Giudice and Rasà, 1992).

Ragalna fault system. It is located in the southwestern flank of the volcano and comprises two linked structures, the main one extending for as much as 5 km in a roughly N direction towards the summit area of the volcano (Rust and Neri, 1996; Azzaro et. al., in press). Examination of the fault system in the field indicates dominantly dip-slip extensional displacement (Rust and Neri, 1996). The active faults of this system bound a triangular structure like horst (Rust and Neri, 1996).

The volcano is also characterized by a peculiar arrangement of the eruptive fissures that diverge from a radial distribution typical of stratovolcano edifices. The fissures are mainly concentrated on three sectors of the volcano named NE Rift, South Rift and West Rift, as previously indicated by several authors (Kieffer, 1975 and 1985; Lo Giudice et al., 1982; Mcguire and Pullen, 1989; **Fig. 2**).

NE Rift. It is located on the northeastern flank of the volcano and from the summit forms a 5-km-long, 2-km-wide topographic ridge made up of eruptive fissures, pit craters and pyroclastic cones. The swarm of eruptive fissures have dispersion axes ranging from 15°E to 50°E showing a gradual clockwise rotation along the rift towards NE (Tibaldi and Groppelli, 2002). The northeastern flank shows another smaller swarm of fissures and cones from the northern slope of the Valle del Bove, with dispersion axes ranging from 70°E to 90°E and a main ENE trend.

South Rift. The southeastern flank is characterized by a more scattered distribution of the eruptive fissures and cones. Over a 12 km wide sector the dispersion axes of the fissures range from 200°E to 140°E. The main belt of the rift develops between the SEC and the southwestern rim of the Valle del Bove along a SSE direction, and then continues southeastward as the rim swings to an easterly direction. On the southern slope of the volcano it forms a more diffuse set of N-S to SSW-NNE striking fissures extending from the Montagnola area to Nicolosi, at a distance of about 10 km.

West Rift. On the west flank eruptive fissures and cones are more radially distributed, even if a concentration of these elements appear over a 4.5 km wide sector between 245°E and 280°E marking the so-called West Rift characterized by WSW and W main trends of the eruptive axis (Bellotti et al., 2010).

2.2 Geological history and origin of volcanism

According to Branca et al. (2004), the beginning of volcanism in Etnean region is due to the northward migration of the Plio-Pleistocene Hyblean magmatic source. Volcanism began at about 500 ka ago through submarine eruptions on the Gela–Catania foredeep basin. About 300 ka ago fissure-type eruptions occurred on the ancient alluvial plain of the Simeto River forming a lava plateau. From about 220 ka ago, the eruptive activity was localized mainly along the Ionian coast where fissure-type eruptions built a shield volcano. Between 129 and 126 ka ago volcanism shifted westward toward the central portion of the present volcano (Branca et al., 2007). This change caused a variation in the volcanic chemical composition (from subalkaline to purely alkaline) as well as in the type of volcanism, which from fissural became central and shifted westward. The stabilization of the plumbing system marked the beginning of the construction of small polygenic edifices (e.g. Trifoglietto volcano) in Valle del Bove from 107 ka to 65 ka ago (De Beni et al., in press). About 57 ka ago, another westward shift of the plumbing system started the building of the stratovolcano (De Beni et al., in press) that represents the main bulk of the Mt. Etna edifice. This volcanic center reached its maximum areal expansion about 40 ka ago, proceeding up to 15 ka when four plinian eruptions formed a large summit caldera, historically named Ellittico Crater (Coltelli et al., 2000). The final evolution of this process took place during the Holocene, when eruptive activity resumed inside the caldera and expanded outside to cover the previous Ellittico edifice forming the volcanic succession of the present active volcanic center (Branca et al., 2004).

The complex geological history and tectonic setting of Mt. Etna have given rise to a great number of models to interpret its origin and the peculiar features for a very active basaltic volcano that is so unusually located in front of an active thrust belt:

- Rittmann (1973) interpreted the intersection of three main fault systems, trending ENE, NNW and WNW, as the mechanism that created a weakness zone for magma uprising.
- Tanguy et al. (1997) proposed how the upwelling of the asthenosphere first caused extensive melting of a mantle diaper, allowing tholeitic magma to accumulate near the mantle-crust interface. Then, increasingly alkaline basalt was generated and fed the entire volcanism of Mt. Etna by undergoing continuous but limited differentiation in a subcrustal reservoir.
- Monaco et al. (1997) infer that the magmatism at Mt. Etna can be related to the dilatational strain on the footwall of an east-facing, crustal scale normal fault located along the Ionian shore. In fact, on the basis of structural, seismological and volcanological studies of 2001 and 2002-2003 eruptions, Monaco et al. (2005) state that the conditions of magma ascent are strongly dominated by extensional structures related to this dilatational strain.
- Gvirtzman and Nur (1999) advanced the idea of the "suction" of asthenospheric material from under the neighboring African plate to cause the voluminous melting under Mt. Etna. Such lateral flow is expected when descending slabs migrate backwards in the mantle. A similar model was also developed by Doglioni et al. (2001). According to these Authors, the right lateral transfer along the Malta escarpment is a transtensional "window" between the Sicilian and Ionian segments of the Apennines slab.
- According to some Authors, Mt Etna's magmatism is related to the instability of the eastern flank of the volcano. Indeed, deformation measurements carried out by GPS, SAR and so on, suggest that the eastern flank of the volcano is sliding toward the sea (e.g. Froger et al., 2001; Lundgren et al., 2003; Palano et al., 2008). Some authors believe this sliding motion may cause the decompression of the plumbing system, facilitating the uprise of magma to the surface (Branca et al, 2003; Neri et al., 2004). The location of the sliding surface is open to debate. Lo Giudice and Rasà (1992) postulate a shallow slip surface (0-1 km a.s.l.) consistent with the very shallow seismicity (depth < 1.5 km). Borgia et al. (1992) and Rust and Neri (1996) suggest a detachment as deep as about 5 km occurring within weak sediments of the Gela-Catania Foredeep. Bousquet and Lanzafame (2001) envisage a decollement between the volcanic pile and the sedimentary substratum (1-2 km a.s.l.). Finally, Tibaldi and Groppelli (2002) suggest that both a shallow and a deep decollement surface can characterise, at the same time, the eastward sliding of the volcano. The unstable zone is confined by the Pernicana

fault to the northwest, by the NE rift and the fissure systems in the summit area, and by the Ragalna fault system to the southwest. The Trecastagni-Mascalucia fault system is likely originated by differential movements within the collapsing sector of the volcano (Rust and Neri, 1996).

- Chiocci et al. (2011), by studying the marine geological and geophysical data of the continental margin facing the volcano, found a large bulge offsetting the margin that is deeply affected by widespread semicircular steps, interpreted as evidence of large-scale gravitational instability. Such features extend inshore to the mobile eastern flank where the larger ground deformations are measured. Both submarine instability and subaerial flank sliding are bounded by two regional tectonic lineaments to accommodate the basinward movement of this large sector of the continental margin topped by the Etna volcanic pile. The Authors infer that the instability process involving the Sicilian continental margin facing Etna volcano during the last 0.1 Ma may be considered a very large mass-wasting phenomenon. This is due to the magmatic intrusion rather than any tectonic process related to a late-orogenic phase of the Apennine Chain thrusting this portion of the continental margin. Indeed, the bulge has no trace of any compressive structures, as previously expected by Borgia et al. (1992) and Rust et al. (2005). Conversely, it is pervaded by extensional and transtensional structures representing the brittle response to a large-scale and long-lasting gravitational instability affecting the continental margin. This model implies that an extensional tectonics induced by the sliding of the volcano eastern flank has been acting continuously over the last 0.1 Ma since the bulge collapse effects are propagating upslope. The continuous decompression at the volcano summit favors the ascent of basic magma without lengthy storage in the upper crust, as one might expect in a compressive tectonic regime. This may be the cause or one of the main contributory causes of the growth of a very active basaltic volcano on top of such an active thrust belt as the Apennine Chain in Sicily.

2.3 The recent volcanic activity since 2000

Two main types of volcanic activity may be distinguished into persistent activity at the summit craters and periodic flank eruptions. The former is characterized by phases of degassing alternating with mild strombolian activity, occasional lava fountains, and lava overflows. Flank eruptions occur from lateral vents usually located along fracture systems. The past decade at Mt. Etna was characterized by different kinds of activity. From 2001 to 2003, two large eruptions characterized by very intense explosive activity took place in the southern and northeastern flanks of the volcano. Successively, Etna remained quiet for about 20 months up to September 2004 when an eruption, differing significantly from the two previous, erupted essentially degassed magma from two vents within Valle del Bove (e.g. Di Grazia et al., 2006). After a 15-month-long period mainly characterized by degassing, the eruptive activity resumed on the eastern flank of SEC in late 2006 with strombolian activity, lava fountaining and lava overflows. During 2007, six episodes of intense lava fountaining/strombolian activity took place at SEC. Finally, after a lava fountain occurring on 10 May 2008 at SEC, a new eruption took place on 13 May from an eruptive fissure that opened east of the summit area (EF; Cannata et al., 2009c; **Fig. 10**). This eruption, ending on 6 July 2009, was characterized by a strong Hawaiian activity at its beginning and by a long phase of gradually decreasing strombolian activity and lava flows during the following months (Aloisi et al., 2009).

3. Mt. Etna's deep plumbing system: tomographic analysis

Since 1977, seismic surveys and seismological studies have progressively improved our knowledge of Etna's structure, but there is still no clear evidence for the presence of a large magma chamber in the crust. In the last two decades, tomographic inversions of P- and S-wave arrival times from local earthquakes have been performed with various techniques allowing a good definition of the P-wave velocity structure beneath the volcano down to 18-24 km depth, more detailed down to 10 km depth (Chiarabba et al., 2000; Laigle et al., 2000; Patanè et al., 2002; Chiarabba et al., 2004; Patanè et al., 2006). However, none of these studies have evidenced the presence of a large anomalous region of relatively low-velocity in the upper crust beneath the volcano. One of the recent ideas about the deep structure of Mt. Etna is the presence of a melted lens capping a mantle upward beneath the volcano (Hirn et al., 1997) with a Moho transition at less than 20 km deep (Nicolich et al., 2000). Following, the main features revealed by Mt. Etna velocity and attenuation tomographies are reported and discussed.

3.1 Active seismic surveys and tomographic studies

The first noteworthy seismic investigation of the crust and upper mantle in Sicily was performed in 1968 by seismic refraction surveys and covered the whole island with only one profile near Etna, located just to the north (Cassinis et al., 1969). The interpretation of seismic sections revealed a low velocity zone in the eastern Sicily continental crust, close to Mt. Etna between 9 and 24 km depth, interpreted as a region with high temperatures due to the proximity of a deep magma chamber under the volcano. Following this early active seismic exploration, only in 1977 a deep seismic sounding focused on Etna's structure, with both a detailed survey and deployment of a temporary seismic array (Colombi et al., 1979). Based on data acquired during these experiments, Sharp et al. (1980) investigated Mt. Etna's structure and the physical properties of the low-velocity anomaly previously observed near the area. They modelled this anomaly as a low-velocity, tri-axial ellipsoid body extending under the entire volcanic area at midcrustal depths (15-25 km), interpreted as a large partially molten magma chamber.

Since the 90's, seismological studies have progressively improved our knowledge of Etna's structure and in the last two decades, tomographic inversions of P- and S-wave arrival times from local VT earthquakes have been performed with various techniques (Hirn et al., 1991, 1997; Cardaci et al., 1993; De Luca et al., 1997; Laigle et al., 2000; Chiarabba et al., 2000; Patanè et al., 2002; Patanè et al., 2003; Chiarabba et al., 2004). None of these tomographic studies showed the presence of a large anomalous region of relatively low-velocity within the crust beneath the volcano interpretable as a large magma chamber. Conversely, the most important feature is the presence of a wide central high-velocity body (HVB) embedded in the pre-Etnean sediments, interpreted as a main solidified intrusive body (cooled batches of magmatic intrusions), which is also an almost aseismic volume surrounded by an active seismic region. This HVB shows a roughly ellipsoidal shape in the upper crust (depth ~10 km) with a NNW-SSE horizontal axis and a vertical axis extending between 0 and 9 km below sea level. However, basalt melt rising through the continental intermediate crust may not produce a slow anomaly and the clear large high Vp body observed in the tomographic images can be related to the volumes where the magma is stored in the crust before the eruption. This seems to be supported by the existence of the wide, elongated aseismic zone located just beneath the summit craters (Chiarabba et al., 2000; Patanè et al., 2004) and by results regarding the spatial distribution of b-values (Murru et al., 1999). De Gori et al. (2005)

tried to yield insights into the physics of the volcanic plumbing system by determining the three-dimensional Qp structure. This attenuation tomography evidenced the presence of a low Qp body located at shallow depth (0–3 km b.s.l.) beneath the south and southwestern sides of the edifice, where the magma was likely stored during 1994–2001. Since attenuation is a physical parameter sensitive to the thermal state of the crustal volume traveled by seismic wave, this interpretation of the low-Qp anomaly, also confirmed by Martinez-Arevalo et al. (2005) for the 2001 eruption, is consistent with the intense recent volcanic activity (2001 and 2002–2003) that concentrated in the southern part of the summit area. Finally, the Patanè et al. (2006) inversion allowed the improvement of even the most recent velocity tomographic results (Patanè et al., 2002; Patanè et al., 2003; Chiarabba et al., 2004) and a better definition of the shallow structure, down to 7 km depth, and shape and geometry of the upper portion of high-velocity Vp volume. However, the most notable result of this work concerns the detection of anomalous zones with low Vp/Vs values located in the central-southern and northeastern part of the volcanic edifice, where geodetic data modeled the dike intrusions feeding the 2002–2003 eruption, located beneath the eruptive fracture systems.

3.2 Shape and geometry of the intermediate and deep plumbing system

Mt. Etna's tomographic models contribute significantly to clarify whether and how tectonic control of magma ascent works at Mt. Etna, revealing a broad complex of intrusive meshes in the upper and middle crust. In particular, we analyze the results obtained by the last tomographic study performed by Patanè et al. (2006) integrating it with previous results and new unpublished data. In summary, the main features revealed by Mt. Etna Vp tomography (Patanè et al. 2006, **Fig. 3a**) are:

- A shallow high Vp anomaly (Vp ranging between 3.5 and 5.5 km/s) beneath the southern craters, the South Rift and mostly beneath the central-southern sector of the Valle del Bove, between 0 and -1 km depth, is interpreted as a solidified intrusive complex (**Fig. 3a**). Contours for 3.5-5.0 km/s at 0 km show that the high velocity anomaly aligns with the present-day south and northeastern Rifts. The presence of the old shallow plumbing system feeding the past Mt. Etna eruptive centers (e.g. Trifoglietto volcano), located along the central-southern part of the Valle del Bove, is also evidenced both at 0 and -1 km a.s.l.. The analysis of the isosurface image at Vp of 3.5 km/s (**Fig. 3b**), reveals a complex pattern of solidified magma chambers and conduits with variable dimensions in the very shallow crust (between 0 and -1 km a.s.l.). These higher velocity volumes can be linked to: i) the wide plutonic body mainly located beneath the Valle del Bove (Patanè et al., 2003, 2006); ii) the solidified magma reservoirs feeding the S, NE and ENE Rift zones.
- A clear high Vp body (Vp ranging between 5.5 and 6.7 km/s), NNW-SSE to NS trending located beneath the central craters extended toward S and SSE, between -2 and -7 km a.s.l. (wide 5-7 km in longitude and 8-10 km in latitude) is interpreted as high density cumulates, fractionated by the magma during its ascent, stocked and congealed at depth (**Fig. 3 a**).

Considering now the two different inversions by Patanè et al. (2003) and Chiarabba et al. (2004), extended also to the deep structure although at lower resolution, the main features observed at major depth are:

- A narrow high Vp body (Vp ranging between 6.8 and 7.5 km/s), 4-6 km laterally wide, beneath the central-southern part of the volcano between 8 and at least 18 km depth (**Fig. 4a**), interpreted as the deeper part of the plumbing system.

- The presence of a melted lens capping a mantle upward beneath the volcano with a Moho transition at depth less than 20 km (**Fig. 4a, b**), as suggested by Nicolich et al. (2000), seems to be supported by recent tomographic results at regional scale (Barberi et al., 2006).

Therefore the high Vp intrusion is the main structural feature of the volcano, testifying to its intense past history, and revealing the accumulation of a very large volume of non-erupted volcanic material. Seismicity seems to occur at its borders and defines a main aseismic volume (**Fig. 3b**).

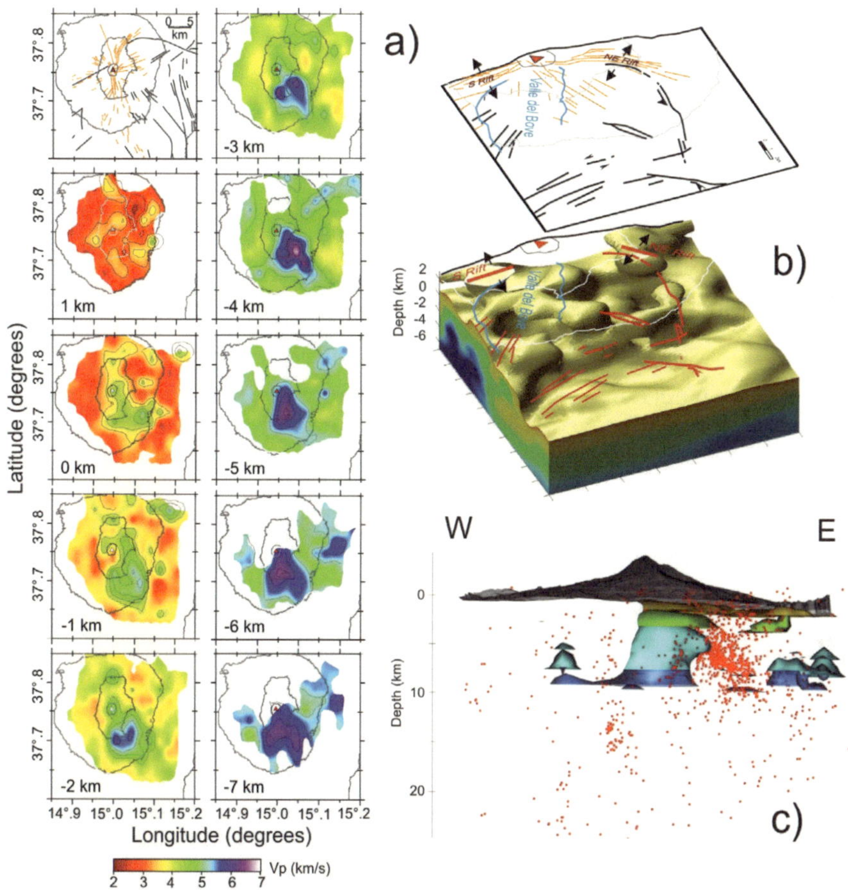

Fig. 3. (a) Mt. Etna's Vp velocity model (Patanè et al. 2006), between 1 a.s.l. and 7 km b.s.l., in the well-resolved regions of the model . The gray lines are elevation isolines (every 1000 m). In the top left square the historical eruptive fissures (orange lines) and major faults (black lines) are shown. b) Isosurface image at Vp of 3.5 km/s for the central-eastern and northern sectors of the volcano. A complex pattern of solidified magma chambers and conduits with variable dimensions is recognizable. At the top, historical eruptive fissures (orange lines) and major faults (black lines) are shown. Major faults are also projected in red in the 3D block. c) Cumulative isosurfaces for different velocities Vp showing the 3D geometry of the HVB down to 10 km depth. The seismicity occurring during 2001-2003 is also shown (red dots), evidencing how the HVB is almost an aseismic volume surrounded by an active seismic region

The bulk of the high Vp body, located to the southeast of the central craters, suggests that the Valle del Bove has been the main site for magma accumulation in the past as confirmed by the presence of the old eruptive centers (e.g. Trifoglietto). The high Vp body NNW-SSE to NS trending between -1 and -5 km depth appear rooted at greater depth. At present, the ascent of magma is controlled by the pervasive high Vp intrusion and seems to occur at its western border. Very shallow dike emplacement at the border of the intrusive body occurs mostly on NNW-trending fracture system, such as those of the 2001 and 2002-2003 eruptions.

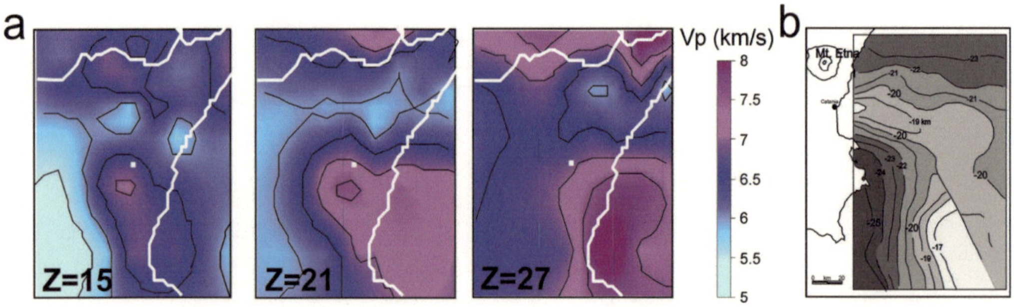

Fig. 4. a) Regional Vp model of the lower crust and uppermost mantle from Barberi et al. (2006). b) Moho topography (km b.s.l) of the northwestern part of the Ionian basin (redrawn from Nicolich et al., 2000)

4. Mt. Etna shallow plumbing system: seismo-volcanic signal analysis

Although the VT earthquakes are the key to tomographic studies in volcanoes, they cannot provide precise information about the location and geometry of the shallow magma conduits (Almendros et al., 2002). In fact, the understanding of the complex velocity structure in the shallow part of the volcano requires estimation of both P- and S- waves variations with a spatial resolution of the order of several hundred meters, which is still not yet available at Mt. Etna.

A more useful approach consists of investigating the seismo-volcanic signals, whose variations and features are often closely related to eruptive activity. Indeed, they are generally considered as an indicator of the internal state of activity of volcanoes (Neuberg, 2000). For this reason their investigation can be very useful for both monitoring and research purposes. Because of the peculiar characteristics of the seismo-volcanic signals, different from the tectonic and VT earthquakes in terms of both waveforms and source mechanisms, new techniques have been developed to investigate their features. In **Fig. 5** examples of VT earthquake, volcanic tremor, LP and VLP events recorded at Mt. Etna are shown.

According to Murray (1990), shallow reservoirs at Mt. Etna are temporary and are occupied by magma only during short periods preceding a single eruption or an eruptive cycle. Patanè et al. (2008) demonstrated how the locations of the tremor sources and of the long-period seismic events can be used at Mt. Etna to constrain both the area and the depth range of magma degassing, highlighting the geometry of the shallow conduits feeding the central craters. In this work the Authors, by a careful analysis of the seismo-volcanic signals recorded during two powerful lava fountaining episodes taking place on 4–5 September and on 23–24 November, 2007 from SEC, discover the magma pathway geometry feeding the eruptive activity at SEC. The imaged conduits consist of two connected resonating dike-like bodies, NNW-SSE and NW-SE oriented, extending from sea level to the surface. In addition, we show how tremor, long-period (LP), and very-long period (VLP) event locations and signatures reflect pressure fluctuations in the plumbing system associated with the ascent/discharge of gas-rich magma linked to the lava fountains.

Thus, in this section we will focus on the most recent eruptive activity of this volcano taking place in 2008-2009, showing the inferences about the volcano dynamics and the shallow system feeding this eruption drawn by the seismo-volcanic signal investigation. **Fig. 6** shows a digital elevation model of Mt. Etna with the locations of the stations used to record the seismic signals during such an eruption. In **Fig. 7** the seismic signal acquired by the vertical component of ECPN during 1-13 May 2008, together with the number of LP events and the seismic RMS, is shown.

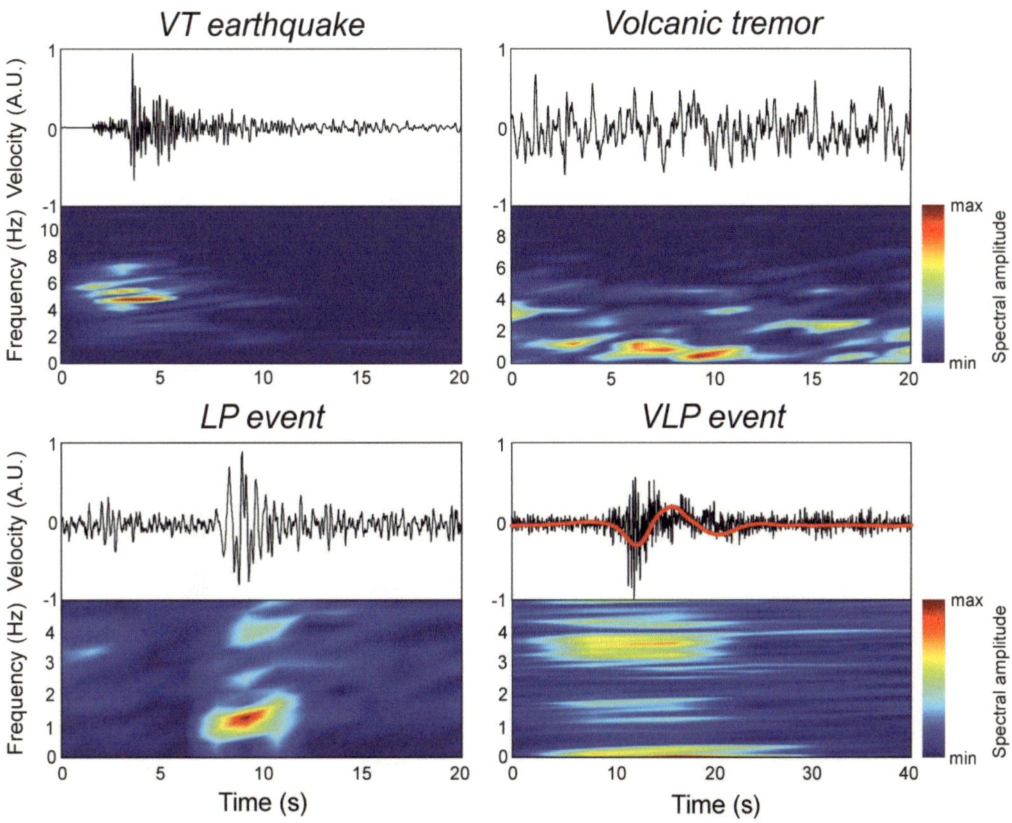

Fig. 5. Waveforms and spectrograms of VT earthquake, volcanic tremor, LP event and VLP events recorded at Mt. Etna. The thick red line plotted over the VLP waveform shows the signal filtered below 0.15 Hz

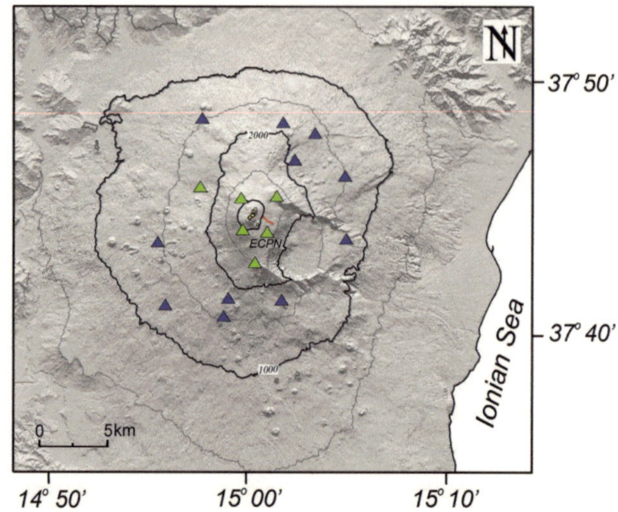

Fig. 6. Digital elevation model of Mt. Etna with the location of the seismic stations used to investigate seismo-volcanic signals (green and blue triangles). The green triangles indicate the stations used to study both volcanic tremor and LP events, while the blue triangles only the volcanic tremor. The red line indicate the eruptive fissure active during 2008-2009

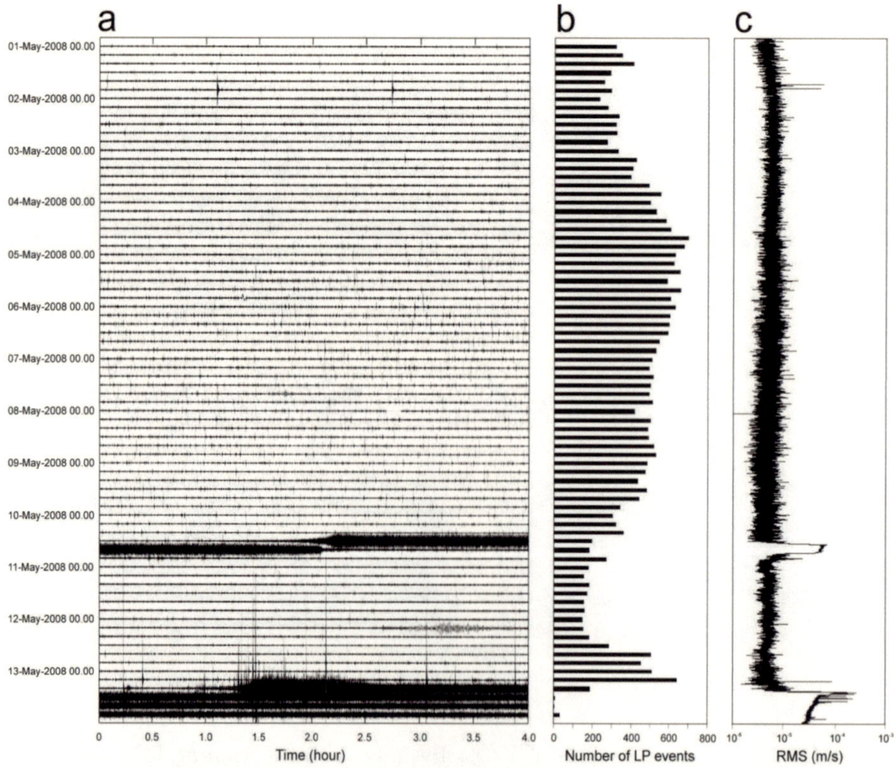

Fig. 7. (a) Seismogram of the vertical component of ECPN station, (b) histogram of the number of LP events in 4-hour-long windows and (c) RMS time series

4.1 Volcanic tremor and LP events

A peculiar aspect of volcanic tremor at Mt. Etna is its continuity in time, as also observed at other basaltic volcanoes with persistent activity such as Stromboli (Italy; Langer and Falsaperla, 1996). Most of the energy of volcanic tremor at Mt. Etna is radiated below 5 Hz (e.g., Lombardo et al., 1996; Falsaperla et al., 2005; Cannata et al., 2008, 2009a). Another interesting feature of the volcanic tremor is its close relationship to eruptive activity, highlighted by variations in amplitude, spectral content, wavefield features, and source

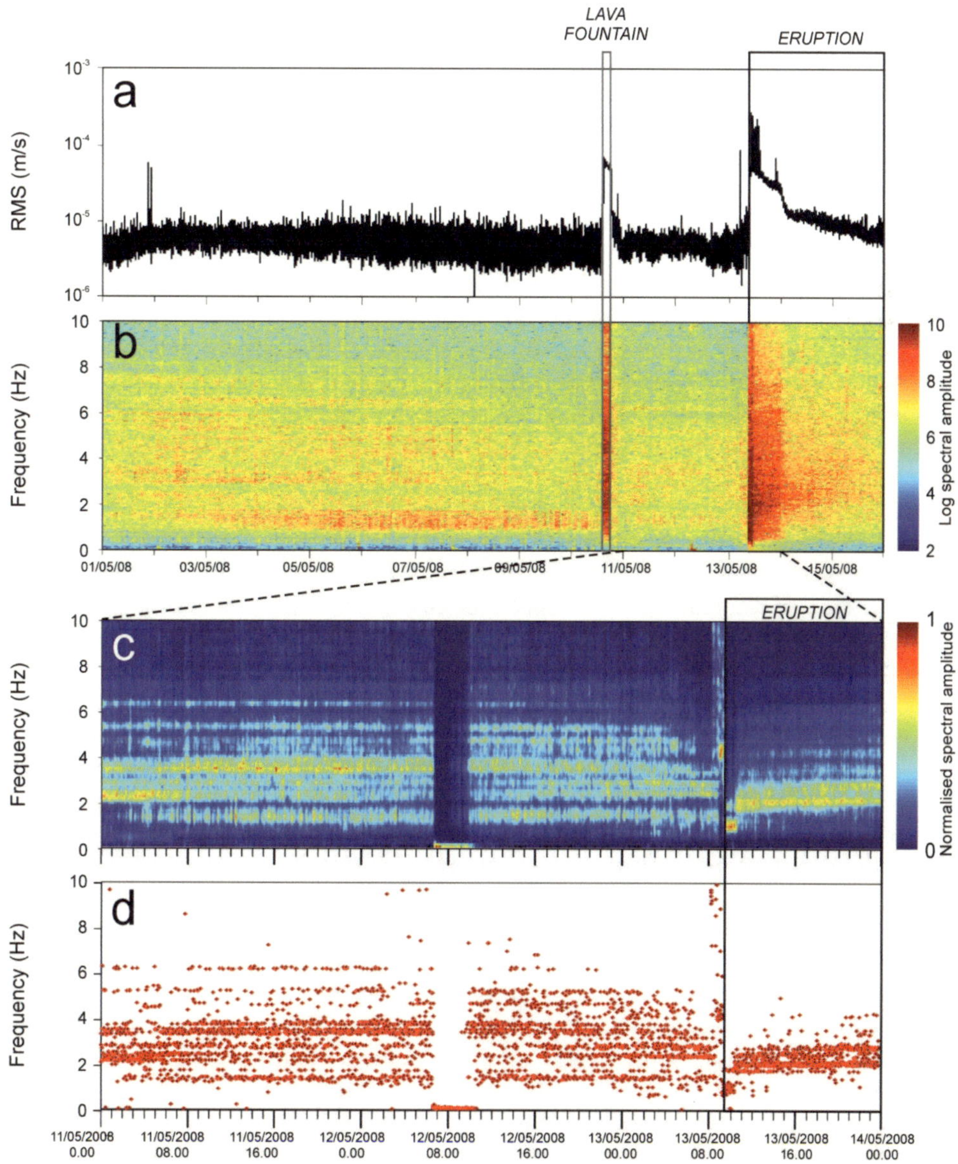

Fig. 8. (a) RMS time series and (b) spectrogram of the seismic signal recorded at the vertical component of ECPN station. (c) Normalised spectrogram and (d) dominant frequencies of the seismic signal recorded at the vertical component of ECPN station

location of volcanic tremor at the same time as changes in volcanic activity (e.g., Gresta et al., 1991; Lombardo et al., 1996; Di Grazia et al., 2006; Alparone et al., 2007; Cannata et al., 2008; Patanè et al., 2008; Cannata et al., 2009a). The volcanic tremor recorded during 1–15 May 2008 was investigated by performing several analyses. First of all, in order to get information about the time changes of tremor energy, the RMS of the seismic signals recorded at the vertical component of ECPN station (see **Fig. 6**) was calculated within 1-minute-long sliding windows (**Fig. 8a**). Successively, since changes of source location and/or mechanism of volcanic tremor are generally accompanied by variations of its spectral content, the Short Time Fourier Transform (STFT) was performed. We calculated a spectrum by 40-secong-long sliding windows of the signal recorded at the vertical component of ECPN station. Then, the spectrogram was plotted in **Fig. 8b**. Moreover, the normalized spectrogram of 11-14 May, together with the dominant frequencies, was also computed and plotted in **Fig. 9c,d**. Finally, since the seismo-volcanic signals are generally related to dynamics of fluid inside the volcanic edifice, the location of their source is basic information for monitoring of volcanoes. Then, the tremor source locations were retrieved by following the approach described by Patanè et al. (2008) and Di Grazia et al. (2009), inverting the spatial distribution of tremor amplitude in 18 stations (green and blue triangles in **Fig. 6**) using a grid-search approach (**Figs. 9,10**). We considered the RMS amplitudes of the 25th percentile instead of average values. This enables us to efficiently remove undesired transients in the signal and consider continuous recordings (Patanè et al., 2008). The source location of tremor is found on the basis of the goodness of the linear regression fit (hereafter referred to as R^2) obtained for each point of a 3-D grid centered underneath the craters (Di Grazia et al., 2006). For this grid, we consider a $6 \times 6 \times 6$ km^3 volume with a spacing between nodes of 250 m. The centroid position of all the 3-D grid points, whose R^2 values do not differ by more than 1% from the maximum R^2, was considered the tremor source location. LP and VLP events, whose sources, similar to volcanic tremor, are related to fluid processes (such as vibration or resonance of fluid-filled cracks; Chouet, 2003), are also recorded at Mt. Etna. A number of papers deal with the relation between eruptive activity and LP events at Mt. Etna (Patanè et al., 2008; Di Grazia et al., 2009; Cannata et al., 2010): it was shown how occurrence rate, energy, spectral content and/or source location of LP events often change before, during and after eruptive activities. LP events recorded during 1 – 15 May 2008 were investigated obtaining several parameters: i) occurrence rate; ii) peak-to-peak amplitude; iii) source location. About 33,000 LP events were detected during the analysed period by STA/LTA algorithm (short time average/long time average; e.g., Withers et al., 1998). Similar to all the triggering algorithms based on dynamic thresholds, the event detection by STA/LTA is affected by the variation of the background noise level: for instance, if the background noise level increases, the threshold in turn will increase, and then the events with lower amplitude will be lost. The LP hourly number and the peak-to-peak amplitudes were calculated and plotted in **Fig. 11a,b**, respectively. Moreover, since the frequency and damping of a resonant system is strongly influenced by the nature of liquid and gas content (Chouet, 2003 and references therein), also the study of the spectral evolution of LP events in volcanic areas provides very useful information for monitoring purposes. Thus, a value of frequency and quality factor for each LP event were obtained by Sompi analysis (Kumazawa et al., 1990) (**Fig. 11c,d**). A moving median over 200 samples was calculated for both frequency and quality factor. Indeed, the median values are less affected by outliers than the average values. Finally, a subset of 1700 LP events with high signal to noise ratio at all the six stations nearest to the summit area (green triangles in **Fig. 6**) was selected to perform location analysis. LP events were located by following a new grid-search method based on the joint computation of two different functions: i) semblance, used to measure the similarity among signals recorded by two or more stations (e.g. Neidell and Taner, 1971; Cannata et al., 2009b); and R^2 values, calculated on the basis of the spatial distribution of seismic amplitude (Patanè et al., 2008; Di Grazia et al., 2009). The 3-D grid of possible locations was 6 km×6 km×3.25 km, centered on the volcanic edifice, and with a vertical extent from 0 km a.s.l. to the top of the volcano. The horizontal and vertical grid spacing was 250 m. The space distributions of both semblance and R^2 values were determined, the two grids of values were normalised by subtracting the minimum value and dividing by the maximum one. Thus, the values belonging to two grids ranged from 0 to 1, and the same weights were assigned to semblance and R^2. Then, the two normalised grids were summed node by node. The source was finally located in the node with the largest composite semblance-R^2 value. This joint method takes advantage of both LP waveform comparison among the different stations and space amplitude distribution. The LP location results are reported in **Figs. 12** and **13**.

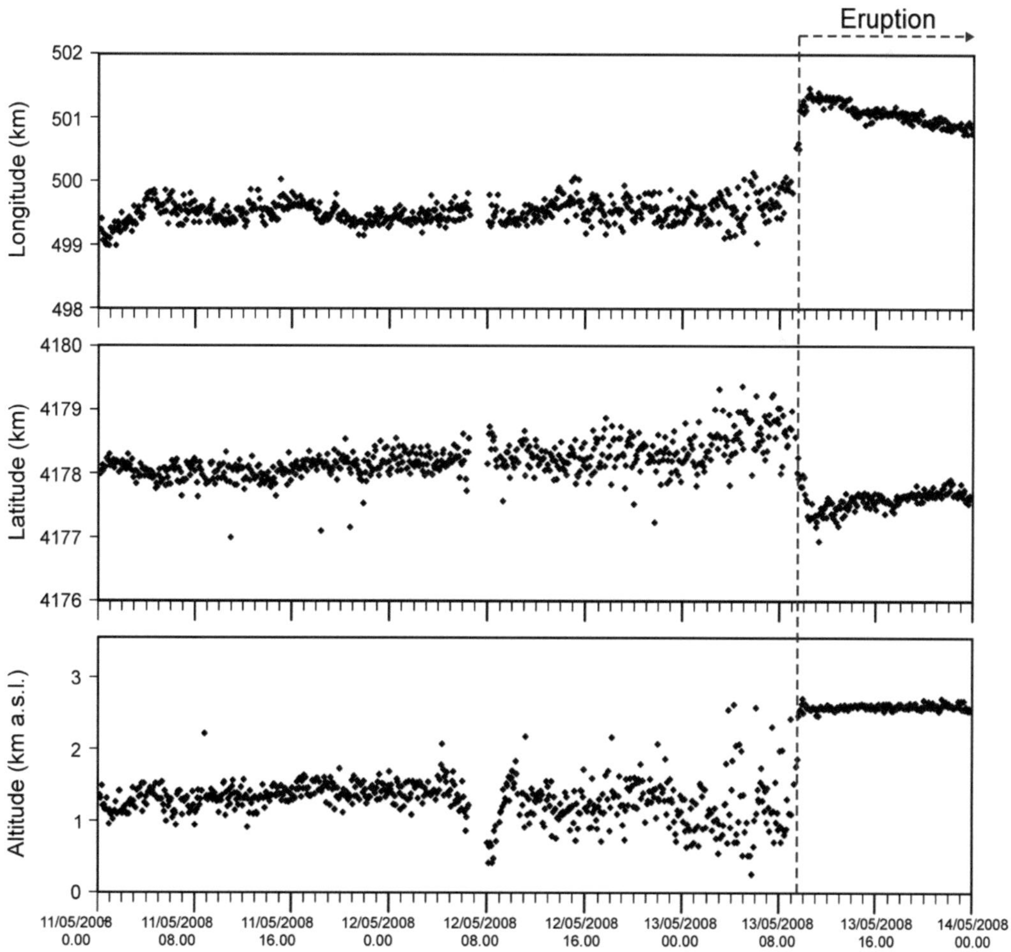

Fig. 9. Time variations of the location of volcanic tremor recorded during 11-14 May 2008

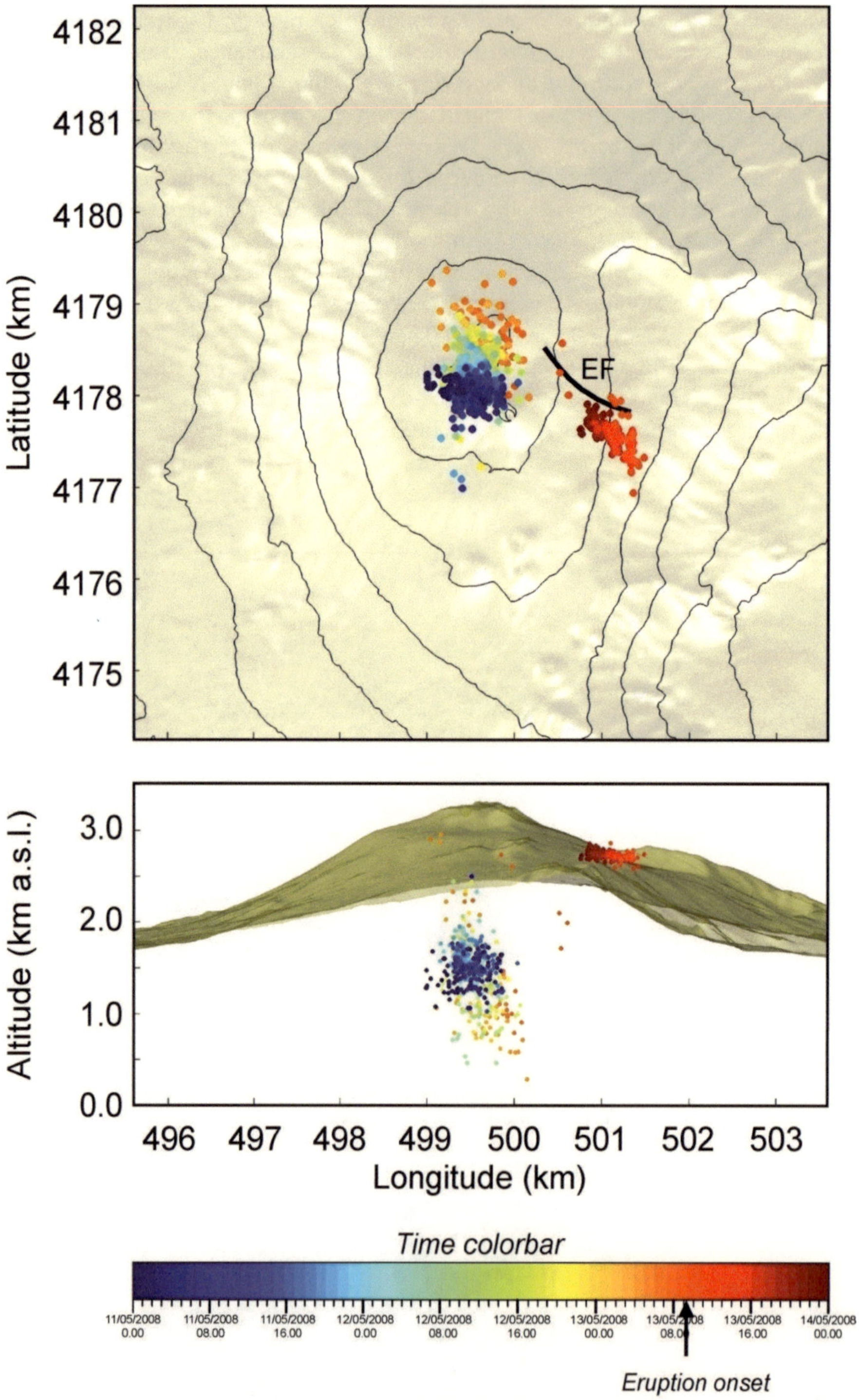

Fig. 10. Map and section of Mt. Etna with the locations of volcanic tremor recorded during 11-14 May 2008

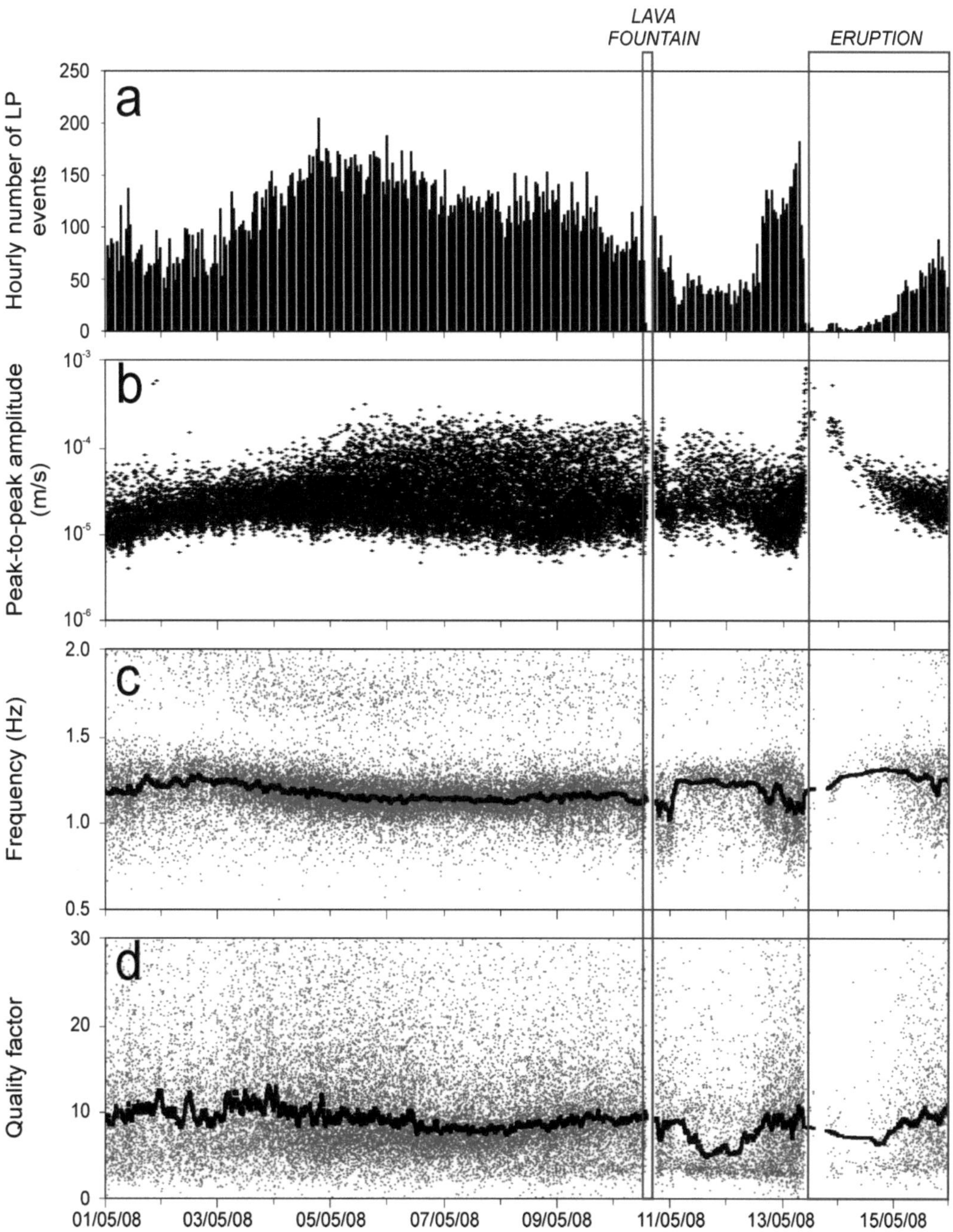

Fig. 11. (a) Histogram of the number of LP events in 1-hour-long windows. (b) Peak-to-peak amplitude of the LP events. (c,d) Frequency and quality factor of the LP events, calculated by Sompi method with autoregressive order of 2

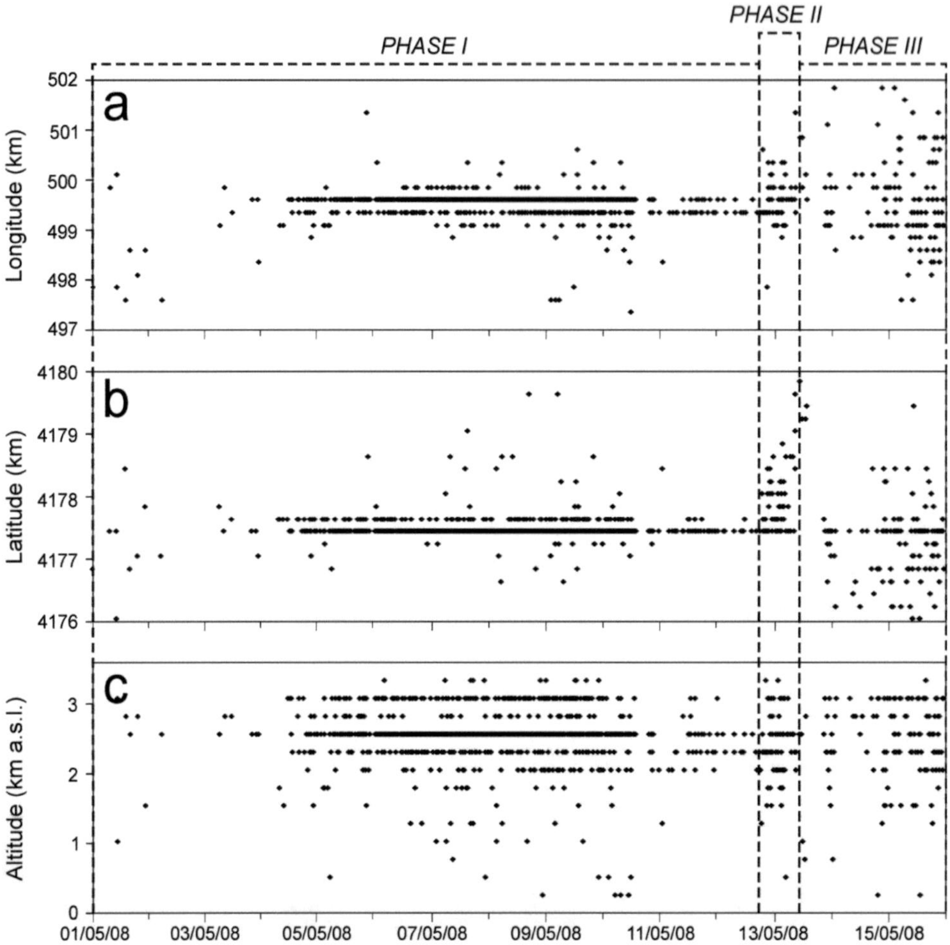

Fig. 12. Time variations of the location of LP events recorded during 1-15 May 2008

Finally, in order to understand the LP source mechanisms, moment tensor inversions were performed. On several volcanoes Moment Tensor Inversion has been achieved to quantify the source processes of LP events (e.g. Ohminato et al., 1998; Chouet et al., 2003; Lokmer et al., 2007). Indeed, due to the link between LP activity and fluid dynamics (Chouet, 2003), the characterization of the LP source mechanism becomes a fundamental tool for understanding processes in magmatic systems. Several studies propose the excitation and resonance of fluid-filled resonator systems as the cause of the source mechanism of these particular events. Different geometries of the resonators were investigated, such as a crack, a spherical inclusion or a conduit simplified to a cylinder. The seismic moment-tensor is a representation of a seismic source by a system of equivalent forces acting at a source point, including the force couples and single forces resulting from mass transport. We performed a moment tensor inversion in the frequency domain (Auger et al., 2006; Lokmer et al., 2007). The Green's Functions (GF) are calculated for a homogeneous velocity model with DEM Etna topography. Synthetic seismograms were generated by Discrete Elastic Lattice algorithm described by O'Brien and Bean (2004), based on discrete particle scheme. The topography model of Mt. Etna covers an area of 19.2 x 16 x 7 km, with a grid node of 50 m, and its origin (x,y,z) is centred on the volcano summit. In order to avoid reflections from the model boundaries, we employed at the bottom and at the edges of the model 4.8 km wide absorbing boundaries. Moreover we used for the GF computation (i) a Gaussian pulse as source function (STF), with a 7.5 Hz cut-off frequency; (ii) velocities for P and S waves of 2000 ms^{-1} and 1175 ms^{-1} respectively, as found by Patanè et al. (2006) and Montellier et al. (2009) in their recent tomographic study of Mt. Etna. Several authors demonstrated how the topography and the velocity model play an important role to correctly reconstruct the moment tensor (e.g. Bean et al., 2008; O'Brien and Bean, 2009). For LP events, which are characterized by frequencies above 5 seconds, uncertainties in the GF can be introduced by a poor knowledge of the velocity structures. This problem can be resolved by using several seismic stations installed very close to the source positions (Bean et al., 2008; Kumagai et al., 2010; De Barros et al., 2011). For this reason, in order to compute the LP moment tensor inversion, we used the LP database recorded by an exceptionally high-density network of 30 temporary broadband stations, installed during the 2008-2009 Etna eruption (De Barros et al., 2011). In order to determine the most reliable mechanism type (crack, pipe, or explosion), the source of the LP is initially modelled performing an unconstrained inversion. Next, starting from the mechanism so obtained, we have constrained the subsequent reversals solution (inversions) found to confirm and refine its characteristics. Once the stability of our results is verified, it was possible to reconstruct the source mechanisms of the LP in 2008 Etna eruption.

4.2 Results and interpretation

The analyses described in the **section 4.1** provided information about the volcano dynamics and on the shallow part of the plumbing system involved in magma movements before and during the first days of the eruption.

The 10 May lava fountain and the following 13 May eruption onset were preceded by a change in the volcanic tremor spectral content (from polychromatic to monochromatic with a spectral peak at 1-2 Hz) and by an increase in LP activity (increases in both occurrence rate and amplitude of LP events) taking place roughly on 4 May. Such an energy increase of LP events can be interpreted as increasing overpressure inside the shallow part of the plumbing system. Increases of amplitude and occurrence rate of LP events preceded eruptive activities also at many other volcanoes such as Redoubt (Chouet et al., 1994), St. Helens (Moran et al., 2008) and Colima (Varley et al., 2010). Before the eruptive activity the volcanic tremor was located below the summit area at depth ranging from 1 to 2 km a.s.l. (**Figs. 9,10**), suggesting an important magma storage volume in this location, as also suggested by previous studies (Allard et al., 2005; Aiuppa et al., 2010). The LP events were located roughly below Bocca Nuova at 2-3 km a.s.l., consistent with the LP location obtained in other papers (Patanè et al., 2008; Cannata et al., 2009b). The lava fountain was accompanied by a sharp increase in the tremor RMS, as observed during other lava fountain activities at Mt. Etna (e.g. Cannata et al., 2008), as well as at other volcanoes (e.g. McNutt, 1994). Because of this increase of tremor amplitude and then of the background noise level, only a few LP events were detected during the lava fountain (**Fig. 11a**). It is also worth noting that during the whole of 10 May the LP frequency peak decreased, then suddenly increasing on the following day (**Fig. 11c**). After the lava fountain, the sudden decrease in the seismic dominant frequencies observed during 06.00-11.00 on 12 May was due to the arrival of teleseismic waves of the Sichuan earthquake (M=7.9) that, according to Cannata et al. (2010), played an important role in the 13 May eruption because of the dynamic stress transfer. Focusing on the 13 May eruption, it was preceded by a few hours by a further increase in the LP activity, accompanied by changes in LP spectral content (decrease of frequency peak and increase of quality factor; **Fig. 11c,d**), and by a northward shift of the sources of both volcanic tremor and LP events (**Figs. 10,13**). Such a seismo-volcanic source migration was consistent with the dyke intrusion in the northern part of the summit area (towards the NE rift zone; **Fig. 2**), inferred by earthquake swarm hypocenters (Di Grazia et al., 2009), ground deformation (Aloisi et al., 2009) and a dry fracture field (Bonaccorso et al., 2011). Finally, the onset of the 13 May eruption was characterised by a sharp RMS increase reaching the maximum values of the whole analysed period (**Fig. 8a**), together with significant changes in spectral content (**Figs. 8c,11**) and shift of tremor and LP sources moving roughly below the eruptive fissure (**Figs. 9,10,12,13**). Also in this case the LP number drastically decreased because of the increase of the background noise level (**Fig. 11a**). All these data suggest the intrusion of a gas-rich magma batch east of the summit area.

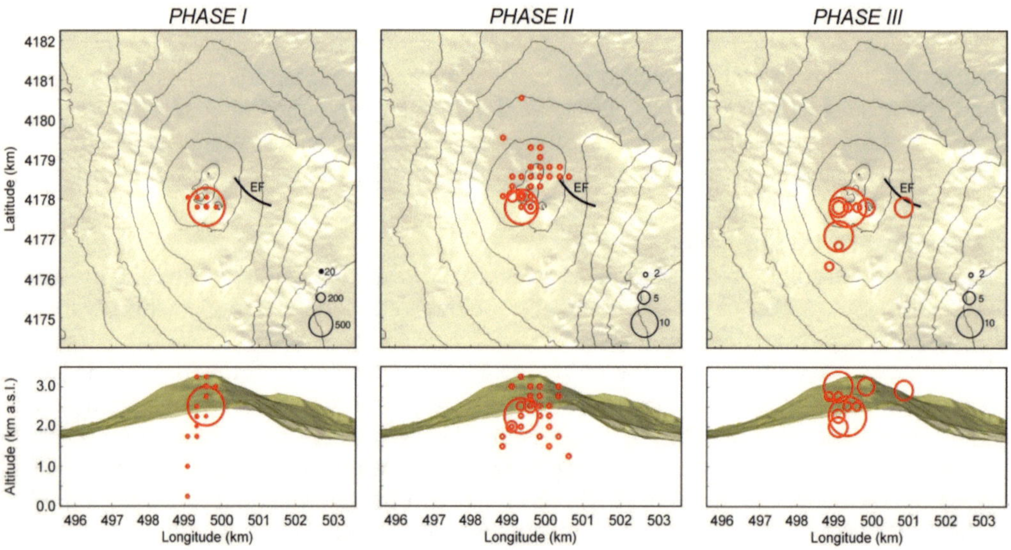

Fig. 13. Map and section of Mt. Etna with the locations of LP events recorded during three phases (see dashed black lines and "phase I", "phase II" and "phase III" in Fig. 12). The radii of the red circles are proportional to the number of the locations in each grid node (see black circles and numbers reported in the right lower corner of the maps)

Between 18 June and 3 July 2008, about 30 temporary broadband stations were deployed on Mt. Etna very close to the summit craters. This high-density network permitted to better investigate on the LP activity. De Barros et al. (2009), classifying more than 500 LP events by cross-correlation analysis, obtained two different families with a similar number of events. In agreement with previous studies (e.g. Saccorotti et al., 2007; Cannata et al., 2009b), the LP source positions were located close to the summit craters, and were slightly different for both families. The focal depth hypocentres were found shallow below the summit: from 0 to 800 m for family 1 and from 0 to 400 m for family 2 (De Barros et al., 2009). For both families the inversions show mechanisms with high volumetric components, most likely generated by a crack, striking in the SW-NE direction (De Barros et al., 2011). In particular for family 1 (**Fig. 14a**) the MT solution shows a subvertical dike striking SSW-NNE; for family 2 the crack solution lies on a plane inclined of 45° and striking SW-NE (**Fig. 14b**). The orientations of the cracks are consistent with local tectonics, which shows a SW–NE weakness direction, as testified by the orientation of the NE rift. De Barros et al. (2011) hypothesize that these events are not related to the lava flow from the eruptive fracture, instead to the decompression phase following the 10 May lava fountain. The LP events studied here show similar characteristics to the events occurring after a lava fountain in the 2007, analysed by Patanè et al. (2008), which interpreted, in particular, the LP belonging to the family 2 as the response to the volcano deflation. This theory is validated by the temporary cessation of the LP events after 22 June 2008, suggesting a return of equilibrium of the upper part of volcano, where pressure and stress return to a static state. Although the poor knowledge of the velocity model can lead to unambiguous explanations of both moment and forces related to the mechanism found, this study demonstrated how the LP moment tensor inversion is a powerful tool to understand the magmatic processes in the shallow plumbing system of Mt. Etna.

In summary, it was shown how the analysis of seismo-volcanic signals is very effective to reconstruct the geometry of the shallow portion of the plumbing system and to investigate the magma dynamics in it. Such information is very important for both research and monitoring purposes.

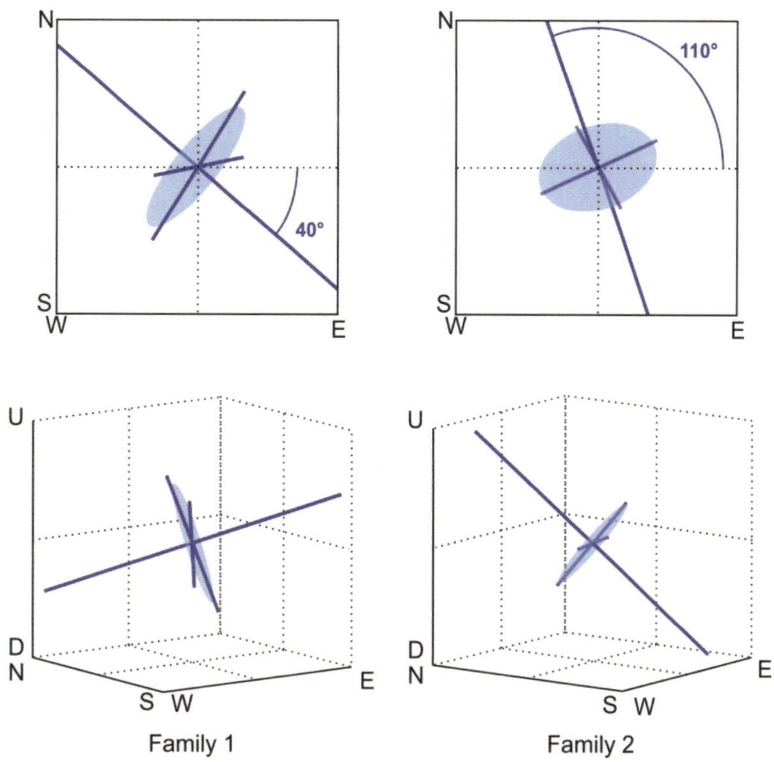

Fig. 14. Map views (up) and 3-D views (down) of the crack source mechanism obtained for the two families of LP events. The circular areas represent the cracks, the normalized eigenvectors are represented by solid lines, and the longest of these are the normal cracks. A subvertical dike striking SSW-NNE is obtained for family 1, and a crack striking SW-NE that lies on a plane inclined of 45° is computed for family 2. Redrawn from De Barros et al. (2011)

5. Summary

Mt. Etna lies in front of the southeast-verging Apennine-Maghrebian fold-and-thrust belt, where the NNW-trending Malta Escarpment separates the Sicilian continental crust from the Ionian Mesozoic oceanic basin, presently subducting beneath the Calabrian arc (Selvaggi and Chiarabba, 1995). Seismic tomographic studies indicate the presence of a mantle plume beneath the volcano with a Moho transition at depth less than 20 km (Nicolich et al.,2000; Barberi et al., 2006). Geophysical and geological evidences suggest that the Mt. Etna magma ascent mechanism is related to the major NNW-trending lithospheric fault (Doglioni et al., 2001). However, the reason for the Mt. Etna mantle plume draining and channeling the magma from the upper mantle source to the surface is not yet clear. All models proposed in literature (Rittmann, 1973; Tanguy et al., 1997; Monaco et al.; 1997; Gvirtzman and Nur, 1999; Doglioni et al., 2001) do not explain why such a mantle plume has originated in this anomalous external position with respect to the arc magmatism and back-arc spreading zones associated with the Apennines subduction. Some ideas on the subduction rollback must be better developed through the comparison with new regional tomographic studies that are being released. Moreover, tomographic studies reveal a complex and large plumbing system below the volcano from -2 to -7 km a.s.l., wide up to 60 km^2 that reduces itself in size down to -18 km of depth close to the apex of the mantle plume. Chiocci et al. (2011) found a large bulge on the underwater continental margin facing Mt. Etna, and suggested that the huge crystallized magma body intruded in the middle and upper continental crust was able to trigger an instability process involving the Sicilian continental margin during the last 0.1 Ma. This phenomenon induces the sliding of the volcano eastern flank observed since the 90s (Borgia et al, 1992; Lo Giudice and Rasà, 1992) because the effects of the bulge collapse are propagating upslope, and the continuous decompression at the volcano summit favors the ascent of basic magma without lengthy storage in the upper crust, as one might expect in a compressive tectonic regime. Taken together, these new evidences (tomographic, tectonic, volcanic) are concerned with the exceptional nature of Mt. Etna and raise the need to explain the origin of the mantle plume that supplies its volcanism. The lower crust and the uppermost mantle need to be better resolved in future experiments and studies. The use of regional and teleseismic events for tomography and receiver function analyses is required to explore a volume that has only marginally been investigated to date. The relation between the magma source in the mantle and the upper parts of the system, as well as the hypothesis above reported on the relation between tectonics and volcanism and the role of lithospheric faults, could be resolved only by applying seismological techniques able to better constrain broader and deeper models.

Finally, although the recent tomographic inversions have progressively improved our knowledge of Etna's shallow structure, highlighting a complex pattern of magma chambers and conduits with variable dimensions, the geometry of the conduits and the dimensions and shapes of small magmatic bodies still require greater investigation. Their precise definition is crucial to delineate a working model of this volcano in order to understand its behaviour and evolution. For this purpose, at least within the volcanic edifice, the precise locations of the seismo-volcanic signals can be considered a useful tool to constrain both the area and the depth range of magma degassing and the geometry of the shallow conduits. In this work, we furnish evidences that the tremor and LP locations allowed to track magma migration during the initial phase of the 2008-2009 eruption and in particular the initial northward dike intrusion, also confirmed by other geophysical, structural and volcanological observations (Aloisi et al., 2009; Bonaccorso et al., 2011), and the following fissure opening east of the summit area at the base of SEC. All these evidences, obtained by the marked improvement in the monitoring system together with the development of new processing techniques, allowed us to constrain both the area and the depth range of magma degassing, highlighting the geometry of the magmatic system feeding the 2008-2009 eruption.

6. References

Achauer, U., Evans, J.R., & Stauber, D.A. (1988). High-resolution seismic tomography of compressional wave velocity structure at Newberry volcano, Oregon Cascade range. *J. Geophys. Res.*, 93, pp. 10135-10147.

Acocella, V., & Neri, M. (2005). Structural features of an active strike-slip fault on the sliding flank of Mt. Etna. *J. Struct. Geol.*, 27, pp. 343-355.

Acocella, V., Neri, M., & Sulpizio, R. (2009). Dike propagation within active central volcanic edifices: constraints from Somma-Vesuvius, Etna and analogue models. *Bull. Volcanol.*, 71, 219-223.

Aiuppa, A., Cannata, A. , Cannavò, F., Di Grazia, G., Ferrari, F., Giudice, G., Gurrieri, S., Liuzzo, M., Mattia, M., Montalto, P., Patanè, D., & Puglisi, G. (2010). Patterns in the recent 2007-2008 activity of Mount Etna volcano investigated by integrated geophysical and geochemical observations. *Geoch., Geophys., Geosys.*, 11, doi:10.1029/2010GC003168.

Allard, P., Burton, M., & Murè, F. (2005). Spectroscopic evidence for lava fountain driven by previously accumulated magmatic gas. *Nature*, 433, pp. 407-409.

Almendros, J., Chouet, B., Dawson, P., & Bond, T. (2002). Identifying elements of the plumbing system beneath Kilauea Volcano, Hawaii, from the source locations of very-long period signals. *Geophys. J. Int.*, 148, pp. 303-312.

Aloisi, M., Bonaccorso, A., Cannavò, F., Gambino, S., Mattia, M., Puglisi, G., & Boschi, E. (2009). A new dyke intrusion style for the Mount Etna May 2008 eruption modelled through continuous tilt and GPS data. *Terra Nova*, 21, pp. 316-321.

Alparone, S., Cannata, A., & Gresta, S. (2007). Time variation of spectral and wavefield features of volcanic tremor at Mt. Etna (January-June 1999). *J. Volcanol. Geotherm. Res.*, doi:10.1016/j.jvolgeores.2006.12.012.

Auger, E., D'Auria, L., Martini, M., Chouet, B.A., & Dawson, P. (2006). Real-time monitoring and massive inversion of source parameters of very long period seismic signals: An application to Stromboli Volcano, Italy. *Geophys. Res. Lett.*, 33(4), L04301, doi:10.1029/2005GL024703.

Azzaro, R., Branca, S., Giammanco, S., Gurrieri, S., Rasà, R., & Valenza, M. (1998). New evidence for the form and extent of the Pernicana Fault System (Mt. Etna) from structural and soil-gas surveying. *J. Volcanol. Geotherm. Res.*, 84, pp. 143-152.

Azzaro, R. (1999). Earthquake surface faulting at Mount Etna volcano (Sicily) and implications for active tectonics. *J. Geodyn.*, 28, pp. 193-213.

Azzaro, R., Barbano, M.S., Antichi, B., & Rigano, R. (2000). Macroseismic catalogue of Mt. Etna earthquakes from 1832 to 1998. *Acta Vulcanol.*, 12, pp. 3-36.

Azzaro, R., 2004. Seismicity and active tectonics in the Etna region: constraints for a seismotectonic model. In: Bonaccorso, A., Calvari, S., Coltelli, M., Del Negro, C.,

Falsaperla, S. (Eds.), Mt. Etna: Volcano Laboratory. *AGU Geophys. Monogr.*, vol. 143, pp. 205–220.

Azzaro, R., Branca, S., Gwinner, K., & Coltelli, M. The volcano-tectonic map of Etna volcano, 1:100.000 scale: morphotectonic analysis from high-resolution DEM integrated with geologic, active faulting and seismotectonic data. In press in *Journal of Geosciences*.

Barberi, G., Patanè, D., Scarfì, L., & Zhang, H. (2006) Tomographic images of Northeastern Sicily and Southern Calabria crust by using TomoDD algorithm, General Assembly of the European Geosciences Union 2006, EGU06-A-04278; SM1-1WE5P-0434

Bean, C., Lokmer, I., & O'Brien, G. (2008). Influence of near-surface volcanic structure on long-period seismic signals and on moment tensor inversions: Simulated examples from Mount Etna. *J. Geophys. Res.*, 113, B08308, doi:10.1029/2007JB005468.

Bellotti, F., Branca, S., & Groppelli, G. (2010). Geological map of Mount Etna West Rift (Italy). *J. Maps*, pp. 96–122, doi:10.4113/jom.2010.1115.

Bonaccorso, A., Bonforte, A., Del Negro, C., Di Grazia, G., Ganci, G., Neri, M., Vicari, A., & Boschi, E. (2011). The initial phases of the 2008-2009 Mt. Etna eruption: a multi-disciplinary approach for hazard assessment. *J. Geophys. Res.*, doi:10.1029/2010JB007906.

Bonforte, A., Guglielmino, F., Coltelli, M., Ferretti, A., & Puglisi, G., (2011). Structural assessment of Mt. Etna volcano from Permanent Scatterers analysis. *Geochemistry, Geophysics, Geosystems*, 12, doi:10.1029/2010GC003213.

Borgia, A., Ferrari, L., & Pasquarè, G. (1992). Importance of gravitational spreading in the tectonic and volcanic evolution of Mount Etna. *Nature*, 357, pp. 231- 235.

Bousquet, J.C., & Lanzafame, G. (2001). Nouvelle interprètation des fracture des éruption latérales de l'Etna: conséquences pour son cadre tectonique. *Bull. Soc. Geol. Fr.*, 172, pp. 455- 467.

Bousquet, J.C., & Lanzafame, G. (2004) The tectonics and geodynamics of Mt. Etna: synthesis and interpretation of geological and geophysical data. In: *Mt. Etna: Volcano Laboratory*, Bonaccorso, A., Calvari, S., Coltelli, M., Del Negro, C., Falsaperla, S., pp. 29-47, American Geophysical Union, Geophysical Monograph 143.

Branca, S., Carbone, D., & Greco, F. (2003). Intrusive mechanism of the 2002 NE-Rift eruption at Mt. Etna (Italy) inferred through continuous microgravity data and volcanological evidences. *Geophys. Res. Lett.*, 30, doi:10.1029/2003GL018250.

Branca, S., Coltelli, M., & Groppelli, G. (2004). Geological evolution of Etna volcano. In: *Mt. Etna: Volcano Laboratory*, Bonaccorso, A., Calvari, S., Coltelli, M., Del Negro, C., Falsaperla, S., pp. 49-63, American Geophysical Union, Geophysical Monograph 143.

Branca, S., Coltelli, M., De Beni, E., & Wijbrans, J. (2007). Geological evolution of Mount Etna volcano (Italy) from earliest products until the first central volcanism (between 500 and 100 ka ago) inferred from geochronological and stratigraphic data. *International Journal of Earth Sciences*, doi:10.1007/s00531-006-0152-0

Cannata, A., Catania, A., Alparone, S., & Gresta, S. (2008). Volcanic tremor at Mt. Etna: Inferences on magma dynamics during effusive and explosive activity. *J. Volcanol. Geotherm. Res.*, doi:10.1016/j.jvolgeores. 2007.11.027.

Cannata, A., Giudice, G., Gurrieri, S., Montalto, P., Alparone, S., Di Grazia, G., Favara, R., & Gresta, S. (2009a). Relationship between soil CO2 flux and volcanic tremor at Mt.

Etna: implications for magma dynamics. *Env. Earth Sci.*, doi:10.1007/s12665-009-0359-z.

Cannata, A., Hellweg, M., Di Grazia, G., Ford, S., Alparone, S., Gresta, S., Montalto, P., & Patanè, D. (2009b). Long period and very long period events at Mt. Etna volcano: Characteristics, variability and causality, and implications for their sources. *J. Volcanol. Geothherm. Res.*, 187, pp. 227–249.

Cannata, A., Montalto, P., Privitera, E., Russo, G. & Gresta, S. (2009c). Tracking eruptive phenomena by infrasound: May 13, 2008 eruption at Mt. Etna.-*Geophys. Res. Lett.*, 36, L05304, doi:10.1029/2008GL036738.

Cannata, A., Di Grazia, G., Montalto, P., Aliotta, M., Patanè, D., & Boschi, E. (2010). Response of Mount Etna to dynamic stresses from distant earthquakes. *J. Geophys. Res.*, 115, B12304, doi:10.1029/2010JB007487.

Cassinis, R., Finetti, I., Giese, P., Morelli, C., Steinmetz, L. & Vecchia, O. (1969). Deep seismic refraction research on Sicily. *Boll. Geof. Teor. Appl.*, 11, pp. 140-160.

Cardaci, C., Coviello, M., Lombardo, G., Patanè, G., & Scarpa, R. (1993). Seismic tomography of Etna Volcano. *J. Volcanol. Geotherm. Res.*, 56, pp. 357–368.

Catalano, R., Doglioni, C., & Merlini, S. (2001). On the Mesozoic Ionian basin. *Geophys. J. Int.*, 144, pp. 49-64.

Cernobori, L., Hirn, A., McBride, J.H., Nicolich, R., Petronio, L., Romanelli, M., & STREAMERS/PROFILES Working Groups (1996). Crustal image of the Ionian basin and its Calabrian margins. *Tectonophysics*, 264, 175-189.

Chiarabba, C., Amato, A., Boschi, E., & Barberi, F. (2000). Recent seismicity and tomographic modeling of the Mount Etna plumbing system. *J. Geophys. Res.*, 105, pp. 10923–10938.

Chiarabba, C., De Gori, P., & Patanè, D. (2004). The Mount Etna plumbing system: the contribution of seismic tomography. In *Mount Etna Volcano Laboratory*, Vol. 143, eds Bonaccorso, A., Calvari, S., Coltelli, M., Del Negro, C. & Falsaperla, S., American Geophysical Union Monography Series.

Chiocci, F.L., Coltelli, M., Bosman, A., & Cavallaro, D. (2011). Continental margin large-scale instability controlling the flank sliding of Etna volcano. *Earth Planet. Sci. Lett.*, doi:10.1016/j.epsl.2011.02.040.

Chouet, B., Page, R.A., Stephens, C.D., Lahr, J.C., & Power, J.A. (1994). Precursory swarms of long-period events at Redoubt Volcano (1989–1990), Alaska: their origin and use as a forecasting tool. *J. Volcanol. Geotherm. Res.*, 62, pp. 95–135.

Chouet, B. (1996). Long-period volcano seismicity: Its source and use in eruption forecasting. *Nature*, 380, pp. 309-316.

Chouet, B. (2003). Volcano seismology. *Pure Appl. Geophys.*, 160, pp. 739–788.

Chouet, B., Dawson, P., Ohminato, T., Martini, M., Saccorotti, G., Giudicepietro, F., De Luca, G., Milana, G. & Scarpa, R. (2003). Source mechanisms mechanisms of explosions at Stromboli Volcano, Italy, determined from moment□tensor inversions of very□long□period data. *J. Geophys. Res.*, 108(B1), 2019, doi:10.1029/2002JB001919.

Clemens, J.C., & Mawer, C.K. (1992). Granitic magma transport by fracture propagation. *Tectonophysics*, 204, pp. 339–360.

Colombi, B., Guerra, I., Luongo, G., & Scarascia, S. (1979). Profilo sismico a rifrazione Acireale-Termini Imerese. CNR Prog. Fin. Geodinamica, pub. 235, pp. 155-170, ed. Giannini, Napoli.

Coltelli, M., Del Carlo, P., & Vezzoli, L. (2000). Stratigraphic constraints for explosive activity in the last 100 ka at Etna volcano, Italy. *Int. J. Earth Sci.*, 89, pp. 665-677.

Continisio, R, Ferrucci, F., Gaudiosi, G., Lo Bascio, D., & Ventura, G. (1997). Malta escarpment and Mt. Etna; early stages of an asymmetric rifting process? Evidences from geophysical and geological data. *Acta Vulcanologica*, 9, 45-53.

De Barros, L., Bean, C., Lokmer, I., Saccorotti, G., Zuccarello, L., O'Brien, G., Métaxian, J.-P., Patanè, D. (2009). Source geometry from exceptionally high resolution long period event observations at Mt. Etna during the 2008 eruption. *Geophys. Res. Lett.*, 36, L24305, doi:10.1029/2009GL041273.

De Barros, L., Lokmer, I., Bean, C.J., O'Brien, G.S., Saccorotti, G., Métaxian, J.-P., Zuccarello, L., & Patanè, D. (2011). Source mechanism of long-period events recorded by a high-density seismic network during the 2008 eruption on Mount Etna. *J. Geophys. Res.*, 116, B01304, doi:10.1029/2010JB007629.

De Beni, E., Branca, S., Coltelli, M., Groppelli, G., & Wijbrans, J. (2011) - 40Ar/39Ar isotopic dating of Etna volcanic succession. In press in *It. J. Geosci.*

De Gori, P., Chiarabba, C., & Patanè, D. (2005). Qp structure of Mount Etna: Constraints for the physics of the plumbing system. *J. Geophys. Res.*, 110, B05303, doi:10.1029/2003JB002875.

De Luca, G., Filippi, L., Patanè, G., Scarpa, R., & Vinciguerra, S. (1997). Three dimensional velocity structure and seismicity of Mt. Etna volcano, Italy. *J. Volcanol. Geotherm. Res.*, 79, pp. 123- 138.

Di Grazia, G., Falsaperla, S., & Langer, H. (2006). Volcanic tremor location during the 2004 Mount Etna lava effusion. *Geophys. Res. Lett.*, 33, L04304. doi:10.1029/2005GL025177.

Di Grazia, G., Cannata, A., Montalto, P., Patanè, D., Privitera, E., Zuccarello, L., & Boschi, E. (2009). A new approach to volcano monitoring based on 4D analyzes of seismo-volcanic and acoustic signals: the 2008 Mt. Etna eruption. *Geophys. Res. Lett.*, 36, L18307, doi:10.1029/2009GL039567.

Doglioni, C., Innocenti, F., Mariotti, & G. (2001). Why Mt. Etna? *Terra Nova*, 13, pp. 25-31.

Falsaperla, S., Lanzafame, G., Longo, V., & Spampinato, S. (1999). Regional stress field in the area of Stromboli (Italy): insights into structural data and crustal tectonic earthquakes. *J. Volcanol. Geotherm. Res.*, 88, pp. 147-166.

Falsaperla, S., Alparone, S., D'Amico, S. , Di Grazia, G., Ferrari, F., Langer, H., Sgroi, T., & Spampinato, S. (2005). Volcanic tremor at Mt. Etna, Italy, preceding and accompanying the eruption of July–August, 2001. *Pageoph*, 162, pp. 1–22.

Froger, J.L., Merle, O., & Briole, P. (2001). Active spreading and regional extension at Mount Etna imaged by SAR interferometry. *Earth Planet. Sci. Lett.*, 148, pp. 245–258.

Gresta, S., Montalto, A., & Patanè, G. (1991). Volcanic tremor at Mt. Etna (January 1984–March 1985): its relationship to the eruptive activity and modelling of the summit feeding system. *Bull. Volcanol.*, 53, pp. 309–320.

Gresta, S., Peruzza, L., Slejko, D., & Distefano, G. (1998). Inferences on the main volcano-tectonic structures at Mt. Etna (Sicily) from a probabilistic seismological approach. *J. Seismol.*, 2, pp. 105- 116.

Gvirtzman, Z., Nur, A. (1999). The formation of Mount Etna as the consequence of slab rollback. *Nature*, 401, pp. 782–785.

Gudmundsson, A. (2006). How local stress control magma-chamber ruptures, dyke injections, and eruptions in composite volcanoes. *Earth Sci. Rev.*, 79, 1-31.

Hirn, A., Nercessian, A., Sapin, M., Ferrucci, F., & Wittlinger, G. (1991). Seismic heterogeneity of Mt. Etna: structure and activity. *Geophys. J. Int.*, 105, pp. 139- 153.

Hirn, A., Nicolich, R., Gallart, J., Laigle, M., & ETNASEIS Scientific Group (1997). Roots of Etna volcano in faults of great earthquakes. *Earth Planet. Sci. Lett.*, 148, pp. 171-191.

Kieffer, G. (1975). Sur l'existence d'une rift zone à l'Etna. *CR Acad Sci Paris*, 280, pp. 263-266.

Kieffer, G. (1985). Evolution structurale et dynamique d'un grand volcan polygénique: stades d'édification et activité actuelle de l'Etna (Sicile). *Thèse doctoral,* Université de Clermont-Ferrand, pp. 1-410.

Kumagai, H., Nakano, M., Maeda, T., Yepes, H., Palacios, P., Ruiz, M., Arrais, S., Vaca, M., Molina, I., & Yamashima, T. (2010). Broadband seismic monitoring of active volcanoes using deterministic and stochastic approaches. *J. Geophys. Res.*, 115, B08303, doi:10.1029/2009JB006889.

Kumazawa, M., Imanishi, Y., Fukao, Y., Furumoto, M., & Yamamoto, A. (1990). A theory of spectral analysis based on the characteristic property of a linear dynamic system. *Geophys. J. Int.*, 101, pp. 613-630.

Laigle, M., Hirn, A., Sapin, M., Lepine, J.C., Diaz, J., Gallart, J., & Nicolich, R. (2000). Mount Etna dense array local earthquake P and S tomography and implications for volcanic plumbing. *J. Geophys. Res.*, 105, pp. 21633- 21646.

Langer, H., & Falsaperla, S. (1996). Long-term observation of volcanic tremor on Stromboli Volcano (Italy): A synopsis. *Pageoph*, 147, pp. 57-82.

Lees, J.M., & Crosson, R.S. (1990). Bayesian ART versus conjugate gradient methods in tomographic seismic imaging: an application at Mount St. Helens, Washington. *Spatial Statistics and Imaging*: Proc. 1988 AMS-IMS-SIAM Summer Res. Conf., pp. 1-18.

Lentini, F., Carbone, S., & Guarnieri, P. (2006). Collisional and postcollisional tectonics of the Apenninic-Maghrebian orogen (southern Italy), in Dilek, Y., and Pavlides, S., eds., *Postcollisional tectonics and magmatism in the Mediterranean region and Asia*, Geological Society of America Special Paper 409, p. 57-81, doi:10.1130/2006.2409(04).

Lo Giudice, E., Patanè, G., Rasà, R., & Romano, R. (1982). The structural framework of Mount Etna. In: Romano R (ed) Mount Etna Volcano, a Review of Recent Earth Sciences Studies. *Memorie della Società Geologica Italiana*, 23, pp. 125-158.

Lo Giudice, E., & Rasà, A. (1992). Very shallow earthquakes and brittle deformation in active volcanic areas: the Etnean region as an example. *Tectonophysics*, 202, pp. 257-268.

Lokmer, I., Bean, C.J., Saccorotti, G., & Patanè, D. (2007). Moment-tensor inversion of LP events recorded on Etna in 2004 using constraints obtained from wave simulation tests. *Geophys. Res. Lett.*, 34, L22316, doi:10.1029/2007GL031902.

Lombardo, G., Coco, G., Corrao, M., & Gresta, S. (1996). Features of seismic events and volcanic tremor during the preliminary stages of the 1991-1993 eruption of Mt. Etna. *Ann. Geophys.*, 39, pp. 403-410.

Lundgren, P., Berardino, P., Coltelli,M., Fornaro, G., Lanari, R., Puglisi, G., Sansosti, E., & Tesauro, M. (2003). Coupled magma chamber inflation and sector collapse slip observed with SAR interferometry on Mt. Etna volcano. *J. Geophys. Res.*, 108, pp. 2247-2261.

Marsh, B. D. (2000). Magma Chambers. In *Encyclopedia of Volcanoes*, Sigurdsson, H., pp. 191-206, Academic Press.

Martinez-Arevalo, C., Patanè, D., Rietbrock, A., & Ibanez, J.M. (2005). The intrusive process leading to the Mt. Etna 2001 flank eruption: Constraints from 3-D attenuation tomography. *Geophys. Res. Lett.*, 32, L21309, doi:10.1029/2005GL023736.

McGuire, W.J., Pullen, A.D. (1989). Location and orientation of eruptive fissures and feeder dykes at Mount Etna: influence of gravitational and regional stress regime. *J. Volcanol. Geotherm. Res.*, 38, pp. 325-344.

McNutt, S.R. (1994). Volcanic Tremor Amplitude Correlated with Eruption Explosivity and its Potential Use in Determining Ash Hazards to Aviation. *U.S. Geological Survey Professional Paper*, 2047, pp. 377-385.

Monaco, C., Tapponnier, P., Tortorici, L., & Gillot, P.Y. (1997). Late Quaternary slip rates on the Acireale–Piedimonte normal faults and tectonic origin of Mt. Etna (Sicily). *Earth Planet. Sci. Lett.*, 147, pp. 125– 139.

Monaco, C., Catalano, S., Cocina, O., De Guidi, G.,Ferito, C., Gresta, S., Musumeci, C., & Tortorici, L. (2005). Tectonic control on the eruptive dynamics at Mt. Etna Volcano (Sicily) during the 2001 and 2002–2003 eruptions. *J. Volcanol. Geotherm. Res.*, 144, pp. 211-233.

Monteiller, V., Got, J.-L., Patanè, D., Barberi, G., & Cocina, O. (2009). Double-difference tomography at Mt Etna volcano: Preliminary results. In *The VOLUME Project. Volcanoes: Understanding Subsurface Mass Movement*, Bean, C. J., et al., pp. 74–84, VOLUME Consort./Jaycee, Dublin, Ireland.

Moran, S.C., Malone, S.D., Qamar, A.I., Thelen, W.A., Wright, A.K., & Caplan-Auerbach, J. (2008). Seismicity associated with the renewed dome-building eruption of Mount St. Helens 2004-2005. In *A volcano rekindled: the renewed eruption of Mount St. Helens, 2004-2006*, Sherrod, D.R., Scott, W.E., and Stauffer, P.H., chapter 2, U.S. Geological Survey Professional Paper.

Murray, J. B. (1990). High-level magma transport at Mount Etna volcano, as deduced from ground deformation measurements. In: *Magma Transport and Storage*, Ryan M.P., pp. 357-383, Wiley, Chichester.

Murru, M., Montuori, C., Wyss, M., & Privitera, E. (1999). The locations of magma chambers at Mt. Etna, Italy, mapped by b-values. *Geophys. Res. Lett.*, 26, pp. 2553–2556.

Neidell, N., & Taner, M.T. (1971). Semblance and other coherency measures for multichannel data. *Geophysics*, 36, pp. 482-497.

Neri, M., Acocella, V., & Behncke, B. (2004). The role of the Pernicana Fault System in the spreading of Mt. Etna (Italy) during the 2002–2003 eruption. *Bull. Volcanol.*, 66, pp. 417-430.

Neuberg, J. (2000). External modulation of volcanic activity. *Geophys. J. Int.*, 142, pp. 232–240.

Nicolich, R., Laigle, M., Hirn, A., Cernobori, L., & Gallart, J. (2000). Crustal structure of the Ionian margin of Sicily: Etna volcano in the frame of regional evolution. *Tectonoph.*, 329, pp. 121-139.

O'Brien, G.S., & Bean, C.J. (2004). A 3D discrete numerical elastic lattice method for seismic wave propagation in heterogeneous media with topography. *Geophys. Res. Lett.*, 31, L14608, doi:10.1029/2004GL020069.

O'Brien, G. S., & Bean, C. J. (2009). Volcano topography, structure and intrinsic attenuation: Their relative influences on a simulated 3d viscoelastic wavefield. *J. Volcanol. Geotherm. Res.*, 183(1–2), 122–136.

Ohminato, T., Chouet, B., Dawson, P., & Kedar, S. (1998). Waveform inversion of very long period impulsive signals associated with magmatic injection beneath Kilauea Volcano, Hawaii. *J. Geophys. Res.*, 103, pp. 23839-23862.

Palano, M., Puglisi, G., & Gresta, S. (2008). Ground deformation patterns at Mt. Etna from 1993 to 2000 from joint use of InSAR and GPS techniques. *J. Volcanol. Geotherm. Res.*, 169, pp. 99-120.

Patanè, D., Chiarabba, C., Cocina, O., De Gori, P., Moretti, M., & Boschi, E. (2002). Tomographic images and 3D earthquake locations of the seismic swarm preceding the 2001 Mt. Etna eruption: Evidence for a dyke intrusion. *Geophys. Res. Lett.*, 29, 1497, doi:10.1029/2001GL014391.

Patanè, D., De Gori, P., Chiarabba, C., & Bonaccorso, A. (2003). Magma ascent and the pressurization of Mount Etna's volcanic system. *Science*, 299, pp. 2061-2063.

Patanè, D., Cocina, O., Falsaperla, S., Privitera, E., & Spampinato, S. (2004). Mt. Etna volcano: a seismological framework. In: S. Calvari, A. Bonaccorso, M. Coltelli, C. Del Negro and S. Falsaperla (Eds.), The Mt. Etna Volcano. AGU, Washington, D.C., 147–165.

Patanè, D., Mattia M. & Aloisi M., (2005). Shallow intrusive processes during 2002-2004 and current volcanic activity at Mt. Etna. *Geophys. Res. Lett.*, 32, L06302, 10.1029/2004GL021773.

Patanè, D., Barberi, G., Cocina, O., De Gori, P., & Chiarabba, C. (2006). Time-resolved seismic tomography detects magma intrusions at Mount Etna. *Science*, 313, pp. 821–823.

Patanè, D., Di Grazia, G., Cannata, A., Montalto, P., & Boschi, E. (2008). The shallow magma pathway geometry at Mt. Etna volcano. *Geochem. Geophys. Geosyst.*, 9, Q12021, doi:10.1029/2008GC002131.

Petford, N., Cruden, A.R. , McCaffrey, K.J.W., & Vigneresse, J.L. (2000). Granite magma formation, transport and emplacement in the Earth's crust. *Nature*, 408, pp. 669-673.

Rasà, R., Azzaro, R., & Leonardi, O. (1996). Aseismic creep on faults and flank instability at Mt. Etna volcano, Sicily. In: "Volcano Instability on the Earth and Other Planets", W.C. McGuire, A.P. Jones & J. Neuberg (ed.). Geological Society Special Publication 110, 179-192.

Rittmann, A. (1973). Structure and evolution of Mount Etna. *Philos. Trans. R. Soc. Lond.*, 274, 5-16.

Rust, D., & Neri, M. (1996). The boundaries of large-scale collapse on the flanks of Mount Etna, Sicily. In: *Volcano instability on the Earth and other planets.* McGuire, W.J., Jones, A.P., Neuberg, J., pp. 193–208, Geol. Soc. Lond. Spec. Pub. 110.

Rust, D., Behncke, B., Neri, M., & Ciocanel, A. (2005). Nested zones of instability in the Mount Etna volcanic edifice, Sicily. *J. Volcanol. Geotherm. Res.*, 144 , pp. 137–153.

Saccorotti, G., Lokmer, I., Bean, C., D. Grazia, G., & Patanè, D. (2007). Analysis of sustained long-period activity at Etna Volcano, Italy. *J. Volcanol. Geotherm. Res.*, 160, 340–354.

Selvaggi, G., & Chiarabba, C. (1995). Seismicity and P-wave velocity image of the Southern Tyrrhenian subduction zone. *Geophys. J. Int.*, 121, pp. 818 - 826.

Sharp, A.D.L., Davis, P.M., & Gray, F. (1980). A low velocity zone beneath Mount Etna and magma storage. *Nature*, 287, pp. 587-591.

Tanguy, J.C., Condomines, M., & Kieffer, G. (1997). Evolution of Mount Etna magma: constraints on the present feeding system and eruptive mechanism. *J. Volcanol. Geotherm. Res.*, 75, pp. 221- 250.

Tibaldi, A., & Groppelli, G. (2002). Volcano-tectonic activity along structures of the unstable NE flank of Mt. Etna (Italy) and their possible origin. *J. Volcanol. Geotherm. Res.*, 115, pp. 277-302.

Torelli, L., Grasso, M., Mazzoldi, G., & Peis, D. (1998). Plio-Quaternary tectonic evolution and structure of the Catania foredeep, the northern Hyblean Plateau and the Ionian shelf (SE Sicily). *Tectonophysics*, 298, pp. 209-221.

Varley, N., Arámbula-Mendoza, R., Reyes-Dávila, G., Stevenson, J., & Harwood, J. (2010). Long-period seismicity during magma movement at Volcán de Colima. *Bull. Volcanol.*, doi:10.1007/s00445-010-0390-7.

Vigneresse, J.L. (1999). Intrusion level of granitic massifs along the Hercynian belt: balancing the eroded crust. *Tectonophysics*, 307, pp. 277–295.

Withers, M., Aster, R., Young, C., Beiriger, J., Harris, M., Moore, S., & Trujillo, J. (1998). A comparison of select trigger algorithms for automated global seismic phase and event detection. *Bull. Seism. Soc. Am.*, 88, pp. 95-106.

5

Transtensional Tectonics Played a Key Role during the Variscan Cycle in the Sardinia-Corsica Massif

Nello Minzoni
Dipartimento di Scienze della Terra; Università di Ferrara
Italy

1. Introduction

Analysis, interpretation and synthesis of the Variscan (Hercynian) chain in the whole Sardinian-Corsican Massif (SCM) are relatively recent. Previous studies were carried out only in SW Sardinia, characterized by important ore deposits and abundant fossils.
Afterward, in SE Sardinia (Sarrabus region) Calvino, 1961, recognized terrigenous deposits ("San Vito" Fm.) and intrusive magmatics ("Porfidi bianchi" and "Porfidi grigi"), assigned to a generic early Paleozoic.
Bosellini & Ogniben, 1968, discovered, in central Sardinia (Sarcidano region) volcanoclastics ("Manixeddu" Fm), volcanics ("Serra Tonnai" Fm), middle Ordovician in age and a thick terrigenous turbiditic succession ("Bruncu su Pizzu" Fm) late Ordovician in age
Minzoni, 1975, identified, in the Sarcidano and Barbagia regions (central-northern Sardinia) a terrigenous succession ("Solanas "Fm), middle Cambrian -early Ordovician in age, covered by sub-aerial acidic volcanics ("Monte Corte Cerbos "Fm.), middle Ordovician in age, in turn covered by the Manixeddu volcanoclastics and by Serra Tonnai volcanics and Bruncu su Pizzu turbidites.. The Solanas Fm was considered equivalent to the San Vito Fm and the effusive Monte Corte Cerbos and Serra Tonnai volcanics were considered equivalent to the intrusive Porfidi bianchi and Porfidi grigi.
Structural analysis indicated that from late Devonian .to early Carboniferous the Variscan orogenesis piled up, in central-northern Sardinia, several south- vergent nappes. North Sardinia was considered the internal zone of the chain, source of the nappes, that were stacked in central-northern Sardinia (Nappe zone) and transported toward southern Sardinia (external chain and foreland).
The Variscan geodynamic cycle was considered completely ensialic, involving continental crust only. North Sardinia was assumed as the portion of the crust most deeply subducted along an intra-continental shear zone and North Corsica was regarded as backland of the chain (Carmignani et al., 1980; 1981).
Afterward, Carmignani et al., 1992, discovered, in northern Sardinia, along a narrow belt (the "Asinara -Posada Line"), small mafic bodies of amphibolite, bearing granulite-eclogite relics. Geochemical data indicated a MORB affinity for the protolith and radiometric dating (Sm / Nd on whole rock; Cappelli et al., 1992), yielded an age of 960 ± 93 Ma. The granulite-eclogite relics were related to subduction processes of oceanic lithosphere, retrogressed in

the amphibolite facies during the Variscan metamorphism. Using these data, this model was proposed: opening in North Sardinia of an ocean during the pre Cambrian.; subduction, during the middle Ordovician, of the oceanic lithosphere and development, in central Sardinia, of a volcanic arc on continental crust; further subduction, during the Silurian, of the oceanic lithosphere; continental collision of the paleo margins in north Sardinia during late Devonian-early Carboniferous and oceanic suture ("Asinara -Posada Line"); gravitative collapse of the thickened orogenic wedge, with ascent of the deepest metamorphic cores; crustal extension and emplacement of granitoids.

Rossi et al., 2009 proposed this geodynamic cycle: Cambrian -Early Ordovician, opening of the Rheic Ocean in central Europe; Middle Ordovician, subduction of the Rheic Ocean and development of an Andean-type volcanism in Sardinia; Late Ordovician -Early Silurian, back arc spreading and rifting on the North Sardinia-central Corsica and opening of the "South Armorica Ocean"; Silurian -Devonian, subduction of the Armorican Ocean; Late Devonian - early Carboniferous, collision of the passive margins; Carboniferous-Early Permian, extension and emplacement of the batholith.

In both the models, the middle Ordovician magmatics in Sardinia represent an arc generated by subduction of oceanic lithosphere.

2. Aim of this chapter

Aim is to propose, reviewing the Paleozoic events in the SCM, a new geodynamic model, in the frame of the European Variscan Realm.

3. The Variscan cycle in the Sardinia - Corsica Massif: data and interpretation

Figure 1 represents the main features of the Variscan chain in the SCM.

3.1 Foreland

Data. SW Sardinia is the external zone and the foreland of the Variscan chain. Near Capo Spartivento, a metamorphic complex ("Settiballas complex", Minzoni, 1981) is cut and metamorphosed by contact by a granitoid (" Orthogneis " or "Porphyroid", in literature). These rocks are covered by the "Bithia Fm.", consisting in arkoses, quarzites, conglomerates, shales and siltites, with thin dolomitic intercalations. Few badly preserved Acritarchs have been assigned to the early Cambrian.

The Bithia Fm grades into early-middle Cambrian terrigenous deposits, with Archeocyat and Trilobite-bearing carbonate levels ("Nebida Fm.").

The middle Cambrian is represented by a thick carbonate succession ("Gonnesa Fm.").

The early-middle Cambrian deposits reach a thickness of about 2000 m.

The deposition in shallow marine conditions of a siliciclastic succession ("Cabitza Fm."), showing a thickness of about 400 m, characterizes the late Cambrian-early Ordovician (Barca et al., 1987).

SW Sardinia, experienced a mild tectonic phase that developed E-W directed folds ("Sardic" Phase).

A conglomerate ("Puddinga") represents the middle-late Ordovician transgression, after a short-lived emergence and erosion (Martini et al., 1991). This conglomerate grades upward into a thin (50-100 m) succession constituted by siltites and marly shales, late Ordovician in age ("Tuviois" Fm).

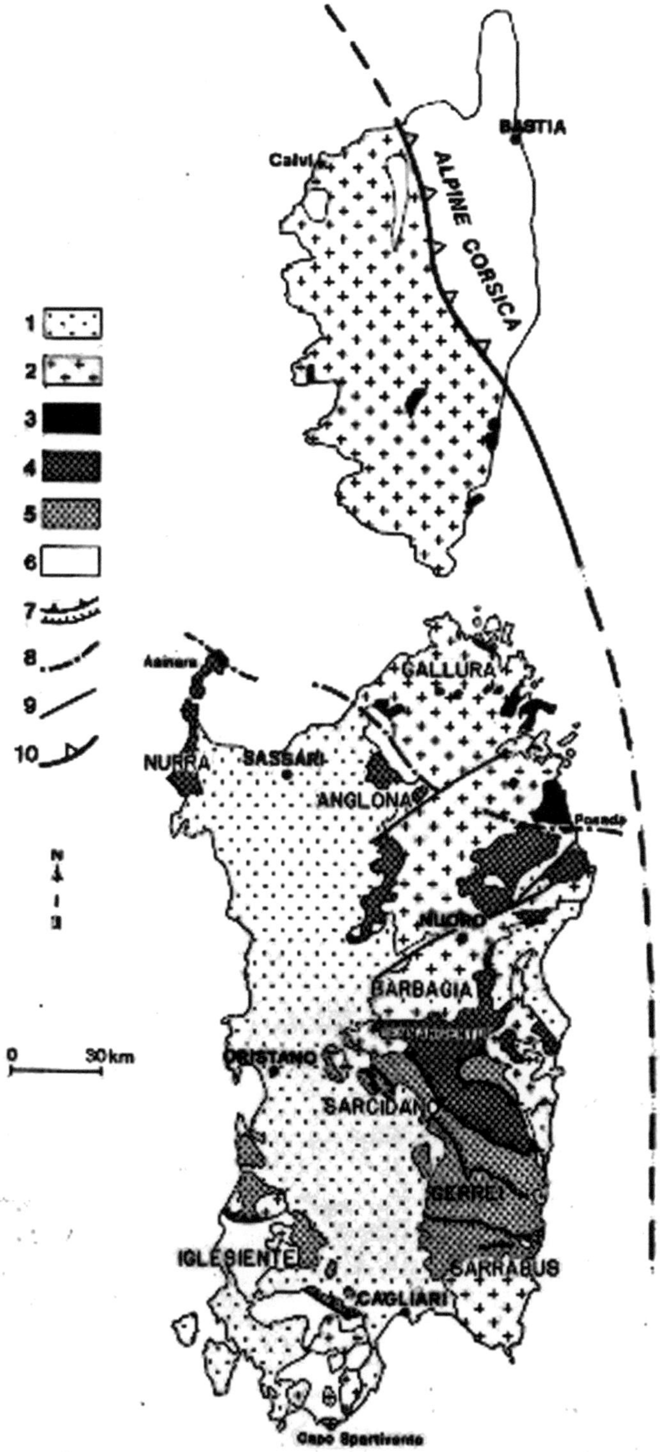

1-Post Variscan rocks.
2-Late Variscan granites.
3-Metamorphic rocks of the inner chain (northern Sardinia and central-southern Corsica).
4-Upper–internal nappes (central- northern Sardinia).
5-Sarcidano- Sarrabus nappe covering, along the Gerrei tectonic window, the lower - external nappes.
6-Paleozoic deposits in the foreland (SW Sardinia) and in the backland (NW Corsica).
7- Overthrusts.
8-Asinara –Posada Line.
9-Main Alpine faults.
10-Front of the Alpine chain

Fig. 1.

The Silurian and the Devonian are represented by the well known pelagic black shales and Tentaculite-bearing limestones.

The main Variscan tectonic phase generated N-S directed folds which, interfering with E-W folds of the Sardic phase, formed a "dome-basin" structural setting. Downward, the Variscan folds become progressively recumbent- isoclinal in type, characterized by sub-horizontal pervasive cleavage and by syn tectonic matamorphism increasing downward from the Anchizone to the green-schist facies. The Variscan cleavage cuts the schistosity of the Settiballas metamorphic complex; the granitoid was transformed in the Orthogneiss – Porphyroid.

Interpretation. The Settiballas metamorphic complex and the granitoid (Orthogneiss-Porphyroid), both represent a pre-Cambrian (pan –African) Basement. It is generally accepted (Cocozza, 1980; Vai & Cocozza, 1986; Gandin, 1987; Gandin et al., 1987;) that the overlying deposits formed in epi-continental domain, evolving from a delta oolitic lagoonal-tidal flat system to open marine conditions. Their thickness (2400 m) suggests that the sedimentation was linked to slow subsidence controlled by long-lived syn-sedimentary normal faults.

The Sardic Phase formed structures locally emerging from a shallow sea. The emergence is pre-dated by the Arenig deposits at the top of the Cabitza Fm and post-dated by the Caradocian deposits of the Tuviois Fm (Vai & Cocozza, 1986) This last consists in fluvio-lacustrine-shallow marine sequences, formed after short-lived emergence.

The Silurian-Devonian succession is usually considered the result of a worldwide climatic transgression (Vai, 2001).

3.2 Nappe zone

In central-northern Sardinia, the main Variscan orogenetic phase piled up several southward- verging nappes (Nappe Zone) characterized by isoclinal- recumbent folds, pervasive slaty cleavage and by northward increasing (from greenschist-to- amphibolite facies) metamorphism.

NE Sardinia represents, in this way, internal zone of the Variscan chain, source of the nappes which were piled up in northern-central Sardinia and transported toward SW (external chain and foreland).

In this frame, the lower nappes represent external domains adjacent to SW Sardinia (lower-external nappes) and the upper nappes represent internal domains adjacent to NE Sardinia (upper-internal nappes).

The Sarcidano-Sarrabus tectonic unit , sandwiched between the lower and the upper nappes, represents an area linking the external with the internal domains.

3.2.1 Lower-external nappes

Data. These small-sized nappes outcrop in the "Gerrei window", beneath the Sarcidano-Sarrabus tectonic unit (Fig.1). They consist of the pre Cambrian Orthogneiss-Porphyroid covered by arkoses (10-50m), siliciclastic sequences (30-80 m) middle Cambrian –early Ordovician in age (Cuili Biringoni Fm), siltites and limestones (10-60 m) Caradocian-Ashgillian in age (Tuviois Fm) and black shales -limestones (80-100 m) Silurian-Devonian in age.

Interpretation. The non deposition on the pre Cambrian Basement of the early-middle Cambrian successions indicates that until the middle Cambrian the lower nappes constituted a terrestrial area extended North of SW Sardinia, this last being affected by the syn sedimentary normal faults controlling the deposition of the Bithia- Nebida-Gonnesa Fms.

The terrestrial condition was interrupted from late Cambrian with the deposition of the Cuili Biringoni Fm. This corresponds, in age but not in thickness, to the Cabitza Fm in SW Sardinia. Therefore, the lower nappes constituted, during the late Cambrian-early Ordovician, an Horst (characterized by the deposition of the condensed Cuili Biringoni Fm) adjacent to a Graben in SW Sardinia (characterized by the deposition of the thick Cabitza Fm).
Instead, during the late Ordovician the sedimentation developed homogeneously in the SW Sardinia and in the area of the lower nappes, being represented, everywhere, by the Tuviois Fm.

3.2.2 Sarcidano-Sarrabus nappe

Data This enormous tectonic unit extends from the Sarcidano -Sarrabus to the Iglesiente (Barca et al., 1984; Albani et al., 1985) thus covering all the lower external nappes and reaching the external zone of the chain. This unit starts with the late Cambrian-early Ordovician San Vito-Solanas Fms showing a thickness of about 400 m. Both correspond, in thickness and in age, to the Cabitza Fm in SW Sardinia.
Emplacement of felsic calc-alkaline -to-basic alkaline magmatics ("Porfidi bianchi" - "Porfidi grigi" and Monte Corte Cerbos-Serra Tonnai volcanics) characterizes the middle-late Ordovician. (Fig.2)

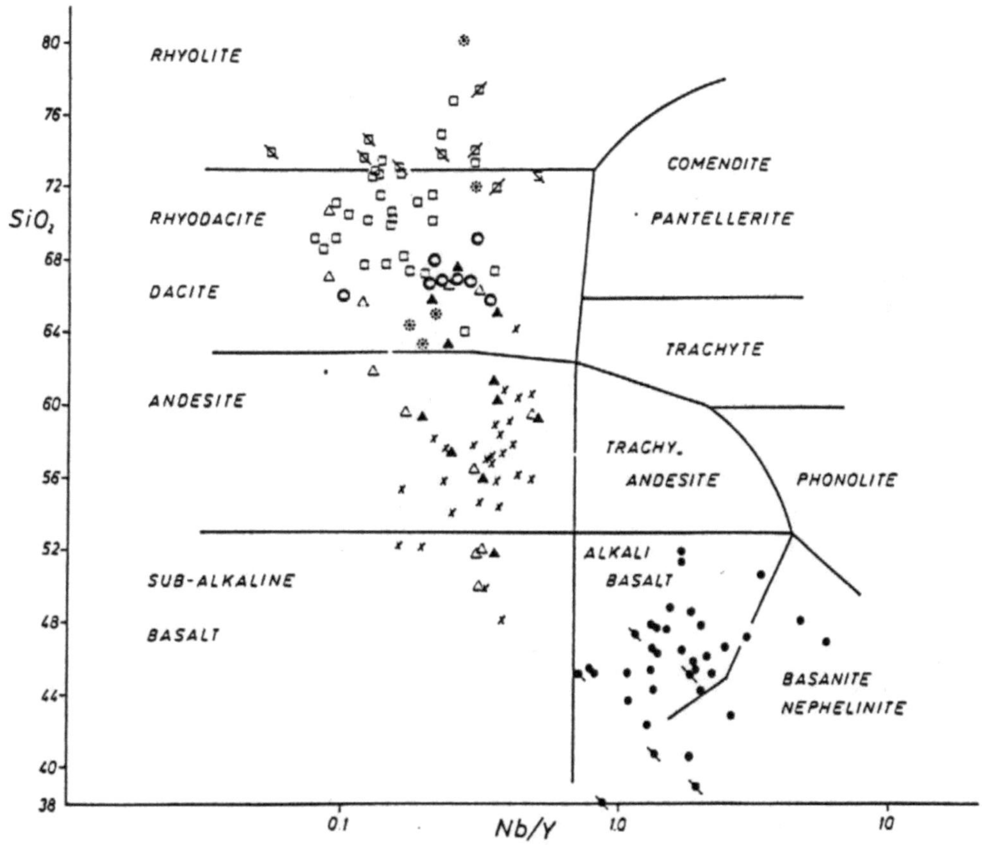

Fig. 2. Middle-to- late Ordovician magmatics in central-northern Sardinia

The erosion of the Monte Corte Cerbos volcanic edificies caused the deposition of the Manixeddu volcanoclastics. Until the early Ordovician the sedimentation developed in similar conditions in the Sarcidano and in the Sarrabus with the deposition of the San Vito and Solanas Fms. The conditions changed during the late Ordovician, with deposition of the thin -condensed Tuviois Fm in Sarrabus and of the thick-turbiditic Bruncu su Pizzu Fm in Sarcidano The Silurian and Devonian consist, in the whole nappe of the black shale ad carbonate sediments.

Interpretation. The thickness (400 m) of the San Vito- Solanas Fms indicates that the Sarrabus and Sarcidano constituted, from late Cambrian to early Ordovician, a Graben controlled by syn sedimentary normal faults. Therefore, the area of the lower nappes, characterized by the thin-condensed Cuili Biringoni Fm, constituted an Horst between the two Grabens, in SW Sardinia and in the Sarrabus-Sarcidano areas.

During the emplacement of the nappe, the syn sedimentary normal faults were inverted and transformed in thrust surfaces, covering the Horst which was dismembered in the small lower nappes.

The emplacement in central-northern Sardinia of abundant magmatics argues in favor of the development during the middle Ordovician of an intra continental rifting, controlled by extensional tectonics. On the other hand, these tectonics developed at the same time of the compressional tectonics in SW Sardinia ("Sardic phase"). Such heterogeneous distribution of the deformation and its spatial variation indicate that during the middle Ordovician, the SCM was affected by wrench dynamics, causing contemporaneously transpressional and transtensional phenomena.

3.2.3 Upper internal nappes

Data. These nappes occur in central-northern Sardinia (Fig.1). In the Gennargentu, the late Ordovician Bruncu su PIzzu Fm reaches a thickness of about 1000 m (Dessau et al., 1982). Isoclinal recumbent folds, well-developed metamorphic cleavage and strong stretching lineations characterize these nappes.

Interpretation. The northward increasing thickness of the late Ordovician turbiditic sequences indicates that only from the late Ordovician an internal domain, extended in central –northern Sardinia (from the Sarcidano to the Gennargentu to the northern Sardinia; that is the areas characterized by the sedimentation of the turbiditic Bruncu su Pizzu Fm) was opposed to an external domain, extended in central-southern Sardinia (from SW Sardinia to the lower external nappes to the Sarrabus; that is the areas characterized by the deposition of the epicontinental Tuviois Fm). Therefore, in the strict sense only from the late Ordovician the true Variscan cycle began to develop.

3.3 Asinara-Posada line

Data. The Asinara-Posada Line (APL) in NE Sardinia, is a km-sized belt in the amphibolite facies, characterized by occurrence of small bodies of amphibolites, containing granulitic or eclogitic relics (Ghezzo et al., 1980), representing remnanys of oceanic crust, trapped between two collided passive margins.

Interpretation. The transtensional tectonics that developed during the late Ordovician with increasing intensity toward NE, reached their acme along the APL, thus opening of an oceanic basin.

3.4 Innermost chain

The part of the Variscan chain from the APL to the central Corsica consists (Rossi et al., 2009, references therein) of intermediate pressure amphibolite facies, migmatite and migmatitic gneisses.

This part of the chain represents the innermost area of the Variscan belt.

3.5 Backland

Data. The backland of the chain outcrops in northern Corsica.

In NW Corsica, a metamorphic complex ("Argentella complex") is covered by a Paleozoic succession. This starts (Baudelot et al., 1977) with thin terrigenous deposits, late Cambrian-early Ordovician in age, affected by a mild compressional phase during the middle Ordovician. The transgression is documented by a conglomerate followed by a thin terrigenous formation late Ordovician in age, covered by the Silurian-Devonian sediments. The Carboniferous consists of carbonate levels ("Calcaire de Capitello").

Interpretation. The Argentella complex represents the pre Cambrian Pan African Basement. The non deposition of the early-middle Cambrian sediments indicates that NW Corsica rconstituted, during this interval of time, a terrestrial area.

The overlying transgressive siliciclastic sequences , late Cambrian-early Ordovician in age, correspond, in thickness and in age, to the Cuili Biringoni Fm in Sardinia; this is in favor of the Horst-Graben structural setting during late Cambriam-early Ordovician times..

The compressional middle Ordivician tectonics indicate that the Sardic Phase also affected NW Corsica. The conglomerate at the base of the late Ordovician succession corresponds to the "Puddinga" in SW Sardinia.

The late Ordovician sequences correspond to the Tuviois Fm in central-southern Sardinia. Therefore, central-southern Sardinia and northern Corsica (both characterized by epicontinental deposits) constituted the external areas opposed to the internal zone in north Sardinia-central Corsica, (characterized by the turbiditic deposits and overall by the ophiolite-bearing successions).

4. Discussion

The various events that, starting from the early Cambrian, progressively affected the SCM, offer a great deal of information about the development of the Variscan geodynamic cycle in this southern segment of the European Variscan Realm. Different thickness of the sediments, their lateral and vertical changes, variations in deposition rate, break in sedimentation, angular unconformities and magmatism: all of them record changes in crustal dynamics and allow to trace the space-time evolution of the orogenetic belt.

Figure 3, summarizes the most representative events.

The sedimentation began from the early Cambrian covering a pre-Cambrian (Pan-African) Basement, represented by metamorphics and by a granite ("Othogneiss"- "-Porphyroid").

Radiometric datings yielded for these last a middle Ordovician age. However, the overlying arkoses interbedded with the early Cambrian deposits are in favor of a pre Cambrian age. Therefore, the granite represents a magmatism concluding a pre-Cambrian orogenetic cycle.

The early -middle Cambrian sediments (Bithia, Nebida and Gonnesa Fms) were deposited only in SW Sardinia. Syn sedimentary long-lived extensional tectonics controlled their thickness.

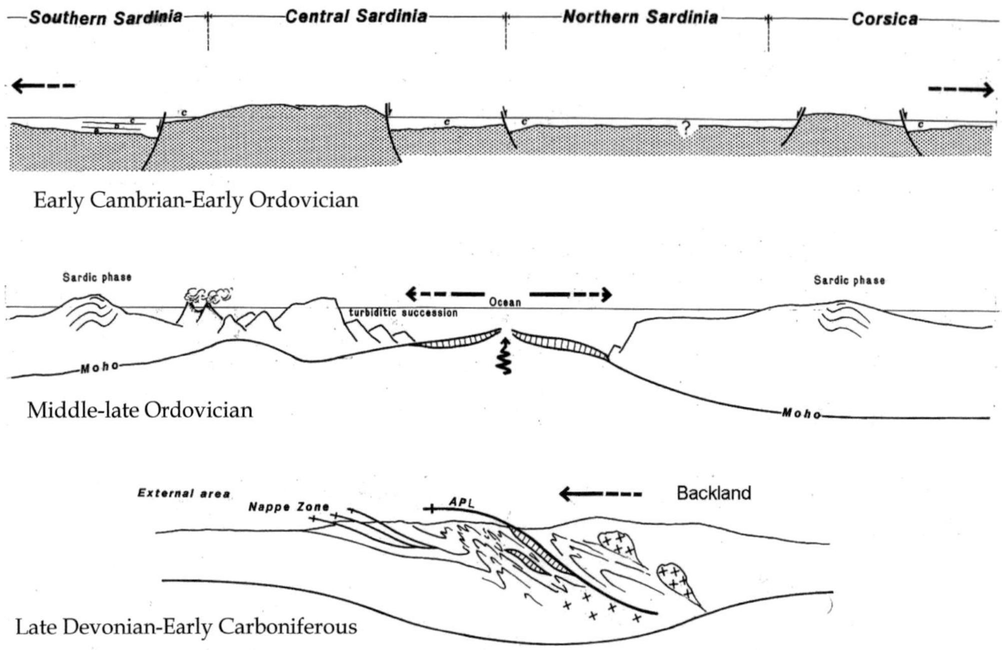

Fig. 3. Variscan geodynamic cycle in the Sardinia-Corsica Massif.
Early-middle Cambrian: sedimentation , in SW Sardinia, of a) Bithia and Nebida Fms. b) Gonnesa Fm.
Late Cambrian- early Ordovician: deposition in the whole Sardinia-Corsica Massif of c) Cabitza- San Vito -Solanas - Cuili Biringoni Fms.
Middle Ordovician: Transpressional and transtensional tectonics, development of "Sardic Phase" and emplacement of felsic calc alkaline magmatics.
Late Ordovician: deposition, of a thick turbiditic succession (Bruncu su Pizzu Fm) and emplacement of basic alkaline magmatics.
Late Ordovician –Silurian: opening the Asinara-Posada Ocean
Late Devonian-early Carboniferous. Orogenesis and formation of the chain. APL: Asinara-Posada Line

From late Cambrian to early Ordovician, the extensional tectonics spread throughout the whole SCM, with the deposition of the Cabitza-San Vito-Solanas Fms (Grabens) and of the Cuili Biringoni Fm (Horsts). Similar deposits also occur in the Kabilies (Boullin et al., 1987) and in Calabria (Minzoni, on going .studies). They correspond to the European "Grès Armoricain". The extend of these deposits indicates a continuity between the SCM and the Variscan Europe until the early Ordovician. Therefore, although there were local signs of incipient rifting, until the early Ordovician the geodynamic regime was governed by slow extensional tectonics, controlled by long-lived normal faults with deposition of shallow water successions and effusion of little or nothing volcanics.

This slow extensional regime drastically changed starting from the middle-late Ordovician. Wrench dynamics caused the sudden development of transpressional and transtesional tectonics.

The transpressional tectonics played a subordinate role. They developed, in fact, only during the middle Ordovician; ("Sardic Phase"). According to Zwart & Dornsiepen, 1980, the "Sardic Phase" constituted local event (Fig. 4), linked to the Caledonian orogenesis.

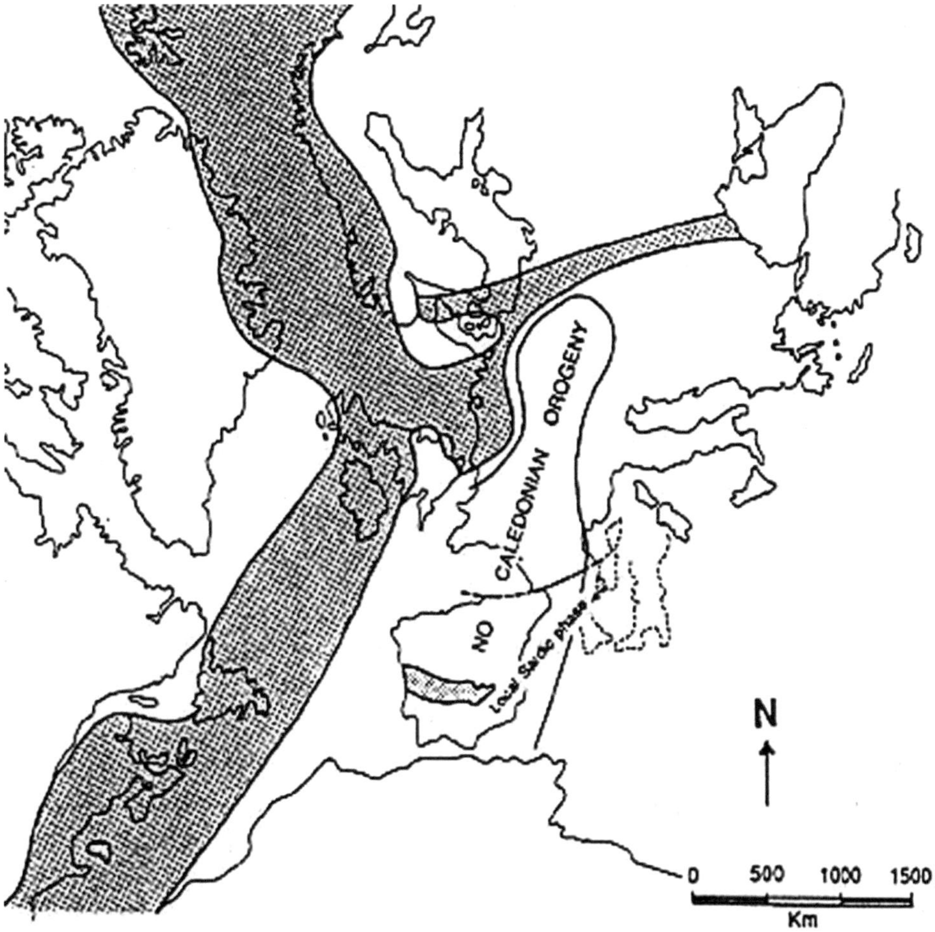

Fig. 4. Area affected by the Caledonian orogenesis and "Sardic Phase"

The transtensional tectonics played the big role. They developed from the middle Ordovician to the Silurian, affecting the zone from central Sardinia to central Corsica. Sinistral strike-slip faults caused the opening of an S-shaped pull apart basin (Fig.5) evolving, in a sequence of closely related stages, from incipient intracontinental rifting to oceanic arm, floored by oceanic crust. The intracontinental rifting stage was characterized, from middle to late Ordovician, by emplacement of the magmatics, ranging from calc-alkaline rhyolite to alkaline basalt.

The transtensional tectonics developed with increasing intensity northward, that is toward the internal areas of the basin, characterized by the occurrence of ophiolltes.. The area extended from northern Sardinia to central Corsica may be regarded as the zone affected by maximum in transtension, with opening of the Oceanic arm: Asinara-Posada Ocean, in this article; South Armorican Ocean of Rossi et al., 2009.

The Variscan orogenesis developed from the late Devonian to the early Carboniferous. The sinistral strike-like master faults, that allowed to the opening of the ocean, now acted as dextral faults thus closing the ocean (Fig.5).

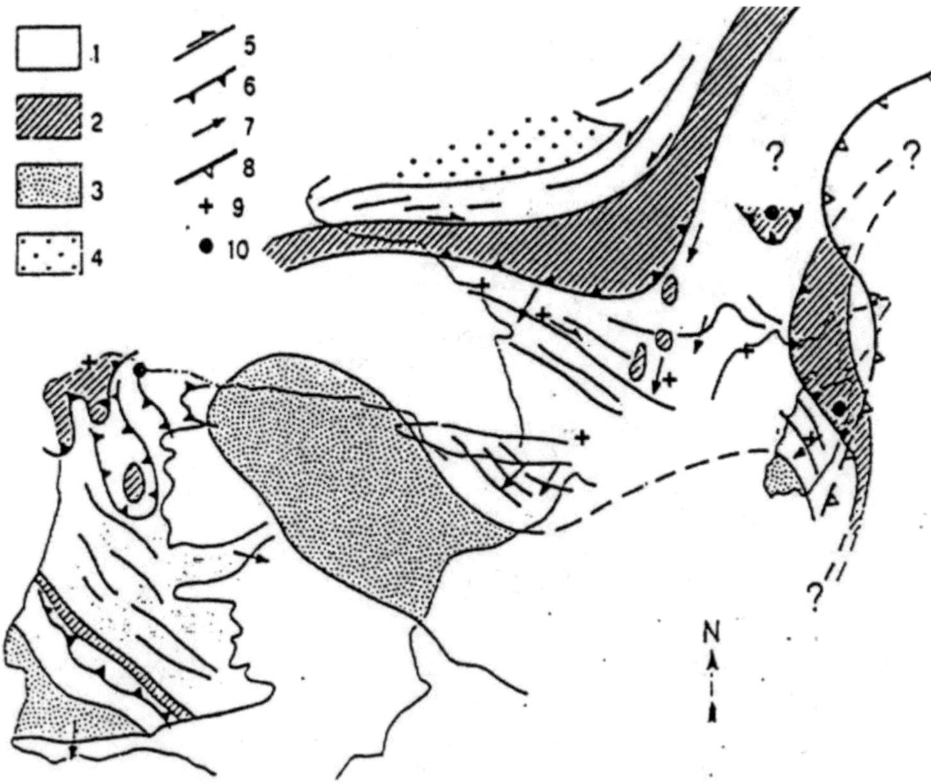

Fig. 5. The Variscan chains in Central-southern Europe
1-Nappe and external zone. 2- Innermost zone, with ophiolitic remnants. 3- Outer foredeep basin. 4-Crystalline basement. 5-Strike-slip faults. 6- Main overthrusts. 7- Tectonic transport. 8- Front of Alpine chain. 9- Early-middle Ordovician bimodal magmatics. 10- Ophiolitic suites

Northward subduction of the oceanic lithosphere was followed by the collision of the two passive margins. The northern margin overrode the southern one, (Elter & Pandeli, 2005; Carosi et al., 2006) with crust-mantle detachment, crustal stacking, oceanic suture along the APL, development of the parauthoctonous nappes in the southern paleomargin (central - northern Sardinia), progressive migration of deformation and metamorphism from the suture zone toward the foreland.
The ensialic model proposed by Helbing, 2003 not accounts for the occurrence of ophiolites (Cruciani et al., 2002) and of the magnetic anomaly (Fig. 6) in NE Sardinia, (Cassano et al., 1979; Fais et al., 1994; Cassano et al., 2001). The magnetic anomaly can be related to an important buried mafic body, suggesting a crustal-scale structure, formed by convergence and collision of two passive margins. The ophiolites, exposed at the surface along the APL, may be considered as fragments of oceanic crust, trapped between the two passive margins (Fig.6).

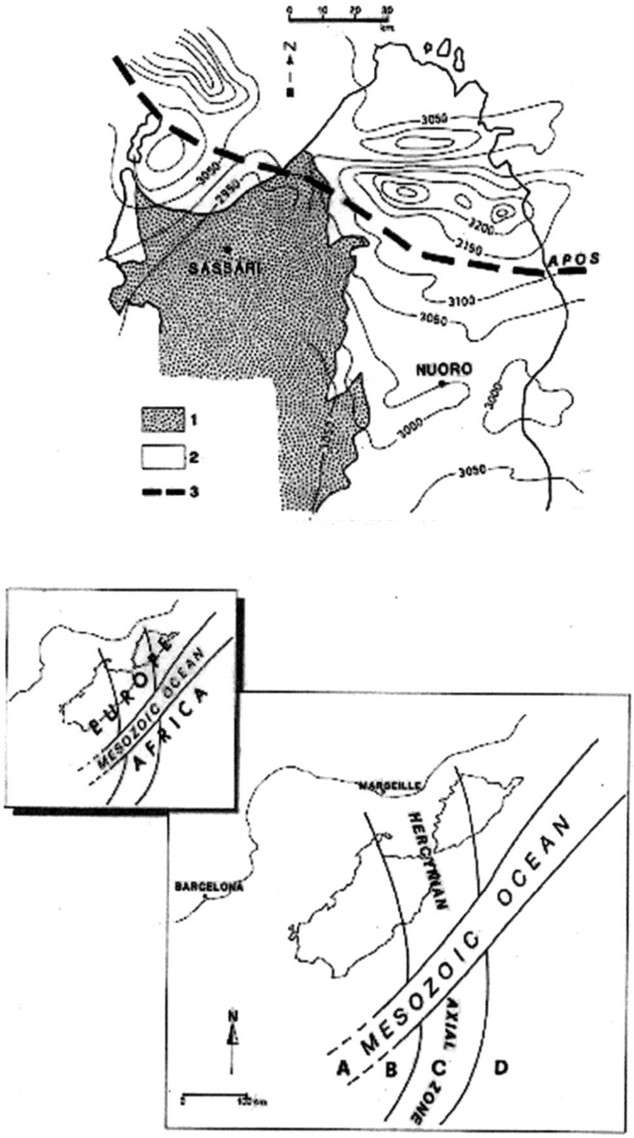

Fig. 6. Aeromagnetic map of northern Sardinia . Contour interval: 50 y.
1- Mesozoic and Cenozoic sedimentary cover.
2- Variscan (Hercynian) crystalline rocks.
3-Asinara-Posada Oceanic Suture (APOS).
The Mesozoic Ocean opened East of the Sardinia-Corsica Variscan Massif (Europe) and West of the Variscan Calabria (Africa).
ABCD represent the palinspasthic restoration of the present Alpine west-verging nappes in Calabria. A- ophiolitic nappes, derived from the Mesozoic Ocean. BCD: crystalline nappes, derived from different tectono-metamorphic zones of the Variscan chain in Calabria. These zones were piled up and transported westward, by the Alpine tectonics. The front of the Alpine chain is shown in Fig.1 and in the Fig. 5

Oceanic sutures are well documented in several European Variscan segments (Fig 5). Metabasic rocks, with high-pressure relics, occurring along the axial zone of the Variscan chain, have been believed as fragments of oceanic crust, trapped along an oceanic suture, between two passive margins (Burg & Matte, 1978; Beher et al. 1984; Matte, 1986; Franke, 1989; Pin, 1990; Pin & Marini, 1993; Matte, 1991; Matte, 2001).
The metabasic rocks often display ocean-floor affinity (Bodinier et al., 1986; Bouchardon et al., 1989; Pin, 1990). The most frequent radiometric datings on the protoliths indicate ages ranging from late Cambrian to early Ordovician (Pin, 1990; Rossi et al., 2009). In some areas, U/Pb datings suggested a pre Cambrian age (Peucat et al., 1982; Paquette et al.. 1985). However, more recent studies reinterpreted these data in terms of zircon inheritance through crustal contamination (Paquette, 1987). The ages of the HP-HT metamorphism in the metabasic suites range from Silurian to early Devonian (Peucat et al., 1982; Peucat & Cogné, 1977; Pin & Lancèlot, 1982; Paquette 1987; Quadt & Gebauer, 1988; Matte, 2001), thus indicating subduction of oceanic lithosphere
In the SCM, the Variscan chain was born and shaped along the Paleozoic domains. They were progressively incorporated in the chain, moving from north southward and from late Devonian to early Carboniferous.
The internal zone of the chain was born reutilizing the normal faults which caused the opening, from late Ordovician to Silurian, of the Asinara-Posada ocean. The northern margin overrode, in this way, the southern one with the development of an accretionary wedge.
In the southern margin, the Sarcidano-Sarrabus nappe took place reworking the syn sedimentary normal faults that, from late Cambrian to early Ordovician, allowed to the deposition of the San Vito-Solanas Fms. These faults once inverted, became an enormous sub horizontal thrusting surface which transported the nappe as far as to the north Iglesiente. The underlying Horst was enveloped and fragmented in several small exotic horses (the present Lower external Nappes) in a stacked imbricate antiform.
In the southernmost Sardinia, the whole Paleozoic succession was folded and detached from the pre-Cambrian basement utilizing the syn sedimentary normal faults that allowed to the deposition of the early Cambrian-middle Cambrian formations. This explains the tectonic-metamorphic characteristics of these areas. In fact, deformation and metamorphism increase downward and reach the green schist facies along a sub horizontal shear surface, located between the lowermost Paleozoic deposits and the underlying pre Cambrian basement.

6. Conclusions

The transtensional tectonics played during the development of the Variscan geodynamic cycle, a key role causing, from the middle Ordovician to the Silurian, crustal thinning, initiating rifting, emplacement of abundant magmatics, breaking of the continental crust and opening of an oceanic arm (Asinara-Posada Ocean). By the activity of the transtensional tectonics, the true Variscan geodynamic cycle began to develop.
Extensional or transtensional tectonics, reaching their acme during the Ordovician-early Silurian and causing intracontinental rifting and opening of oceanic basins, have been identified in several areas of the European Variscan Realm (Montaigne Noire, Armorican Massif, Pyrenees and Alps). Emplacement of "Bimodal Magmatics" accompanies these tectonics. They are usually related, as in the SCM, to subduction of oceanic lithosphere and

considered an Andean-type arc. Alternatively, they may be linked (Briand et al., 2002) to plume mantle activity prior to the onset of rifting and breaking of the continental lithosphere. Their chemical characteristics may be interpreted in terms of partial melting at different depths of a progressively depleted mantle, during the progressive upwelling within intra continental rifting.

These phenomena developed by several pulsating and space -time evolving episodes. In fact, they migrated from central Europe (with the opening, from late Cambrian to early Ordovician, of the Reich Ocean) to the SCM (with the opening , from the late Ordovician to the early Silurian, of the Asinara- Posada Ocean). It is worthy of note that in the SCM the Variscan orogenesis migrated progressively from north siuthward and from the late Devonian to the early Carboniferous.

This model is consistent with the characteristics of the Variscan chain in the whole Europe. The sutures are characterized by scarcity of oceanic remnants. This may be due to wrench dynamics that led to the opening, by transtensional tectonics, of several pull apart basins. This may also explain the extraordinary complexity of the Variscan orogeny in Europe, its unusual width and its space-time evolution

The SCM and the on going studies in Calabria argue in favor of this model. Their connection with the Variscan Europe is not the result of fortuitous coincidence.

7. Acknowledgments

This article took advantage by the collaboration, in computer processing, of my son Christian and by scientific comments of my son Marcello (geologist at Shell Oil Company, Houston, Texas) and of Rebecca ("Becky"), Marcello's wife.

8. References

Albani R., Di Milia A., Minzoni N., & Tongiorgi M. (1985)- Nuovi dati palinologici e considerazioni geologiche sull'età delle Arenarie di Solanas (Cambro- Ordoviciano, Sardegna centrale). *Atti Soc, Tosc. Sc, Nat. Mem.*, 91, 1-20.

Barca S., Cocozza T., Del Rio M. Pillola G.L. & Pittau Demelia P. (1987)- Datation de l'Ordovicien inférieur par Didtyonema flabelliformis dans la partie supérieur de la formation de Cabitza (SW de la Sardaigne, Italy). Coséquences géodinamiques. *C.R. Acad. Sc. Paris*, 305, 1109-1113.

Barca S., Del Rio M., MinzonI N. & Pittau P. (1984)- Presenza di Tremadociano ad Acritarchi in unità tettoniche erciniche a sud del Lago di Mulargia (Sardegna centrale). *Riv. Paleont. Strat.* 89, 315-334. Baudelot S., Doubinger J., Durand-Delga M., & Vellutini P. (1977)-Caractéristiques et age des cinq cycles Paléozoiques du Nord-Ovest de la Corse . *Bull. Soc . Géol. France* , 18, 1221- 1228.

Beher M.J., Engel W., Franke W., Giese P.& Weber K. (1984)- The Variscan Belt in central Europe: main structures, geodynamic implications, open questions. *Tectonophisics,* 109, 15-40

Bodinier J. L., Giraud A., Dupuy C., Leyreloup A. & Dostal J. (1986)- Caractérisation géochimique des métabasites associées a la suture hercynienne . Massif Céntral Francaise et chamrouse (Alpes), *Bull.Soc. Géol. France,* 1, 115-123.

Bosellini A. & Ogniben G. (1968)- Ricoprimenti ercinici in Sardegna . *Ann. Univ. Ferrara,* 9,1,1, 11-15.

Bouchardon J.L., Santallier D., Briand B., Menot R. & Piboule M.C.(1989)- Eclogites in the French Paleozoic orogen: geodynamic significance. *Tectonophisics,* 169, 317-322.

Bouillin J. P., Majeste'-Menjoulas C., Baudelot S., Cygan C. & Fournier-Vinas C.(1987) - Les formations paléozoiques de l'Arc Calabro-Peloritaine dans leur cadre structural. *Boll. Soc. Geol. It.,* 106, 683-698.

Briand B., Bouchardon J.L.,;Capiez P., Piboule M., 2002)-Felsic (A-type)-basic (plume mantle) Early Paleozoic bimodal magmatism in the Maures Massif (Southeastern France).*Geologi.cal Magazine,* 139; 3, 291-311.

Burg J.P. & Matte P: (1978)- A cross section through the Frenc Massif and the scope of its Variscan evolution. *Z. Dtsch.Geol. Ges.,* 129, 429-440.

Calvino F. 1961)- Lineamenti strutturali del Sarrabus- Gerrei (Sardegna sud –orientale), *Boll. Serv. Geol. It.,* 81, 489-556.

Cappelli B., Carmignani L., Castorina F., Di Pisa A., Oggiano G. & Petrini R. (1992)- A Hercynian suture in Sardinia: geological and geochemical evidence. *Geodynamica acta,* 5, 101-118.

Carmignani L, Barca S., Cappelli B., Di Pisa A., Gattiglio M., Oggiano G., & Pertusati P.C. (1992)- A tentative geodynamic model for the Variscan basement of Sardinia. IGCP 276, *Newsletter* 5, 61-82.

Carmignani L., Cocozza T., Minzoni N. & Pertusati P.C. (1981) -Structural and paleogeographic lineaments of the Variscan cycle in Sardinia. *Geologie en Miynbown,* 60, 171-181.

Carmignani L., Cocozza T., Minzoni N., Pertusati P.C. & Ricci C.A (1980)- E' la Corsica il retropaese della catena ercinica in Sardegna? *Mem. Soc. Geol. It. ,* 20, 47-75

Carosi R., Frassi C., Montomoli C. (2006)-Le metamorfiti dell'Anglona e della Gallura sud-occidentale (Sardegna settentrionale, Italia):nuove implicazioni sull'evoluzione post-collisionale della catena sarda. *Rend. Soc.Geol. It., 2, Nuova Serie,* 108-109

Cassano E., Anelli L. , Cappelli V. & Latorre P. (2001) - Magnetic and gravity analysis of Italy. In: G.B. Vai and I.P. Martini (eds.), *Anatomy of an Orogen: the Apennines and Adjacent Mediterranean Basins,* 53-64, Kluver Academic Publishers, Dordrecht/Boston/London

Cassano E., Marcello A., Nannini R., Pretti S., Ranieri G., Salvadori R. & Salvadori I. (1979)- Rilievo aereomagnetico della Sardegna e del mare circostante,. *Ente Minerario Sardo,* 3/4,7-30.

Cocozza T. (1980) -The Cambrian of Sardinia. *Mem.Soc.Geol.It.,* 20, 163-187.

Cruciani G., Franceschelli M., Marchi M. & Zucca M. (2002)- Geochemistry of metabasites from NE Sardinia, Italy: nature of the protoliths, magmatic trend and geotectonic setting. *Mineralogy and petrology.* 74, 25-47.

Dessau G., Duchi G.,Moretti A., & Oggiano G.,(1982)-Geologia del Valico di Correboi (Sardegna centro-orientale) *Boll.Soc. Geol. It.* 101, 62-130.

Elter F. & Pandeli E. (2005)-Structural –metamorphic correlation between three Variscan segments in southern: Maures Massif (France), Corsica(France)-Sardinia(Italy), and northern Appennines (Italy). *Journal of the Virtual Explorer,* 19, 29-36.

Fais S., Klingele E. E., & Tocco R. (1994)- Geophisical interpretation of the Gallura magnetic anomaly (north-east Sardinia,Italy). *Tectonophisics,* 233, (1-2), 125-144

Franke W. (1989)- Variscan plate tectonics in central Europe; current ideas and open questions. *Tectonophisics,* 169, 221-228

Gandin A., Minzoni N. & Courjolt-Rade P. (1987)- Shelf- to -basin transition in the Cambrian-lower Ordovician of Sardinia (Italy). *Geol. Rund.*, 76/3, 827-833.

Gandin A. (1987)- Depositional and paleogeographic evolution of Cambrian in S-W Sardinia: IGCP 5, *Newsletter* 7,151-166.

Ghezzo C., Memmi I. & Ricci,C.A. (1980) - Un evento granulitico nella Sardegna nord-orientale. *Mem. Soc. Geol. It.* 20, 23-38.

Helbing H. (2003)- No suture in the Sardinian Variscides: a structural , petrological and geochronological analysis. *Tübinger Geow. Arbeiten,* 68,1-90.

Martini I.P., Tongiorgi M., Oggiano G. & Cocozza T. (1991)-Ordovician alluvial fan to marine shelf transition in sw Sardinia, western Mediterranean sea: tectonically ("Sardic phase") influenced clastic sedimentation. *Sedimentary Geology,* 72, 97-115.

Matte P. (2001)- The Variscan collage and orogeny (480-290 Ma) and the tectonic definition of the Armorica microplate : a rewiew. *Terra Nova* , 13, 122-128.

Matte P. (1986) -La chaine Varisique parmi les chaines Palèozoiqes pèri- Atlantiques, Modële d'évolution et position des blocs continentau au permo-carbonifëre, *Bull.Soc. France,* 8, 9-24.

Matte P.(1991)-Accretionary history and crustal evolution of the Variscan belt in western Europe. *Tectonophysics,* 196, 309-337.

Memmi I., Barca S., Carmignani L.,Cocozza T., Elter F., Franceschelli M., GAttiglio M., Ghezzo C., Minzoni N., Naud G., Pertusati P.C. & Ricci C.A. (1982)- Further geochemical data on the pre-Hercynian igneous activities of Sardinia and their geodynamic significance. IGCP 5, *Newsletter* 5, 87-91.

Minzoni N. (1981)- Il pre-Cambriano del Sulcis meridionale (Sardegna). *Miner. Petrogr. Acta,* 24, 51-56.

Minzoni N. (1975) - La serie delle successioni paleozoiche a sud del Gennargentu . *Boll.Soc. Geol. It.,* 94, 347-392

Minzoni N. (1995)- The Variscan geodynamic cycle in the Sardinian-Corsican Massif, according to the most recent data. J.P. Hogan and M. C. Gilbert eds. Basement tectonics, 12, 155-168.Kluwer Acad. Publ. Netherland.

Paquette J.L (1987)- Comportement des systemes isotopiques U/ Pb et Sm/ Nd dans le métamorphisme éclogitique . Chaine hercynienne et chaine alpine. *Mem. Doc. Cent. Armoricain Etud Struct. Socles.* 14, 1-130

Paquette J.L., Peucat J.J., Bernard-Griffiths J. & Marchand J. (1985)- Evidence for old pre Cambrian relics shown by u-pb zircon dating of eclogites and associated rocks in the Hercynian belt of south Brittany. France. *Chem. Geol.* 52, 203-216.

Peucat J. & Cogne' J. (1977) -Geocronology of some blueschists from Ile de Groix (France). *Nature* 28, 131-132.

Peucat J., Vidal P., Godard G. & Postaire B. (1982)-Precambrian U-Pb zircon ages of eclogites and garnet pyroxenites from Brittany (France): an old oceanic crust in the w European Hercynian belt. *Earth* Plan. *Sc. Lett.* 60, 70-78.

Pin C. & Lancélot J. (1982)- U-Pb dating of an early Paleozoic bimodal magmatism in the French Massif Central and its further metamorphic evolution. *Contr. Miner. Petrolog.* 79, 1-12.

Pin C. (1990) - Variscan oceans: ages, origins and geodynamic implications inferred from geochemical and radiometric data. *Tectonophysics* , 177, 215-227.

Pin C.& Marini F. (1993) - Early Ordovician continental break-up in Variscan Europe: Nd-Sr isotope and trace element evidence from bimodal igneous associoations of the southern Massif Central, France. *Lithos* 29, 177-196.

Quadt A. & Gebauer D. (1988) - Sm/Nd, U/Pb and Rb/Sr dating of high-pressure ultramafic to felsic rocks from the Moldanubian area of NE Bavaria (FRG) and the Saxonian granulite massif (GDR). Abstract, Conf. Bohemian Massif, (Prague), 71.

Rossi P., Oggiano G. & Cocherie A.(2009)- A restored section of the "Southern Variscan realm" across the Sardinia-Corsica microcontinent. *C.R. Geosciences,* 341, 224-238

Vai G. B. & Cocozza t. (1986)- Tentative schematic zonation of the Hercynian chain in Italy . *Bull. Soc. Geol. France,* 2, 95-114

Vai G.B. (2001)- Basement and early (pre-Alpine) history. In: G.B. Vai and I.P. Martini (eds.), Anatomy of an Orogen: the Apennines and Adjacent Mediterranean Basins, 121-150, Kluver Academic Publishers, Dordrecht/Boston/London

Zwart H. J. & Dornsiepen U. F. (1980) - The Variscan and pre-Variscan tectonic evolution of central and western Europe; a tentative model. *Sociétè Géologique du Nord,* C6, 26 CGI. 227-232.

6

Relationships between Lithospheric Flexure, Thrust Tectonics and Stratigraphic Sequences in Foreland Setting: the Southern Apennines Foreland Basin System, Italy

Salvatore Critelli, Francesco Muto, Vincenzo Tripodi and Francesco Perri
Dipartimento di Scienze della Terra, Università della Calabria
Italy

1. Introduction

We discuss here tectonics and sedimentation processes occurring during continent-continent collision and relationships between accretionary processes on overplate, flexural lithosphere on underplate and related controls on clastic sedimentation in developing foreland basin systems. This paper focuses on and clastic sedimentation developed during the sequential history of an orogenic system, in the Mediterranean Region. These clastic trends, covering a large time span from Early Mesozoic to the present, may contribute: (1) to the paleogeographic and paleotectonic reconstructions of the southern Italy portions of the western Mediterranean orogen, and (2) to the general models of complex relationships between clastic sedimentation and paleotectonic history of other major orogens.

The evolutionary record of Earth's processes preserved in the form of sedimentary rocks has been pivotal in paleogeographical and paleotectonic reconstructions of source/basin systems. Compositional trends of clastic strata through space and time are used to infer the structural history of adjacent mountain belts and to monitor the key geodynamic changes during orogenic processes (*e.g.* Dickinson, 1985, 1988; Critelli & Ingersoll, 1994; Critelli, 1999).

The controls on the composition and dispersal pathways of clastic strata along the convergent plate margins have long been debated (*e.g.* Dickinson, 1988; Ingersoll *et al.*, 1995). Clastic infilling of sedimentary basins in orogenic systems have been used as important indicators of tectonic activity and climatic changes. In the orogenic systems, clastic sedimentation may record the accretionary processes, the accomodation of the thrust units, and the flexural features of the foreland plate.

The development of an orogenic wedge during continental collision results in thickening of the crust. The excess mass of this thickened crust acts as a load on the underthrust plate, causing it to be flexed downwards close to the load, so developing a foreland basin (*e.g.* Beaumont, 1981; Sinclair and Allen, 1992). During plate convergence, the vertically acting load of the mountain belt migrates over the foreland plate, thus resulting in the migration of the associated foreland basin.

The foreland is the region between the front of a thrust belt and the adjacent craton (*e.g.* Dickinson, 1974; Bally and Snelson, 1980; Allen *et al.*, 1986; Miall, 1995). Large volumes of clastic sediment are derived from erosion of the thrust belt and deposited in the foreland basin. The foreland basin generally is defined as an elongate trough that forms between a linear contractional orogenic belt and the stable craton, mainly in response to flexural subsidence caused by thrust-sheet loading in the orogen (fig. 1).

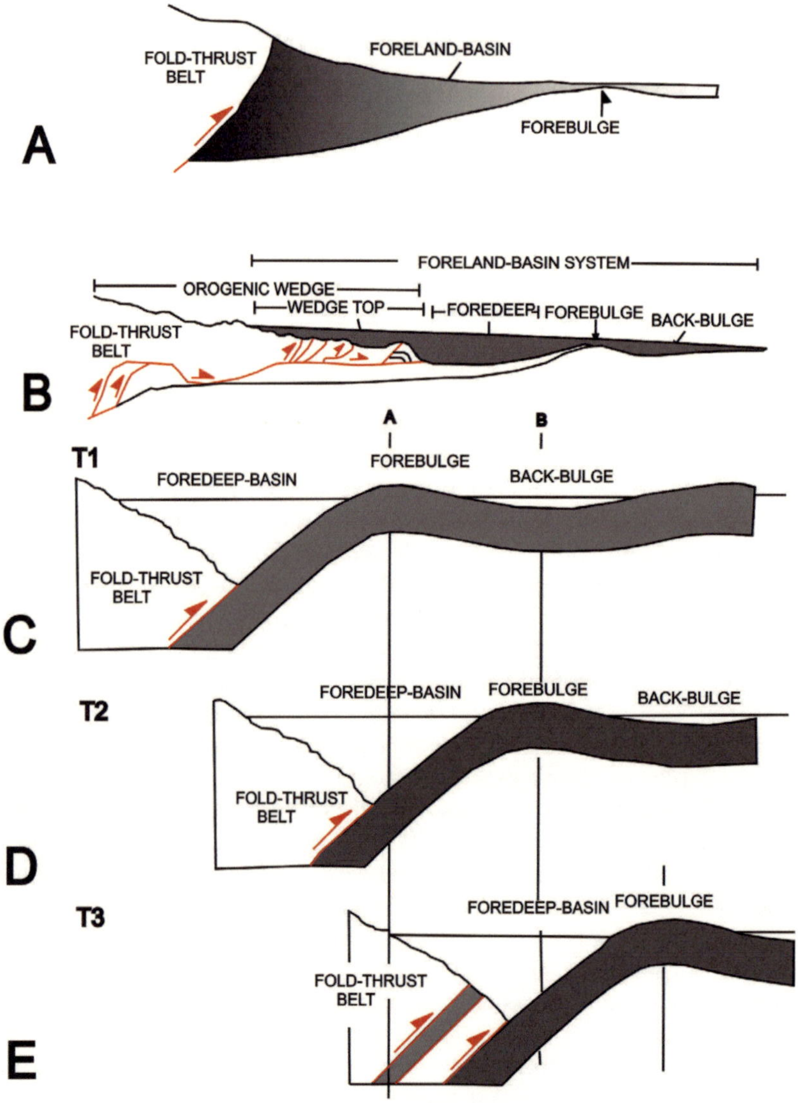

Fig. 1. Diagrammatic cross sections showing the generally accepted notion of foreland-basin geometry(*a*, *b*) and the relationships of lithospheric flexure to accomodation space in foreland systems (*c* to *e*) (*e.g.*Giles and Dickinson, 1995; De Celles and Giles, 1996). *a*) General relationship between fold-thrust-belt, foreland basin and forebulge; *b*) foreland-basin geometry and depozones: wedge-top, foredeep, forebulge and back-bulge depozones; *c* to *e*) relationship of the flexural features in times; *c*) is the initial (Time1) foreland system; *d*) foreland evolution during accretion of the fold-thrust-belt at Time 2; forebulge is migrated cratonward; *e*) previous forebulge is assembled within the fold-thrust-belt. Modified after Giles and Dickinson (1995), and DeCelles and Giles (1996)

Foreland basin stratigraphy records tectonic, eustatic, and climatic changes at convergent plate margins (*e.g.* Miall, 1995). The formation of unconformities is the results of the interplay of temporal variations in the erosion and lateral progradation rates of the orogenic wedge, as well as tectonic and eustatic sea-level changes (*e.g.* Beaumont, 1981; Jordan, 1981; Schedl and Wiltschko, 1984; Peper *et al.*, 1995).

2. The clastic infill of Foreland basin systems

2.1 Depositional zones and geometries

In foreland settings, subsidence and uplift are profoundly affected by lithospheric flexure. Foreland basin subsidence is primarily controlled by downflexing of the lithosphere in response to thrust accommodation and loading (e.g. Jordan, 1981, 1995; Beaumont, 1981). Subsidence rate gradually decreases away from the thrust front producing an asymmetrical depression. Flexure uplift (forebulge) occurs as an isostatic response to downwarping and forms the distal margin of the foreland basin. Cratonward of the forebulge flexure, a broad shallow downwarp or intrashelf basin forms, the back-bulge basin (Fig. 1; e.g. Quinlan and Beaumont, 1984; DeCelles and Giles, 1996).

The dimension and amount of flexural subsidence and uplift produced by the flexural features (i.e., foreland basin, forebulge, back-bulge basin) primarily depend on the geometry and density of the tectonic load, rheology of the lithosphere, density and volume of the sediment infill, and amount of thrust wedge and forebulge erosion (e.g. Beaumont, 1981; Jordan, 1981; Vai, 1987; DeCelles and Giles, 1996; Sgrosso, 1998). The interrelationships between lithospheric flexure, single thrust accomodation within the accretionary wedge and flexural subsidence experiences geometrically complexes entities within the foreland region. The foreland basin system may be diveded into four depozones, the wedge-top, the foredeep, the forebulge, and the back-bulge depozones (Fig. 1; e.g. DeCelles and Giles, 1996). Boundary between depozones may shift laterally through time following the deformation propagation. The longitudinal dimension of the foreland basin system is roughly equal to the length of the adjacent fold-thrust belt (e.g. DeCelles and Giles, 1996).

Wedge-Top Depozone - Large amounts of syntectonic sediment cover the frontal part of the fold-thrust-belt. The sediment that accumulates on top of the frontal part of the orogenic wedge constitutes the wedge-top depozone. Its extent toward the foreland is defined as the limit of deformation associated with the frontal tip of the underlying orogenic wedge. The main distinguishing characteristics of wedge-top deposits are the abundance of progressive unconformities (e.g., Riba, 1976) and various types of growth structures (folds, faults, cleavages; Boyer and Elliot, 1982; Cello and Nur, 1988; Srivastava and Mitra, 1994; DeCelles and Giles, 1996; Zuppetta and Mazzoli, 1997). Aerially extensive aprons of alluvial sediment or shallow shelf deposits commonly drape the upper surface of the orogenic wedge during periods the wedge is not deforming in its frontal part, and large, long-lived feeder canyons may develop and fill in the interior parts of orogenic wedges (e.g. Ori et al., 1986; DeCelles and Giles, 1996). Sediments of the wedge-top depozone typically reflect the erosion and unroofing of the thrust-belt (e.g. Critelli and Le Pera, 1994, 1995a, 1998; Trop and Ridgway, 1997).

Foredeep Depozone - It is the mass of sediment that accumulates between the frontal tip of the orogenic wedge and the forebulge. Foredeeps are typically 100-300 km wide and 2-8 km thick (e.g., DeCelles and Giles, 1996). Sediment is derived predominantly from the fold-thrust-belt, with minor contributions from the forebulge and craton (e.g. Schwab, 1986; DeCelles and Hertel, 1989; Critelli and Ingersoll, 1994; Critelli and Le Pera, 1998). Foredeep depozones have frequently recorded transitions from early deep-marine sedimentation

("*Flysch*") to late coarse-grained, nonmarine and shallow-marine sedimentation ("*Molasse*") (e.g. Covey, 1986; Crook, 1989; Sinclair and Allen, 1992). The transition from "Flysch" to "Molasse" most likely reflects the fact that foreland basin systems originate as oceanic trenches (or remnant ocean basin) and later become shallow marine or nonmarine as continental crust enters the subduction zone (e.g. Ingersoll et al., 1995; DeCelles and Giles, 1996).

Forebulge Depozone.- It consists of the region of potential flexure uplift along the craton side of the foredeep. Because of forebulge depozone is a positive, and potentially migratory, feature, which may be eroded, its potential of preservation is low. One signal of the presence of a ancient forebulge may be the erosional unconformity surface. The forebulge generally is considered to be a zone of nondeposition or erosion, and the resulting unconformity may be used to track its position through time (e.g. Vai, 1987; Bosellini, 1989; Crampton and Allen, 1995; DeCelles and Giles, 1996; Sgrosso, 1998; Critelli, 1999).

In subaerial foreland basin systems (in which foredeep is not filled to the crest of forebulge) the forebulge may be a zone of erosion, with streams draining both toward and away from the orogenic belt (Crampton and Allen, 1995). If sediment derived from thrust-belt progrades into the forebulge, a thin condensed fluvial and aeolian sediment is deposited (DeCelles and Giles, 1996; Critelli, 1999).

In subaqueous foreland basin systems (in which foredeep is not filled up to the crest of forebulge), local carbonate platforms may develop in the forebulge depozone; extensive forebulge carbonate platforms and ramps can connect the foredeep with the back-bulge depozone (Giles and Dickinson, 1995; Critelli, 1999).

Back-bulge Depozone.- It constitutes the sediment that accumulates between the forebulge depozone and the craton. Sediment contributions from the craton and development of carbonate platforms may be significant in submarine systems (e.g., DeCelles and Giles, 1996). Stratigraphic units in the back-bulge are generally much thinner than those in the foredeep, and consist of dominantly shallow-marine and nonmarine sediments.

2.2 Compositional signatures of Foreland clastics

Lithologic Provenance Models. — Foreland regions are one of the typical setting in which huge volumes of clastic sediments are rapidly accumulated. Provenance studies in this tectonic setting have long been used to contribute the complex history of the basin evolution, sediment dispersal pathways, dating major thrust events, and the unroofing history of the thrust-belt (*e.g.* Wiltschko and Dorr, 1983; Graham *et al.*, 1986; Dickinson, 1988; Jordan *et al.*, 1988; Steidtmann and Schmitt, 1988; Critelli, 1999). Tectogenic sediments may be shed as alluvial fans in front of rising thrust sheets, and the age of these sediments may indicate the time of motion on some faults. In this setting, the uplift-erosion-transport-deposition system are genetically and intimately related to the style of deformation in thin-skinned, thrusted terrains. Transport of clastic sediment in the same direction as tectonic transport is the commonly assumed setting for the clastic-wedge/thrust association (*e.g.* Graham *et al.*, 1986; Jordan *et al.*, 1988; Steidtmann and Schmitt, 1988), that is named as «synthetic dispersal». However, opposite sediment dispersal pathways with rispect the tectonic transport is named as «anthitetic dispersal» (*e.g.* Steidtmann and Schmitt, 1988). The results is that sediment dispersal pathways in foreland basin systems are controlled by geometries within the thrust sheet system, as frontal ramps, lateral ramps, diverse hanging-wall beds dip. In settings where distinct source-rock compositions are eroded sequentially, as in the case of predominantly vertical uplift of a stratigraphic section, «unroofing sequences» are commonly formed in the resultant clastic wedge (*e.g.* Graham *et al.*, 1986; DeCelles, 1988; Steidtmann and Schmitt, 1988). This erosional inverted clast stratigraphy can provide valuable information about the evolving source and the identification of specific source areas (*e.g.* DeCelles, 1988; Colombo, 1994; Critelli and Ingersoll, 1994; Critelli *et al.*, 1995; Critelli and Le Pera, 1998). In the case of thin-skinned thrusted terrains (where horizontal transport dominates), layered rocks having different lithologies are exposed to erosion as they pass over a ramp providing a blended clastic dispersal of the exposed rock types. The resulting clastics may show no unroofing sequences, but include the same blended clast composition for relatively great thickness. These blended clastics may indicate that the source rocks were formed by tectonic transport over a ramp (*e.g.* Steidtmann and Schmitt, 1988). In thin-skinned thrust belts, both «unroofing sequences» and «blended clastics» can result in combinations.

Large and Regional-Scale Models based on Sand(stone) Petrofacies. – Numerous studies have demonstrated that sand (stone) from foreland basins are characterized by high framework percentages of quartz and unstable sedimentary and metamorphic lithic fragments, and the mean composition is quartzolithic (*e.g.* Dickinson, 1985, 1988; Schwab, 1986; DeCelles and Hertel, 1989; Critelli and Ingersoll, 1994; Critelli, 1999; Critelli et al., 2003). These studies provide a basis for interpretations of tectonic setting from sand (stone) composition. Sandstone petrofacies can be considered a reliable general guide to the overall tectonic settings of most sediment provenances; and although many processes may modify the composition of sedimentary detritus, the fundamental imprint of provenance tectonics is preserved in the final sedimentary products. The key petrofacies of various sedimentary basins occur most typically when transport is short and direct. In other cases sandstone petrofacies in many sedimentary basins have multiple sources showing complex paleotectonic and paleogeographic relationships to the basins (*e.g.* Dickinson, 1988). The foreland basin systems are a typical basin-setting in which multiple sources can be active in the same time, and the derivative sandstones may show mixed petrofacies (fig. 2; Dickinson *et al.*, 1986; Schwab, 1986; Critelli, 1999). Schwab (1986), in a general statement of foreland-basin sandstone petrofacies, testifies the complex pattern of provenance relationships during the foreland basin evolution. Quartzose sand is typical during the early stage of foreland infill, where the thrust-belt has low elevation and consequently supplies low amounts, whereas cratonal region is flexing and supplies more amounts (Dickinson *et al.*, 1986; Cazzola and Critelli, 1987). Subsequent petrofacies is typical quartzolithic, when the thrust-belt is growing. Local provenances from magmatic arcs, uplifted subduction complexes or uplifted carbonate rocks of the forebulge only represent small amounts of the clastic record within foreland basin system. If thrust belt has severe uplift exposing the crustal basement, petrofacies can evolve to quartzofeldspathic sand (*e.g.* Critelli and Ingersoll, 1994; Garzanti *et al.*, 1996; Critelli and Le Pera, 1998; Critelli and Reed, 1999; Critelli, 1999; Critelli et al., 2007; Barone et al., 2008).

2.3 Present-day morphotectonic zones, Foreland basin system

From south to north the Southern Italy is subdivide into the following morphotectonic belts (Ippolito et al., 1975): (1) The northern Calabrian Arc, including ophiolites, crystalline basement rocks and Mesozoic sedimentary sequences; (2) the Cilento and Calabro-Lucanian Ranges, having ophiolitic, metasedimentary and sedimentary rocks. The Ranges include a Paleogene Subduction Complex (the Calabro-Lucanian Flysch Unit or Liguride Complex of southern Italy), the middle Miocene foreland strata of the Cilento Group and younger sequences, and the Mesozoic to Miocene carbonate platform and slope (inner platform or Alburni-Cervati-Pollino Units and the Monti della Maddalena Unit); (3) the Campano-Lucanian Ranges, including Mesozoic to upper Miocene deep-sea sequences of the Lagonegro and Sicilide units, the outer platform sequences (Monte Alpi Unit), and the

Miocene foreland strata; (4) the Lucanian-Apulia lowland, including the Pliocene to Quaternary foreland clastics; and (5) the Apulian Swell, a Mesozoic to Quaternary carbonate platform (external platform).

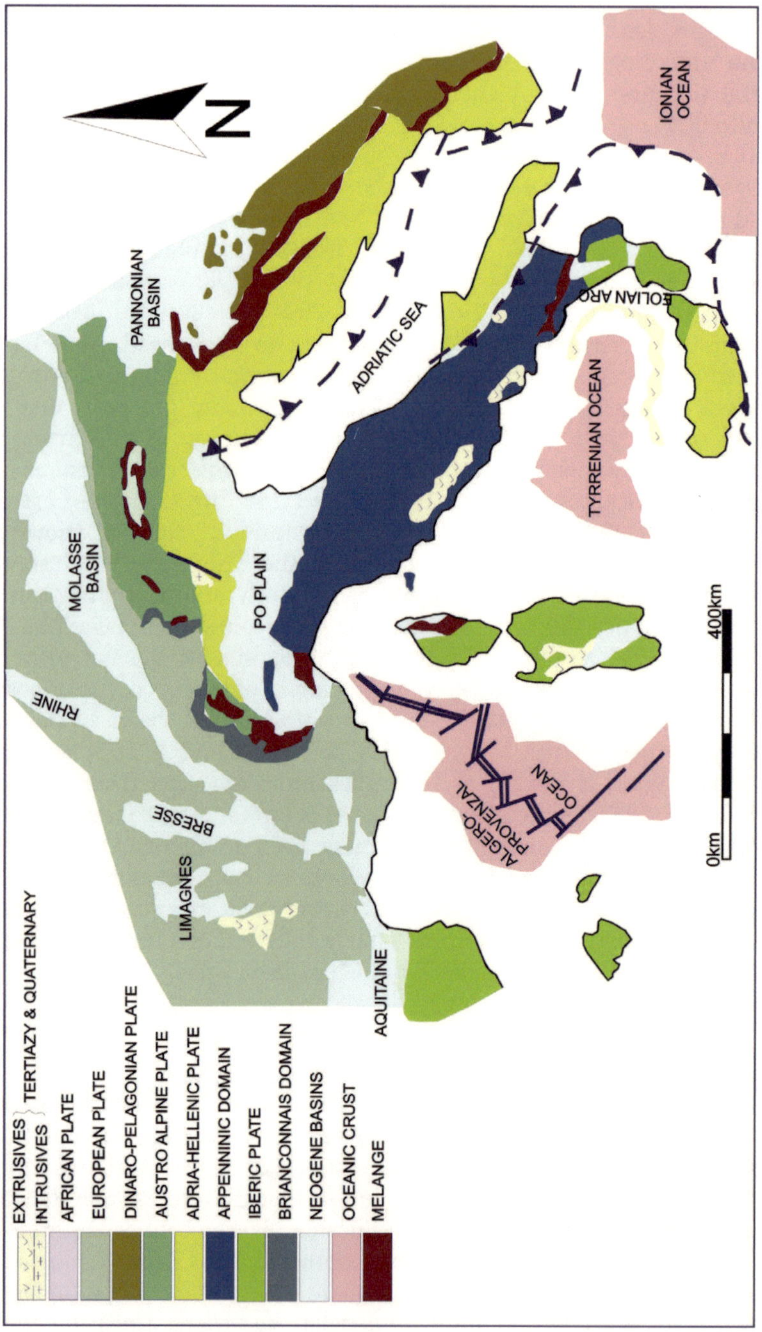

Fig. 2. Present day distribution of the main geodynamic domains of the Alpine region. Modified after Stampfli and Marchant (1997)

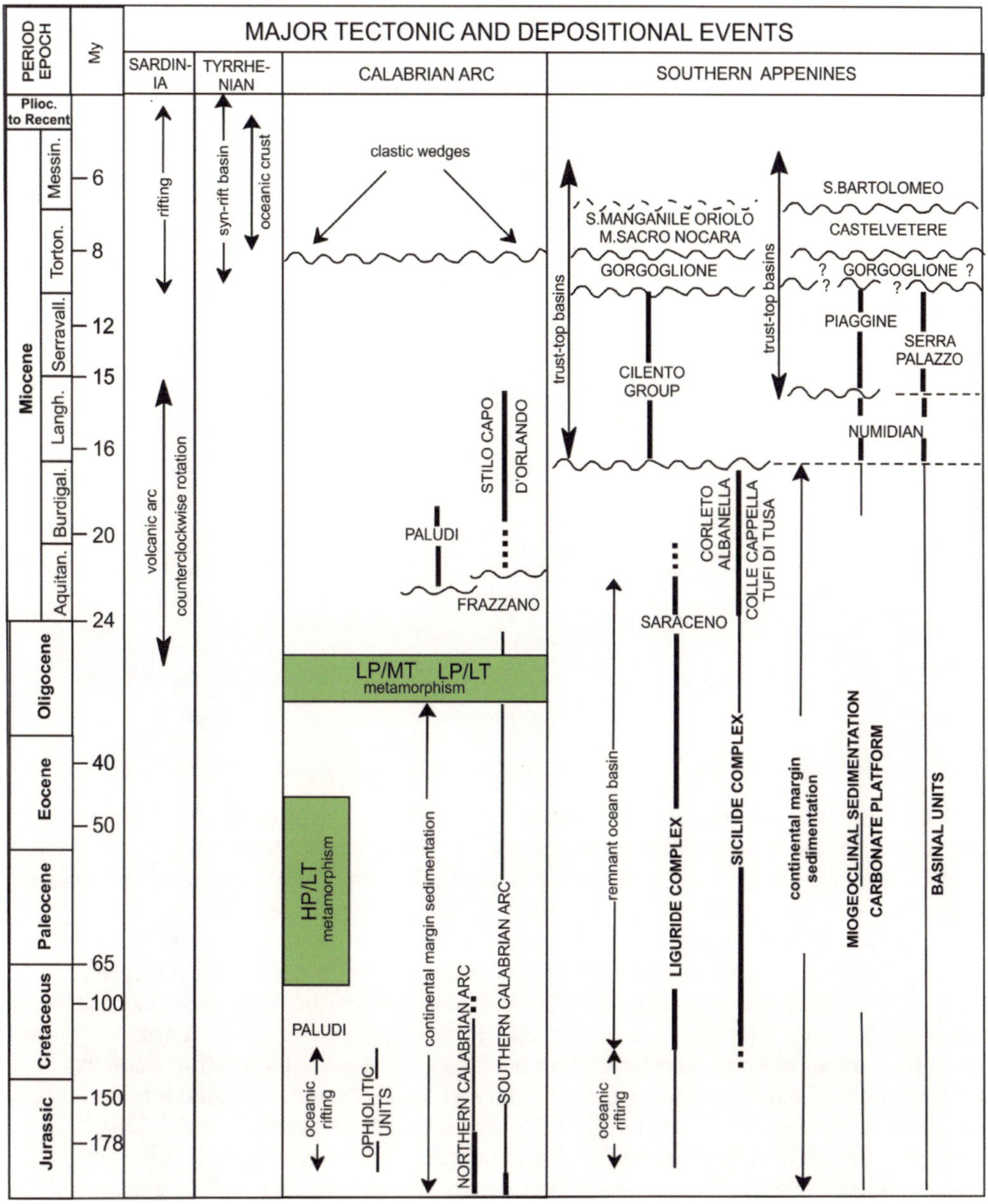

Fig. 3. Chart showing major Mesozoic-Cenozoic tectonic and depositional events in southern Italy sedimentary assemblages. Modified after Critelli and Le Pera (1995a), Critelli et al. (1995b) and Critelli (1999)

2.4 Approach and scope

This paper presents results of regional, structural, stratigraphic and provenance relationships that constraint the post-Oligocene tectonic history of the southern Apennines foreland basin system (Fig. 2). The paper focus on the effects of tectonic deformation during sequential history of the growing orogen in southern Italy.

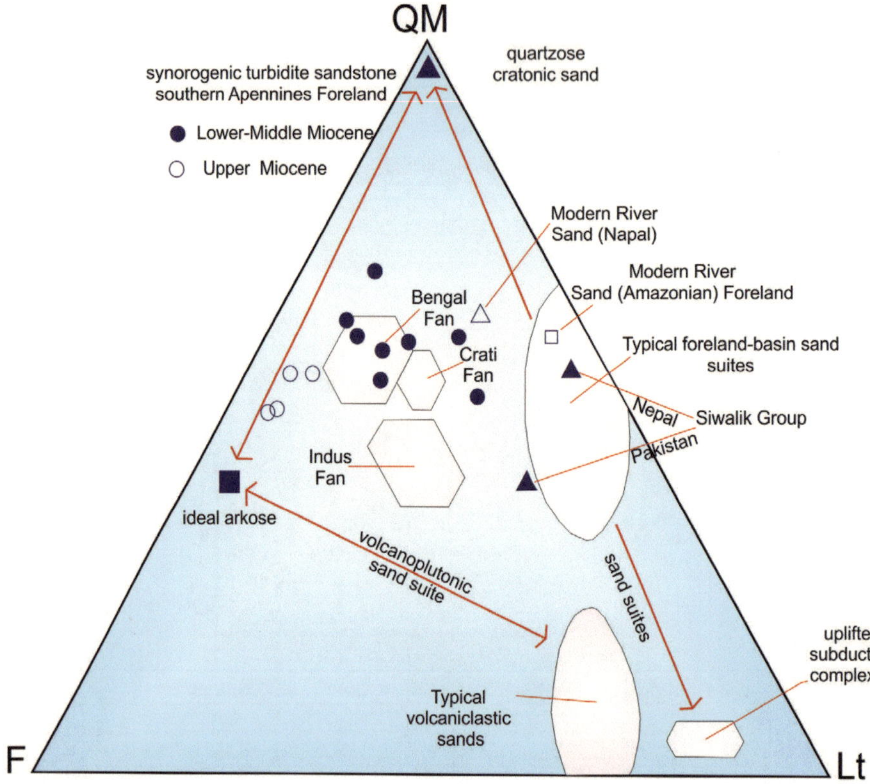

Fig. 4. QmFLt (Qm = monocrystalline quartz, F = feldspars, Lt = aphanitic lithic fragments) diagram to illustrate concept of mixing detritus from different provenance types to produce detrital modes reflecting mixed provenance (from Dickinson, 1988). Typical foreland-basin sand suites were derived from uplifted fold-thrust belts exposing sedimentary and metasedimentary strata. The mixed provenance relations are also typical of some foreland basin systems and remnant ocean basins (i.e. southern Apennines foreland, Indus and Bengal fans of the Himalayan belt). During early stage of foreland infill, sand may derive from cratonal areas, generating quartzose sand. Subsequent petrofacies is quartzolithic, and during final foreland infill, petrofacies may be mixed and quartzofeldspathic. Foreland sandstone detrital modes reference data plotted are: the modern Amazonian foreland (open square; data from DeCelles and Hertel, 1989), the Himalayan foreland [Siwalik (filled triangle) and modern rivers (open triangle); data from Critelli and Ingersoll, 1994], the Bengal and Indus Fan (polygons; data from Garzanti et al., 1996), and the southern Apennines foreland (filled and open circles for Miocene sandstone, and polygon for Holocene Crati Fan; data from Critelli and Le Pera, 1994, 1995). The arrows, symbols (filled triangle and square) and specific fields within the diagram show the sand suite trends from different generic types of provenance terranes (e.g. Dickinson, 1985, 1988; Critelli, 1999)

The entire stratigraphic, structural and compositional data set are interpreted using new general models of sequential evolution of foreland basin systems.

The Calabrian terranes form an arcuate mountain belt that lies between the thrust belts of the Apennines to the north and the Maghrebides to the west (Fig. 2). The study area is a transect across the Calabria block and Apulia platform (Figs.5,11).

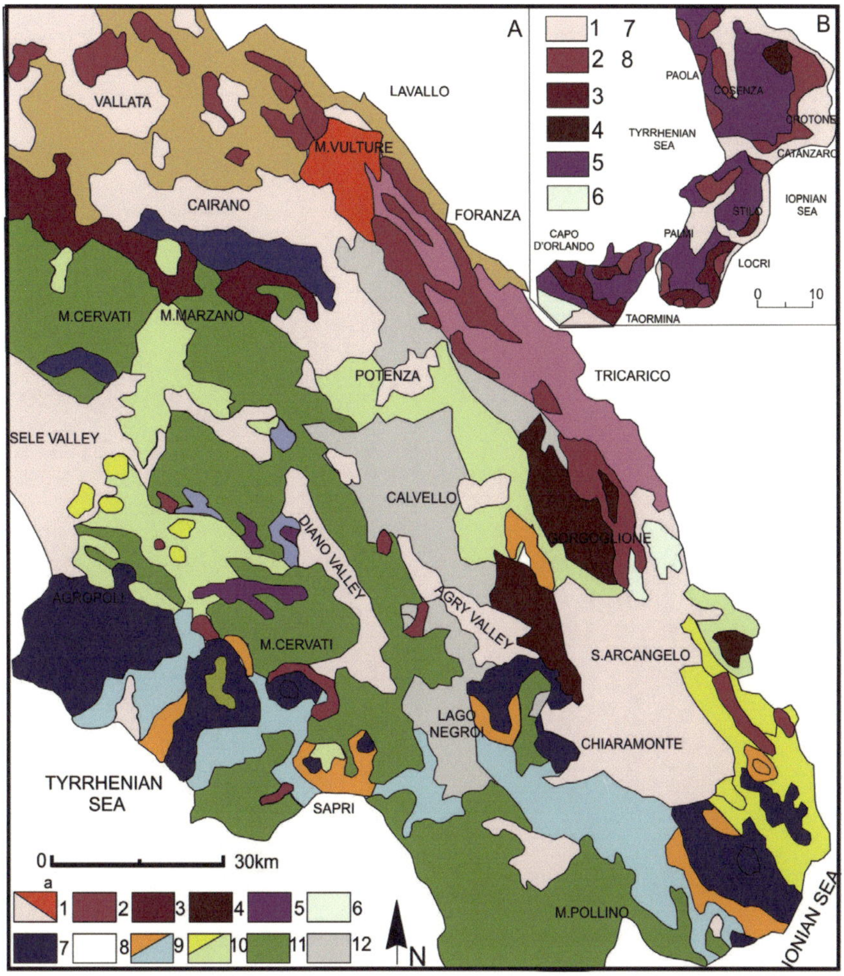

Fig. 5. Geological sketch map of the main tectonostratigraphic units of the Southern Apennines (A) and the Calabria-Peloritani Arc (B). A] 1) Lower Messinian to Holocene sediments (a: Monte Vulture volcanic and volcaniclastic rocks); 2) San Bartolomeo Formation (Messinian); 3) Castelvetere, Oriolo,Monte Sacro, Nocara, Serra Manganile formations (upper Tortonian to lower Messinian); 4) Gorgoglione Formation (Tortonian); 5) Piaggine Formation (Serravallian to Tortonian); 6) Serra Palazzo Formation (Langhian to Tortonian); 7) Cilento Group (Langhian to Tortonian); 8) Numidian Sandstone Formation (Langhian); 9) Liguride Complex (Cretaceous to early Miocene; a: Saraceno Formation); 10) Sicilide Complex (Jurassic to early Miocene; a: Albanella, Corleto, Colle Cappella, Tufiti di Tusa formations); 11)Shallow-water to deep-water carbonate units (Triassic to middle Miocene); 12) Deep-water pelagic sediments (Triassic to early Miocene; Lagonegro, Molise, Sannio units). B] 1) Pliocene to Holocene sediments and recent volcanic deposits; 2) Upper Tortonian to Messinian sediments; 3) Stilo-Capo d'Orlando Formation (early Miocene); 4) Longobucco Group (Jurassic); 5) Jurassic to Cretaceous Ophiolitiferous units,and Paleozoic metamorphic and plutonic units; 6) Maghrebian units (Mesozoic to Tertiary); 7) Paludi Formation (Late Oligocene to early Miocene); 8) Frazzan`o Formation (Oligocene to early Miocene). Modified after Critelli et al. (1995a,b) and Critelli (1999)

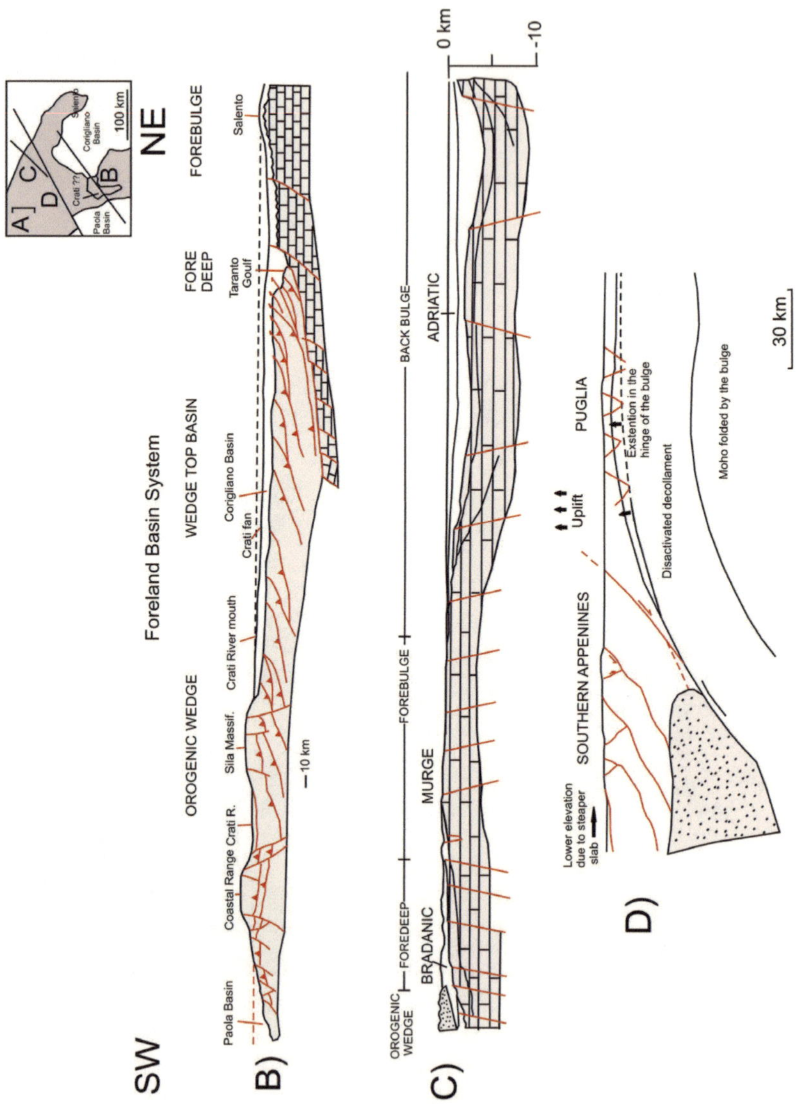

Fig. 6. Cross sections from the Adriatic Sea to the Tyrrhenian Sea, crossing the main depozones of the modern southern Apennines foreland basin system (Critelli, 1999); locations of the cross sections are in a). b) Cross section from the Paola Basin to the eastern Apulia, showing the Paola slope basin (Eastern Tyrrhenian margin), the Calabrian thrust-belt (orogenic wedge), the southern Apennines foreland region (wedge-top and foredeep basins) and the flexed Apulia foreland [Modified after Cello et al. (1981), and Pescatore and Senatore (1986)]. c) Cross section from the outer thrust front of the southern Apennines orogenic wedge to the southern Adriatic Sea; locations of the modern subaerial foredeep (Bradanic trough), forebulge (Murge), and back-bulge (southern Adriatic Sea) depozones are shown [Modified after Ricchetti (1980), and Ricchetti and Mongelli (1980)]. d) General schematic deep cross-section of the southern Apennines orogenic wedge showing formation of the Puglia bulge. During middle Pleistocene, the Bradanic trough was inverted from subsidence to uplift. Modified after Doglioni et al. (1994)

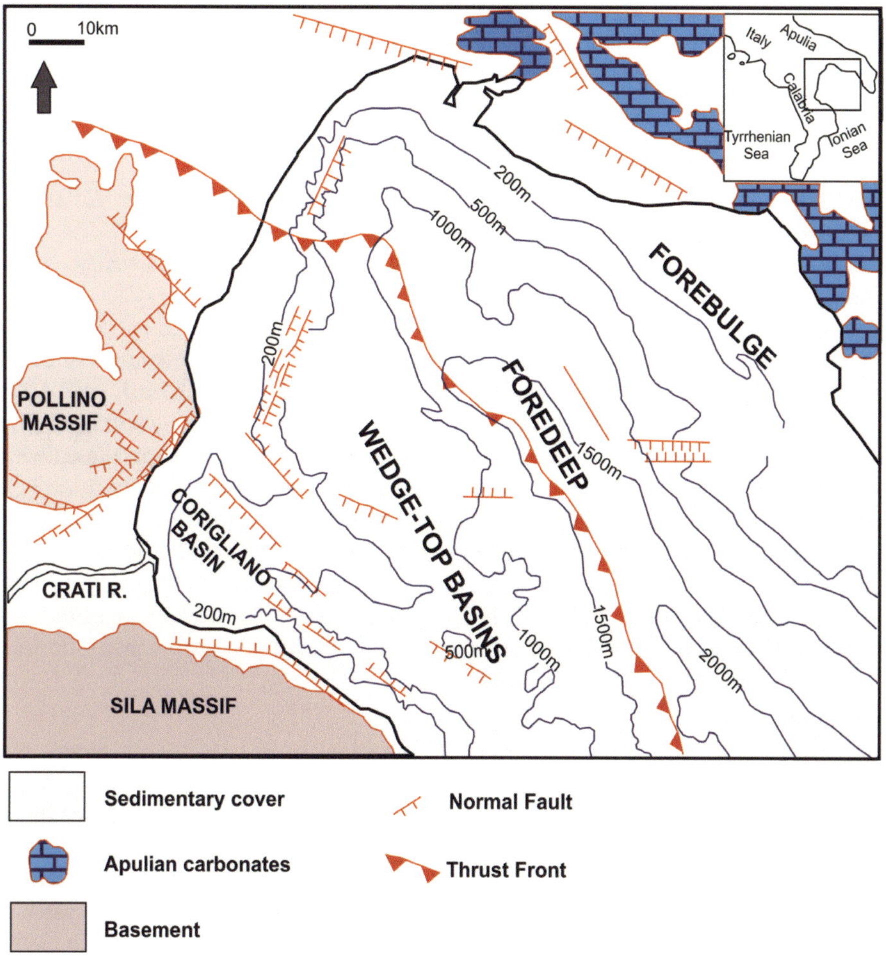

Fig. 7. Bathymetric and structural map of the Gulf of Taranto, which is the deep-marine part of the modern foreland basin system along the south-east side of the Apennines orogenic belt. The Corigliano basin represents the wedge-top depozone of the foreland region, in which the Crati Fan has been developing during the last 6000 y. Modified after Critelli and Le Pera (1998, 2003) and Critelli (1999)

The modern physiography and geology of Calabria are the results of post-30 Ma geodynamic processes in which synchronous accretionary processes were active along the eastern flank (northern Ionian Sea), and rifting processes along the western flank (Eastern Tyrrhenian Margin).

The subduction plane, as such as the southern Apennines and Calabrian accretionary prism, have migrated eastward or southeastward causing the roll-back of the subduction (e.g. Malinverno and Ryan, 1986; Royden et al., 1987; Doglioni, 1991; Gueguen et al., 1997, 1998). The roll-back of the subduction hinge (rate of hinge retreat is 6 cm/y; Royden et al., 1987; Patacca et al., 1993) appears to have been slowed and buckled during the Late Pleistocene by the interference of the thick continental lithosphere of the Adria Plate (Apulian swell) at the front of the belt (e.g. Doglioni et al., 1994, 1996).

The frontal active accretionary wedge, below sea-level, whereas the main elevated ridge to the west is in uplift and extension instead. The modern basin configuration of this thrust belt is represented by the wedge-top depozone (Corigliano-Amendolara basins), the marine and subaerial foredeep depozone (Gulf of Taranto and the Bradano river basin, respectively), the forebulge (the Gallipoli Basin) and the back-bulge (southern Adriatic Sea) (Figs. 6,7; e.g. Critelli and Le Pera, 1998).

Several Pliocene-Pleistocene basins cross-cut the Apennines and northern Calabria thrust pile, the most important are the Vallo di Diano, Val d'Agri, Potenza Basin, Mercure Basin and Crati Basin (e.g. Turco et al., 1990; Cinque et al., 1993; Colella, 1994; Tortorici et al., 1995; Schiattarella, 1998; Tavarnelli and Pasqui, 1998).

On the backarc area similar fault-controlled Pliocene-Pleistocene basins (Tortorici et al., 1995), as such as the Paola Basin and Gioia Basin, represent the synrift troughs of the eastern Tyrrhenian margin (e.g., Savelli and Wezel, 1980; Barone et al., 1982; Sartori, 1982, 1990).

Respect of low elevation, some calculations (Doglioni et al., 1996; Gueguen et al., 1998) show that the thickness of sedimentary strata in the Apennines exceed 20-25 km, the entire crustal thickness is about 30 km, and a thick pile of synorogenic sediment accumulation (up to 10 km from Miocene to modern) suggesting a delamination of the lithospheric mantle during Apenninic subduction (e.g. Channel and Mareschal, 1989; Doglioni et al., 1996).

The Calabrian ranges is peculiar for their high uplift rates that are 1 mm yr-1 (e.g. Cosentino and Gliozzi, 1988; Sorriso-Valvo, 1993; Westaway, 1993), where the maximum uplift is toward the frontal part of the accretionary prism (Ionian side) (Cosentino and Gliozzi, 1988).

2.5 Plate-tectonic evolution

The study area shows rocks which experienced a large series of geodynamic events occurred between early to middle Paleozoic orogenesis to actual.

The key geodynamic events into the Mediterranean region can be summarized as follow:

a) The Mesoalpine (Eocene to early Oligocene) tectonic phase in southern Italy corresponds with the subduction of the Adria-Ionian oceanic lithosphere beneath the Iberia plate (Fig. 8). This tectonic stage is responsible for the initial flexure, a general erosional processes of both the inner platform (Alburni-Cervati-Pollino-Bulgheria; Boni, 1974; D'Argenio, 1974) and outer platform (Monte Alpi-Apulia).

The Mesoalpine tectonic phase caused regional metamorphism at around 38 Ma (e.g. Steck and Hunziker, 1994), and intra-orogenic magmatism along the Periadriatic zone.

The Middle Oligocene (32-30 Ma) is characterized by intense magmatic activity, part of which is directly linked to the Algero-Provencal rift (Provence and Sardinia), part along the Insubric line and part along the periadriatic domain. In the Alps and northern Apennines, the Eocene and Oligocene siliciclastic sedimentary sequences record provenance from (a) Iberic plate (Corsica-Sardinia-Brianconnais), (b) Adria plate (austroalpine domain), (c) European plate, (d) syneruptive magmatic activity, and from (e) both European and Adria forebulges.

In the southern Italy domain, the Calabro-Lucanian Flysch Unit and the Sicilide Complex strata represent deposition in the remnant ocean basin related to the western subduction of the Adria oceanic lithosphere beneath the Iberia plate (Fig. 8; e.g. Knott, 1988; Dewey et al., 1989; Critelli, 1993; Guerrera et al., 1993; Critelli and Le Pera, 1998; Critelli, 1999). The subduction has been active for all the Paleogene and lower Miocene, producing an accretionary prism, the calabro-Lucanian Flysch Unit and the Sicilide p.p. Complex, and a diffuse calcalkaline volcanism in Sardinia. The Liguride Complex records the accretionary processes along the Adria margin and the consumption of the oceanic crust.

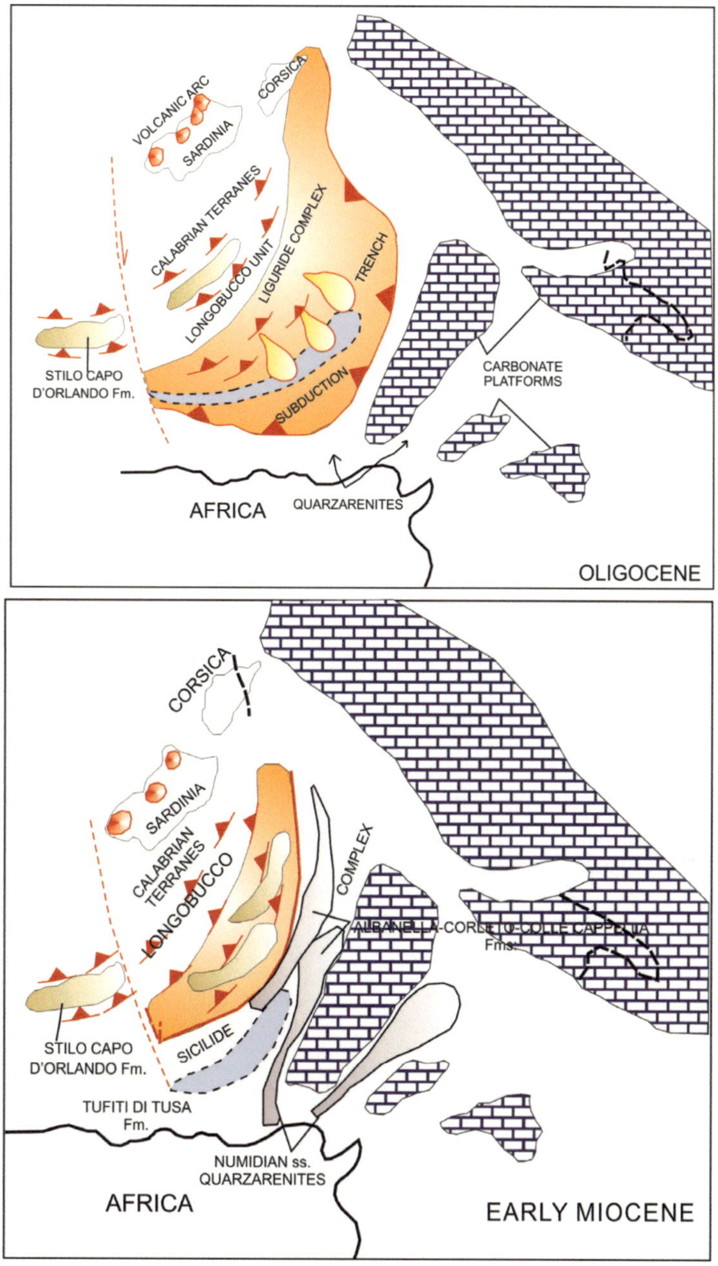

Fig. 8. Paleogeographic reconstructions (Oligocene to early Miocene) of the western Mediterranean during progressive closure of the remnant ocean basin and onset of collision in southern Apennines. Modified after Dewey et al. (1989), Critelli (1993, 1999) and Perri et al. (2011)

f) During early (Fig. 8) to middle (Fig. 9) Miocene the Apenninic domain is the place where immense volume of turbiditic sedimentation is in response of E-NE accretionary processes along the Adria plate (e.g. Ricci Lucchi, 1986; Patacca and Scandone, 1987; Boccaletti et al.,

1990). Here, the foreland basin system is developed over deformed Liguride Complex, during the early-middle Miocene, over Sicilide, Lagonegro and inner platform units during the upper Miocene (Fig. 10), over the previous units and the western margin of the Apulia platform during the Pliocene to Quaternary. The foreland basin system (wedge-top, foredeep, forebulge, back-bulge depozones) migrated in time, and siliciclastic and carbonatoclastic deposits, filling the wedge-top and the foredeep, where derived from progressive unroofing of the Calabrian crustal block or from erosion of the forebulge (e.g. Critelli and Le Pera, 1998).

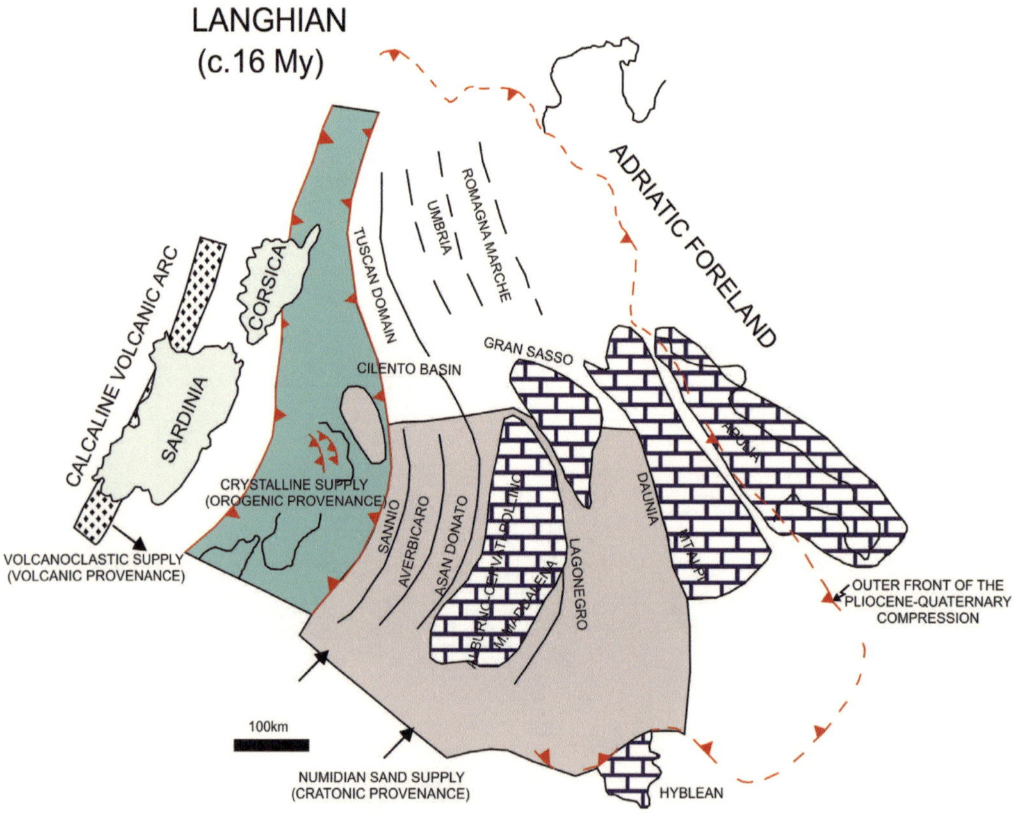

Fig. 9. Palinspastic restoration of the Apenninic domains during Langhian. Modified after Patacca et al.(1992) and Critelli (1999)

g) the geodynamic events of the last 10 My, in the western-central Mediterranean is named the Tyrrhenian phase (15-0 Ma) (Fig. 10). The Tyrrhenian phase (or back-arc extension) is responsible for the fragmention and dispersion of pieces of the Iberian and European plates (Calabria, Sardinia, Corsica), increased the displacement of the accretionary prism over the Adria plate, the eastward migration of the magmatic arcs, and the roll-back of the Adriatic lithosphere (Malinverno and Ryan, 1986; Patacca et al., 1990, 1993; Argnani et al., 1995; Doglioni et al., 1996; Gueguen et al., 1997, 1998). The Tyrrhenian backarc basin migrated eastward (northeastward in the northern Apennines and southeastward in Calabria and Sicily) at velocities of up to 5-7 cm/yr in the most arcuate parts of the arc (Doglioni, 1991; Gueguen et al., 1998).

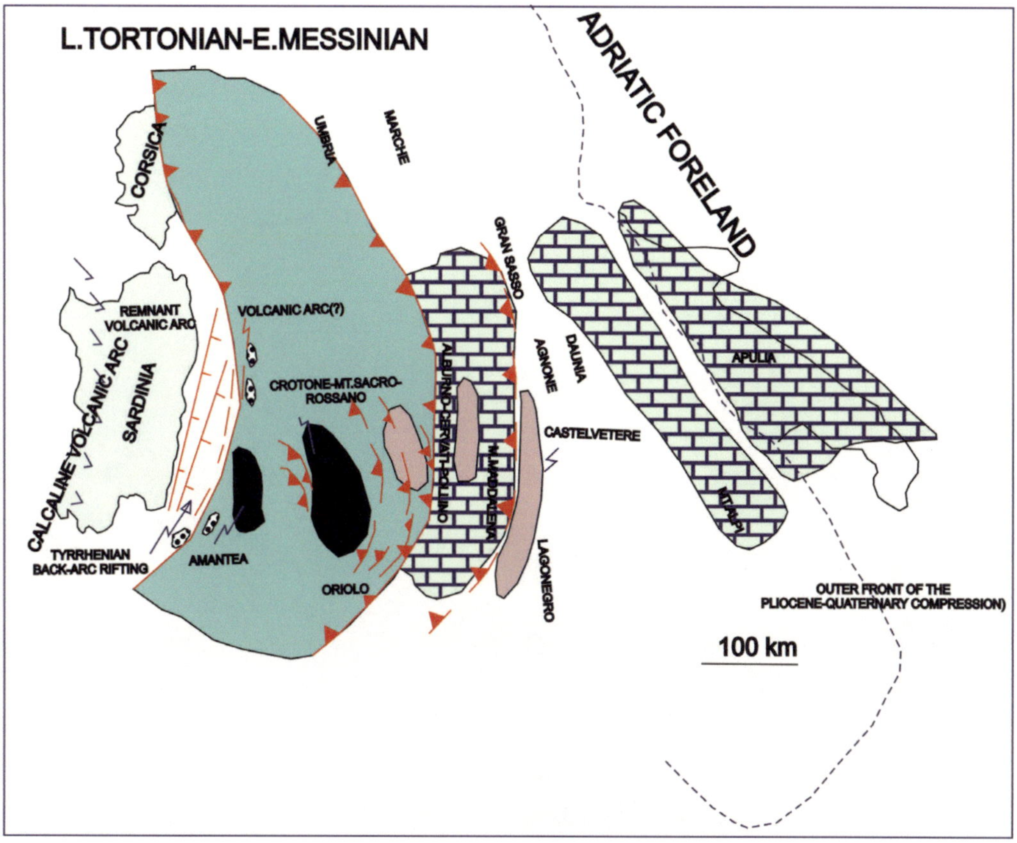

Fig. 10. Palinspastic restoration of the Apenninic domains during the late Tortonian to early Messinian (time interval 8-6 my). Initial back-arc rifting of the Tyrrhenian Sea, and diverse wedge-top and foredeep depozones in the southern Apennine foreland region. Modified after Patacca et al. (1990) and Critelli (1999)

2.6 Structural evolution and rise of the Calabrian terranes

The Paleozoic metamorphic and plutonic terranes of the Calabrian Arc represent the remnants of Caledonian, Hercynian and Alpine orogens (e.g. Amodio Morelli et al., 1976; Schenk, 1981; Zanettin Lorenzoni, 1982; Atzori et al., 1984; Del Moro et al., 1986; Zeck, 1990; Messina et al., 1994), that are drifted from the southern Iberic plate and accreted since upper Oligocene over the Adria-Africa lithosphere. They are a key tectonic element of the southern Italy orogen.

However, other authors consider the Calabrian basement terranes as a part of the Austroalpine domain of the African Plate (e.g. Haccard et al., 1972; Alvarez et al., 1974; Alvarez, 1976; Amodio Morelli et al., 1976; Scandone, 1979, 1982; Bonardi et al., 1982, 1993; Dercourt et al., 1986). In other alternative interpretations, the nappes of the Calabrian Arc originated from a microcontinent located between the European and African continents (e.g. Wildi, 1983; Guerrera et al., 1993; Critelli and Le Pera, 1998; Critelli, 1999; Mongelli et al., 2006; Perrone et al., 2006; Critelli et al., 2008; Perri et al., 2008; 2010) or the Calabrian-Arc terranes are the result of the amalgamation of three "crustal microblocks" (e.g. Vai, 1992).

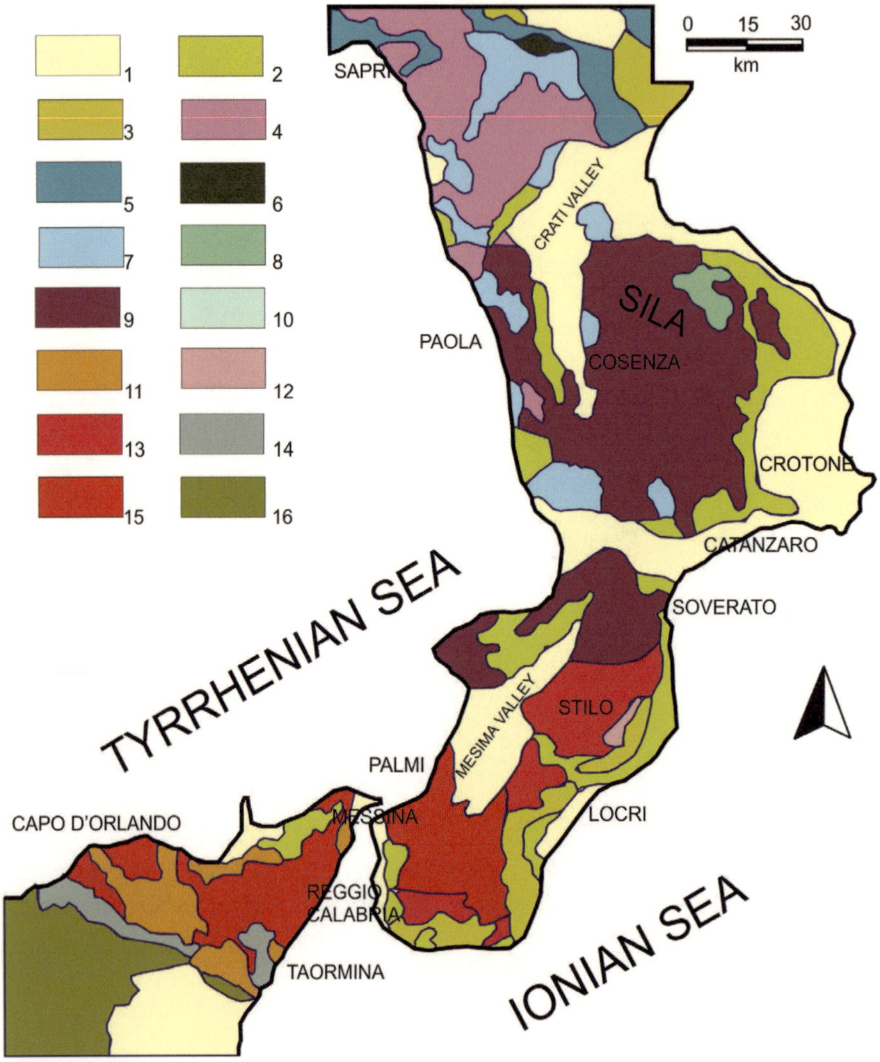

Fig. 11. Tectonic sketch map of the Calabria-Peloritani Arc. 1) Pliocene to Holocene sediments, and volcanic and volcaniclastic rocks; 2) Upper Tortonian to Messinian clastics and evaporites; 3) Cilento Group (Middle Miocene); 4) San Donato, Verbicaro and Pollino Units (Triassic to Miocene); 5) to 7) Liguride Complex: 5. Calabro-Lucanian Flysch Unit (Upper Jurassic to Upper Oligocene); 6. Ophiolitiferous blocks and M`elange; 7. Frido Unit (Upper Jurassic to Upper Oligocene); 8) Longobucco and Caloveto Groups (Lower Lias to Lower Cretaceous) and Paludi Formation (Upper Oligocene); 9) Sila, Castagna and Bagni basement Units (Paleozoic); 10) Malvito, Diamante-Terranova, Gimigliano Ophiolitiferous units (upper Jurassic to Lower Cretaceous); 11) Floresta Calcarenite (Middle Miocene), Stilo-Capo d'Orlando Formation (Lower Miocene); 12-13) Stilo Unit: 12. Carbonate rocks of the Stilo Unit (Upper Triassic? to Cretaceous) and 13. Basement rocks (Paleozoic); 14) Sedimentary Cover of the Longi-Taormina Unit (Upper Triassic to Oligocene); 15) Basement rocks (Paleozoic) of the Aspromonte, Africo, Mandanici, Fondachelli, Longi, Taormina units; 16) Sedimentary units of the Maghrebian Chain. Modified after Bonardi et al. (1993), and Critelli (1999)

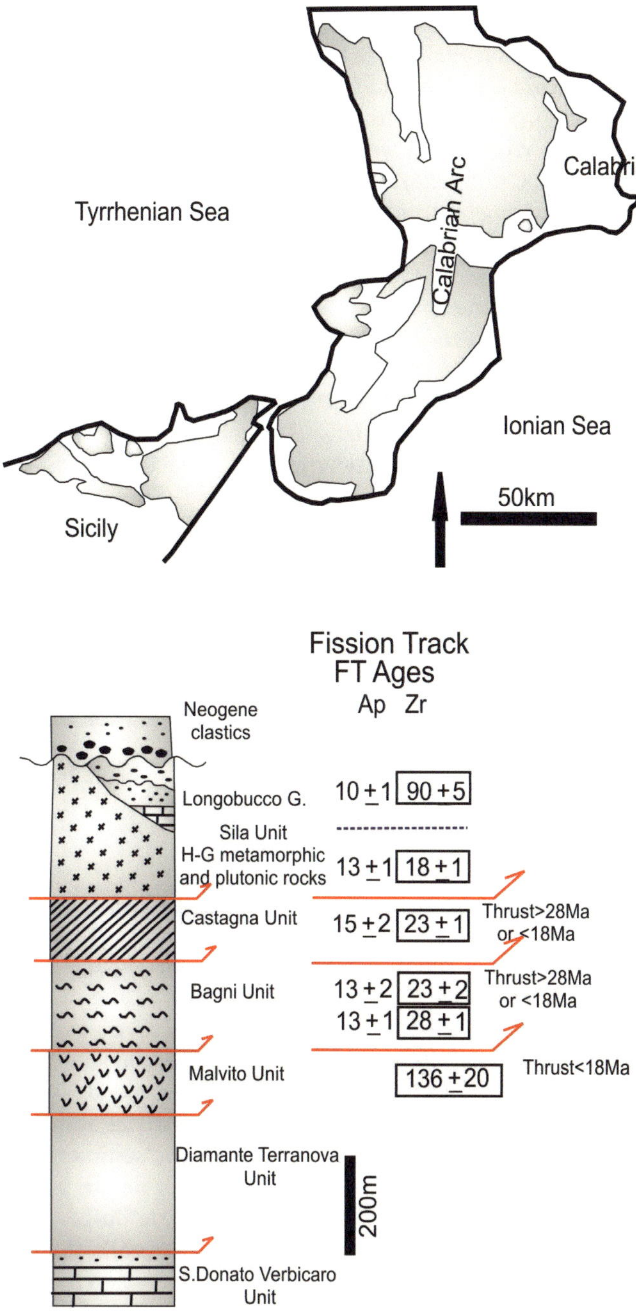

Fig. 12. Sketch map of the Calabria-Peloritani Arc, showing the outcrop of the ophiolitiferous and crystalline basement rocks, and simplified tectonostratigraphic terranes of the northern Calabrian Arc.Simplified tectonostratigraphy of the nappe sequence is modified after Amodio Morelli et al. (1976). Data of Apatite and Zircon Fission-Track ages is from Thomson (1998). Data on fission-track ages of the key thrusted terranes constraints times of tectonic accomodation of the nappe piles. See Thomson (1998) for further details

The northen Calabrian Arc can be divided into three stacked tectonostratigraphic assemblages (Figs. 11, 12; e.g. Amodio Morelli et al., 1976; Bonardi et al., 1976; Scandone, 1979, 1982; Cello et al., 1981; Bonardi et al., 1982; Colonna and Compagnoni, 1982; Colonna, 1998). The lowest is made of mainly carbonate rocks of Mesozoic age (D'Argenio et al., 1973; Ietto and Barillaro, 1993; Iannace et al., 1995; Perrone, 1996; Ietto and Ietto, 1998) that were originally deposited on the continental margins of the Apulia/Adria plate (Channel et al., 1979). These sediments were stripped from their basement during the Early Miocene collision of Calabria with Africa and Adria, and now form part of the Africa-verging Apennine fold-thrust belt. The middle tectonic units are composed of two nappes (Diamante-Terranova, and Malvito units; Fig. 12) of Mesozoic to Cenozoic metasedimentary and ophiolitic rocks, which can be interpreted as the remnains of the neo-Tethyan Ocean and related accretionary wedge (De Roever et al., 1974; Amodio Morelli et al., 1976; Lanzafame et al., 1979; Guerrera et al., 1993; Cello et al., 1996). The uppermost tectonic units consists of thrust sheets of Paleozoic igneous and metamorphic rocks (Bagni, Castagna and Sila units) and Mesozoic (Longobucco Group) to Cenozoic sediments (Figs. 11, 12), considered to be the basement and cover, respectively, of the former Iberian/Europe margin of Neotethys (e.g. Ogniben, 1969, 1973; Bouillin, 1984; Bouillin et al., 1986; Knott, 1987, 1988; Dietrich, 1988; Dewey et al., 1989; Thomson, 1998).
Thomson (1998), with fission track studies, demonstrates that the emplacement of continental basement rocks with Alpine metamorphism over ophiolitic rocks is constrained as a thrust contact of lower-to-middle Miocene age (<23 Ma), and the other major thrust contact of the diverse alpine basement units may be <18 Ma (Fig. 12). The relative cooling ages range from 35 to 15 Ma, where most of this phase of accelerated cooling can be attributed to increased erosion and progressive exhumation since 23 Ma to about 10 Ma (Thomson, 1994, 1998).

3. Sequential history of growing orogen in Southern Italy and clastic sediments in space and time

3.1 Pre-collisional and earliest collisional clastic units (Late Cretaceous to early Miocene)

In this large time occurred the main Alpine tectonic phases and final closures of the Piemontese-Ligurian oceanic basin to the north, whereas in the southern Italy, a more external remnant oceanic basin, the Lucanian oceanic basin, divides the Adria margin from the Mesomediterranean Microplate (e.g. Channel and Mareschal, 1989; Dewey et al., 1989; Guerrera et al., 1993; Critelli and Le Pera, 1998; Guerrera et al., 2005; Perrone et al., 2006; Critelli et al., 2008). The Adria plate experienced abrupt surficial and lithospheric changes, as changing nature of the pelagic basins, onset of siliciclastic sedimentation, huge volumes of cratonic quartzose sediments, emersion and erosion of carbonate platform domains, and deformation of the inner carbonate platform to form a forebulge (Patacca et al., 1992; Sgrosso, 1998). The oceanic lithosphere was subducting beneath the Mesomediterranean Microplate, with the Liguride basin representing the oceanic accretionary wedge and a diffuse calcalkaline volcanism was located in Sardinia.
Mesomediterranean Microplate- We include in the Mesomediterranean Microplate, the Calabria-Peloritani (CP, Fig. 13), Betic Cordillera (Alpujarrides, Rondaides and Malaguides) and Rif (Sebtides, Tetouanides and Ghomarides) (AL, Fig. 13), Tellian Maghrebides (Grande and Petite Kabilies, Eudough, Algers Massif) (KB, Fig. 13) (Wildi, 1983; Bouillin et al., 1986; Guerrera et al., 1993; 2005; Mongelli et al., 2006; Perrone et al., 2006; Critelli et al., 2008; Perri, 2008; Perri et al., 2008, 2010).

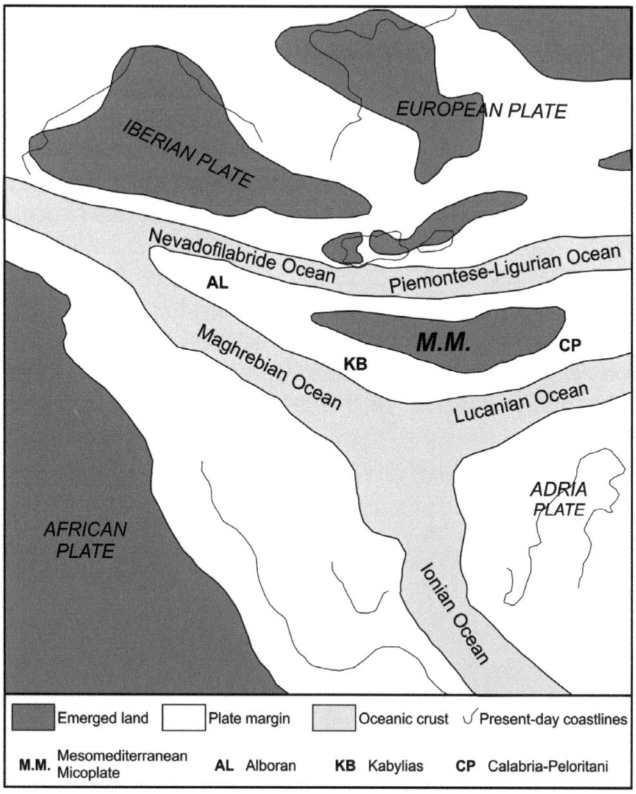

Fig. 13. Simplified Early Cretaceous palaeogeographical sketch of the Western Mediterranean Region (after Perrone et al., 2006 and Critelli et al., 2008, modified). Oceanic basins and margins around the Mesomediterranean Microplate are shown

During Paleogene to early Miocene, sediments over the Calabria-Peloritani Arc include the Upper Oligocene to lower Miocene Frazzanò Formation (de Capoa et al., 1997), in the Peloritani sector of the arc, and the upper Oligocene to lower Miocene Paludi Formation (northern sector) and Stilo Capo d'Orlando Formation (southern sector). These sandstones are quartzofeldspathic (Fig. 14; Zuffa and De Rosa, 1978; Puglisi, 1987; Cavazza, 1989; Nigro and Puglisi, 1993; Critelli et al., 1995b) and reflect their local provenance from crystalline rocks of the Calabrian terranes. The tectonic setting of these basins is complex; the sequences suturing some crystalline thrust units could represent a wedge-top deposition on advancing calabrian thrust-belt (e.g. Weltje, 1992; Patacca et al., 1993; Wallis et al., 1993) or may represent deposition in a forearc setting (e.g. Cavazza et al., 1997). An alternative interpretation is that they could represent remnants of deposition in foreland setting related to the back-thrust belt of the Betics-Alps orogen (e.g. Doglioni et al., 1997; Gueguen et al., 1997, 1998).
The oceanic area (Liguride basin) experienced deformation and accretion, involving in a remnant ocean basin (Fig. 8). A tectonic melange (Northern-Calabrian Unit; Critelli, 1993, 1999; Mongelli et al., 2010) was formed in this time frame, including olistholiths and broken formations of oceanic sequences (both basement and its pelagic sedimentary cover) and crystalline rocks (gneiss and granite) (Spadea, 1982). The subduction of the Adria oceanic lithosphere beneath the European plate, producing along the southern-end of the European plate a continental-margin calcalkaline volcanic arc in Sardinia (e.g. Scandone, 1982;

Malinverno and Ryan, 1986; Channel and Mareschal, 1989; Dewey et al., 1989). Cretaceous to Eocene quartzose sandstone (Crete Nere Formation; Bonardi et al., 1988; Monte Soro Unit; Barbera et al., 2011), and late Paleogene (upper Eocene to upper Oligocene) quartzofeldspathic and volcanolithic sandstones are tectonically assembled within the tectonic mélange (Northern-Calabrian Unit) (Figs. 14). Quartzofeldspathic sandstone was derived from mixtures of ophiolitiferous detritus and neovolcanic detritus. Syneruptive volcanolithic sandstones, having basaltic and andesitic fragments, reflect climax of activity of the Sardinia volcanic arc during its initial arc volcanism (late Oligocene, 32-30 Ma; Critelli, 1993). The overlying sequence, the Saraceno Formation (that caps the Liguride Complex), is unconformably over the Northern-Calabrian Unit, is mixture of siliciclastic and carbonatoclastic strata, that are hybrid arenites, lithic and quartzolithic sandstones (Figs. 14). These sandstones reflect a provenance evolution from sedimentary-dominant (both carbonate and siliciclastic fragments) detritus to metamorphic and sedimentary mixtures. This provenance evolution testifies the initial signal of accretion and unroofing of the frontal thrust system of the northern Calabrian terranes (Sila Unit; e.g. Messina et al., 1994).

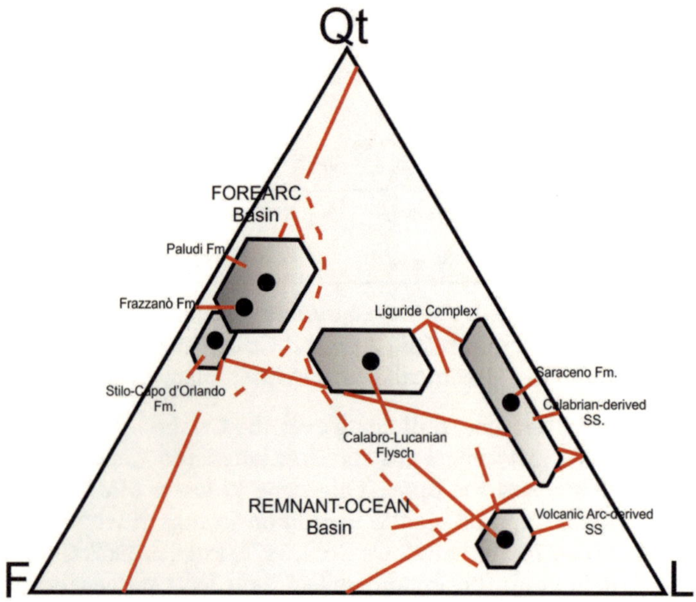

Fig. 14. QtFL plot (with superposed provenance fields of Dickinson, 1985) summarizing sandstone compositions from Paleogene to early Miocene sedimentary assemblages deposited during closure of the Liguride remnant ocean basin. Assemblages include the Calabro-Lucanian Flysch Unit (Tectonic Melange) and the Saraceno Formation of the Liguride Complex (remnant-ocean basin), and episutural (forearc basin) assemblages of the Paludi, Frazzan`o and Stilo-Capo d'Orlando formations. Data from Liguride Complex are from Critelli (1993); data from Paludi, Frazzan`o and Stilo-Capo d'Orlando Formations are from Zuffa and De Rosa (1978), Puglisi (1987), Cavazza (1989), Nigro and Puglisi (1993)

Adria Margin.- Siliciclastic sediments are rare or absent within basins located on Adria continental crust. The Paleogeographic scenario of this continental margin includes the outer carbonate domain (Apulia and Monte Alpi platforms), the Lagonegro basin, the inner carbonate platform, the Alburno-Cervati-Pollino units and the Sannio and Sicilide p.p.

basins (Fig. 9). These paleogeographic domains are characterized by thick (several thousands of meters in thickness) carbonate platforms, and their slopes, and pelagic basins. The pelagic basins receive dominatly Cretaceous to Upper Oligocene clays, marls and resedimented carbonate gravity flows (Patacca et al., 1992). There are no important traces of siliciclastic turbidite sandstones, the carbonate platforms record repeated emersions during the Cretaceous (Albian, and Albian-Cenomanian; e.g. Carannante et al., 1988a), showing pervasive karsification and deposition of bauxite deposits (locally up to 10 m). The bauxite deposits seems to be related to an intense weathering of original fine volcanic deposits (Carannante et al., 1988a).

During Paleogene, carbonate platforms and their slopes are characterized by thin stratigraphic sections, repeated emersions and non deposition (hiatus) intervals, that include the Eocene-Oligocene in the Alburno-Cervati-Pollino-Bulgheria units, all the Paleogene in the Monti della Maddalena unit and Apulia platform unit, and the Cretaceous to Early Miocene in the Monte Alpi Unit (Marsella et al., 1995).

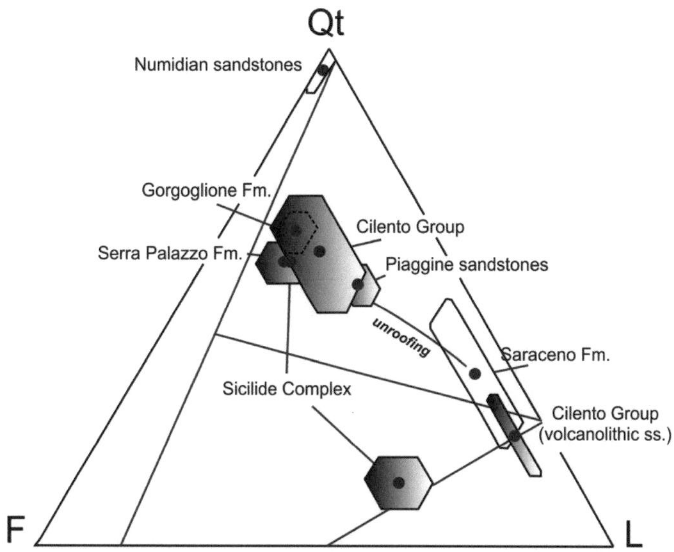

Fig. 15. QtFL plot (with superposed provenance fields of Dickinson, 1985) summarizing sandstone compositions from Burdigalian to Tortonian sedimentary assemblages deposited within the southern Apennines foreland basin system. Detrital modes of the Saraceno Formation shown as reference point of the onset of the unroofing history of the Calabrian terranes. The volcanic provenance from the volcanic arc is also recorded within the foreland basin sequences (Cilento Group and Sicilide Complex). Data from: Sicilide Complex is from Critelli et al. (1990, 1995b) and Fornelli and Piccarreta (1997); Cilento Group and Gorgoglione Formation is from Critelli and Le Pera (1994); Piaggine Sandstones and Serra Palazzo Formation is from Critelli and Le Pera (1995a)

Basinal sequences are dominantly pelagic, consisting of resedimented carbonate gravity flows (turbidite calcarenite to conglomerate debris flows and grain flows), representing carbonate slope aprons in the Lagonegro basin (Flysch Rosso Formation; Pescatore et al., 1988) and the Sicilide basin (Monte Sant'Arcangelo Formation; Lentini, 1979), interbedded with siliceous clays and shales. Rare siliciclastic or hybrid arenites are in the Paleocene sections of the Sicilide Complex (Monte Sant'Arcangelo Formation; Selli, 1962; Lentini, 1979). The sandstone is quartzolithic, including abundant quartz and metamorphic and sedimentary lithic fragments. Eocene to Oligocene Colle Cappella Sandstone Formation (the lower portion of the Nocara Flysch; Ogniben, 1969; Zuppetta et al., 1984), is a turbidite system including abundant sandstones that are quartzolithic (Critelli and Le Pera, 1998). These sandstones have very abundant low-grade metamorphic fragments, suggesting initial erosion and accretion of the Calabrian terranes. However, the Colle Cappella Sandstone Formation could reasonable be younger than proposed ages, and be considered as early Miocene in age (Aquitanian to Burdigalian; Figs. 8,15) (Critelli et al., 1994, 1995b).

During early Miocene an abrupt paleogeographic and geodynamic change occur along the Adria margin (Figs. 8, 9). Transgressive shallow-water calcarenite sediments were deposited on carbonate platform domains (Fig. 16; Selli, 1957; Carannante et al., 1988b; Patacca et al., 1992; Sgrosso, 1998). Within the Sicilide and Lagonegro basins a thick (up to 1000 m) quartzose turbidite sand, the Numidian Sandstone Formation (Patacca et al., 1992) represent the key signal of a mature quartzose (cratonic) provenance from the northern Africa continental margin (Fig. 9; Wezel, 1970a, 1970b; Patacca et al., 1992). This widespread quartzose material was deposited, during upper Burdigalian (?) to Langhian, within the nascent foredeep of the Sannio-Sicilide, on the forebulge of the Alburno-Cervati-Pollino units, and on the back-bulge Lagonegro depozones (Patacca et al., 1992). Syneruptive andesitic volcaniclastic layers are interbedded with the shallow-water calcarenites (Fig. 16), and quartzose sandstones testifying the volcanic activity on the Sardinia Arc (Patacca et al., 1992), as such as arkosic debris flows (cf. Carbone et al., 1987) recording signals of provenance from accreted crustal block of the Calabrian terranes.

The active volcanic source and the crystalline sources of the Calabria-Peloritani Arc are recorded within the Sicilide foredeep, forming distinct early Miocene siliciclastic turbidite systems having sand compositions ranging from volcanolithic (Tufiti di Tusa Formation) to quartzolithic and quartzofeldspathic (Albanella, Corleto and Colle Cappella Formations) (Fig. 15; e.g. Critelli et al., 1994; Fornelli and Piccarreta, 1997; Critelli and Le Pera, 1998; Perri et al., 2011). Volcaniclastic detritus, interbedded with quartzose sandstone strata, seem to be also deposited within the back-bulge Lagonegro depozone (Pescatore et al., 1988).

3.2 Early collisional clastic units (Late Burdigalian to Tortonian)

Final closure of the Liguride remnant ocean basin and onset of continental collision in the southern Apennines are dated as early Miocene (Burdigalian). The provenance of the detrital constituents of the Miocene foreland sandstones was dominantly from the Calabrian Arc terranes, the active growing front of the fold-thrust belt (Fig. 8). Nevertheless, folded and thrusted remnant oceanic sequences, active volcanics, and the forebulge of the flexed Adria margin were in time and space important detrital sources of the southern Apennines foreland basin system (Fig. 8).

The Source Areas of the Southern Apennines Foreland Basin System.- The key sources of clastics deposited within the foreland basin system include different present-day realms (morphotectonic zones), that are:

a. the Basement rocks of the northern Calabrian Terranes. Initial signals of the provenance is during final closure of the Liguride Complex (Saraceno Formation). During accretionary processes of, thrust units of the Calabrian terranes along the Adria margin (Figs. 11, 12).
b. the uplifted subduction complex (the Calabro-Lucanian Flysch Unit), during the mid-late Miocene (Figs. 15,).

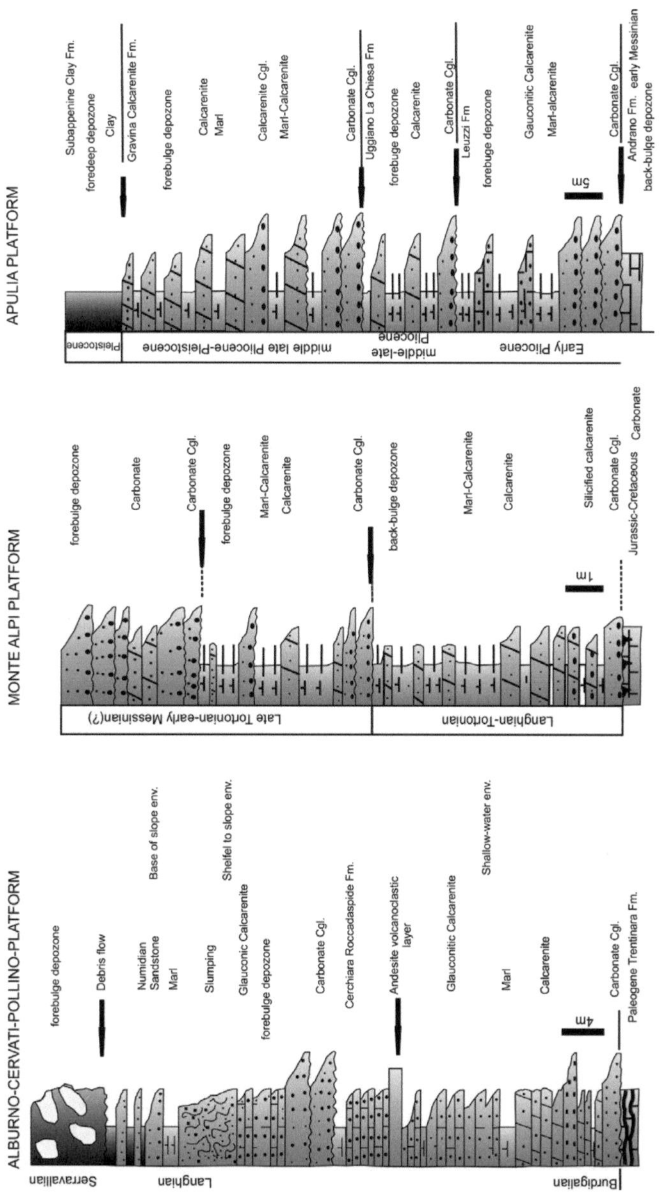

Fig. 16. Schematic columnar sections of the upper portions of the Adria carbonate platform domains, involved in flexural features during southern Apennines foreland basin system. The three sections correspond with the forebulge depozone sedimentation during (a) Burdigalian to early Tortonian (Alburno-Cervati-Pollino units or inner platform domain), (b) late Tortonian to Messinian (Monte Alpi unit), and (c) early Pliocene to the present (Apulia unit). Reference stratigraphic data for: the Alburno-Cervati-Pollino platform are from Carannante et al. (1988b), Critelli (1991), Patacca et al. (1992); the Monte Alpi Unit are from Sgrosso (1988), Taddei and Siano (1992), and personal unpublished data; the Apulia platform are from Ricchetti (1981), Ciaranfi et al. (1988), Ricchetti et al. (1988). Modified after Critelli (1999)

c. the Mesozoic to Tertiary Apulia/Adria basinal and platform domains (Lagonegro, Sannio, Sicilide units, the Alburno-Cervati-Pollino units, Verbicaro-San Donato-Bulgheria-Monti della Maddalena units, Monte Alpi unit, and the Apulia unit). The forebulge sources to the foredeep depozones, were the Alburno-Cervati-Pollino-Monti della Maddalena units from Burdigalian to Tortonian; the Monte Alpi unit, from early Messinian to lower Pliocene, and the Apulia unit, since Pliocene (Fig. 16).
d. An additional source of sediment is volcanic, that is mainly related to the calcalkaline volcanic arcs (between 32 and 11 Ma) was widespread along the western side of Sardinia (Cherchi and Montadert, 1982; Assorgia et al., 1986), or more recently to the intraorogenic alkaline volcanism in the Oligocene to Miocene.

3.3 Burdigalian-early Langhian Foreland basin system

The upper Sicilide Complex represents the oldest deposits of the foredeep basin (Critelli et al., 1995b). In the forebulge and back-bulge depozones, a widespread quartz arenite, the Numidian sandstone, as well as shallow-water calcarenite and thin volcaniclastic layers were deposited (Selli, 1957, 1962; Perrone, 1987; Carbone et al., 1987; Carannante et al., 1988b; Santo and Sgrosso, 1988; Patacca et al., 1992; Sgrosso, 1998) (Fig.16).
Siliciclastic strata of the Sicilide Complex (Fig. 8) include quartzolithic, volcanolithic and quartzofeldspathic sandstones. Thick (up to 300m) volcanolithic strata of the Tufiti di Tusa include syneruptive (e.g. Critelli and Ingersoll, 1995) andesite and basaltic andesite sandstones, recording climax of volcanic activity of the calcalkaline volcanic arc (Critelli et al., 1990; Fornelli and Piccarreta, 1997). Interbedded with carbonatoclastic sequences of the Adria forebulge (the Capaccio-Roccadaspide, Cerchiara Formations; Carannante et al., 1988a), similar andesitic volcanolithic sand testifies to wide dispersal of the neovolcanic detritus in the forebulge and back-bulge depozones (Pieri and Rapisardi, 1973; Perrone, 1987; Pescatore et al., 1988; Critelli, 1991; Patacca et al., 1992). Metamorphiclastic quartzolithic and quartzofeldspathic sandstones occur in the lower portions of the Tufiti di Tusa below the volcaniclastic strata (Critelli et al., 1990), and characterize the Corleto, Colle Cappella and Albanella formations (Fig. 15; Critelli et al., 1994; Fornelli and Piccarreta, 1997). They are derived from low to middle grade metasedimentary terranes, and are partly derived from ophiolitic rocks (Fornelli & Piccarreta, 1997). Interbedded thick carbonatoclastic (calcarenite-marl) strata within Tufiti di Tusa, Albanella and Corleto Formations, testify a provenance from the forebulge.
Langhian to Tortonian Foreland Basin System.- Since Langhian time, elongate turbidite basins have formed on top of advancing thrust-sheet systems (Fig. 9). The Cilento Group, Serra Palazzo, Piaggine, Gorgoglione, Sorrento, Castelvetere, Oriolo, San Bartolomeo formations, Monte Sacro, Serra Manganile and Nocara Conglomerate formations (Figs. 3) are the main turbiditic successions that were deposited after the completion of rotation of Corsica and Sardinia (19±1 Ma; Dewey et al., 1989), in progressively shifting wedge-top and foredeep depozones of the growing foreland basin system. Except for the Serra Palazzo Formation and the Sorrento Sandstone, the base of each succession is everywhere marked by an unconformable contact (Patacca et al., 1990; Sgrosso, 1998).
The Cilento Group (Fig. 17) ("*Cilento Flysch*" according to Ietto et al., 1965), Langhian to Tortonian in age (Amore et al., 1988; Russo et al., 1995; Zuppetta and Mazzoli, 1997), ranging from 2000 to 1200 m thick, rests unconformably on the Liguride Complex, and in turn it is unconformably overlain by the upper Tortonian Gorgoglione Formation, and the upper Tortonian to lower Messinian (?) Monte Sacro, Oriolo and Serra Manganile formations (Critelli et al., 1995b). The Cilento Group consists of different turbidite

depositional systems (Valente, 1993). In addition to siliciclastic turbidite beds, the Cilento Group includes numerous carbonatoclastic megabeds (ranging from few meters to 65 m thick; olistostrome beds (ranging from ten to hundreds of meters thick), and coarse volcaniclastic debris flows and turbidites. Sandstones (Fig. 15) of the Cilento Group are quartzolithic, volcanolithic and quartzofeldspathic (Critelli and Le Pera, 1994). Hybrid arenites and calcarenites characterize the carbonatoclastic megabeds. Sandstone strata of the lower portions are metamorphiclastic quartzolithic and quartzofeldspathic, resting on quartzolithic sandstone of the Liguride Complex.

Proportions of volcanic and plutonic detritus increase upward in the upper Pollica Formation and lower San Mauro, Torrente Bruca and Albidona formations. A volcaniclastic interval in the lower San Mauro Formation includes abundant felsic (rhyodacite to rhyolite) calcalkaline volcanic clasts (Critelli and Le Pera, 1994).

Sandstone of the upper Cilento Group is plutoniclastic quartzofeldspathic, consisting of abundant phanerites of plutonic and metamorphic fragments. In the upper Cilento Group thick carbonatoclastic and olistostroma megabeds record major tectonic events on both active thrust belt and forebulge (e.g. Critelli and Le Pera, 1994, 1998). Carbonatoclastic megabeds record huge volumes of sand-sized and mud derived from flexed Adria margin. These beds have impressive volumes and basinal lateral continuity (Colella and Zuffa, 1988; Cieszkowski et al., 1995). Olistostroma beds are siliciclastics, and include mountain-sized blocks of Calabrian terranes and Liguride Complex terranes (including also oceanic crust rocks; Ietto et al., 1965; Cocco and Pescatore, 1968; Carrara and Serva, 1982; Di Girolamo et al., 1992; Valente, 1991, 1993). Liguride-derived detritus appear only in the middle-upper Cilento Group, suggesting initial signals of the Liguride Complex emersion. Clear signals of the Liguride Complex emersion and erosion are recorded within the Piaggine Sandstone (Fig. 15; Serravallian to Tortonian; Sgrosso, 1981, 1998; Castellano et al., 1997). Quartzolithic sandstone of the Piaggine is derived from abundant Liguride Complex detritus (over than 50%;, suggesting that near the Serravallian-Tortonian boundary, the Liguride Complex was in a subaerial position, probably representing the emerged frontal thrust of the mountain belt (Critelli and Le Pera, 1995a, 1998).

The Serra Palazzo Formation (Selli, 1962; Ogniben, 1969) has been interpreted as the foredeep basin of the Langhian to Serravallian (Tortonian?) southern Apennines foreland region (Patacca et al., 1990). It has quartzofeldspathic sandstone (Fig. 15), hybrid arenite, and calcarenite, suggesting provenance from both thrust belt and forebulge. The middle-upper sedimentary succession, includes an olistostromal bed of carbonate clasts (olistoliths) (Loiacono and Sbarra, 1991) recording abrupt flexure along the passive margin.

Sandstone of the Gorgoglione Formation (Selli, 1962), upper Tortonian in age (Patacca et al., 1990), is quartzofeldspathic having similar provenance to that of the upper Cilento Group sandstone. The Cilento and Gorgoglione sandstone modes record accretionary processes of the Calabrian terrane, and initial unroofing of the crystalline terranes (Critelli and Le Pera, 1994, 1995a, 1998).

3.4 Late collisional clastic units (Late Tortonian to Messinian)

Southern Apennines Foreland Basin- The history of deep erosion of the Calabrian terranes is clearly recorded by Upper Tortonian to Messinian clastics (Fig. 10).

These foreland clastics, including Castelvetere, Monte Sacro, Oriolo, Serra Manganile, Nocara, and San Bartolomeo formations (Fig. 3), abruptly shift sand composition toward "ideal arkose" (e.g. Dickinson, 1985) or continental-block-derived sandstone (Fig. 18), suggesting deeply eroded Calabrian terranes. The previous forebulge of the Alburno-

Cervati-Pollino-Monti della Maddalena units, during the late Tortonian were assembled within the fold-thrust-belt, and the new forebulge of the foreland basin system might be located on the Monte Alpi Unit (Fig. 10; Patacca et al., 1992).

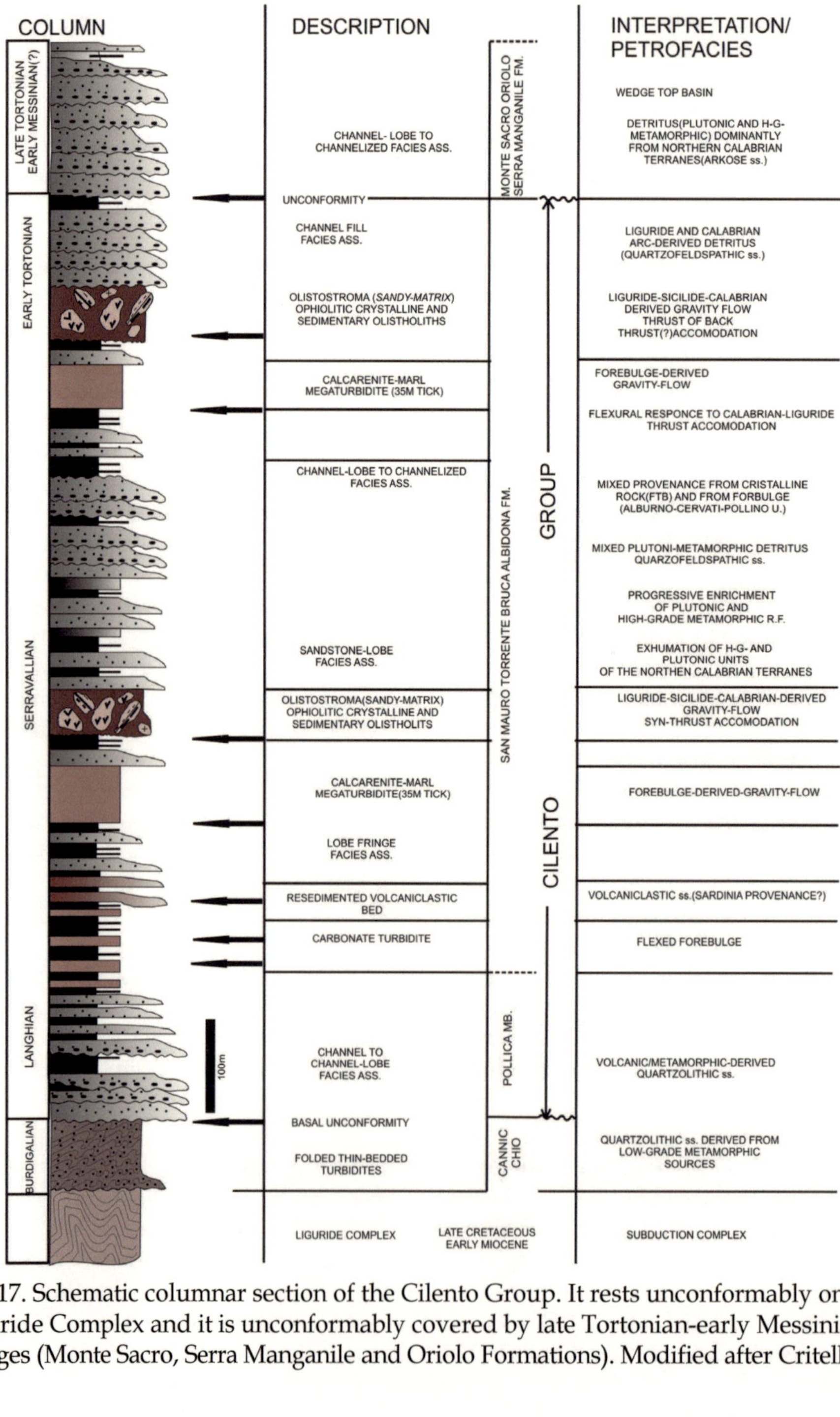

Fig. 17. Schematic columnar section of the Cilento Group. It rests unconformably on Liguride Complex and it is unconformably covered by late Tortonian-early Messinian clastic wedges (Monte Sacro, Serra Manganile and Oriolo Formations). Modified after Critelli (1999)

Forebulge sedimentation of the Monte Alpi Unit (cf. Sgrosso, 1988b, 1998; Taddei and Siano, 1992) is thin (20m to about 100m in thickness) and consists of shallow-water to coastal arenite, marl and carbonate conglomerate (Fig.16). Arenite and rudite of the Monte Alpi are dominantly composed of carbonate detritus. Arenites of this forebulge sequence, are pure to impure calclithite, composed of ancient extrabasinal carbonate grains (e.g. Zuffa, 1987) having Cretaceous to early Miocene tests, and the siliciclastic detritus includes rounded to subrounded quartz, plagioclase, radiolarian chert, fine grained quartz-siltite and quartzite, and rare serpentinite/serpentine schist and volcanic lithic fragments. Rare quartzolithic sandstones, having abundant quartz, carbonate lithic grains and plagioclase, are interbedded with the calcilithite strata. Plutonic and metamorphic detritus is absent in these arenites.

On the Alburno-Cervati-Pollino units, locally, thin arkosic sandstone strata unconformably overlain the Miocene forebulge sequence or are directly on Cretaceous to Paleogene carbonates (cf. Patacca et al., 1992; Sgrosso, 1998). These arkosic strata crop out on the Alburno-Cervati Mountains (Tempa del Prato Sandstone) and on the Pollino Mountains (Civita Sandstone) (Patacca et al., 1992; Sgrosso, 1998;, and include abundant plutonic and high-grade metamorphic detritus, as such as extrabasinal carbonate detritus.

Monte Sacro, Serra Manganile, Oriolo and Nocara Conglomerate formations (Figs. 5, 17) unconformably covering Liguride and Sicilide Complexes, and the Cilento Group, represent the wedge-top depozone sequences (Fig. 10, 17).

The Castelvetere Formation (Pescatore et al., 1970) has been interpreted as the foredeep basin (Patacca et al., 1990; Critelli and Le Pera, 1995b; Fig. 10). The Castelvetere has a thick olistostrome bed in the basal portions (Fig. 19), including mountain-block carbonate olistoliths (Pescatore et al., 1970; Pescatore, 1978; Carrara and Serva, 1982), that record involvement of the Langhian to Tortonian passive margin (e.g., the Alburno-Cervati Unit) within the thrust belt. Castelvetere sandstone modes are plutoni-metamorphiclastic, with up-section increases of sedimentary detritus (Critelli and Le Pera, 1995b). Sedimentary detritus is carbonate dominant in the lower Castelvetere; up-section increases of siliciclastic detritus suggests progressive erosion of older clastic wedges. Interbedded with quartzofeldspathic turbidite sandstone, the upper Castelvetere has a thick olistostrome bed composed of clastic detritus derived from Sicilide/Sannio and Liguride complexes, and a 1m thick volcaniclastic layer (Fig. 19). The siliciclastic olistostroma may be the signal of the syn-thrust accomodation of the Sicilide/Sannio Complex and possibly of the Liguride Complex. The syneruptive volcaniclastic layer consists of pyroclast fragments (pumice and shards) having felsic subalkaline composition (dacite) (Critelli and Le Pera, 1995b).

Messinian sandstones of the wedge-top and foredeep basins (Patacca et al., 1990, 1992; Sgrosso, 1998), the Monte Sacro, Serra Manganile, Oriolo, Nocara, Tempa del Prato, Caiazzo, San Bartolomeo, Agnone formations have homogeneous quartzofeldspathic compositions similar to the Castevetere sandstone (Fig. 18).

Northeastern Calabria Foreland Basin- The Tortonian to Messinian strata of the northeastern Calabria represent the more proximal portions of the southern Apennines foreland basin system. This sequence is directly unconformably over the Paleozoic plutonic and metamorphic rocks, or over the upper Oligocene to lower Miocene turbidite strata of the Paludi Formation (Fig. 19; Roda, 1967; Roveri et al., 1992), and it represents the basin fill of a wedge-top depozone (Critelli, 1999). These strata crop out along the piedmont of the Sila Massif, from the Trionto River (Rossano-Cariati zone) to south of the Neto River (Crotone zone) (cf. Cotecchia, 1963; Roda, 1964, 1967; Ogniben, 1973; Di Nocera et al., 1974; VanDijk, 1990; Van Dijk and Okkes, 1991; Roveri et al., 1992; Barone et al., 2008).

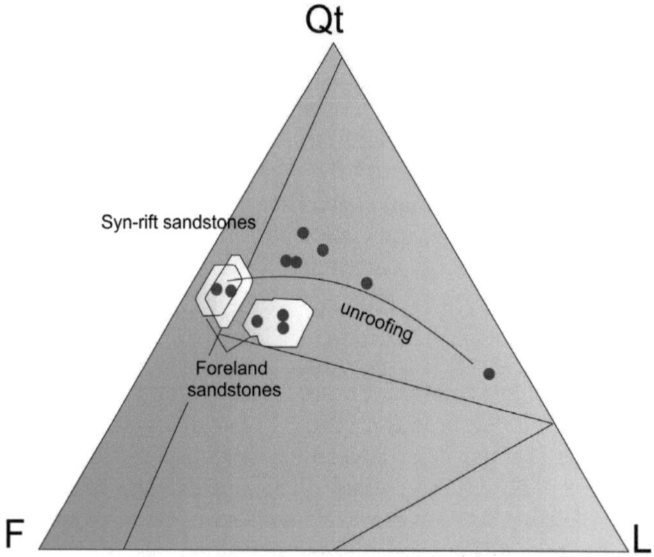

Fig. 18. QtFL plot (with superposed provenance fields of Dickinson, 1985) summarizing sandstone compositions from late Tortonian to early Messinian sedimentary assemblages deposited within the southern Apennines foreland basin system, and within the intermontane basins (northwestern Calabria) related to the backarc rifting of the Tyrrhenian Sea. Detrital modes of the Saraceno Formation, and Burdigalian to lower Tortonian foreland sandstones (only average values) are plotted as reference point of the unroofing history of the Calabrian terranes. Modified after Critelli and Le Pera (1998) and Critelli (1999)

Unconformably conglomerate and sandstone strata having rich macro-fauna (Clypeaster sandstone Formation, Cotecchia, 1963; or San Nicola dell'Alto Formation, Ogniben, 1955; Roda, 1964) represent the onset of the foreland basin system on advancing Calabrian thrust belt. These strata include diverse sedimentary facies associations, representing a depositional sequence (Roveri et al., 1992) and they are interpreted as a turbiditic system, having an overall fining and thinning upward trend, in the Crotone basin (where it is over 1000m in thickness; Roveri et al.,1992).

It represents the main reservoir of dry gas (Roveri et al., 1992) In the other areas, these strata include also continental strata (alluvial fans), nearshore and shallow-water deposits (area between Bocchigliero and Campana). These strata are overlain by fine-grained turbiditic systems and, toward the thrust culminations of the Sila Massif, by shelfal deposits. These strata, correspond with the Ponda Formation (Roda, 1964) of the Crotone basin, or the "Argilloso-marnosa Formation" of the Rossano basin (Ogniben, 1955) and may represent deposition during low-stand systems tract (Roveri et al., 1992).

The Rossano wedge-top depozone, during late Tortonian-early Messinian, abruptly receives huge volumes of Sicilide-derived olistostroma "Argille Scagliose Formation" (Ogniben, 1955, 1962) composed by variegated clay matrix and large blocks (olistholiths) of Cretaceous-Oligocene limestone, Miocene quartzolithic (similar to the Albanella-Colle Cappella sandstones) and quartzose sandstones (Numidian sands). These gravity flow deposits may be related to an out of sequence thrust accomodation or to a back-thrust of the Sicilide unit (Critelli, 1999). Infact, at the same time interval, within the foredeep depozone, the Castelvetere Formation has a similar olistostrome layer (Fig. 19); Critelli and Le Pera, 1995b).

On the successions of Rossano (north) and Crotone (south) Basins rests tectonically a sedimentary allocthonous succession defined named "Cariati Nappe" (Roda, 1967a; Ogniben, 1973). This succession is made up of turbiditic bodies with thinning-upward trend of Langhian-Serravallian in age, involved in backthrusting starting late Tortonian and involving the evaporitic and post evaporitic units in the Rossano Basin (Barone et al., 2008). The CN includes a Middle to Upper Miocene clastic succession unconformably covering an Oligocene to Burdigalian siliciclastic flysch. The Miocene and post Messinian emplacement of the so-called "Cariati Nappe" (CN) in the central sector of the area interrupts the lateral continuity and affects the sedimentary supply of a such configured wedge-top basin.

The Messinian sequence is characterized by evaporite deposits which record the Mediterranean salinity crisis. The evaporites consist mainly of gypsum and halite, followed by a thin mudstone interval, and thin clastic and evaporite beds (Ogniben, 1955; Roda, 1964; Romeo, 1967; Di Nocera et al., 1974). Overlying the evaporite sequence, an erosional unconformity marks the base of a Late Messinian to Pliocene depositional sequence within the Crotone Basin (Roveri et al., 1992). This depositional sequence consists of a basal conglomerate and sandstone strata with fining-upward trend (transgressive systems tract; Carvane Conglomerate Formation; Roda, 1964), overlain by basin-wide marine shales (high-stand systems tract; Marne argillose dei Cavalieri Formation; Roda, 1964) (Roveri et al., 1992).

The juxtaposition of authocthonous basinal successions (Rossano and Crotone successions) and allochthonous (Cariati Nappe) would suggest the detection, during the Serravallian-Tortonian, of the sedimentary basins developed in different contest; A basin on the inner set of the Arco Calabro Units which the western edge is well outcropping, and an outer external basin set on Sicilide units and Albidona formation.

Therefore, the Cariati Nappe would give the meaning of a backthruts of Tortonian age, related to the upper-middle Miocene accretionary phases that sharing the Foreland Basin system of the intersection of southern Apennines-Calabrian terrane.

Because of its sedimentary succession, the Cariati Nappe would include many tettonostratigrafic similarities with the sedimentary successions of the Upper Ionian Calabria and Lucania, which identify the area of the Montegiordano-Nocara-Rocca Imperiale ridge (Zuppetta et alii, 1984; Mostardini & Merlini, 1986; Patacca & Scandone, 1987, 2001; Carbone & Lentini, 1990; Cinque et alii, 1993; Critelli, 1999) where the successions of the Albidona and the hight portion of the Sicilidi Units, posed by the Argille Scagliose formation and Colle Cappella Sandstones, rests conglomeratic and arenaceous turbiditic successions belonging Serravallian-Tortonian of Oriolo Formation and Nocara Conglomerates Formation.

Synchronously with major tectonic events in the foreland thrust-belt, extensional tectonic activity affected the Tyrrhenian margin, just after the Tortonian compressive event; thereafter, evolution of the Tyrrhenian basin strongly influenced peripheral deformation of the Apennines foreland region (e.g., Malinverno and Ryan, 1986; Royden et al., 1987; Kastens et al., 1988; Lavecchia, 1988; Patacca et al., 1990; Sartori, 1990).

Since late Tortonian, the Calabrian terranes have provided abundant detritus to both the foreland region and intermontane basins of the backarc region (Figs. 3, 10) characterized by similar detrital provenances. Sandstones are "ideal arkose", and are identical in composition with the distal deep-marine upper Tortonian to Messinian foreland strata (e.g., Critelli and Le Pera, 1995a; Critelli et al., 1995b) (Fig. 18).

Marginal syn-rift strata of the Coastal Range of western Calabria (Amantea Basin) has Upper Tortonian to Messinian arkose, hybrid arenites and calcarenite, however, are similar to sandstone strata of the Crotone Basin (peri-ionian area) (Critelli, 1999; Barone et al., 2008).

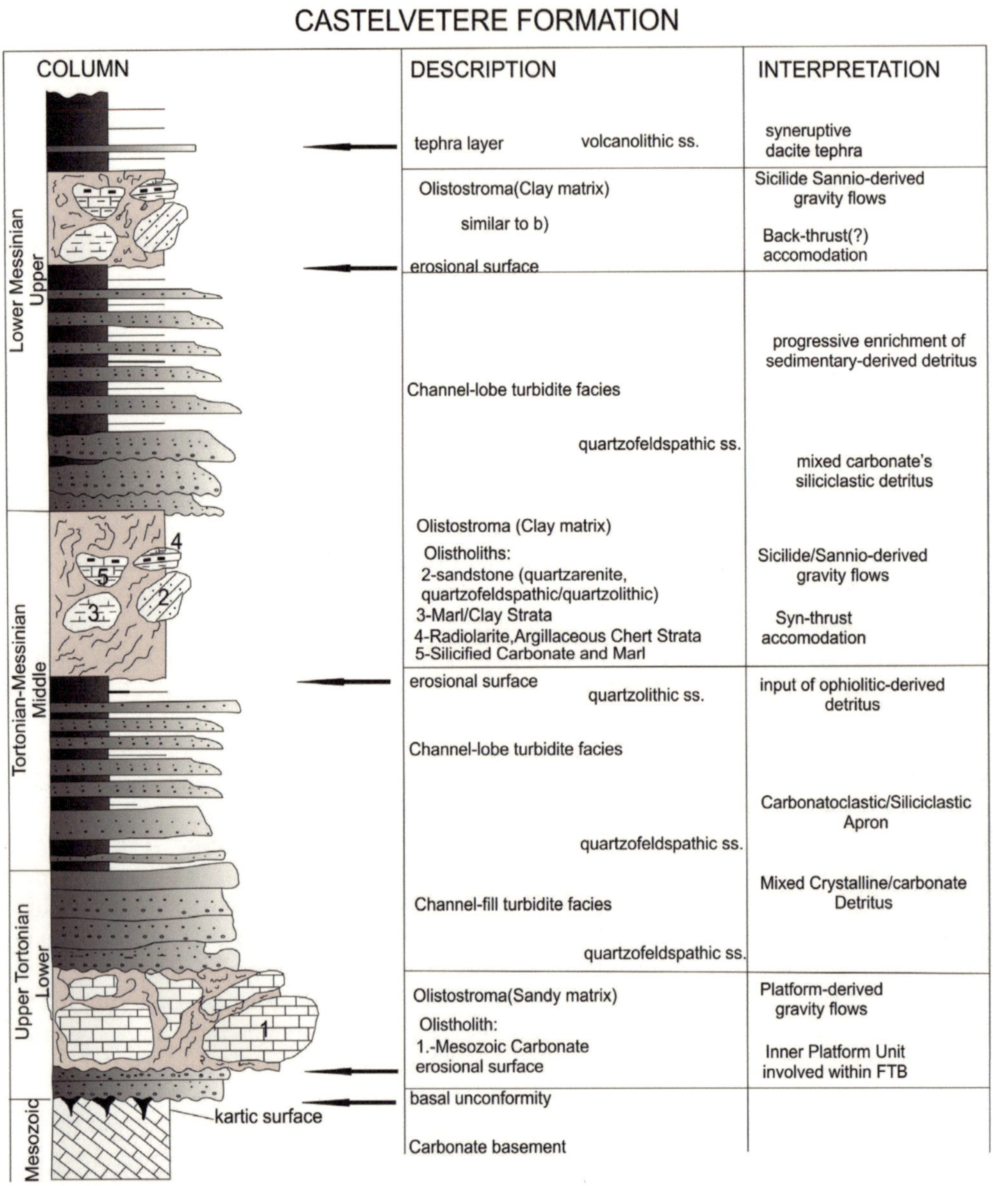

Fig. 19. Schematic columnar section of the Castelvetere Formation. It rests unconformably on Mesozoic carbonate platform unit. Reconstructed stratigraphy is from personal data and from Pescatore et al. (1970), Cocco et al. (1974), and Sgrosso (1998). Modified from Critelli (1999)

3.5 Pliocene to Quaternary clastic units

Southern Apennines.- During Pliocene, loading of the lithosphere by eastward thrusting of the Southern Apennines thrust-belt over the Adriatic plate resulted in flexural warping of the Apulia platform forming a downwarp, the Bradanic foredeep, an upwarp at the Apulia western edge, the Murge-Salento forebulge (Figs. 6, 16). The Bradanic foredeep basin was initially filled by early Pliocene pelagic and turbiditic sediments, and from middle-upper Pliocene to Pleistocene by hemipelagic clays (Argille Subappennine; Fig. 16), shallow-water calcarenite (Gravina Calcarenite Formation; Fig. 14), deltaic, coastal and alluvial clastics (e.g. Casnedi et al., 1982; Casnedi, 1988; Pieri et al., 1996). Clastics of the bradanic foredeep depozones reflect the erosion of the previous accreted clastic and carbonate thrust units of the southern Apennines fold-thrust-belt. The Sant'Arcangelo Basin was one of the wedge-top depozone of the Pliocene-Pleistocene foreland basin system (e.g. Hyppolite et al., 1994a, 1994b). Sediments of this basin show blended clasts from sedimentary thrust units, the uplifted subduction complex (Liguride Complex) and also from Calabrian arc. Most of the typical northern Calabrian provenances were trapped in piedmont areas of the Crotone and Rossano Basin, or within the Crati Basin. Because of the strike-slip movements were also active during Pliocene and Pleistocene in the southern Apennines realm, diverse fault controlled small basins were developed (e.g. Turco et al., 1990, Van Dijk et al., 2000, Tansi et al., 2007).

Estimated uplift rates are almost equal to late Quaternary denudation rates. Tectonics and climate have had a strong effect on the landforms of the Calabrian mountain ranges, resulting in the higher accumulation rates.

Additional Quaternary sediment sources for the Paola and Corigliano basins are active volcanic centres bordering the Paola Basin, and submarine structural highs, such as Amendolara embankment (Romagnoli and Gabbianelli, 1990), bordering the Corigliano Basin, producing reworking intrabasinal detritus. Quaternary sedimentation of both basins is strongly influenced by glacio-eustatic changes (Chiocci, 1994; Trincardi et al., 1995).

The Corigliano trough represents the Holocene submarine wedge-top depozone of the southern Apennines and northern Calabria foreland region (Pescatore and Senatore, 1986) (Fig. 7). It is morphologically characterized by a restricted shelf area, numerous gullies and canyons, and a submarine fan, the Crati Fan, developed during Holocene and connected with the torrential-type Crati delta on the shelf (Ricci Lucchi et al., 1984; Romagnoli and Gabbianelli, 1990).

The Crati River drains both the Calabrian crustal block to the west, east and south, and the southern Apennines Mesozoic to Tertiary sedimentary terranes to the north.

The Tyrrhenian margin of northern Calabria consists of diverse small coastal drainages, draining both Calabria continental block and the southern Apennines thrust belt, supplying sediments to the deep-marine Paola Basin. The basement of the basin consists of crystalline rocks of the Calabrian terranes or the upper Tortonian to Messinian sedimentary sequences. The basal unconformity is early Pliocene in age, and sediments of this age are bathyal (Fabbri et al., 1981). The main Pliocene to early Pleistocene unconformities seem to be related to the abrupt uplift of the Calabrian Coastal Range ("Catena Costiera"; e.g. Ortolani, 1978; Fabbri et al., 1981; Barone et al., 1982; Wezel, 1985).

Modern beach and fluvial sands of the Tyrrhenian margin of northern Calabria have three distinct petrofacies from north to the south, namely, (1) the calclithitic Lao petrofacies (at the northern end of the Paola Basin drainage area), having a provenance from the south-western flank of the southern Apennines slope, including dominantly Mesozoic carbonate rocks, (2) the quartzolithic Coastal Ranges petrofacies (in the central portion of the Paola Basin drainage area), having a provenance from dominantly metamorphic terranes (dominantly phyllite and schist, and gneiss) of the Coastal Ranges, and (3) the quartzofeldspathic Santa Eufemia Gulf petrofacies (at the southern end of the Paola Basin drainage area), having a provenance from metamorphic (dominantly gneiss, and phyllite and schist) and plutonic terranes of the Sila and Serre Mountains and from sedimentary terranes of the Catanzaro

Graben. They represent the actualistic petrofacies of the mainland areas of the deep-marine Paola Basin (e.g. Le Pera and Critelli, 1997; Critelli and Le Pera, 2003).

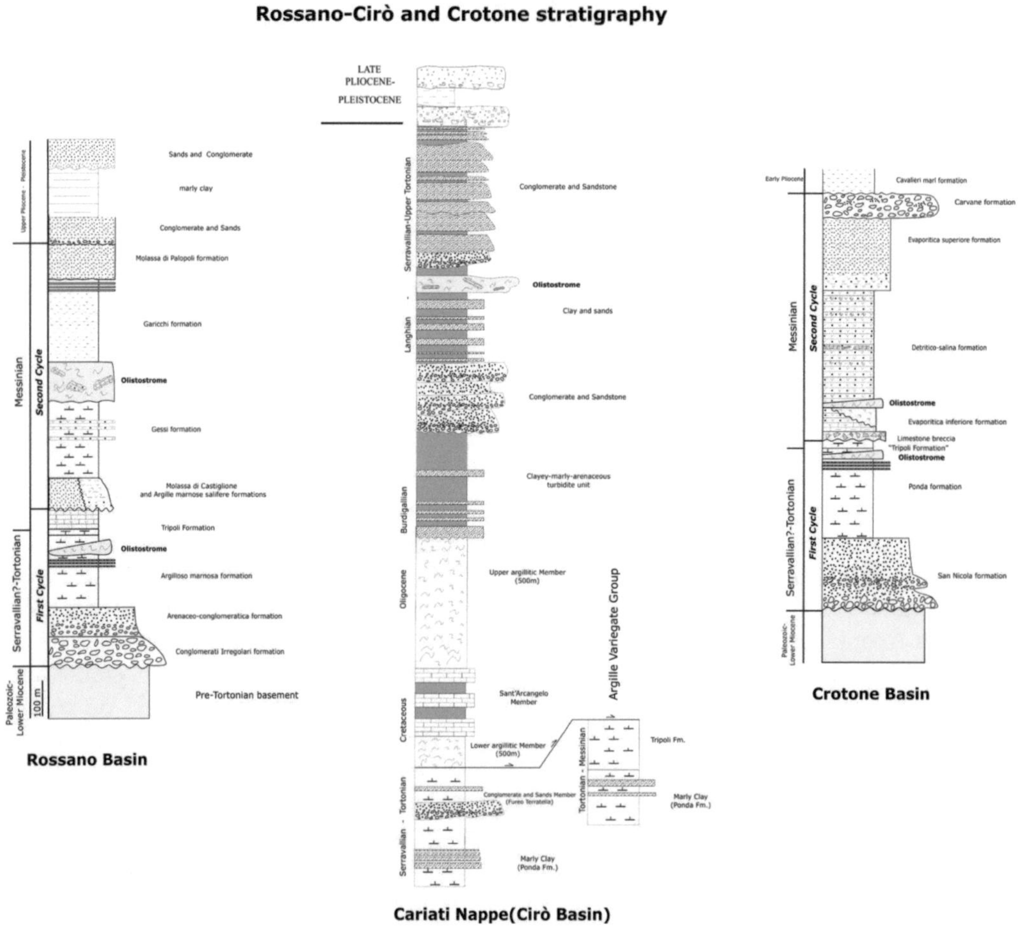

Fig. 20. Schematic columnar sections of the Crotone-Cirò and Rossano basins. The sedimentary successions rest unconformably on Paleozoic plutonic and metamorphic rocks of the northern Calabrian Arc (Sila Unit), or on Oligocene to lower Miocene Paludi Formation. These sequences represent the more proximal late Tortonian to Pliocene strata (wedge-top depozone) of the southern Apennines foreland basin system. Reconstructed stratigraphy is from personal data and from Ogniben (1962), Roda (1964, 1967), Romeo (1967), and Roveri *et al.* (1992). Modified after Critelli (1999)

Late Quaternary turbidite sands of the Paola Basin have distinct petrofacies (Fig. 21), that are: (a) a quartzolithic petrofacies, including also calclithitic turbidite sands, and (b) a volcanic-rich petrofacies, including distinctive syneruptive volcaniclastic sands (Critelli, 1999).

The quartzolithic sand petrofacies widely occurs into the Paola Basin and it is strictly related to the composition of the Coastal Range littoral province. At the northern end of the Paola Basin, distinct sedimentaclastic (calclithitic) turbidite sands, reflects a provenance from the Lao littoral province (Le Pera and Critelli, 1997; Le Pera, 1998; Critelli and Le Pera, 2003).

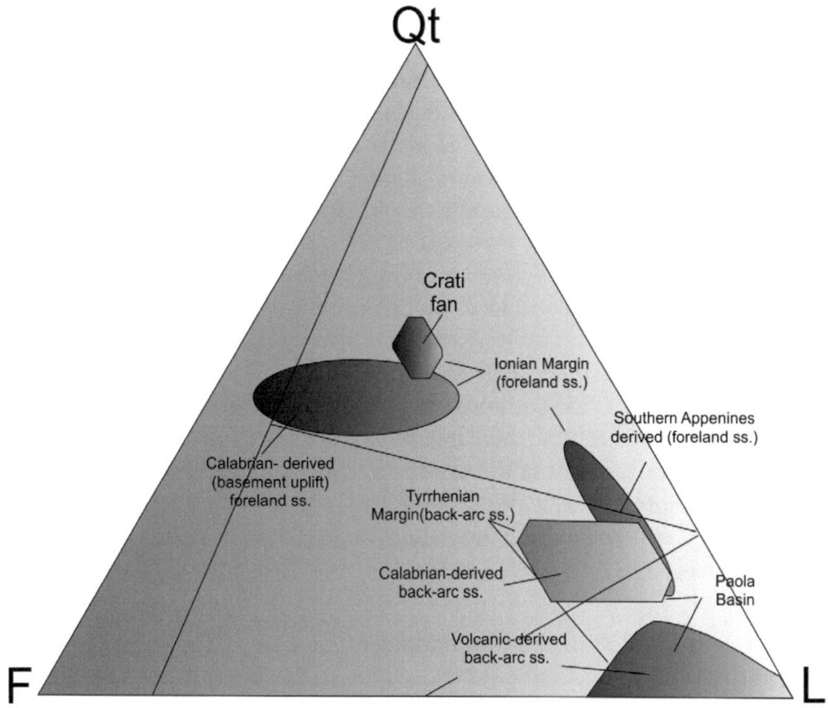

Fig. 21. QtFL plot (with superposed provenance fields of Dickinson, 1985) summarizing sand compositions for upper Pleistocene to modern sands of the Crati Fan, the Paola Basin (turbidite sands), and beach and fluvial sands of both Ionian and Tyrrhenian margins of northern Calabria. Sands from Ionian margin include the Crati Fan turbidites, the Crati River and Delta, the drainages derived from the Sila Massif and from the southern Apennines thrust belt. Sands from Tyrrhenian margin include Paola Basin turbidites, tephra layers interbedded with turbidites, coastal and fluvial systems of the Coastal Range. Data from Le Pera (1998) and Critelli and Le Pera (2003). Modified after Critelli (1999)

The volcanic-rich sand petrofacies also well represented within the Late Quaternary stratigraphic sequence of the Paola Basin. This petrofacies includes two main syneruptive volcaniclastic turbidites, one is located close to a datum plane at 20,000 y (calcalkaline volcanic provenance), and the upper one (alkaline volcanic provenance) is at the top of the basin-fill.

3.5 The climax of accretionary processes and evolution of Foreland basin system

The unroofing history of the Calabrian terranes, started during final closure of the Liguride basin, abruptly increased during accretionary processes over the Adria margin, occurred during early Miocene (Fig. 8). Increasing detrital feldspars and metamorphic detritus in the early Miocene sandstones (lower Cilento Group; upper Sicilide Complex) suggest dissection of the frontal terranes of the northern Calabrian arc. Local huge arrivals of volcaniclastic detritus testifies the climax of activity of the calcalkaline volcanic arc of Sardinia (Fig. 8).

In middle Miocene (Serravallian to Tortonian) sandstone detrital modes recorded a major change from lower Cilento Group to upper Cilento Group and Gorgoglione Formation (Fig.

14), with marked increased in detrital feldspars, medium- to high-grade metamorphic and plutonic rock fragments (Critelli and Le Pera, 1994). This compositional change, related to rapid northeastern movement of the Calabrian terranes and thrust accomodation of the high-grade Hercynian metamorphic rocks (Fig. 12), reflects rapid rise of Calabria and sharp increase in denudation rates, as documented also by fission-tracks (e.g. Thomson, 1998).
At this stage, the Adria (Alburno-Cervati-Pollino-Monti della Maddalena units) forebulge was involved in tectonic deformation and assembled within the orogenic belt (Fig. 10); the Liguride subduction complex, that was part of the deep duplex system, locally emerges producing abundant detritus to the foreland (Piaggine Sandstone and Olistostroma beds of the upper Cilento Group; Critelli and Le Pera, 1995a, 1998) (Figs. 15, 17).
In upper Miocene (late Tortonian to Messinian) sandstone detrital modes recorded an other major change (increasing feldspars, high-grade metamorphic and plutonic fragments), and the composition shift toward "ideal arkose" (Fig. 17; Critelli and Le Pera, 1995a). This time is also marked by the onset of the Tyrrhenian rifting on the back of the orogenic belt, causing an increased eastern displacement of the thrust system (e.g., Cello et al., 1981, 1989; Carbone and Lentini, 1990; Patacca et al., 1990; Lentini et al., 1994; Sgrosso, 1998). This other compositional change of the foreland sandstones reflects an increasing of the uplift rates, and deep erosion levels into mid-crustal rocks along the core of the Calabrian thrust belt.

3.6 Discontinuous migration of flexural features

The syntectonic Miocene stratigraphic succession indicates episodic, eastern migration of the forebulge. The position of the flexural forebulge did not progress continuously eastward through time but appeared to stall at its initial position from Langhian to Tortonian during deposition of the Cilento Group, Gorgoglione Formation and “Cariati Nappe” succession (Fig. 9). Only during the upper Tortonian to Messinian the forebulge moved rapidly eastward during deposition of Castelvetere and San Bartolomeo formations (Fig. 10). Possible reasons for discontinuous migration of flexural features that may apply to the southern Apennines foreland are episodic migration of the thrust load and inhomogeneitis within the lithosphere (e.g. Boyer and Elliot, 1982; Waschbusch and Royden, 1992; Giles and Dickinson, 1995; Critelli, 1999).
The buildup and migration of an accretionary prism includes progressive cratonward outstepping of the thrust front incorporating new material into the thrust load as it migrates (e.g. Giles and Dickinson, 1995). Thrust loads may also build up by almost vertical stacking of thrust sheets along ramps within the hinterland, producing the critical taper needed for the thrust system to migrate (e.g. Boyer and Elliot, 1982). In this case, flexural features would not continuously migrate cratonward because the thrust load itself is not continuously migrating cratonward. Waschbusch and Royden (1992) suggest that discontinuous migration of flexural features may also be caused by inhomogeneities within the lithosphere that fix the position of the forebulge to a weak segment of the foreland lithosphere.
Flexed features of the southern Apennines foreland basin system change discontinuously during the last 23 my. The forebulge did not migrate from early Miocene to the Tortonian and it may have been fixed to a weak zone corresponding to the former miogeoclinal shelf margin of the Alburno-Cervati-Pollino inner platform domain (Figs. 9,16; Patacca et al., 1992). The Calabrian allochthon terranes and associated subduction zone progressively migrated toward the fixed forebulge until stresses reached a threshold during late

Tortonian-early Messinian (Fig. 10), the time of deposition of the Castelvetere Formation (Critelli and Le Pera, 1995b).

At this time the forebulge migrated rapidly eastward causing deformation of the former back-bulge basin strata (Lagonegro basin). The inferred new forebulge may be the inner Apulia platform unit (or Monte Alpi Unit; Figs. 10, 16). The Monte Alpi Unit has an unconformable lower Messinian carbonatoclastic sequence (Figs. 16), that is transitional to shallow-water (cf. Sgrosso, 1988a, 1988b, 1998; Taddei and Siano, 1992), representing remnants of deposition on the forebulge. The last eastward forebulge migration occur during Pliocene, the new forebulge is the Apulia platform (Fig. 16), and the foredeep is the Bradanic trough (Fig. 6; e.g. Ricchetti, 1980; Ricchetti and Mongelli, 1980; Casnedi et al., 1982; Critelli, 1999).

4. Conclusions

In foreland settings, subsidence and uplift are profoundly affected by lithospheric flexure. Foreland basin subsidence is primarily controlled by downflexing of the lithosphere in response to thrust accommodation and loading. The interrelationships between lithospheric flexure, single thrust accomodation within the accretionary wedge and flexural subsidence experiences geometrically complexes entities within the foreland region (e.g. Critelli, 1999).

This chapter has examined clastic sediments and interpreted many stratigraphic sequences that were deposited in the southern Apennines foreland basin system during the complex orogenic history of the western Mediterranean, suggesting that interplay of lithospheric flexure and thrust accomodation were important factors in controlling accomodation trends.

From late Paleogene to the present the siliciclastic sedimentary sequences of southern Italy filled basins that are directly related to this convergent setting, causing consumption of the oceanic lithosphere, and subsequent accretion of the Calabrian allochthonous terranes over the Adria-Africa plate generating post-Oligocene foreland basin systems.

Earliest onset of continental accretion on Adria margin occurred during late Burdigalian-early Langhian; foreland clastic strata of the upper Sicilide Complex were derived from dominantly metasedimentary and related sedimentary covers rocks of the frontal Calabrian terranes (e.g. Critelli, 1999). Sudden influx of neovolcanic detritus suggests continuing provenance from the active volcanic arc that was possibly hundred kilometers distant (Fig. 8).

Thick foreland clastic sequences formed during Langhian (Fig. 9) over accreted Liguride and Sicilide Complexes.

Langhian to Serravallian detrital modes of the Cilento Group (Fig. 17) have abrupt changes from quartzolithic (phyllite and schist source rocks) sandstone to quartzofeldspathic (plutonic and gneissic source rocks) sandstone. This change in detrital modes was accompanied by interbedded carbonatoclastic detritus derived from abrupt flexure of the Alburno-Cervati-Pollino forebulge (Figs. 17). These major changes in foreland clastic deposition occurred *c.*16-15 Ma (Critelli and Le Pera, 1994), the time of exhumation of the northern Calabria crystalline rocks (Thomson, 1994, 1998). Signals of sandstone detrital modes and fission tracks on source rocks indicate increasing uplift between 15 and 10 Ma, after which deep dissection affected the northern Calabrian Arc.

During Serravallian to lower Tortonian (15 to about 10 Ma), abundant ophiolitic and pelagic detritus within the foreland basin (upper Cilento Group and Piaggine Sandstone) record emersion and erosion of the Liguride and Sicilide complexes. At *c.*11 Ma, volcanism of western Sardinia abruptly terminated, rendering Sardinia into a remnant arc.

During Late Tortonian to Early Messinian, abrupt changes in the southern Apennines foreland region occurred. The former forebulge (e.g., Alburno-Cervati-Pollino units) was involved in tectonic deformation and incorporated within the thrust belt, and an eastward shifting of the foredeep depozone occurred. Foredeep and wedge-top sandstone strata shifted in composition toward continental block or arkose, suggesting major uplift of the Calabrian thrust belt. Coeval thrust accomodation of Sicilide-Sannio units, recorded as large gravitative deposits within foredeep and wedge-top depozones (Figs. 19, 20) and recycling of older sedimentary sequences are important signals of accretionary processes within the thrust-belt. Signals of post-8 Ma volcanism are recorded in the foreland sequences, but sources are unknown.

Since Late Tortonian (10-8 Ma; Fig. 10), backarc rifting has produced the Tyrrhenian Sea. At this time, the northern Calabrian Arc has been the western border of the northern Ionian foreland region and the eastern margin of the Tyrrhenian backarc basin (Fig. 10). Intermontane and syn-rift basins of the western Calabria and proximal and distal foreland basins of eastern Calabria and southern Apennines have identical sand composition, plotting within ideal arkose or continental block provenance field (e.g., Dickinson, 1985, 1988) (Fig. 18). The maximum rate of foreland thrust advancement (8 cm yr^{-1}) occurred in late Tortonian to Messinian (Patacca et al., 1990); major changes in uplift rate in the northern Calabrian Arc correspond with the abrupt change in sandstone detrital modes (Critelli and Le Pera, 1995a; Critelli, 1999; Barone et al. 2006).

Upper Tortonian to Messinian nonmarine to shallow-marine and deep-marine successions, cropping out on the western and eastern Calabrian Arc, representing synrift clastic wedges related to backarc rifting in the peri-Tyrrhenian area (western sequences), or foreland clastic wedges in the peri-Ionian area (eastern sequences; Barone et al., 2008).

Pliocene and Quaternary of the northern Calabrian Arc, represented by foredeep and related wedge-top basins on the eastern side (Gulf of Taranto and Corigliano Basin), and a slope basin on the western side, the eastern Tyrrhenian margin (Paola Basin). These receive detritus primarily from deep erosion of northern Calabria. The modern deep-marine basins of offshore northern Calabria have many similarities to the middle to upper Miocene clastic sequences in both foreland and backarc regions of the southern Italy. The type of sedimentary provenance of the southern Italy foreland basin system, providing an example of the changing nature of the orogenic belt through time, may contribute to general understanding and application of other major orogens.

Quaternary erosional rates for same areas of Calabria are over 200 mm/Ka with maximum values exceeding 800 mm/Ka (Ibbeken and Schleyer, 1991). The enormous sediment production of Calabria crosses the river-mouth areas and the beaches, and is spread over shelf and slope or is transferred via submarine canyons to lower bathyal plains of the Ionian seas. The Corigliano Basin, in the Ionian Sea, record the sedimentary processes acting on the northen Calabrian Arc terranes and can be considered as a modern analogue of /thrust-belt/foreland transect. During Quaternary, in the southern Apennines thrust-belt, other major thrust accomodations, as such as normal faulting, sinistral strike-slip movement, block rotation and strong uplift occur defining the morphotectonic zones of the orogenic belt. The northern Ionian Sea and the Bradanic River basin represent the submarine and subaerial foredeep depozone, respectively. The Corigliano-Amendolara submarine basins, and the Sant'Arcangelo basin represent the wedge-top depozones,. The forebulge is finally located to the western Apulia platform, while the back-bulge depozone is located to the southern Adriatic Sea (Critelli, 1999).

The flexed foreland lithosphere and its forebulge has played an important rule to the development of the foreland basin system. The initially formed forebulge is interpreted to have been in the Alburno-Cervati-Pollino domain, and the final position of the forebulge is the present Apulia platform domain (Figs. 6, 7, 16). Thus the forebulge migration distance was over 150 km. Forebulges contributed to the sediment supply within the foreland basin, even if it is minor with respect to the orogenic provenance. Forebulges have produced istantaneous huge volumes of single carbonatoclastic megabeds testifying major forebulge instability (Critelli, 1999).

An estimation of detrital supply from Calabria during the last 25 my suggests that at least 5 to 8 km of crust has been removed from the Calabrian orogenic belt and deposited in the marine basins (Critelli and Le Pera, 1998; Critelli, 1999). Erosional budgets and accumulation rates document the immense volume of detrital sediments transferred from deeply weathered crystalline rocks of the Calabrian Arc to marine basins.

The clastic compositions of the southern Apennines foreland basin strata reflect the changing nature of the thrust belt through time, recording the history of accretion of the Calabria microplate over the Adria margin. The type of sedimentary provenance analysis, providing an example of the close relations between clastic compositions and growing orogen in southern Italy orogenic system, may contribute and have general application to other major orogens.

5. References

Alvarez W., 1976. A former continuation of the Alps. *Geol. Soc. Am. Bull.*, 87: 891-896.

Alvarez W., Cocozza T., Wezel F.C., 1974. Fragmentation of the Alpine orogenic belt by microplate dispersal. *Nature*, 248: 309-314.

Amodio Morelli L., Bonardi G., Colonna V., Dietrich D., Giunta G., Ippolito F., Liguori V., Lorenzoni S., Paglionico A., Perrone V., Piccarreta G., Russo M., Scandone P., Zanettin-Lorenzoni E., Zuppetta A., 1976. L'arco calabro-peloritano nell'orogene appenninico-maghrebide. *Mem. della Soc. Geol. It.*, 17: 1-60.

Amore F.O., Basso C., Ciarcia S., Di Nocera S., Matano F., Torre M., Ciampo G., Di Donato V., Esposito P., Staiti D., 1998. Nuovi dati stratigrafici sul Pliocene affiorante tra il Fiume Ufita e il Torrente Cervaro (Irpinia, Appennino meridionale). *Boll. Soc. Geol. It.*, 117: 455-466.

Argnani A., Marani M., Savelli C., Galassi B., 1995. Migrazione del vulcanismo di arco cenozoico nel contesto geodinamico intraorogenico del Mar Tirreno meridionale: un riesame. In: R. Polino, R. Sacchi (eds.), *Rapporti Alpi-Appennino*. Accademia Nazionale delle Scienze, Roma, 377-396.

Assorgia A., Barca S., Onnis G., Secchi F.A.G., Spano C., 1986. Episodi sedimentari e vulcanici oligo-miocenici nel settore occidentale dell'Arcuentu e loro contesto geodinamico (Sardegna SW). *Mem. Soc. Geol. It.*, 35: 229-240.

Atzori P., Ferla P., Paglionico A., Piccarreta G., Rottura A., 1984. Remnants of the Hercynian orogen along the "Calabrian-Peloritan arc", southern Italy: a review. *J. Geol. Soc.* London, 141: 137-145.

Bally A.W., Snelson S., 1980. Realms of subsidence. *Mem. Can. Soc. Petrol. Geol.*, 6: 9-75.

Barbera G., Critelli S., Mazzoleni P., 2011. Petrology and geochemistry of Cretaceous Sedimentary Rocks of the Monte Soro Unit (Sicily, Italy): constraints on weathering, diagenesis and provenance. *Journal of Geology*, v. 119: 51-68.

Barone A., Fabbri A., Rossi S., Sartori R., 1982. Geological structure and evolution of the marine areas adjacent to the Calabrian Arc. *Earth-Evolution Sc.*, 3: 207-221.

Barone M., Critelli S., Di Nocera S., Le Pera E., Matano F., Torre M., 2006. Stratigraphy and detrital modes of Upper Messinian post-evaporitic sandstones of the Southern Apennines, Italy: evidence of foreland-basin evolution during the Messinian Mediterranean salinity crisis. *International Geoloy Review*, v. 48, p. 702-724.

Barone M., Dominici R., Muto F.,Critelli S., 2008. Detrital modes in a late miocene wedge-top basin, northeastern calabria, italy: Compositional record of wedge-top partitioning. *Journal of Sedimentary Research*, 2008, v. 78, 693–711.

Beaumont C., 1981. Foreland Basins. Geoph. *J. Royal Astr. Soc.*, 65: 291-329.

Boccaletti M., Ciaranfi N., Cosentino D., Deiana G., Gelati R., Lentini F., Massari F., Moratti G., Pescatore T., Ricci Lucchi F., Tortorici L., 1990. Palinspastic restoration and paleogeographic reconstruction of the peri-Tyrrhenian area during the Neogene. *Paleogeogr. Paleoclim. Paleoecol.*, 77: 41-50.

Bonardi G., Cello G., Perrone V., Tortorici L., Turco E., Zuppetta A., 1982. The evolution of the northern sector of the Calabria-Peloritani Arc in a semiquantitative palinspastic restoration. *Boll. Soc. Geol. It.*, 101: 259-274.

Bonardi G., Giunta G., Messina A., Perrone V., Russo S., 1993. The Calabria-Peloritani Arc and its correlation with Northern Africa and Southern Europe. Field Trip Guidebook. In: A. Messina, S. Russo (eds.), The Calabria-Peloritani Arc and its Correlation with Northern Africa and Southern Europe. IGCP Project n. 276, *Newsletter*, 6: 27-90.

Boni M., 1974. Le argille rosse continentali del passaggio Paleocene-Miocene nella piattaforma carbonatica campano-lucana. *Boll. Soc. Geol. It.*, 93: 1059-1094.

Bosellini A., 1989. Dynamics of tethyan carbonate platforms. In: P.D. Crevello, F.F. Sarg, J.F. Read (eds.), Controls on Carbonate Platform and Basin Development. SEPM (Soc. for Sedim. Geol.) spec. pub. 44: 3-14.

Bouillin J.P., 1984. Nouvelle interprétation de la liason Apennin-Maghrébides en Calabre; consequences sur la paléogéographie téthysienne entre Gibraltar et les Alpes. *Rev. Géol. Dyn. Géogr. Phys.*, 25: 321-338.

Bouillin J.P., Durand Delga M., Olivier P., 1986. Betic-Rifain and Tyrrhenian arcs: distinctive features, genesis and development stages. In: F.C. Wezel (ed.), *The Origin of Arcs*, 281-304. Elsevier, Amsterdam.

Boyer S.E., Elliot D., 1982. Thrust systems. Am. Ass. Petrol. Geol. Bull., 66: 1196-1230.

Carannante G., D'Argenio B., Dello Iacovo B., Ferreri V., Mindszently A., Simone L., 1988a. Studi sul carsismo Cretacico dell'Appennino Campano. *Mem. Soc. Geol. It.*, 41: 733-759.

Carannante G., Matarazzo R., Pappone G., Severi C., Simone L., 1988b. Le calcareniti mioceniche della Formazione di Roccadaspide (Appennino campano-lucano). *Mem. Soc. Geol. It.*, 41: 775-789.

Carbone S., Lentini F., 1990. Migrazione neogenica del sistema catena-avampaese nell'Appennino meridionale: problematiche paleogeografiche e strutturali. *Riv. It. Paleont. Stratigr.*, 96: 271-296.

Carbone S., Lentini F., Sonnino M., De Rosa R., 1987. Il Flysch Numidico di Valsinni (Appennino Lucano). *Boll. Soc. Geol. It.*, 106: 331-345.

Carrara S., Serva L., 1982. I ciottoli contenuti nei flysch cretacico-paleogenici e miocenici e nei depositi post-tortoniani dell'Appennino meridionale. Loro significato paleotettonico. *Boll. Soc. Geol. It.*, 101: 441-496.

Casnedi R., 1988. La Fossa Bradanica: origine, sedimentazione e migrazione. *Mem. Soc. Geol. It.*, 41: 439-448.

Casnedi R., Crescenti U., Tonna M., 1982. Evoluzione dell'avanfossa adriatica meridionale nel Plio-Pleistocene, sulla base di dati di sottosuolo. Mem. Soc. Geol. It., 24: 243-260.

Castellano M.C., Putignano M.L., Sgrosso I., 1997. Sedimentology and stratigraphy of the Piaggine Sandstones (Cilento, southern Apennines, Italy). *Gior. Geologia,* 59: 273-287.

Cavazza W., 1989. Detrital modes and provenance of the Stilo-Capo d'Orlando Formation (Miocene), southern Italy. *Sedimentology*, 36: 1077-1090.

Cavazza W., Blenkinsop J., DeCelles P.G., Patterson R.T., Reinhardt E.G., 1997. Stratigrafia e sedimentologia della sequenza sedimentaria oligocenico-quaternaria del bacino calabro-ionico. *Boll. Soc. Geol. It.*, 116: 51-77.

Cazzola C., Critelli S., 1987. Litostratigrafia e petrologia delle quarzoareniti torbiditiche oligomioceniche di Asilah (Catena del Rif, Marocco nord-occidentale). *Mineralogica et Petrographica Acta*, 30: 203-226.

Cello G., Nur A., 1988. Emplacement of foreland thrust systems. *Tectonics*, 7: 261-271.

Cello G., Tortorici L., Turco E., Guerra I., 1981. Profili profondi in Calabria settentrionale. *Boll. Soc. Geol. It.*, 100: 423-431.

Cello G., Martini N., Paltrinieri W., Tortorici l., 1989. Structural styles in the frontal zones of the southern Apennines, Italy: an example from the Molise district. *Tectonics*, 8: 753-768.

Cello G., Invernizzi C., Mazzoli S., 1996. Structural signature of tectonic processes in the Calabrian Arc (southern Italy): evidence from the oceanic-derived Diamante-Terranova unit. *Tectonics*, 15: 187-200.

Channell, J.E.T. and Mareschal, J.C., 1989. Delamination and asymmetric lithospheric thickening in the development of the Tyrrhenian Rift. In: Coward, M.P., Dietrich, D., Park, R.G. (eds.), Alpine Tectonics. Geological Society Spec. Pub. 45: 285-302.

Cherchi A., Montadert L., 1982. Il sistema di rifting oligo-miocenico del Mediterraneo occidentale e sue conseguenze paleogeografiche sul Terziario sardo. *Mem. Soc. Geol. It.*, 24: 387-400.

Chiocci F.L., 1994. Very high-resolution seismics as a tool for sequence stratigraphy applied to outcrop scale - examples from Eastern Tyrrhenian Margin Holocene/Pleistocene deposits. *Am. Ass. Petrol. Geol. Bull.*, 78: 378-395.

Ciaranfi N., Pieri P., Ricchetti G., 1988. Note alla carta geologica delle Murge e del Salento (Puglia centromeridionale). *Mem. Soc. Geol. It.*, 41: 449-460.

Cieszkowski M., Oszczypko N., Pescatore T., Slaczka A., Senatore M.R. and Valente A., 1995. Megatorbiditi calcareo-marnose nelle successioni flyscioidi dell'Appennino meridionale (Cilento, Italia) e dei Carpazi settentrionali (Polonia). *Boll. Soc. Geol.* It., 114: 67-88.

Cinque A., Patacca E., Scandone P., Tozzi M., 1993. Quaternary kinematic evolution of the southern Apennines. Relationships between surface geological features and deep lithospheric structures. *Ann. Geofisica*, 36: 249-260.

Cocco E., Cravero E., Ortolani F., Pescatore T., Russo M., Torre M., Coppola L., 1974. Le unit`a irpine nell'area a nord di Monte Marzano, Appennino meridionale. *Mem. Soc. Geol. It.*, 13: 607-654.

Cocco E., Pescatore T., 1968. Scivolamenti gravitativi (olistostromi) nel Flysch del Cilento (Campania). *Boll. Soc. Nat. Napoli*, 77: 51-91.

Colella A., 1994. Coarse-grained deltas in neotectonic strike-slip and extensional settings: tectonic and sedimentary controls on the architecture of deltas and basin fills (Crati Basin and Messina Strait, Southern Italy). In: *Pre Meeting Fieldtrip Guidebook. Int. Assoc. Sedimentologists, 15th Regional Meeting*, Ischia, Italy, 245-277.

Colella A., Zuffa G.G., 1988. Megastrati carbonatici e silicoclastici della Formazione di Albidona (Miocene, Appennino meridionale): implicazioni paleogeografiche. *Mem. Soc. Geol. It.*, 41: 791-807.

Colombo F., 1994. Normal and reverse unroofing sequences in syntectonic conglomerates as evidence of progressive basinward deformation. *Geology*, 22: 235-238.

Colonna V., Compagnoni R., 1982. Guida all'escursione sulle unità cristalline della Catena Costiera (Calabria). *Rend. Soc. It. Min. Petr.*, 38: 1141-1152.

Colonna V., 1998. Ruolo ed estensione regionale dei thrusts a vergenza settentrionale in Sila Grande (Calabria). *Boll. Soc. Geol. It.*, 117: 249-260.

Cosentino D., Gliozzi E., 1988. Considerazioni sulle velocità di sollevamento di depositi eutirreniani dell'Italia meridionale e della Sicilia. *Mem. Soc. Geol. It.*, 41: 653-665.

Cotecchia V., 1963. I terreni sedimentari dei dintorni di Cariati, Scala Coeli e Pietrapaola, al margine nord-orientale della Sila (Calabria). *Mem. Ist. Geol. Min.* Univ. Padova, 24: 1-40.

Covey M., 1986. The evolution of foreland basins to steady state: evidence from the western Taiwan foreland basin. In: P.A. Allen, P. Homewood (eds.), *Foreland Basins*. Int. Ass. Sediment. spec. pub. 8: 77-90.

Crampton S.L., Allen P.A., 1995. Recognition of forebulge unconformities associated with early stage foreland basin development: example from the North Alpine foreland basin. *Am. Ass. Petrol. Geol. Bull.*, 79: 1495-1514.

Critelli S., 1991. Evoluzione delle mode detritiche delle successioni arenitiche terziarie dell'Appennino meridionale. *Mem. Soc. Geol. It.*, 47: 55-93.

Critelli S., 1993. Sandstone detrital modes in the Paleogene Liguride Complex, accretionary wedge of the southern Apennines (Italy). *J. Sedim. Petr.*, 63: 464-476.

Critelli S., 1999, The interplay of lithospheric flexure and thrust accomodation in forming stratigraphic sequences in the southern Apennines foreland basin system, Italy. *Memorie dell'Accademia Nazionale dei Lincei*, v. 10: 257-326.

Critelli S., Ingersoll R.V., 1994. Sandstone petrology and provenance of the Siwalik Group (northwestern Pakistan and western-southeastern Nepal). *J. Sedim. Res.*, A64: 815-823.

Critelli S., Ingersoll R.V., 1995. Interpretation of neovolcanic versus palaeovolcanic sand grains: an example from Miocene deep-marine sandstone of the Topanga Group (southern California). *Sedimentology*, 42: 783-804.

Critelli S., Le Pera E., 1994. Detrital modes and provenance of Miocene sandstones and modern sands of the southern Apennines thrust-top basins (Italy). *J. Sedim. Res.*, A64: 824-835.

Critelli S., Le Pera E., 1995a. Tectonic evolution of the Southern Apennines thrust-belt (Italy) as reflected in modal compositions of cenozoic sandstone. *J. Geology*, 103: 95-105.

Critelli S., Le Pera E., 1995b. La Formazione di Castelvetere nell'evoluzione petrostratigrafica dell'avanfossa del Tortoniano-Messiniano dell'Appennino meridionale. *Boll. Soc. Geol. It.*, 114: 615-634.

Critelli S., Le Pera E., 1998. Post-Oligocene sediment-dispersal systems and unroofing history of the Calabrian microplate, Italy. *Int. Geol. Rev.*, 40: 609-637.

Critelli S., Le Pera E., 2003. Provenance relations and modern sand petrofacies in an uplifted thrust-belt, northern Calabria, Italy. In: *Quantitative Provenance Studies in Italy* (Ed. by R. Valloni & A. Basu). *Servizio Geologico Nazionale, Memorie Descrittive della Carta Geologica d'Italia*, v. 61, p. 25-38.

Critelli S., Reed W.E., 1999. Provenance and stratigraphy of the Devonian (Old Red Sandstone) and Carboniferous sandstones of Spitsbergen, Svalbard. *European J. Mineralogy*, 11: 149-166.

Critelli S., De Rosa R., Sonnino M., Zuffa G.G., 1990. Significato dei depositi vulcanoclastici della Formazione delle Tufiti di Tusa (Miocene inferiore, Lucania meridionale). *Boll. Soc. Geol. It.*, 109: 743-762.

Critelli S., Arribas J., Le Pera E., Tortosa A., Marsaglia K.M. & Latter K., 2003. The recycled orogenic provenance sand suite from an uplifted thrust-belt, Betic Cordillera, southern Spain and the Alboran Basin. *Journal of Sedimentary Research*, v.73, p.72-81.

Critelli S, Mongelli G, Perri F, Martìn-Algarra A, Martìn-Martìn M, Perrone V, Dominici R, Sonnino M, Zaghloul MN, 2008.Compositional and geochemical signatures for the sedimentary evolution of the Middle Triassic–Lower Jurassic continental redbeds from Western-Central Mediterranean Alpine Chains. *J Geol* 116:375-386

Critelli S., Le Pera E., Galluzzo F., Milli S., Moscatelli M., Perrotta S., Santantonio, M., 2007, Interpreting siliciclastic-carbonate detrital modes in Foreland Basin Systems: an example from Upper Miocene arenites of the Central Apennines, Italy, *in* Arribas J., Critelli S. and Johnsson M., editors, *Sedimentary Provenance: Petrographic and Geochemical Perspectives. Geological Society of America Special Paper 420*: 107-133.

Crook K.A.W., 1989. Suturing history of an allochthonous terrane at a modern plate boundary traced by flysch-to-molasse facies transitions. *Sedim. Geol.*, 61: 49-79.

D'Argenio B., 1974. Le piattaforme carbonatiche Periadriatiche. Una rassegna di problemi nel quadro geodinamico Mesozoico dell'area mediterranea. *Mem. Soc. Geol. It.*, 13: 137-159.

D'Argenio B., Pescatore T., Scandone P., 1973. Schema geologico dell'Appennino meridionale (Campania e Lucania). In: Moderne Vedute sulla Geologia dell'Appennino. Acc. Naz. Lincei, Quad. 183: 49-72.

De Capoa P., Guerrera F., Perrone V., Serrano F., 1997. New biostratigraphic data on the Frazzanò Formation (Longi-Taormina Unit): consequences on defining the deformation age of the Calabria-Peloritani arc southern sector. *Riv. It. Paleont. e Stratigr.*, 103: 343-356.

DeCelles P.G., Hertel F., 1989. Petrology of fluvial sands from the Amazonian foreland basin, Peru and Bolivia. *Geol. Soc. Am. Bull.*, 101: 1552-1562.

DeCelles P.G., Giles K.A., 1996. Foreland Basin Systems. *Basin Research*, 8: 105-123.

Del Moro A., Paglionico A., Piccarreta G., Rottura A., 1986. Tectonic structure and post-Hercynian evolution of the Serre, Calabrian Arc, southern Italy: geological, petrological and radiometric evidences. *Tectonophysics*, 124: 223-238.

Dercourt J. et 18 al., 1986. Geologic evolution of the Tethys belt from the Atlantic to the Pamirs since the Lias. *Tectonophysics*, 123: 241-315.

De Roever E.W.F., Piccarreta G., Beunk F.F., Kieft C., 1974. Blue amphiboles from north-western and central Calabria (Italy). *Per. Mineral.*, 43: 1-37.

Dewey J.F., Helman M.L., Turco E., Hutton D.H.W., Knott S.D., 1989. Kinematics of the western Mediterranenan. In: M.P. Coward, D. Dietrich, R.G. Park (eds.), Alpine Tectonics. Geological Society Special Publication, 45: 265-283.

Dickinson W.R., 1974. Plate tectonics and sedimentation. In: W.R. Dickinson (ed.), Tectonics and Sedimentation. *SEPM (Soc. for Sedim. Geol.*), spec. pub. 22: 1-27.

Dickinson W.R., 1985. Interpreting provenance relations from detrital modes of sandstones. In: G.G. Zuffa (ed.), Provenance of Arenites, 333-361. D. Reidel, Dordrecht.

Dickinson, W.R., 1988. Provenance and sediment dispersal in relation to paleotectonics and paleogeography of sedimentary basins. In: K.L. Kleinspehn, C. Paola (eds.), *New Perspectives in Basin Analysis*, 3-25. New York, Springer-Verlag.

Dietrich D., 1988. Sense of overthrust shear in the Alpine nappes of Calabria (southern Italy). *J. Struct. Geol.*, 10: 373-381.

Di Girolamo P., Morra V. and Perrone V., 1992, Ophiolitic olistoliths in middle Miocene turbidites (Cilento Group) at Mt. Centaurino (southern Apennines, Italy). *Ofioliti*, v. 17, p. 199-217.

Di Nocera S., Ortolani F. Russo M., Torre M., 1974. Successioni sedimentarie messiniane e limite Miocene-Pliocene nella Calabria settentrionale. Boll. Soc. Geol. It., 93: 575-607.

Doglioni C., 1991. A proposal for the kinematic modelling of W-dipping subductions-possible applications to the Tyrrhenian-Apennines system. Terra Nova, 3: 423-434.

Doglioni C., Mongelli F., Pieri P., 1994. The Puglia uplift (SE Italy): an anomaly in the foreland of the Apenninic subduction due to bucling of a thick continental lithosphere. *Tectonics*, 13: 1309-1321.

Doglioni C., Harabaglia P., Martinelli G., Mongelli F., Zito G., 1996. A geodynamic model of the southern Apennines accretionary prism. *Terra Nova*, 8: 540-547.

Doglioni C., Gueguen E., Sàbat F., Fernandez M., 1997. The western Mediterranean extensional basins and the Alpine orogen. *Terra Nova*, 9: 109-112.

Fabbri A., Gallignani P., Zitellini N., 1981. Geologic evolution of the peri-Tyrrhenian sedimentary basins. In: F.C. Wezel (ed.), Sedimentary Basins of Mediterranean Margins. C.N.R. Italian Project of Oceanography. Bologna, Ed. Tecnoprint, 101-126.

Fornelli A., Piccarreta G., 1997. Mineral and chemical provenance in some early Miocene sandstones of the southern Apennines (Italy). *Eur. J. Min.*, 9: 433-447.

Garzanti E., Critelli S., Ingersoll R.V., 1996. Paleogeographic and paleotectonic evolution of the Himalayan Range as reflected by detrital modes of Tertiary sandstones and modern sands (Indus Transect, India and Pakistan). *Geol. Soc. Am. Bull.*, 108: 631-642.

Giles K.A., Dickinson, W.R., 1995. The interplay of eustasy and lithospheric flexure in forming stratigraphic sequences in foreland settings: an example from the Antler foreland, Nevada and Utah. In: S.L. Dorobek, G.M. Ross (eds.), *Stratigraphic Evolution of Foreland Basins*. SEPM (Soc. for Sedim. Geology) spec. pub. 52, 187-211.

Graham S.A. et 14 others, 1986. Provenance modelling as a tecnique for analysing source terrane evolution and controls on foreland sedimentation. In: P.A. Allen, P. Homewood (eds.), Foreland Basins. *Int. Ass. Sediment.*, spec. pub. 8: 425-436.

Gueguen E., Doglioni C., Fernandez M., 1997. Lithospheric boudinage in the western Mediterranean back-arc basin. *Terra Nova*, 9: 184-187.

Gueguen E., Doglioni C., Fernandez M., 1998. On the post-25 Ma geodynamic evolution of the western Mediterranean. *Tectonophysics*, 298: 259-269.

Guerrera F., Martin-Algarra A., Perrone V., 1993. Late Oligocene-Miocene syn-/-late-orogenic successions in Western and Central Mediterranean Chains from the Betic Cordillera to the Southern Apennines. *Terra Nova*, 5,: 525-544.

Guerrera F., Martín-Martín M., Perrone V.,Tramontana M., 2005. Tectono-sedimentary evolution of the southern branch of the Western Tethys (Maghrebian Flysch Basin and Lucanian Ocean): consequences for Western Mediterranean geodynamics. *Terra Nova* 4, 358–367.

Haccard D., Lorenz C., Grandjacquet C., 1972. Essai sur l'evolution tectogénètique de la liason Alpes-Apennines (de la Ligurie à la Calabre). *Mem. Soc. Geol. It.*, 11: 309-341.

Hippolyte J.C., 1992, Tectonique de l'Apennin méridional: structures et paléocontraintes d'un prisme d'accrétion continentail. Université de Paris, [PhD Thesis].

Hippolyte J.C., Angelier J., Roure F., Casero P., 1994a. Piggyback basin development and thrust belt evolution: structural and paleostress analyses of Plio-Quaternary basins in the southern Apennines. *J. Struct. Geol.*, 16: 159-173.

Hippolyte J.C., Angelier J., Roure F., Casero P., 1994b. A major geodynamic change revealed by Quaternary stress patterns in the Southern Apennines (Italy). *tectonophysics*, 230: 199-210.

Iannace A., Boni M., Zamparelli V., 1995. The middle-Upper Triassic of the San Donato Unit Auct. (northern Calabria): stratigraphy, paleogeography and tectonic implications. *Riv. It. Pal. Strat.*, 101: 301-324.

Ibbeken H., Schleyer R., 1991. Source and Sediment. A Case Study of Provenance and Mass balance at an Active Plate Margin (Calabria, Southern Italy). Springer-Verlag, Berlin, 286 pp.

Ietto A., Barillaro A.M., 1993. L'Unità di San Donato quale margine deformato cretacico-paleogenico del bacino di Lagonegro (Appennino meridionale-Arco calabro). *Boll. Soc. Geol. It.*, 112: 1-20.

Ietto A., Ietto F., 1998. Sviluppo e annegamento di un sistema carbonatico piattaforma-bacino nel Trias superiore della Catena Costiera calabrese. *Boll. Soc. Geol. It.*, 117: 313-331.

Ietto A., Pescatore T., Cocco E., 1965. Il Flysch mesozoico-terziario del Cilento occidentale. *Boll. Soc. Natur. Napoli*, 74: 396-402.

Ingersoll R.V., Graham S.A., Dickinson W.R., 1995. Remnant ocean basins. In: C.J. Busby, R.V. Ingersoll (eds.), *Tectonics of Sedimentary Basins*, 363-391. Blackwell Science, Oxford.

Ippolito F., D'Argenio B., Pescatore T., Scandone P., 1975. Structural-stratigraphic units and tectonic framework of Southern Apennines. In: C. Squires (ed.), *Geology of Italy*. Petroleum Exploration Society of Libya, 317-328.

Jordan T.E., 1981. Thrust loads and foreland basin evolution, Cretaceous, western United States. *Am. Ass. Petrol. Geol. Bull.*, 65: 291-329.

Jordan T.E., 1995. Retroarc foreland and related basins. In: C.J. Busby, R.V. Ingersoll (eds.), *Tectonics of Sedimentary Basins*, 331-362. Blackwell Science, Oxford.

Jordan T.E., Flemings P.B., Beer J.A., 1988. Dating thrust-fault activity by use of foreland-basin strata. In: K.L. Kleinsphen, C. Paola (eds.), *New Perspectives in Basin Analysis*, 307-330. Springer-Verlag, New York.

Kastens K. et 20 others, 1988. ODP Leg 107 in the Tyrrhenian Sea: insights into passive margin and back-arc basin evolution. *Geol. Soc. Am. Bull.*, 100: 1140-1156.

Knott S.D., 1987. The Liguride Complex of southern Italy: a Cretaceous to Paleogene accretionary wedges. *Tectonophysics*, 142: 217-226.

Knott S.D., 1988. Structure, sedimentology and petrology of an ophiolitic flysch terrain in Calabria, south Italy. University of Oxford, [PhD Thesis], 152 pp.

Lanzafame G., Spadea P., Tortorici L., 1979. Mesozoic ophiolites of northern Calabrian and Lucanian Apennine (southern Italy). *Ofioliti,* 4: 173-182.

Lavecchia G., 1988. The Tyrrhenian-Apennines system: structural setting and seismotectogenesis. *Tectonophysics,* 147: 263-296.

Lentini F., 1979. Le Unità Sicilidi della Val d'Agri (Appennino Lucano). *Geol. Rom.*, 18: 215-225.

Lentini F., Carbone S., Catalano S., 1994. Main structural domains of the central Mediterranean Region and their Neogene tectonic evolution. *Boll. Geof. Teor.* e Appl., 36: 103-125.

Le Pera E., 1998. Relazioni composizionali tra aree fonti e sabbie fluviali, costiere e marine attuali e recenti in Calabria settentrionale. Università di Bologna, [PhD Thesis], 278 pp.

Le Pera E., Critelli S., 1997. Sourceland controls on the composition of beach and fluvial sand of the northern Tyrrhenian coast of Calabria, Italy: implications for actualistic petrofacies. *Sedim. Geol.*, 110: 81-97.

Loiacono, F. and Sbarra, R., 1991, Caratteri sedimentologici della Formazione di Serra Palazzo nei dintorni di Tricarico (Basilicata). Memorie della Società Geologica Italiana, v. 47, p. 157-166.

Malinverno A., Ryan, W.B.F., 1986. Extension in the Tyrrhenian Sea and shortening in the Apennines as result of arc migration driven by sinking of the lithosphere. *Tectonics*, 5: 227-245.

Marsella E., Bally A.W., Cippitelli G., D'Argenio B., Pappone G., 1995. Tectonic history of the Lagonegro domain and southern Apennine thrust belt evolution. *Tectonophysics*, 252: 307-330.

Messina A., Russo S., Borghi A., Colonna V., Compagnoni R., Caggianelli A., Fornelli A., Piccarreta G., 1994. Il Massiccio della Sila Settore settentrionale dell'Arco Calabro-Peloritano. *Boll. Soc. Geol. It.*, 113: 539-586.

Miall A.D., 1995. Collision-related foreland basins. In: C.J. Busby, R.V. Ingersoll (eds.), *Tectonics of Sedimentary Basins*, 393-424. Blackwell Science, Oxford.

Mongelli G, Critelli S, Perri F, Sonnino M, Perrone V (2006) Sedimentary recycling, provenance and paleoweathering from chemistry and mineralogy of Mesozoic continental redbed mudrocks, Peloritani Mountains, Southern Italy. *Geochem J* 40:197-209

Mongelli G., Critelli S., Dinelli E., Paternoster M., Perri F., 2010. Mn- and Fe-carbonate rich layers in Meso-Cenozoic shales as proxies of environmental conditions: a case study from the southern Apennine, Italy. *Geochemical Journal, 44: 211-223.*

Mostardini F., Merlini S., 1986. Appennino centro-meridionale. Sezioni geologiche e proposta di modello strutturale. *Mem. Soc. Geol. It.*, 35: 177-202.

Nigro F., Puglisi D., 1993. Analisi strutturale e sedimentologico-petrografica del Flysch di Frazzanò (Eocene-Oligocene, Sicilia nord-orientale). Gior. Geologia, 55: 103-116.

Ogniben L., 1955. Le argille scagliose del Crotonese. Mem. e Note Ist. *Geol. Appl. Napoli*, 6: 1-72.

Ogniben L., 1962. Le Argille Scagliose e i sedimenti messiniani a sinistra del Trionto (Rossano, Cosenza). *Geol. Rom.*, 1: 255-282.

Ogniben L., 1969. Schema introduttivo alla geologia del confine calabro-lucano. *Mem. Soc. Geol. It.*, 8: 453-763.

Ogniben L., 1973. Schema geologico della Calabria in base ai dati odierni. *Geol. Rom.*, 12: 243-585.

Ori G.G., Roveri M., Vannoni F., 1986. Plio-Pleistocene sedimentation in the Apenninic-Adriatic foredeep (Central Adriatic Sea, Italy). In: P.A. Allen, P. Homewood (eds.), Foreland Basins. *Int. Ass. Sediment.*, spec. pub. 8: 183-198.

Ortolani F., 1978. Alcune considerazioni sulle fasi tettoniche mioceniche e plioceniche dell'Appennino meridionale. *Boll. Soc. Geol. It.*, 97: 606-616.

Patacca E., Scandone P., 1987. Post-Tortonian mountain building in the Apennines. The role of the passive sinking of a relic lithospheric slab. In: A. Boriani, M. Bonafede, G.B. Piccardo, G.B. Vai (eds.), *The Lithosphere in Italy*. Acc. Naz. Lincei, 80: 157-176.

Patacca E., Sartori R., Scandone P., 1990. Tyrrhenian basin and Apenninic arcs: kinematic relations since Late Tortonian times. *Mem. Soc. Geol. It.*, 45: 425-451.

Patacca E., Sartori R., Scandone P., 1993. Tyrrhenian basin and Apennines. Kinematic evolution and related dynamic constraints. In: E. Boschi et al. (eds.), Recent *Evolution and Seismicity of the Mediterranean Region*, 161-171. Kluwer Acad. Publ.

Patacca E., Scandone P., Bellatalla M., Perilli N., Santini U., 1992. The Numidian-sand event in the southern Apennines. *Mem. Sc. Geol.*, Univ. Padova, 43: p. 297-337.

Patacca E. & Scandone P. (2001) - Late thrust propagation and sedimentary response in the thrust belt-foredeep system of the Southern Apennines (Pliocene-Pleistocene). In Vai G.B. & Martini I.P. Eds. *Anatomy of a mountain: The Apennines and adjacent Mediterranean basins*. Kluwer Academic Publ., 401-440.

Peper T., Van Balen R., Cloetingh S., 1995. Implications of orogenic wedge growth, intraplate stress variations, and eustatic sea-level change for foreland basin stratigraphy - inferences from numerical modeling. In: S.L. Dorobek, G.M. Ross (eds.), *Stratigraphic Evolution of Foreland Basins*. SEPM (Soc. for Sedim. Geology) spec. pub. 52, 25-35.

Perri F, Cirrincione R, Critelli S, Mazzoleni P, Pappalardo A (2008) Clay mineral assemblages and sandstone compositions of the Mesozoic Longobucco Group (north-eastern Calabria): implication for burial history and diagenetic evolution. *Int Geol Rev* 50:1116-1131.

Perri F., Critelli S., Mongelli G., Cullers R.L. (2010) – Sedimentary evolution of the Mesozoic continental redbeds using geochemical and mineralogical tools: the case of Upper

Triassic to Lowermost Jurassic M.te di Gioiosa mudstones (Sicily, southern Italy). *International Journal of Earth Sciences*, (in press).

Perri F., Critelli S., Cavalcante F., Mongelli G., Dominici R., Sonnino M., De Rosa R. (2011) - Geochemical signatures for the Miocene volcaniclastic succession of the Tufiti di Tusa Formation, southern Apennines, Italy. *Geological Magazine* (in press).

Perrone V., 1987. I depositi miocenici della dorsale di Monte Soprano (Appennino campano): segnalazione di vulcanismo andesitico e nuova interpretazione. *Boll. Soc. Geol.* It., 106: 3-12.

Perrone V., 1996. Une nouvelle hypothèse sur la position paléogéographique et l'éevolution tectonique des Unités de Verbicaro et de San Donato (région Calabro-Lucanienne; Italie): implications sur la limite Alpes-Apennin en Calabre. C.R. Acad. Sci. Paris, 322: 877-884.

Perrone V., Martín-Algarra A., Critelli S., Decandia F.A., D'Errico M., Estevez A., Iannace A., Lazzarotto A., Martín-Martín M., Martín-Rojas I., Mazzoli S., Messina A., Mongelli G., Vitale S. & Zaghloul M.N. (2006) - "Verrucano" and "Pseudoverrucano" in the central-western Mediterranean Alpine chains: palaegeographical evolution and geodynamic significance. In: A. Chalouan & G. Moratti (eds) *Tectonics of the Western Mediterranean and North Africa*. Geological Society, London, Special Publications, 262, 1-43

Pescatore T., 1978. The Irpinids: the model of tectonically controlled fan and base-of-slope sedimentation in southern Italy. In: D.J. Stanley, G. Kelling (eds.), *Sedimentation in Submarine Canyons Fans and Trenches*, 325-339. Dowden, Hutchinson and Ross, Inc., Stroudsburg.

Pescatore T. and Senatore M.R., 1986. A comparison between a present-day (Taranto Gulf) and a Miocene (Irpinian Basin) foredeep of the southern Apennines (Italy). In: P.A. Allen, P. Homewood (eds.), Foreland Basins. *Int. Ass. Sediment. spec. pub.* 8: 169-182.

Pescatore T., Sgrosso I., Torre M., 1970. Lineamenti di tettonica e sedimentazione nel Miocene dell'Appennino campano-lucano. *Mem. Soc. Nat. Napoli*, 78: 337-408.

Pescatore T., Renda P., Tramutoli M., 1988. Rapporti tra le unità lagonegresi e le unità sicilidi nella media valle del Basento, Lucania (Appennino meridionale). *Mem. Soc. Geol. It.*, 41: 353-361.

Pieri P., Rapisardi L., 1973. Su alcune grovacche vulcaniche dell'Appennino Dauno. *Boll. Soc. Natur. Napoli*, 82: 191-218.

Pieri P., Sabato L., Tropeano M., 1996. Significato geodinamico dei caratteri deposizionali e strutturali della Fossa Bradanica nel Pleistocene. *Mem. Soc. Geol. It.*, 51: 501-515.

Puglisi D., 1987. Le successioni torbiditiche cretacico-terziarie della Sicilia nord-orientale nel quadro dell'evoluzione del settore meridionale dell'arco calabro-peloritano e della catena maghrebide siciliana. *Gior. Geologia*, 49: 167-185.

Quinlan G.M., Beaumont C., 1984. Appalachian thrusting, lithospheric flexure and Paleozoic stratigraphy of the eastern interior of North America. *Can. J. Earth Sc.*, 21: 973-996.

Riba O., 1976. Syntectonic unconformities of the Alto Cardener, Spanish Pyrenees: a genetic interpretation. *Sed. Geol.*, 15: 213-233.

Ricchetti G., Ciaranfi N., Luperto-Sinni E., Mongelli F., Pieri P., 1988. Geodinamica ed evoluzione sedimentaria e tettonica dell'avampaese apulo. *Mem. Soc. Geol. It.*, 41: 57-82.

Ricchetti G., 1980. Contributo alla conoscenza strutturale della fossa bradanica e delle Murge. *Boll. Soc. Geol. It.*, 99: 421-430.

Ricchetti G., Mongelli F., 1980. Flessione e campo gravimetrico della micropiastra apula. *Boll. Soc. Geol. It.*, 99: 431-436.

Ricci Lucchi F., 1986. The Oligocene to recent foreland basins of the northern Apennines. In: P.A. Allen, P. Homewood (eds.), Foreland Basins. *Int. Ass. Sediment. spec. pub.* 8: 105-139.

Ricci Lucchi F., Colella A., Gabbianelli G., Rossi S., Normark W.R., 1984. Crati fan, Mediterranean. In: A.H. Bouma, W.R. Normark, N.E. Barnes(eds.), *Submarine Fans and Related Turbidite Systems*, 51-57. Springer-Verlag, New York.

Roda C., 1964, Distribuzione e facies dei sedimenti neogenici nel Bacino Crotonese. *Geol. Rom.*, 3: 319-366.

Roda C., 1967. I sedimenti neogenici autoctoni ed alloctoni della zona di Cirò-Cariati (Catanzaro e Cosenza). *Mem. Soc. Geol. It.*, 6: 137-149.

Romagnoli C., Gabbianelli G., 1990. Late Quaternary sedimentation and soft-sediment deformation features in the Corigliano Basin, north Ionian Sea (Mediterranean). Gior. *Geologia*, 52: 33-53.

Romeo M., 1967. Stratigrafia micropaleontologica del Messiniano di Rossano. Paleont. Italica, 63: 1-74.

Roveri M., Bernasconi A., Rossi M.E., Visentin C., 1992. Sedimentary evolution of the Luna Field Area, Calabria, southern Italy. In: A.M. Spencer (ed.), Generation, Accumulation and Production of Europe's Hydrocarbons II. *Special Publication of the European Association of Petroleum Geoscientists* No. 2., 217-224. Berlin, Springer-Verlag.

Royden L., Patacca E., Scandone P., 1987. Segmentation and configuration of subducted lithosphere in Italy: an important control on thrust belt and foredeep basin evolution. *Geology*, 15: 714-717.

Russo M., Zuppetta A., Guida A., 1995. Alcune precisazioni stratigrafiche sul Flysch del Cilento (Appennino meridionale). *Boll. Soc. Geol. It.*, 114: 353-359.

Santo A., Sgrosso I., 1988. La Formazione del Torrente Raganello: secondo ciclo sedimentario miocenico nel Gruppo del Pollino (Calabria). *Boll. Soc. Geol. It.*, 107: 413-424.

Sartori R., 1982. L'arco calabro-peloritano: aspetti di geologia marina. *Rend. Soc. It. Min. e Petr.*, 38: 941-950.

Sartori R., 1990. The main results of ODP Leg 107 in the frame of Neogene to Recent geology of Perityrrhenian areas. In: K. Kastens, G. Mascle et al. (eds.), *Proceedings of the Ocean Drilling Program, Scientific Results*, 107: 715-730. College Station, Texas.

Savelli D., Wezel F.C., 1980. Morphological map of the Tyrrhenian Sea. In: C.N.R., P.F. Oceanografia e Fondi Marini (Theme <<Bacini Sedimentari>>). Firenze, Litografia Artistica Cartografica (colour map scale 1:1.250.000).

Scandone P., 1979. Origin of the Tyrrhenian Sea and Calabrian Arc. *Boll. Soc. Geol. It.*, 98: 27-34.

Scandone P., 1982. Structure and evolution of Calabrian arc. *Earth-Evolution Sc.*, 3: 172-180.

Schedl A., Wiltschko D.V., 1984. Sedimentological effects of a moving terrain. *J. Geology*, 92: 273-287.

Schenk V., 1981. Synchronous uplift of the lower crust of the Ivrea Zone and of southern Calabria and its possible consequences for the Hercynian orogeny in southern Europe. *Earth and Planetary Sc. Lett.*, 56: 305-320.

Schwab F.L., 1986. Sedimentary "signatures" of foreland basin assemblages: real or counterfeit?. In: P.A. Allen, P. Homewood (eds.), Foreland Basins. *Int. Ass. Sediment. spec. pub.* 8: 395-410.

Selli R., 1957. Sulla trasgressione del Miocene in Italia meridionale. *Gior. Geol.*, 24: 1-54.

Selli R., 1962. Il Paleogene nel quadro della geologia dell'Italia centro-meridionale. *Mem. Soc. Geol. It.*, 3: 737-789.

Sgrosso I., 1981. Il significato delle Calciruditi di Piaggine nell'ambito degli eventi del Miocene inferiore nell'Appennino Campano-Lucano. *Boll. Soc. Geol. It.*, 100: 129-137.

Sgrosso I., 1988a. Nuovi elementi per un più articolato modello paleogeografico nell'Appennino centro meridionale. *Mem. Soc. Geol. It.*, 41: 225-242.

Sgrosso I., 1988b. Nuovi dati biostratigrafici sul Miocene del Monte Alpi (Lucania) e conseguenti ipotesi paleogeografiche. *Mem. Soc. Geol. It.*, 41: 343-351.

Sgrosso I., 1998. Possibile evoluzione cinematica miocenica nell'orogene centro-sud-appenninico. *Boll. Soc. Geol. It.*, 117: 679-724.

Sinclair H.D., Allen P.A., 1992. Vertical versus horizontal motions in the Alpine orogenic wedge: stratigraphic response in the foreland basin. *Basin Res.*, 4: 215-232.

Sorriso Valvo M., 1993. The geomorphology of Calabria, a sketch. Geogr. *Fis. Din. Quat.*, 18: 257-273.

Spadea P., 1982. Continental rocks associated with ophiolites in Lucanian Apennines. *Ofioliti*, 7: 501-522.

Srivastava P., Mitra G., 1994. Thrust geometries and deep structure of the outer and lesser Himalaya, Kumaon and Garhwal (India): implications for evolution of the Himalayan fold-and-thrust-belt. *Tectonics*, 13: 89-109.

Stampfli G. M., Marchant R. H., 1997. Geodynamic evolution of the Tethyan margins of the western Alps. In: O. A. Pfiffner, P. Lehner, P. Heitzmann, St. Mueller, A. Steck (eds.), *Deep Structure of the Swiss Alps*, Results of NRP 20. Birkh¨auser Verlag, Basel: 223-239.

Steck A., Hunziker J.C., 1994.The Tertiary structural and thermal evolution of the Central Alps - Compressional and extensional structures in a orogenic belt. *Tectonophysics*, 238: 229-254.

Steidtmann J.R. and Schmitt J.G., 1988. Provenance and dispersal of tectogenic sediments in thin-skinned, thrusted terrains. In: K.L. Kleinsphen, C. Paola (eds.), *New Perspectives in Basin Analysis*, 353-366. Springer-Verlag, New York.

Taddei A., Siano M.G., 1992. Analisi biostratigrafica e considerazioni paleoecologiche sulla successione neogenica del Monte Alpi (Lucania). *Boll. Soc. Geol. It.*, 111: 255-272.

Tansi, C., Muto, F., Critelli, S., and Iovine, G., 2007, Neogene–Quaternary strike-slip tectonics in the central Calabria Arc (southern Italy): *Journal of Geodynamics*, v. 43, p. 397–414.

Tavarnelli E., Pasqui, V., 1998. Transtensional fault growth by isolated segment linkage: an example from the Potenza Basin, southern Apennines, Italy. *Boll. Soc. Geol. It.*, 117: 261-269.

Thomson S.N., 1994. Fission track analysis of the crystalline basement rocks of the Calabrian Arc, southern Italy: evidence of Oligo-Miocene late-orogenic extension and erosion. *Tectonophysics*, 238: 331-352.

Thomson S.N., 1998. Assessing the nature of tectonic contacts using fission-track thermochronology: an example from the Calabrian Arc, southern Italy. *Terra Nova*, 10: 32-36.

Tortorici L., Monaco C., Tansi C., Cocina O., 1995. Recent and active tectonics in the Calabrian arc (southern Italy). *Tectonophysics*, 243: 37-55.

Trincardi F., Correggiari A., Field M.E., Normark W.R., 1995. Turbidite deposition from multiple source: Quaternary Paola Basin (eastern Tyrrhenian Sea). *J. Sedim. Res.*, B65: 469-483.

Trop J.M., Ridgway K.D., 1997. Petrofacies and provenance of a late Cretaceous suture zone thrust-top basin, Cantwell basin, central Alaska range. *J. Sedim. Res.*, 67: 469-485.

Turco E., Maresca R., Cappadona P., 1990. La tettonica plio-pleistocenica del confine calabro-lucano: modello cinematico. *Mem. Soc. Geol. It.*, 45: 519-529.

Vai, G.B., 1987. Migrazione complessa del sistema fronte deformativo-avanfossa-cercine periferico: il caso dell'Appennino settentrionale. Mem. Soc. Geol. It., 38: 95-105.

Vai G.B., 1992. Il segmento calabro-peloritano dell'orogene ercinico. Disaggregazione palinspastica. *Boll. Soc. Geol. It.*, 111: 109-129.

Valente A., 1991. Caratteri sedimentologici di una successione torbiditica nel Cilento orientale (Appennino meridionale). *Mem. Soc. Geol. It.*, 47: 191-196.

Valente A., 1993, Studi sedimentologici sulla successione miocenica di Monte Sacro (Flysch del Cilento). Università di Napoli, [PhD Thesis], 170 pp.

Van Dijk J.P., 1990. Sequence stratigraphy, kinematics and dynamic geohistory of the Crotone Basin (Calabrian Arc, central Mediterranean): an integrated approach. *Mem. Soc. Geol. It.*, 44: 259-285.

Van Dijk J.P., Okkes M., 1991. Neogene tectonostratigraphy and kinematics of Calabrian basins: implications for the geodynamic of the central Mediterranean. *Tectonophysics*, 196: 23-60.

Van Dijk J.P. , Bello M., Brancaleoni G.P., Cantarella G., Costa V., Frixa A.,Golfetto F., Merlini S., Riva M., Torricelli S., Toscano C., Zerilli A.,2000. A regional structural model for the northern sector of the Calabrian Arc (southern Italy.) *Tectonophysics* 324 (2000) 267–320

Wallis S.R., Platt J.P., Knott S.D., 1993. Recognition of syn-convergence extension in accretionary wedges with examples from the Calabrian Arc and the eastern Alps. *Am. J. Sc.*, 293: 463-495.

Waschbusch P.J., Royden L.H., 1992. Episodicity in foredeep basins. *Geology*, 20: 915-918.

Weltje G.J., 1992. Oligocene to early Miocene sedimentation and tectonics in the southern part of the Calabrian-Peloritan Arc (Aspromonte, southern Italy): a record of mixed-mode piggy-back basin evolution. *Basin Res.*, 4: 37-68.

Westaway, R., 1993. Quaternary uplift of Southern Italy. *J. Geophys. Res.*, 98: 21741-21772.

Wezel F.C., 1970a. Geologia del Flysch Numidico della Sicilia nord-orientale. *Mem. Soc. Geol. It.*, 9: 225-280.

Wezel F.C., 1970b. Numidian Flysch: an Oligocene-early Miocene continental rise deposit off the African Platform. *Nature*, 228: 275-276.

Wezel F.C., 1985. Structural features and basin tectonics of the Tyrrhenian Sea. In: D.J. Stanley, F.C. Wezel (eds.), Geological Evolution of the Mediterranean Basin, 153-194. New York, Springer-Verlag.

Wildi W., 1983. La chaîne tello-rifaine (Algérie, Maroc, Tunisie): structure, stratigraphie et évolution du Trias au Miocène. *Rev. Géol. Dyn. Géogr. Phys.*, 24: 201-297.

Wiltschko D.V., Door J.A. Jr., 1983. Timing of deformation in overthrust belt and foreland of Idaho, Wyoming, and Utah. *Am. Ass. Petrol. Geol. Bull.*, 67: 1304-1322.

Zanettin Lorenzoni E., 1982. Relationships of main structural elements of Calabria (southern Italy). *Nues Jabh. Geol. Paläont. Mh.*, 7: 403-418.

Zeck S.E., 1990. The exhumation and preservation of deep continental crust in the northwestern Calabrian arc, southern Italy. University of California, Santa Barbara [PhD Thesis], 277 pp.

Zuffa G.G., 1987. Unravelling hinterland and offshore paleogeography from deep-water arenites. In: J.K. Leggett, G.G. Zuffa (eds.), Deep-Marine Clastic Sedimentology. *Concepts and Case Studies*, 39-61. London, Graham and Trotman.

Zuffa G.G., De Rosa R., 1978. Petrologia delle successioni torbiditiche eoceniche della Sila nord-orientale (Calabria). *Mem. Soc. Geol. It.*, 18: 31-55.

Zuppetta A., Mazzoli S., 1997. Deformation history of a synorogenic sedimentary wedge, northern Cilento area, southern Apennines thrust and fold belt, Italy. *Geol. Soc. Am. Bull.*, 109: 698-708.

Zuppetta A., Russo M., Turco E., Bartoli A., 1984. Nuovi dati sul <<Flysch di Nocara>> (Calabria settentrionale). *Boll. Soc. Geol. It.*, 103: 615-622.

Part 2

Around the Adriatic

7

Two Tectonic Geomorphology Studies on the Landscape and Drainage Network of Chain and Piedmont Areas of the Abruzzi Region (Central Apennines, Italy)

Miccadei Enrico and Piacentini Tommaso
Laboratory of Tectonic Geomorphology and GIS,
Dipartimento di Geotecnologie per l'Ambiente ed il Territorio (DIGAT),
Università degli Studi "G. d'Annunzio" Chieti-Pescara
Italy

1. Introduction

"Landscapes in tectonically active areas result from a complex integration of the effects of vertical and horizontal motions of crustal rocks and erosion or deposition by surface processes. In a sense, many landscapes can be thought of as resulting from competition among those processes acting to elevate the Earth's surface and those that tend to lower it" (Burbank and Anderson, 2001). Extracting information from deforming landscapes with an integrative approach is the main subject of tectonic geomorphology.

The landscape features have different dimensional scales that correspond to different tectonic implications. The major landforms (from continent scale to orogen scale; 10^7-10^4 km) and the intermediate ones (from mountain belt scale to ridge and valley scale; 10^4-10 km) result from the interaction of both endogenic and exogenic processes, with a dominancy of one over the other at different places and times, while the minor ones (at single landform scale; 10-10^{-2} km) are related to tectonics (tectonic landforms) or to erosion processes (i.e. fluvial, slope, glacial landforms) as defined since the beginning of the history of tectonic geomorphology (Gerasimov, 1946; Cailleux and Tricart, 1956; Mescerjakov, 1968; Ollier, 1981, 1999; Morisawa and Hack, 1985; Panizza and Castaldini, 1987; Ascione and Cinque, 1999; Burbank and Pinter, 1999; Burbank and Anderson, 2001; Peulvast and Vanney, 2001; Scheidegger, 2004). At intermediate to small scale, mountain belts, and related piedmont, are one of the main subjects of tectonic geomorphology. At this scale, drainage basins are the key features in the landscape. Basins consist of river channels, hill slopes, crests of interfluves and drainage divides that define the shape of the catchment. Some of these elements will respond more rapidly to changes imposed on them than others, according to the combination of many factors such as lithology, local tectonics, rock uplift/subsidence and climate changes (Morisawa and Hack, 1985; Kühni and Pfiffner, 2001; Twidale, 2004).

In tectonically active landscapes, changes in the incision/aggradation behaviour of the rivers are associated with the variations in climate and tectonics (Schumm, 1969; Bull, 1991; Merrits et al., 1994; Ascione and Cinque, 1999; Burbank and Pinter, 1999; Pazzaglia and

Brandon, 2001; D'Agostino et al., 2001; Pazzaglia, in press). Such changes occur following a specific sequence involving incision, valley widening and aggradation, and tend to form a series of fluvial terraces. Otherwise, local tectonics tend to shape certain landforms such as river bends, linear valleys, beheaded and hanging valleys, knick points, counterflow confluences of streams and alluvial fans (Miccadei et al., 2004; D'Alessandro et al., 2008; Della Seta et al., 2008 and references therein). The analysis and correlation of these features within the drainage basins allows for the discovery and definition of geomorphic markers of tectonics, as well as its timing. A major influence in landscape is certainly due to rock material properties, although at different scales there are different influences of rock material properties on landscape evolution. The geomorphological features of mountain areas shaped on hard rocks are well recorded in the general configuration of topography and in well preserved tectonic landforms. In piedmont areas, or in general in areas developed on soft rocks, the evidence of tectonics in the landscape is less clear. In these contexts only integrative studies based on (a) terrain analysis, (b) morphostructural analysis of the relief, (c) analysis of geomorphic markers such as certain landforms (geomorphological evidence of tectonics) and deposits (developed in continental environment), (d) drainage basins' analysis and morphometry, and (e) dating of deposits and landforms, provide clear indications concerning the role of tectonics in the landscape evolution.

Central Italy is characterized by a recent (Pliocene to present) geomorphological history and in this area several studies have been carried out at both local and regional scale, based on the integrative approach, by means of tectonic geomorphology methods (D'Alessandro et al., 2003; Miccadei et al., 2004; Ascione et al., 2008; D'Alessandro et al., 2008; Della Seta et al., 2008). In this paper two studies are presented on chain areas and piedmont areas in order to outline the methodological approach focused to decipher the role of morphotectonics and selective erosion in the landscape evolution (Fig. 1):

- chain area - escarpment between the Montagna del Morrone ridge and the Sulmona tectonic basin (central Abruzzi);
- piedmont area - dip stream valley (Sangro river valley, south-eastern Abruzzi)

These studies allow for the characteristics of the main morphostructural domains of central Italy (chain area and piedmont area) to be outlined and suggest, in general, the use of a similar methodological approach, but focused also on different geomorphological landscapes.

2. Study area

The Abruzzi region is located on the eastern slope of the central Apennine (central Italy). The geomorphological evolution of the region is related to a complex geological and structural framework developed since the Late Miocene with the formation of the Apennine thrust belt as part of the Mediterranean mountain system. The whole region has been affected since the Pliocene by extensional tectonics, uplift processes and strong morphostructural processes that have induced very active geomorphological processes. These processes have outlined and shaped the major morphostructural domain in the Abruzzi area: Apennine chain, Adriatic piedmont and Adriatic coastal plain (Fig. 1) (Patacca and Scandone, 2007).

At regional scale, the geomorphological analysis and the correlation to geological and structural characteristics allowed the identification, and the morphostructural characterisation, of the major landforms (Fig. 2; D'Alessandro et al., 2003). The results point out the clear coherence of present landforms with the tectonic framework in the Abruzzi area. In the chain area, exhumed thrust ridges and faulted homocline ridges are present (generally NW-SE, NNW-SSE, N-S), separated by tectonic valleys, fault line valleys, tectonic basins and tectonic-karstic basins, partially filled up with continental deposits. In the piedmont periadriatic area the most important morphological elements are represented by homocline relief (gently NE dipping), mesa relief, dip-stream valleys (SW-NE) and alluvial plains. The latter grade eastwards towards the coastal plain.

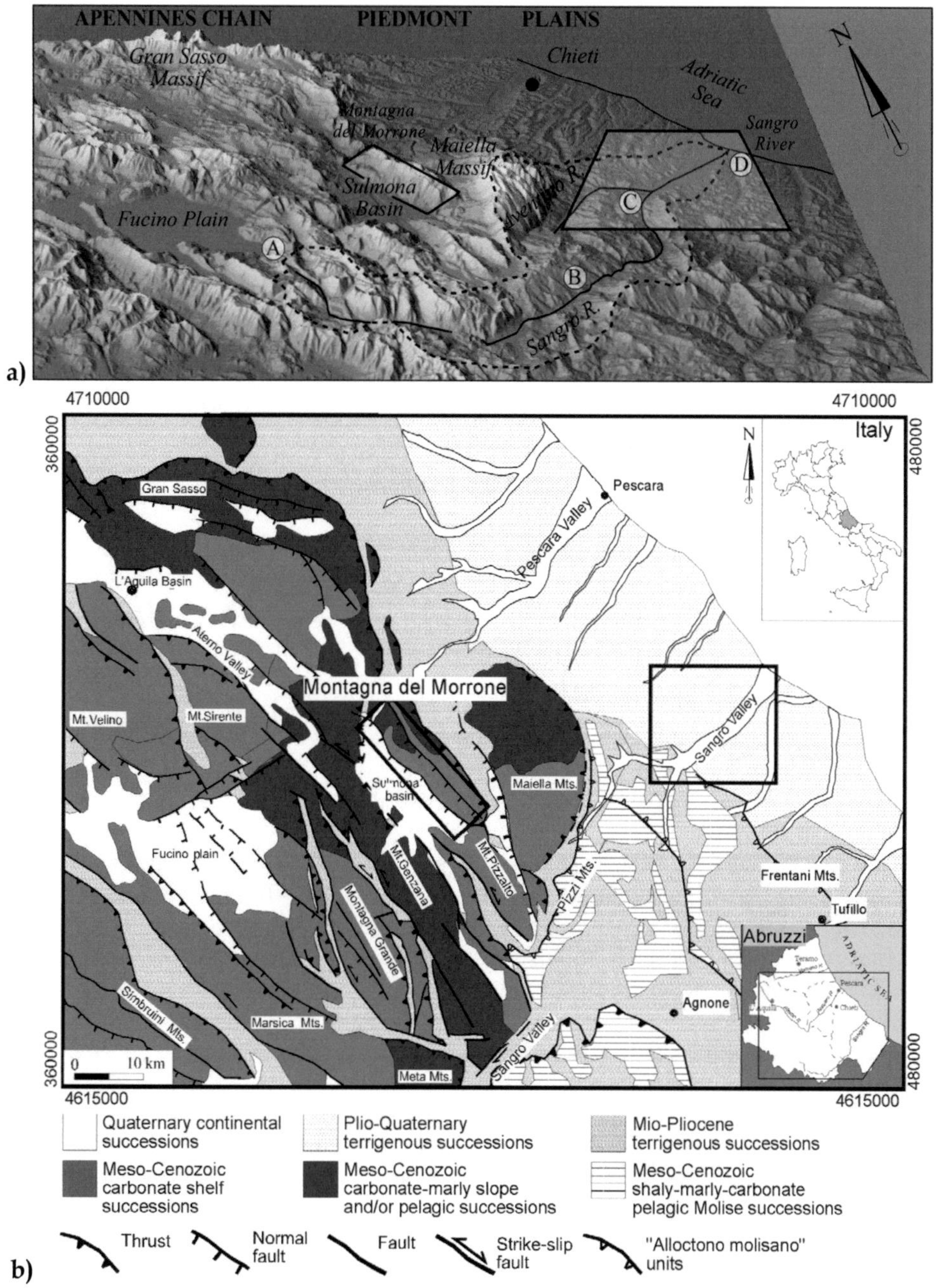

Fig. 1. Shaded relief image (a) and geological (b) map of the Abruzzi area (central Italy). Black boxes indicate the study areas

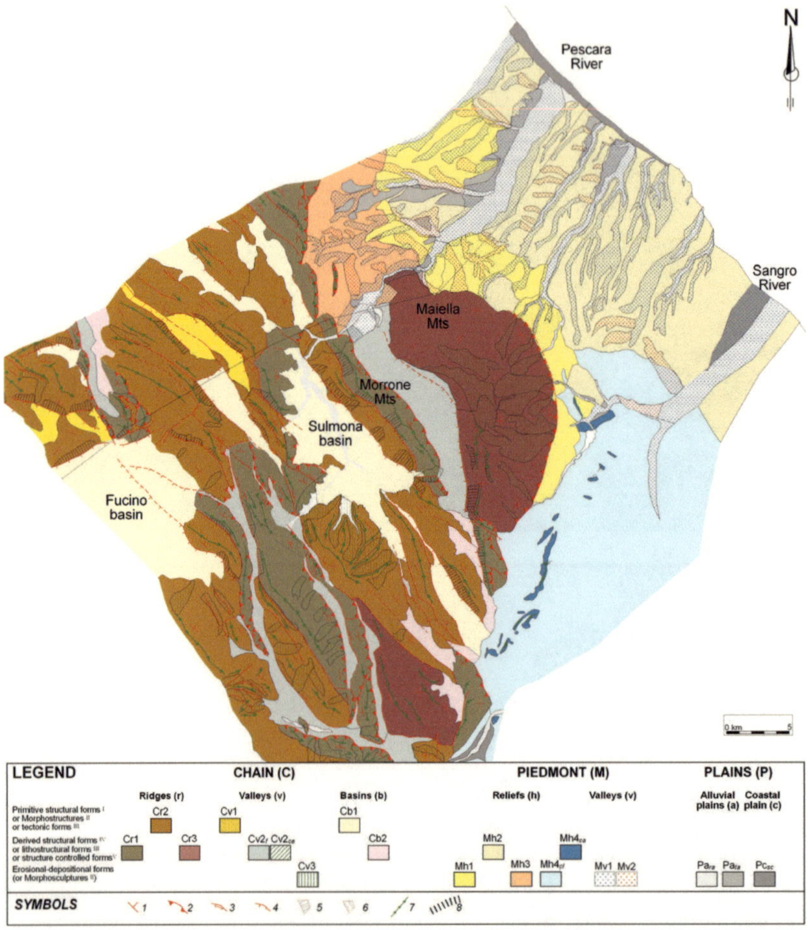

Fig. 2. Map of the morphostructural elements of central-eastern Abruzzi (modified from D'Alessandro et al., 2003).

LEGEND: CHAIN (C). Ridges: Cr1) *Exhumed thrust ridges*; Cr2) *Faulted homoclinal ridges*; Cr3) *Exhumed anticline ridges*. Valleys: Cv1) *Tectonic valleys*; Cv2) *Fault line valley* ($_f$ Neogene arenaceous-clayey foredeep sequences; $_{ca}$ Mezo-Cenozoic carbonate sequences); Cv3) *Transversal and Radial valleys*. Basins: Cb1) *Tectonic basin*; Cb2) *Karst-tectonic basin*. – PIEDMONT (M). Relief: Mh1) *Homoclinal relief* (Plio-Quaternary clayey-sandy sequences); Mh2) *Mesa relief* (Plio-Quaternary clayey-sandy-conglomeratic sequences); Mh3) *Eroded thrust relief* (Neogene arenaceous-clayey foredeep sequences); Mh4) *Hills on a chaotic and folded clayey-calcareous assemblage* ($_{cl}$ *"Argille varicolori"* auctorum complex and Cenozoic marly sequences; $_{ca}$ Meso-Cenozoic carbonate sequences); Mv1) *Dip-stream valleys*; Mv2) *Strike-stream valleys*. – PLAINS (P). Pa) Alluvial plains ($_{ra}$ recent alluvial deposits; $_{ta}$ Pleistocene terraced alluvial deposits); Pc) Coastal plain ($_{sc}$ recent sandy and conglomeratic deposits)

SYMBOLS: *1)* Regional attitude; *2)* Thrust (Miocene-Pliocene activity); *3)* Strike-slip fault (?Pliocene activity); *4)* Normal fault (Upper Pliocene - Quaternary activity); *5)* Major fault scarp; *6)* Major fault related scarp; *7)* Major crest; *8)* Primary drainage divide

[I]) Ascione and Cinque, 1999; Peulvast and Vanney, 2001. [II]) Gerasimov, 1946; Mescerjakov, 1968; Panizza, 1997. [III]) Bartolini, 2002. [IV]) Peulvast and Vanney, 2001. [V]) Ascione and Cinque, 1999

The piedmont of Abruzzi region is characterized by a low relief hill landscape (i.e. cuesta, mesa, plateau reliefs) on Mio-Plio-Quaternary terrigenous deposits, related to sin-, late-orogenic phases of the Apennines, by post-orogenic Quaternary marine regressive deposits and fluvial continental deposits. The transition from marine to continental deposits dates the emersion of the area and the starting point of the drainage evolution at the late Lower Pleistocene - early Middle Pleistocene. The Pleistocene fluvial landscape evolution of the piedmont area and the comprehension of the role of tectonics is an intriguing issue, being a key area for the Apennines' geodynamics, at the transition between compressional active tectonic areas, towards the east (Adriatic) and extensional active tectonic areas towards the west (Apennines chain).

3. Methods

The tectonic geomorphology studies presented in this work are carried out at drainage basin scale by means of: cartographic analysis and morphometry of orography and hydrography (map- and DEM-based), photogeology analysis, Quaternary continental deposits, fluvial terraces and morphotectonic detail field mapping, and morphotectonic cross section drawing. The topographic data in vector and in raster format were provided by Struttura Speciale di Supporto Sistema Informativo Regionale of Abruzzi Region (http://www.regione.abruzzo.it/xcartografia/).

Orography analysis is based on the 40m DEM. Within GIS software slope maps, energy of relief maps and hypsometry maps were realized. Hydrography analysis is based on 40m DEM and scale 1:25,000 topographic maps. Longitudinal profile, drainage density, azimuth of the drainage network, patterns and hydrography parameters were mapped (bifurcation parameters, hierarchic parameters, areal parameters etc.) in order to define the drainage development, to outline the control of morphological alignments and to suggest the tectonic control on basin arrangements (Horton, 1945; Miller, 1953; Schumm, 1956; Strahler, 1957; Avena et al., 1967; Avena and Lupia Palmieri, 1969; Ciccacci et al., 1992, 1995; Keller and Printer, 1996).

Photogeology analysis is performed on scale 1:33,000 aerial photos (Abruzzi Region Flight, 1982-1987) for the preliminarily mapping of main landforms (tectonic and structural landforms, fluvial landforms, slope landforms etc.). Structural geomorphology field mapping is carried out on 1:10.000 scale investigating bedrock geology, Quaternary continental deposits (fluvial, alluvial fan, slope deposits), main tectonic elements and morphotectonic evidence (ridges, slopes, valleys, hydrography forms and fluvial terraces) according to main national guidelines and national and international literature (Ambrosetti et al., 1976; ENEL, 1981; Panizza and Castaldini, 1987; Ciccacci et al., 1986; SGN, 1992, 1994; Leeder and Jakson, 1993; GNGFG, 1994; Centamore et al., 1996; Lupia Palmieri et al., 1996; Molin et al., 2004; Frenkel and Pazzaglia, 2005; APAT, 2007; Bull, 2007; D'Alessandro et al., 2008; Della Seta et al., 2008; ISPRA, 2009; Picotti et al., 2009; Miccadei et al., 2011; Pazzaglia, in press). The coupling of Quaternary continental deposits, fluvial terraces and morphotectonic evidence by means of morphotectonic profiles allowed finally to outline the relationships between forms and deposits.

These methods, taking into account the relationship between forms and deposits, outlined by morphotectonic mapping and morphotectonic profiles, can contribute to defining the main steps of landscape evolution and the major control on it (tectonics, rock properties, climate change etc.), and to estimate the timing of landscape development.

4. Case studies - chain area: tectonic basin and fault escarpment (Montagna del Morrone ridge)

4.1 Introduction

Montagna del Morrone (2061 m a.s.l.) is one of the main central Apennine ridges (central-eastern part of the Abruzzi Apennines; Fig. 1). It is made up of marine Meso-Cenozoic carbonate rocks, forming an asymmetrical anticline fold with a NW-SE axis, NE verging and overthrust onto Neogene terrigenous deposits. The SW slope is broken by several normal fault systems, NW-SE striking and SW dipping, which separate the ridge from the Sulmona tectonic basin (Fig. 1, 2; Miccadei et al., 1999; Doglioni et al., 1998; Miccadei et al., 2004). This slope shows a complex physiography, both longitudinal and transversal, with secondary ridges, scarps, gentle slopes or counter slopes (Fig. 3). The summit is gently undulating in the southern part, while the northern part is a narrow crest. At the base of the slope a sharp slope change joins the wide plain of the Sulmona tectonic basin and corresponds with one of the main normal fault lines of the Abruzzi area. Many ephemeral streams drain the slope down to the basal break forming alluvial fans.

The Sulmona tectonic basin is a half graben with a NW-SE master fault forming its eastern boundary along the Mt. Morrone slope and is filled by a thick sequence of lacustrine, fluvial and slope Middle-Upper Pleistocene deposits (Miccadei et al., 1999; 2002). It shows a peculiar physiographic setting: the lowest mean topographic height (300 m) in central Apennines intramontane basins, a strong relief (2000 m) up to the eastern ridge (Mt. Morrone) and an anomalous triangular shape. Here, a complex fluvial drainage system converges (Aterno River from NW, Sagittario River from SW and S, Gizio River from S, Vella River and Velletta River from SE) and flow into Pescara River.

The collected data (orography, hydrography, Quaternary continental deposits, morphotectonic evidence) allow us to define geomorphic markers of tectonics and to outline Quaternary landscape evolution of the escarpment between the Sulmona basin and Montagna del Morrone. In order to couple deposits and landforms, six morphotectonic sections are presented (three ridge transversal profiles representative of the northern, central and southern sectors; three stream channel and interfluve profiles representative of the drainage basins).

Fig. 3. Panoramic view (from SW) of the Monte Morrone SW escarpment

4.2 Results

The analysis of the morphotectonic features of the area is based on the investigation of geology (bedrock units, superficial deposits, tectonics and neotectonics) and geomorphology (structural, slope, karst and fluvial landforms, and alluvial fans).

4.2.1 Geology

Bedrock units are made up of carbonate rocks of Lias to Paleogene age divided into units according to their resistance to weathering and erosive processes (Fig. 4a): *bedded carbonate rocks* (outcropping in the northern area), *massive carbonate rocks* (outcropping in the central area), *carbonate rocks in thick beds* (outcropping in the southern areas), *dolomite rocks* (outcropping in the lower part of the slope in the northern and central sectors).
Such rock formations, as documented in the relevant literature, can be referred to various Meso-Cenozoic palaeogeographic domains: slope-basin in the northern sector, margin in the central sector and carbonate shelf in the southern sector (APAT, 2006).
Quaternary continental deposits (Fig. 4b) are essentially breccias and conglomerates that can be referred to talus slope and debris cones, to alluvial fans and to eluvial and colluvial covers. There are also chaotic breccias that can be accounted for by paleo-landslide. Based on comparisons with the sector of the Sulmona basin, these deposits can be placed between the Early Pleistocene and the Holocene (Sylos Labini et al., 1993; Carrara, 1998; Miccadei et al., 1999; Lombardo et al., 2001). These deposits are distributed non-homogeneously along all of the slope, but have good continuity at the base and in the mid-slope.
The SW escarpment of Montagna del Morrone is formed by the limb of the anticline structure disarticulated by systems of normal faults, known in the literature as the Monte Morrone fault zone (Vittori et al., 1995; Ciccacci et al., 1999; Miccadei et al., 2002, 2004). The bedrock formations are generally in counter-slope dipping strata, with NW- SE attitude and dipping from 20°NE (in the low part of the slope) to 70°NE (in the high part). Locally, at the base of the slope, there are also SW dipping strata (Fig. 4a).
There are two main normal fault systems with a predominantly N40°-50°W orientation (Fig. 4). These displace the bedrock formations and superficial deposits, and clearly display morphological evidence at different heights on the slope, corresponding to sharp slope breaks or clear fault scarps as described in the following paragraphs.
From the base upwards the main fault systems are as follows:
Basal border fault: this is a system of normal faults with Apennine orientation and SW dip which affect the bedrock formations (displacement higher than 1000 m), as well as the superficial deposits (estimated displacement of 700 m in the breccias of the Lower?-Middle Pleistocene [Miccadei et al., 2002] and up to several tens of metres in the alluvial fans of the Upper Pleistocene). Towards the north, it is made up of fault plains with N20÷30W orientation, 60SW dip and located at heights from 550 m to 650 m a.s.l. (Popoli, Roccacasale). Towards the south, it is made up of faults with attitude of N50÷60/50SW placed at heights between 750 and 800 m a.s.l. (Eremo di Celestino V, Pacentro).
The Schiena d'Asino fault: this is a system of normal faults with a N20÷30W strike and 60°÷50° SW dip (displacement in the bedrock formations over 1000 m) located at heights between 1100 and 1400 m a.s.l.. These fault plains, in the northern and central parts, are characterised by large rock fault scarps. The system continues southwards, but with a clear reduction of displacement and morphological evidence.
In the middle part of the slope, there is a secondary fault system with a N30÷50W orientation, widely covered by surface deposits. Minor faults with a NE-SW orientation and

limited extension are also present transversal to the slope, but mostly in the central sector; they are characterized by thick cataclasite strata. In general they correspond to small valleys, mostly covered by surface deposits. They can be interpreted as transfer elements between the Schiena d'Asino fault and the Basal border fault.

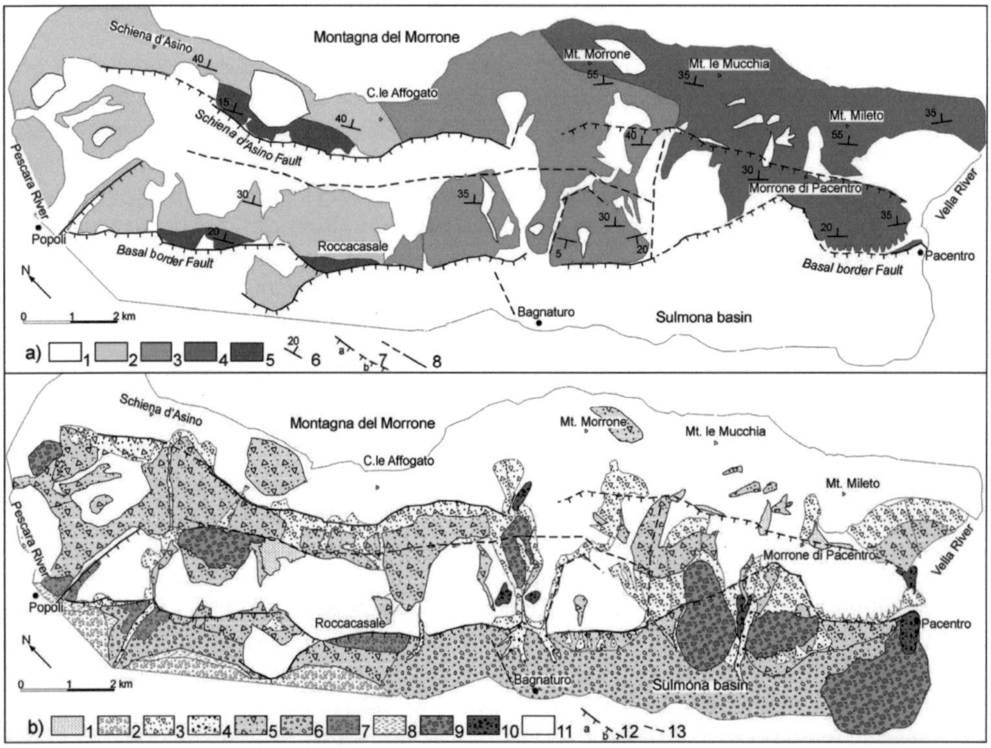

Fig. 4. Geological-structural scheme of the Montagna del Morrone SW slope (Miccadei et al., 2004). a) Bedrock units: 1) Superficial deposits; 2) bedded carbonate rocks; 3) massive carbonate rocks; 4) carbonate rocks in thick beds; 5) dolomite rocks; 6) attitude of strata and dip angle; 7) normal faults (a: visible slickenside; b: invisible slickenside); 8) faults. b) Quaternary continental deposits: 1) eluvial and colluvial deposits (Holocene); 2) sands and gravels, fluvial (Holocene); 3) loose stratified carbonate breccias, slope (Holocene); 4) loose carbonate gravel and breccias, alluvial fan (Holocene); 5) stratified carbonate breccias loose or poorly cemented, slope (Upper Pleistocene); 6) heterometric carbonate gravel and breccias loose or poorly cemented, alluvial fan (Upper Pleistocene); 7) heterometric carbonate gravel and breccias loose or poorly cemented, alluvial fan (late Middle Pleistocene); 8) limestone and clayey-silt, lacustrine (Middle Pleistocene); 9) heterometric and chaotic breccias, paleo landslide (Lower?-Middle Pleistocene); 10) cemented carbonate gravel and breccias, alluvial fan (Lower?-Middle Pleistocene); 11) bedrock formations (Meso-Cenozoic); 12) normal fault (a: visible slickenside; b: invisible slickenside); 13) fault

4.2.2 Geomorphology

This section includes results from orography and hydrography analysis and from geomorphological field mapping. Particular attention has been devoted to the morphometric analysis of the slope, of the drainage network and basins and of the alluvial fan/catchment systems. The main landforms mapped on the slope, both erosional and depositional, have been defined along with their relative morphogenetic agents. These data are described and discussed in the following paragraphs.

4.2.2.1 Orography

The analysis of orography and slope has outlined a NW-SE straight slope 20 km long and up to 1700 m high (Fig. 5) connecting the Sulmona basin (~350 m a.s.l.) and the Morrone ridge (2061 m a.s.l.). The planar and profile form of the slope consists of concave and convex elements, but mostly of planar segments and sharp breaks (Fig. 5). The orography analysis brought to light a strong longitudinal and transversal heterogeneity, and enables us to distinguish three sectors:

- northern sector, a double-ridged slope formed by two relatively down-faulted and uplifted blocks along the two major normal faults. It is made up of two rectilinear steep free faces, gently undulated (concave and convex) in plan, separated by an undulated horizontal or counter slope element.
- central sector, made up of three main units, a rectilinear steep free face upslope, a gently sloping midslope, undulated in plan, and then again a steep lower slope.
- southern sector, a single major uplifted block. The toe-slope is always marked by a sharp junction passing to the alluvial fan area with a high piedmont angle and it is broken by the mouths of narrow transversal valleys.

The Sulmona basin is a wide plain at 350-400 m a.s.l. partly dissected by the main river valleys (Aterno river, Sagittario river) with abrupt 50-100 m high terrace scarps.

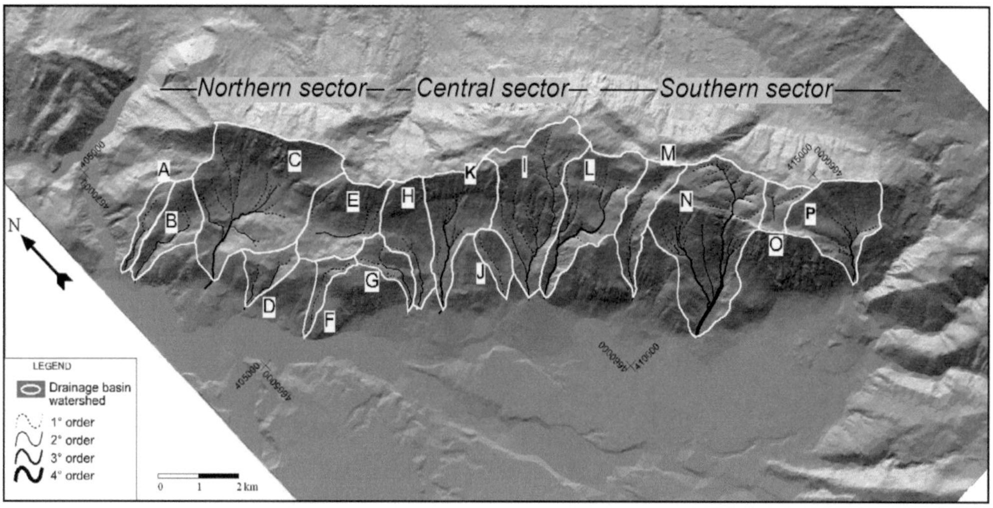

Fig. 5. Orography, drainage basins and network (ordered according to Strahler, 1957) of the Montagna del Morrone SW slope (Miccadei et al., 2004)

4.2.2.2 Hydrography

The drainage network on the steep and heterogeneous escarpment is made up of ephemeral stream channels and the toe-slope break, these stream channels become less defined, forming wide alluvial fans. The slope was subdivided into 16 basins (A-P; Fig. 5; Tab. 1, 2). Of these 16 basins, eight are spread from the line of the crest right to the base of the slope (C, H, I, K, L, M, N and P). Two are endoreic (E and O) on the upstream half of the slope and six develop on the downstream half of the slope (A, B, D, F, G and J). The relief of these basins varies from a minimum of 266 m at the endoreic Basin O, to a maximum of 1510 m at Basin I which extends down from the highest peak (2061 m a.s.l., Mt. Morrone) right to 551 m a.s.l. The total planimetric area of the 16 basins is about 50,0 km², while the total area of the slope is 74,3 km²: the slope is organised into drainage basins over 67% of its planimetric area, while the remaining 33% is made up of areas of interfluve.

Basin	Area (Km²)	Perim. (Km)	H (m)	L (Km)	Re	Rc	Rh	∫ips	ΣNu	ΣL (Km)	D	F
A	0,85	5,60	809	2,59	0,40	0,34	0,31	0,49	3	2,46	2,88	3,51
B	1,92	6,70	792	2,87	0,54	0,54	0,28	0,47	3	3,75	1,96	1,57
C	9,03	13,25	1335	3,85	0,88	0,65	0,35	0,41	20	15,71	1,74	2,21
D	1,15	5,21	533	2,04	0,59	0,53	0,26	0,68	7	3,77	3,27	6,07
E	3,36	7,88	1021	2,83	0,73	0,68	0,36	0,27	4	4,17	1,24	1,19
F	0,76	5,65	667	2,37	0,42	0,3	0,28	0,65	1	2,64	3,45	1,31
Average Northern sect.	*2,85*						*0,31*	*0,50*			*2,42*	*2,64*
G	1,11	5,58	596	2,08	0,57	0,45	0,29	0,67	4	3,26	2,92	3,59
H	2,62	8,20	1360	3,22	0,57	0,49	0,42	0,47	4	2,85	1,09	1,53
I	5,78	12,22	1510	4,50	0,60	0,49	0,34	0,56	16	13,93	2,41	2,77
J	0,93	4,23	588	1,81	0,60	0,65	0,32	0,59	1	1,83	1,97	1,08
K	3,57	9,73	1425	3,84	0,56	0,47	0,37	0,52	14	6,58	1,84	3,92
Average Central sect.	*2,80*						*0,35*	*0,56*			*2,05*	*2,58*
L	4,38	10,85	1464	4,25	0,56	0,47	0,34	0,60	13	9,81	2,24	2,97
M	2,05	8,62	1420	3,45	0,47	0,35	0,41	0,67	3	4,39	2,14	1,46
N	7,75	12,01	1436	4,44	0,71	0,68	0,32	0,64	47	22,67	2,93	6,06
O	0,91	4,68	266	*1,15*	0,94	0,52	0,23	0,45	5	1,93	2,11	5,47
P	3,72	8,43	1230	2,61	0,83	0,66	0,47	0,62	13	6,46	1,74	3,49
Average Southern sect.	*3,76*						*0,35*	*0,60*			*2,23*	*3,89*
Average all basins	*3,12*				*0,62*	*0,52*	*0,33*				*2,25*	*3,01*
Total	*49,89*											

Table 1. Main area and relief geomorphic indices of the basins: H) maximum relief; L) longitudinal length; Re) elongation ratio; Rc) circularity ratio; Rh) relief ratio; ∫ips) hypsometric integral; ΣNu) number of stream segments; ΣL) total stream segment length; D) drainage density; F) drainage frequency

The areal and relief properties of the basins (Tab. 1: Re, elongation ratio; Rc, circularity ratio; Rh, relief ratio; Schumm, 1956; Mayer, 1986; Keller and Pinter, 1996) were analysed, together with the hypsometric data (Tab. 1: ∫ips; Strahler, 1952), not simply to give an indication of the morpho-evolutionary stage, but also to identify the principal situations of disequilibrium and structural control.

Basin	Nu (1°)	Nu (2°)	Nu (3°)	Nu (4°)	ΣNu	Rb (1°-2°)	Rb (2°-3°)	Rb (3°-4°)	Rb aver.	Nd (1°)	Nd (2°)	Nd (3°)	Rbd (1°-2°)	Rbd (2°-3°)	Rbd (3°-4°)	Rbd aver.	R aver.
A	2	1	0	0	3	2,0				2	1	0	2,0				
B	2	1	0	0	3	2,0				2	1	0	2,0				
C	14	5	1	0	20	2,8	5,0		3,9	13	5	1	2,6	5,0		3,8	0,1
D	4	2	1	0	7	2,0	2,0		2,0	4	2	1	2,0	2,0		2,0	0,0
E	3	1	0	0	4	3,0				3	1	0	3,0				
F	1	0	0	0	1	0,0				1	0	0	0,0				
G	3	1	0	0	4	3,0				3	1	0	3,0				
H	3	1	0	0	4	3,0				3	1	0	3,0				
I	12	3	1	0	16	4,0	3,0		3,5	11	3	1	3,7	3,0		3,3	0,2
J	1	0	0	0	1	0,0				1	0	0	0,0				
K	10	3	1	0	14	3,3	3,0		3,2	8	3	1	2,7	3,0		2,8	0,3
L	9	3	1	0	13	3,0	3,0		3,0	6	3	1	2,0	3,0		2,5	0,5
M	2	1	0	0	3	2,0				2	1	0	2,0				
N	34	9	3	1	47	3,8	3,0	3,0	3,3	28	8	3	3,5	2,7		3,1	0,2
O	4	1	0	0	5	4,0				4	1	0	4,0				
P	10	2	1	0	13	5,0	2,0		3,5	7	2	1	3,5	2,0		2,8	0,8
Total	115	35	9	1	160					98	33	9					
Average									3,2							2,9	0,3

Table 2. Geomorphic indices of the drainage network of the 16 basins present on the SW escarpment of Montagna del Morrone. Nu) stream number; Rb) bifurcation ratio; Nd) number of streams flowing into higher order streams; Rbd) direct bifurcation ratio; R) bifurcation index

The northern sector shows elongated (Re=0,4-0,5) and irregular (Re=0,7-0,9) drainage basins with moderately high relief ratios (Rh=0,26-0,36; Tab. 1). The drainage pattern is heterogeneous, sub-dendritic in the upper part and parallel in the lower part. Only Basin C is extended across the whole escarpment, but it shows a clear downstream narrowing (Fig. 5). The downslope interfluves are made of triangular-shaped fault related slopes passing upslope to moderate transversal spur ridges and then to the horizontal undulated mid-slope. In the upper part the valleys are just notched into the uplifted block of the Schiena d'Asino fault. The stream channels have concave-convex profiles with moderate knick points. The hypsometric integrals have values lower than in the southern sector (0,5-0,3) and show concave-convex curves (Fig. 6, Tab. 1).

The southern sector shows elongated (Re=~0,5) and irregular (Re=0,7-0,9) drainage basins with high relief ratios (Rh=0,23-0,47; Tab. 1), separated in the interfluves by wide straight rectilinear slopes (Fig. 5). The drainage patterns are parallel in the lower part of the slope and rectangular on the summit, the drainage density is moderate (1,74-2,93), intermediate between central and southern sectors, the stream channel profiles are convex with sharp knick points, the stream channel/interfluve relief is low, the ipsometric integral show high values (>0,6) and convex curves (Fig. 6, Tab. 1),.

The central sector is in an intermediate situation: the catchments are developed all along the slope, except for a single case (Basin J), but they show a strong downstream narrowing (Re=~0,5; Rh=0,29-0,42). The drainage pattern is parallel, transversal to the slope (Fig. 5), and characterised by the lowest drainage density (1,09-2,92). The stream channel profiles are mostly planar with moderate knick points and the hypsometric integral values are intermediate (0,4-0,6) with moderately convex curves (Fig. 5, Tab. 1).

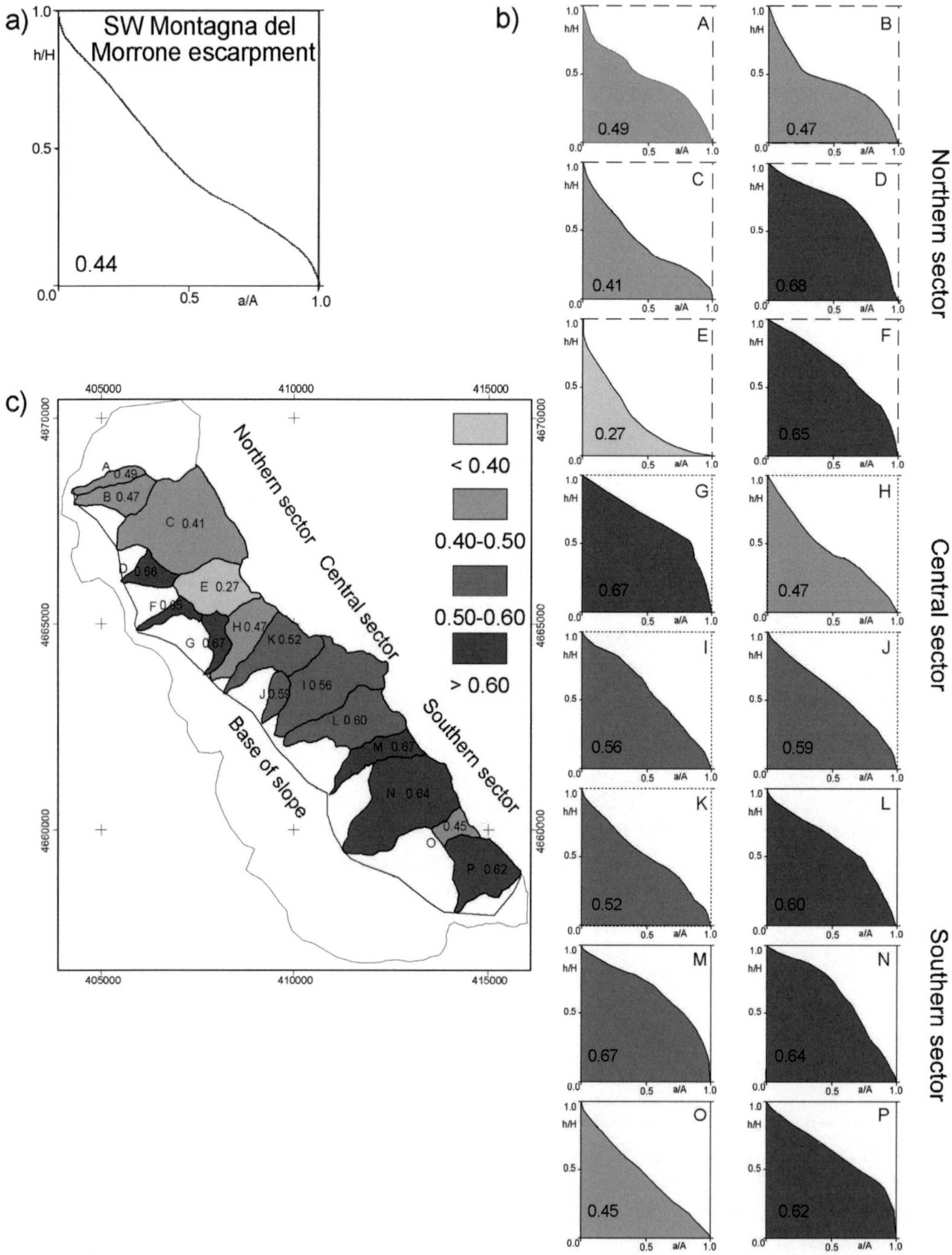

Fig. 6. Ipsometric analysis of the SW escarpment of Montagna del Morrone (Miccadei et al., 2004). a) Ipsometric curves and ipsometric integral value of the whole escarpment.
b) Ipsometric curves and ipsometric integral values of the 16 drainage basins.
c) Planimetric distribution of the ipsometric integral values

4.2.2.3 Structural landforms

The geomorphological surveys allow for the mapping of landforms such as fault scarps, fault related slopes and crests (Fig. 7), essentially controlled by the normal fault systems on the slopes.

Fault scarps are located on the slope, at various heights corresponding to the main fault lines. Three subtypes were distinguished: fault scarps, partly retreated and weathered fault scarps, retreated and weathered fault scarps (Demangeot, 1965; Wallace, 1977; Brancaccio et al., 1978; Bosi et al., 1993; Stewart and Hancock, 1994; Ascione and Cinque, 1997; Peulvast and Vanney, 2001).

The *fault scarps* are made up of rock scarps from some tens of metres to 100 m high, markedly straight, with a basal and a summit part. The basal part is made up of well smoothed scarplets, from a few decimetres to some metres high, 45° - 70° dipping. This typology occurs mainly in the northern sector of the ridge and in the upper parts of the slope. Along the basal fault line these can be identified between Popoli and Roccocasale, at heights from 400 m to 600 m.

The *partly retreated and weathered fault scarps* are rock scarps up to 100 m high, sinuous, 60°-35° dipping. The basal smoothed scarplets corresponding to the fault plain are only locally preserved and partially covered by scree slopes. Upslope the free faces have, to some extent, retreated from the fault line. Inactive talus deposits, and at certain points the apex of inactive alluvial fans, can show evidence of displacement. These features were identified at the base of the slope above all in the southern sector (Pacentro).

The *retreated and weathered fault scarps* can be identified as weak breaks in the slope, often discontinuous and partially or completely covered by surface deposits (talus debris and alluvial fans). These landforms are linked upslope to moderate and weathered rock scarps, which result from retreat of the fault scarps.

Fault related slopes: are made up of generally straight and rectilinear high angle slopes (30° - 60°), in limestone from stratified to massive, generally counter-slope plunging or sub-horizontal, bordered at the base by the different kinds of fault scarps described above. They are present particularly in the upper part of the slope, in the northern sector (Schiena d'Asino, C.le Affogato) and in the lower part (Popoli, Roccacasale, Pacentro) (Fig. 7). Especially in the central and southern sectors, they are incised by gullies and affected by slope processes that have formed talus slopes and debris cones, both inactive and active, at the base. In the lower part of the escarpment the fault related slopes, dissected by the outlets of the drainage basins, have a sub-triangular shape.

Crest lines: develop in a clear, sharp and slightly asymmetrical shape in the northern sector, while in the central and southern sectors they are more discontinuous, set upslope from a less inclined and gently undulating slope unit.

4.2.2.4 Slope landforms

Several slope landforms are mapped in the study area, even though non-homogeneously distributed: landslide scarps, rock slide bodies, talus slopes and debris cones and also evidence of deep seated gravitational slope deformations (Fig. 7).

Landslide scarps: are made up of arched or semi-circular rock scarps on limestone bedrock formations. They show a marked concave profile, but are generally very weathered. These forms are located on the higher parts of the slope, in the northern, central and southern sectors of the ridge.

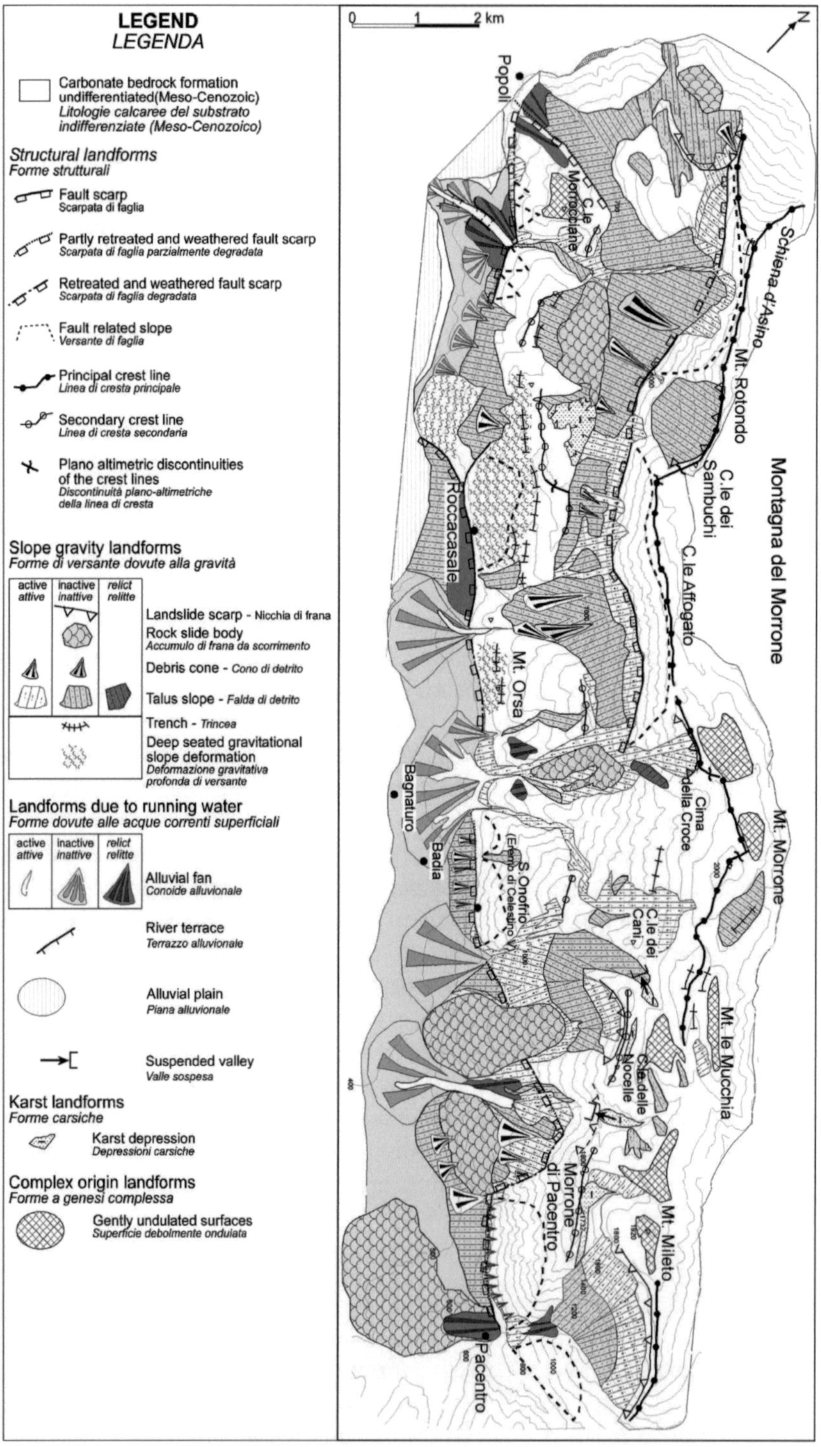

Fig. 7. Geomorphological map of the SW escarpment of Montagna del Morrone (Miccadei et al., 2004)

Rock slide bodies: are made up of limestone in large blocks and of heterometric carbonate breccias (up to boulder size) in a chaotic arrangement with abundant clay-silt matrix, or, in some cases, of considerable volumes of limestone stratified rock that still maintains the original lithostructural arrangement. They show with a surface up to 3 km^2 (length/width ratio of 2:1 to 1:2) and thickness up to hundreds of metres. The longitudinal profile of the slip surface and landslide body is always markedly concave-convex. The movement from the slip surfaces is variable from several hundred metres to several km. The movement is generally complex, however, it is principally attributable to translational and rotational rock slide mechanisms. Similarly to the slip surfaces, the landslide accumulation is also partially covered by talus slopes and debris cones, by active and inactive alluvial fans (Late Pleistocene - Holocene) and by some relict parts, which have been attributed to the Mid-Pleistocene (Miccadei et al., 1999). This suggests an Early?-Mid Pleistocene age for the landslides identified on the slope. They are therefore entirely inactive paleo-landslides; only a few minor ones can be attributed to more recent ages.

Talus slope and debris cones: are formed by bodies of heterometric carbonate breccias. Various inactive forms can be identified from the characteristics of the material, the abundance or absence of matrix, soil and vegetation or from the characteristics of overlooking rock slopes. They are present along the whole escarpment of Montagna del Morrone.

Deep seated gravity slope deformations (D.S.G.S.D.): some areas of the slope are interrupted by elongated trenches and sackung-like features running parallel to the slope itself (NW-SE to NNW-SSE orientation), which are some tens of metres wide and some hundreds of metres long (up to 1000 m). These depressions are in general partly filled with debris and colluvial deposits. These features bring to light the presence of D.S.G.S.D. (Cavallin et al., 1987; Crescenti et al., 1989; Dramis and Sorriso-Valvo, 1994). They are especially evident in the central-northern part of the slope at heights from 1200 m to 500 m, upslope from the principal fault slopes along the Basal border fault (between Popoli and Roccacasale). In some cases they can be identified also upslope from the fault slope of Schiena d'Asino. In the summit area of the southern sector the arrangement of the trenches and karst depressions lead to an elongated NW-SE oriented depression from several tens to hundreds of metres wide and several kilometres. These forms do not display signs of recent movement, but they are very evident indeed and have not be shaped or filled by the geomorphological processes.

4.2.2.5 Karst landforms and complex origin landforms

Mapped landforms are gently undulated surfaces, small karst depressions and suspended valleys (Fig. 7).

Gently undulated surfaces: are areas with gently undulating morphology shaped in the bedrock formation, at a height that ranges from 1800 m to 2000 m, close to the top of the ridge in the central and southern sectors. The occurrence of small dolines and karst valleys suggests that the karst weathering is an important morphogenetic factor.

Karst depressions: are closed depressions with irregular shapes, elongated with a NW-SE or SW-NE orientation, medium in size (length ranging from 500 to 1000 m, width ranging from some tens of metres to 200 m) and filled with residual soils and colluvium. They are located between Mt. Morrone and Morrone di Pacentro at heights of 1500 - 1750 m. Being suspended at these heights, some of these features have been preserved, while others were broken by the incision of the streams along the slope and, also in this case, by intersection with the slip plain of some of the major landslides (Fig. 7).

Suspended valleys: are small valleys, with a flat or concave floor, located in the upper part of the slope (central and southern sectors). They have a very gentle stream channel gradient, abruptly passing downstream to a high stream channel gradient. This creates knick points and strongly convex channel profiles.

4.2.2.6 Fluvial and water erosion landforms

Fluvial and water erosion landforms are mostly present in the lower part of the SW Morrone slope. Major landforms mapped are: fluvial terraces, alluvial plains, alluvial fans (Fig. 7). Alluvial fans are active, inactive or relict and were the subject of morphometric analysis of the fan/catchment systems.

Alluvial fans: all along the join between the slope and the plain several fans are present, ranging in size from several ha to 2,20 km^2 and with slope angles of up to more than 17°. The apex is located close to the Basal border fault, slightly upslope, entrenched in the fault related slopes and in the lower part of the catchments. Only the apex of Basin N is deeply entrenched, possibly because it is located between two wide landslide bodies (Fig. 7). The landforms are mostly inactive. The geometry and the spatial relationship between active and inactive forms indicate a general fan aggradation, except in the northern sector. Basin C, in particular, shows a clear entrenching of three subsequent fans and the formation of two orders of terraces.

Morphometric analysis of fan/catchment systems: the morphometric analysis on the main fan/catchment systems was processed in a GIS and on the DEM, following the most relevant literature (Bull, 1964; Saito, 1982; Blair and McPherson, 1994; Oguchi and Ohmori, 1994; Allen and Hovius, 1998; Allen and Densmore, 2000) and it is summarized in Tab. 3. The relationship between the main parameters and the interpolated functions are shown in Fig. 8. The first graph shows the relationship *fan area vs. catchment area* (Fig. 8a), defined by one of the most widely accepted functions ($A_f = k\ A_b^x$, where A_f = fan area , A_b = basin area, k and x = constant; Allen and Densmore, 2000), also defined by the ϕ ratio (fan area/catchment area; Allen and Hovius, 1998). Note the good alignment of most of the data except for a few anomalies (Basin C and N; triangular symbol in Fig. 8a).

The fan area was compared to the volume eroded from the catchments (EVc, Tab. 3, Fig. 8b), estimated as follows:

EVc = Vmax-Vc-TLVc (*Vmax* = volume of a prism with base corresponding to the catchment area and height to the catchment relief; *Vc* = volume between the catchment surface and a horizontal surface at the minimum height of the catchment; *TLVc* = estimated volume lacking because of the tectonic displacement along the Schiena d'Asino fault). In the third graph the relationship between the estimated fan volume (Vf) and the estimated volume eroded from the catchments (EVc) is shown (Fig. 8c). In both graphs the data distribution is similar to the first graph (Fig. 8a), but much more scattered; the anomalous data of Basin C and N is confirmed. The fourth graph (Fig. 8d) is similar to the first (Fig. 8a), but we must also consider the relief ratio (Rh) of the catchment in order to verify whether not only the dimension, but also the steepness could be an important factor in the geometry of the fans.

So the fan/catchment systems that seem to be anomalous in the previous graphs (Basin C and Basin N, triangular symbol in Fig. 8a,b,c,d) have been considered in detail. They both have a small alluvial fan, compared to the catchment area, and they have a low ϕ ratio value (fan area/catchment area) with respect to the other basins (Tab. 3). In the first case (Basin C) the deeply entrenched fans indicate the occurrence of deposition and erosion pulse, which led to the fan being undersize due to sediment removal. The geometry of the catchment and the distribution of surface deposits in it, indicate the presence of possible internal storage points that could have contributed to the undersizing of the fans, preventing the sediment supply. In the second case (Basin N) the geometry of the drainage pattern, the basin, the stream channel profile and the ipsometric integral (Fig. 6,7) suggest that the upper part of the catchment underwent a capture during the evolution of the slope. Therefore, the morphometric ratios were recalculated eliminating the supposed captured part (N* in Tab. 3). The four graphs of Fig. 8a′,b′,c′,d′ were reprocessed eliminating the anomalous data (Basin C, Basin N) and considering the recalculated data (N*): note the clear increase in the R^2 value of the regression line calculated. Particularly the approximation of the N* value to the tendency line could be an indirect confirmation of the capture process in the upper part of Basin N: the fan morphometry is still in equilibrium with the pre-capture catchment morphometry. Furthermore, note the increase of R^2 in the graph of Fig. 8d′ in relation to the value in the graph of Fig. 8a′ which suggests the influence of catchment steepness in defining the fan area.

Basin	Af (km^2)	Hf (km)	Lf (km)	Sf Hf/Lf	Vf (km^3)	Ac (km^2)	Rc (m)	Lc (km)	Rh c	LVc (km^3)	TLVc (km^3)	EVc (km^3)	∫ ips	ϕ
C (Tot)	1,64	0,160	1,64	0,10	43,7E-3	9,03	1335	3,850	0,35	7,05	2,00	5,05	0,41	0,18
C (pars)	0,46	0,070	0,85	0,08	5,4E-3	9,03	1335	3,850	0,35	7,05	2,00	5,05	0,41	0,05
K	1,44	0,200	1,44	0,14	47,9E-3	3,57	1425	3,836	0,37	2,48	0,30	2,18	0,52	0,40
I+L	2,20	0,320	1,95	0,16	117,3E-3	10,16	1510	4,500	0,34	6,23	1,60	4,63	0,58	0,22
J	0,46	0,225	1,17	0,19	17,1E-3	0,93	588	1,811	0,32	0,23	x	0,23	0,59	0,49
J+I+L	2,66	0,320	1,95	0,16	141,7E-3	11,09	1510	4,500	0,34	6,45	1,60	4,85	0,59	0,24
M	1,17	0,255	1,28	0,20	49,8E-3	2,05	1420	3,448	0,41	0,94	x	0,94	0,67	0,57
N	1,39	0,100	1,16	0,09	23,2E-3	7,75	1436	4,440	0,32	4,12	x	4,12	0,64	0,18
N*	1,39	0,100	1,16	0,09	23,2E-3	4,75	1360	2,950	0,46	2,58	x	2,58	0,60	0,29
N**	1,74	0,350	2,16	0,16	33,7E-3	4,75	1360	2,950	0,46	2,58	x	2,58	0,60	0,37

Table 3. Morphometric parameters of the main alluvial fan and related source catchments (C, northern sector; K, J, I, L, central sector; M, N, southern sector). Af) Fan area; Hf) Fan relief; Lf) Fan length; Sf) Average fan slope; Vf) Estimated fan volume; Ac) Catchment area; Rc) Catchment relief; Lc) Catchment length; Rh c) Catchment relief ratio; LVc) Catchment lacking volume; TLVc) Tectonic lacking volume; EVc) Estimated eroded volume; ∫ ips) Hypsometric integral; ϕ) Fan area/Catchment area; x) negligible; N*) without possible captured upper part of the catchment; N**) considering the entrenched apex

4.3 Discussion

4.3.1 Orography and hydrography

The distribution of slope and relief is irregular in relation to the tectonic setting (Fig. 9): in the southern sector, slope and relief are mostly in the lower part along the wide free face; in the central and particularly in the northern sector, slope and relief are mostly in the upslope, low in the midslope and increase again in the lower part down to the toe-slope break.

The hydrography analysis outlines a poorly developed drainage system with slow denudation processes and strongly controlled by extensional tectonics. The southern sector of the ridge is characterized by a poorly dissected morphology and a clear stage of inequilibrium. This is due to a strong lithological and tectonic control: a single block of resistant rocks relatively uplifted by the activity of the Basal border fault and poorly incised by the drainage network.

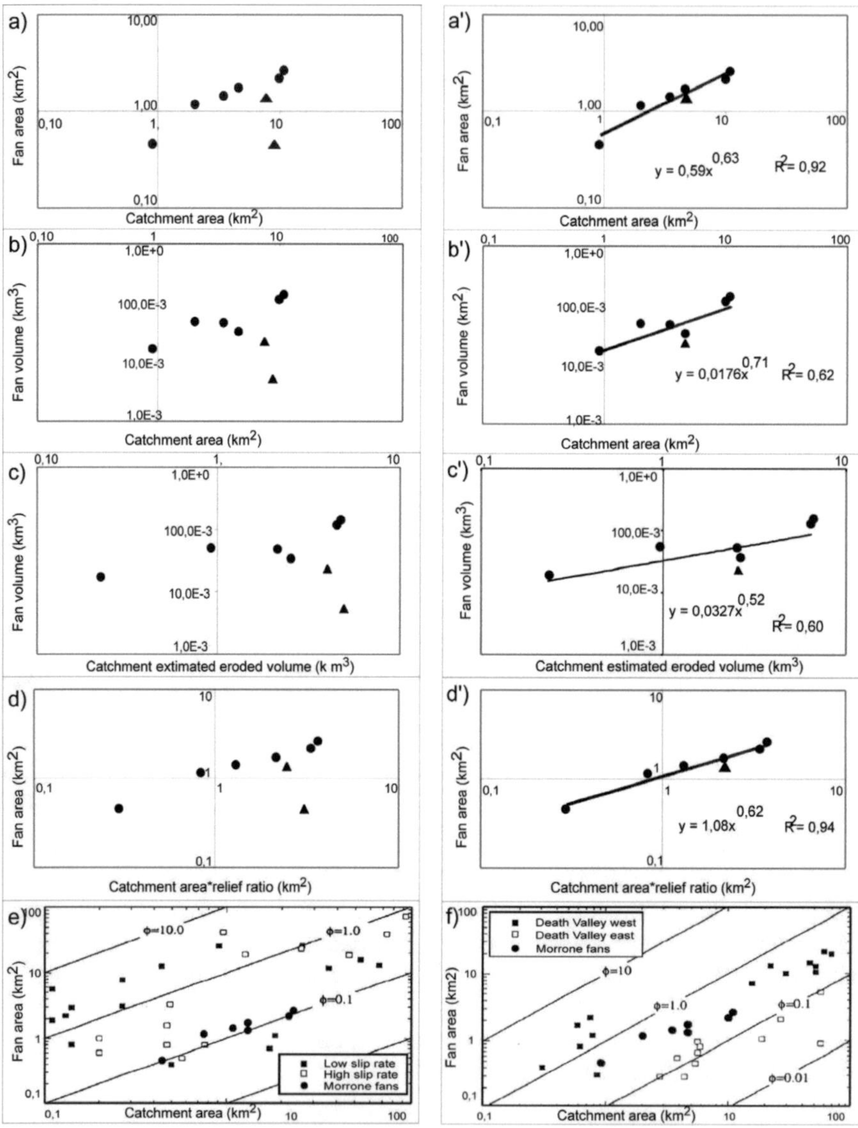

Fig. 8. Graphics illustrating the relationships between some of the main morphometric parameters of alluvial fan/catchment systems (refer to Tab. 3). Note the logarithmic axes; each symbol represents a single fan/catchment pair (Circle: normally developed fan/catchment systems; Triangle: anomalous developed fan/catchment systems; see text for detail). a) Fan area vs. catchment area. b) Fan volume vs. catchment area. c) Fan volume vs. catchment estimated eroded volume. d) Fan area vs. catchment area x relief ratio. a') b') c') d') are the same graphics of a, b, c, d, reprocessed eliminating and recalculating the anomalous data (see text for detail). e) comparison of values of ϕ ratio (fan area/catchment area) calculated on the Montagna del Morrone with value obtained from numerical modelling (Allen and Densmore, 2000). f) comparison of values of ϕ ratio (fan area/catchment area) calculated on the Montagna del Morrone with value calculated in different structural context (Death Valley, Nevada U.S., Allen and Densmore, 2000)

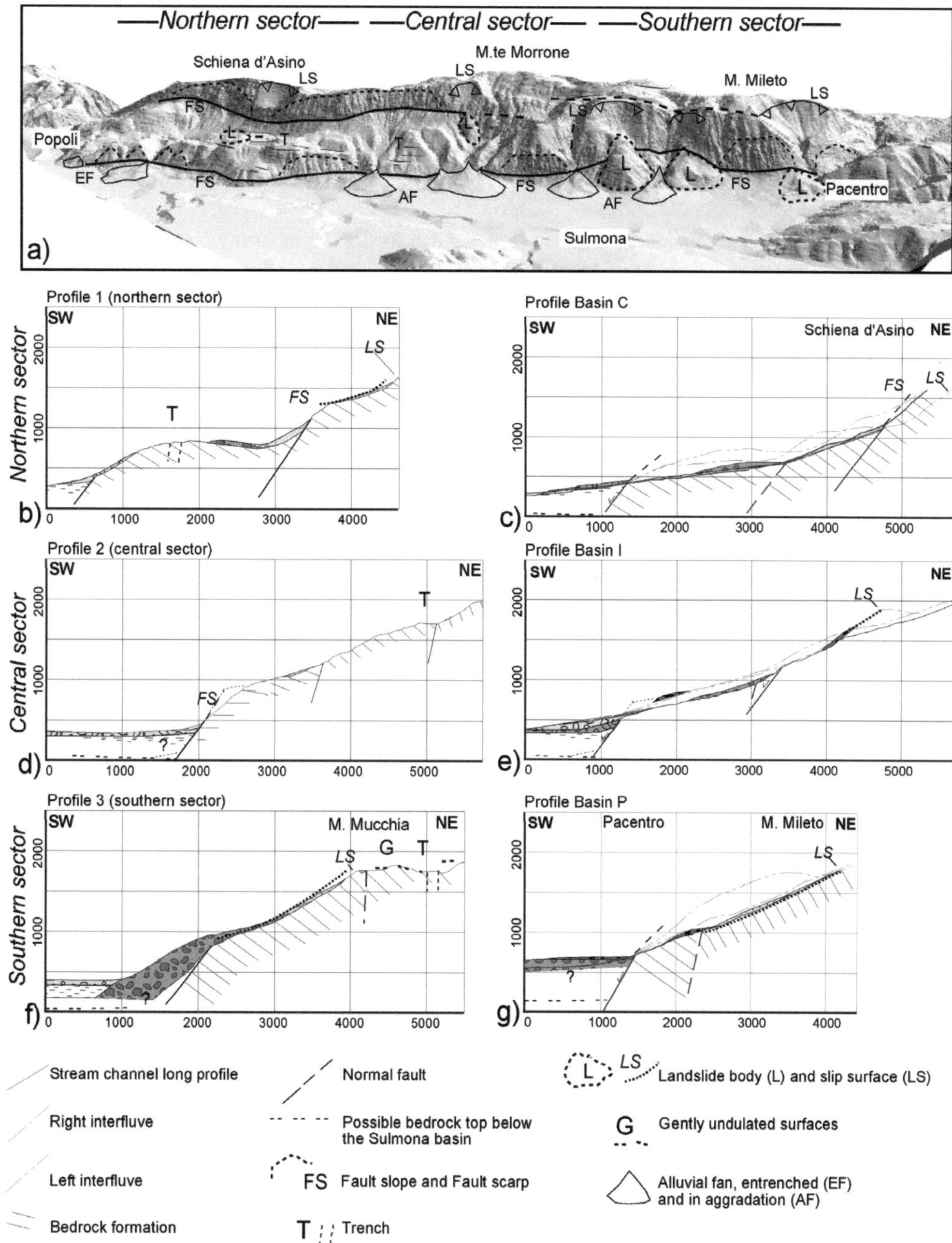

Fig. 9. a) Synthetic 3D morphostructural scheme profiles of the SW escarpment of the Montagna del Morrone (Miccadei et al., 2004). b, c) synthetic transversal and stream channel morphostructural profiles of the northern sector of the escarpment; d, e) synthetic transversal and stream channel morphostructural profiles of the central sector of the escarpment; b, c) synthetic transversal and stream channel morphostructural profiles of the southern sector of the escarpment (for the legend of the deposits see Fig. 4)

The different morphometry of the drainage of the northern sector is thought to be due not to a different development of the erosional processes, but to a different morphostructural setting of this sector of the escarpment. It consists of a double-ridge made up of two different blocks risen in parallel along the Basal border fault and the Schiena d'Asino fault, which have formed two separate fault related slopes with the slightly undulated area in between (Fig. 4, 9). This setting has led to the separation of the catchments between the upper and lower blocks of the slope and the concave-convex hypsometric curve of the basin developed throughout the escarpment (Basin C).
This setting of the central sector is controlled by the interplay of principal faults parallel to the ridge and secondary transversal faults (Fig. 4, 9): the southern termination of the Schiena d'Asino fault, with a reduced morphostructural role, has formed an upper fault related slope, but has not separated upper and lower catchments as in the northern sector. The relative uplift of the lower block along the Basal border fault and the conflicting drainage deepening brought about the downstream narrowing of the catchments. The secondary transversal faults control the development of the parallel drainage network.

4.3.2 Geomorphology

The geomorphological surveys allow for the mapping of structural landforms, slope landforms, karst landforms and fluvial and water erosion landforms.
The processes that have controlled the evolution of the escarpment are highlighted by the characteristics and degree of physical weathering, retreat of fault scarps and fault related slopes, and in particular by the analysis of transversal profiles (mostly rectilinear with more or less evident rock scarps, Fig 9) when compared to the distribution of slope depositional forms (rock landslides and talus slopes) (Fig. 7).
The variable degradation of the fault scarps and the morphology of the fault slopes (according to Brancaccio et al., 1978; Wallace, 1978; Blumetti et al., 1993; Bosi et al., 1993; Stewart and Hancock, 1994; Ascione and Cinque, 1997; Peulvast and Vanney, 2001), suggest a variable balance between the relative tectonic uplift, rejuvenating the fault scarps, and the slope denudation processes. Variability of rock resistance seems to have a control on the development of the geomorphic processes influencing the physical weathering because of the different type of stratification, degree of fracturing and local presence of cataclasite.
Moreover, it is worth noting that the upslope profile of several fault scarps is poliphasic (Fig. 9). This suggests again the cyclic alternation of relief building phases linked to tectonic activity and slope denudation events.
In the northern sector, the slope related to the Schiena d'Asino fault shows a profile made up of a clear fault scarp separating slope segments with different dip angles (Fig. 9b, c). Upslope there are many minor rock cliffs and secondary scarps, while downslope there is a talus slope. On the basis of the models proposed by the literature, particularly for the Apennine area (Demangeot, 1965; Brancaccio et al., 1978; Bosi et al., 1993; Ascione and Cinque, 1997), the slope is thought to be affected by a period of repeated tectonic activity with slope development by replacement with moderate sediment accumulation on the downfaulted block. A possible renewal of the tectonic activity would have formed the present basal fault scarp, which is only partly weathered. On the slope related to the Basal border fault, only triangular shaped fault related slopes, retreated and developed, are preserved (Fig. 9b,c; Brancaccio et al., 1978; Wallace, 1978). This clearly shows the role of drainage downcutting in the geomorphology of the lower part of the northern sector.
In the southern sector, the geomorphological characteristics of the escarpment indicate that the relative uplift has taken place mostly on the Basal border fault (Fig. 9f,g). The basal fault

scarp has in many cases clearly retreated and the fault line is covered by scree (Demangeot, 1965; Ascione and Cinque, 1997). Furthermore, on the fault related slope, there are wide rock landslide bodies and remnants of relict alluvial fans, referable to Early?-Mid Pleistocene age. This suggests an early stage of strong activity on the Basal border fault, leading to slope development by wide and sudden mass movements together with early slope replacement processes on the resistant, but highly jointed rocks. This created a steep slope, mostly planar, and supplied slope deposits along the slope down to the base, which are now preserved in remnants. The continuation of the fault activity, possibly at a reduced rate, has brought about a gradual slope development, shaping the basal fault scarps with a high sediment supply that has partly covered the fault lines, the relative scarplets and the landslide bodies placed on them (Fig. 7, 9a,f,g).

Several slope landforms are mapped in the study area, even though non-homogeneously distributed: landslide scarps, rock slide bodies, talus slopes and debris cones, and also evidence of deep seated gravitational slope deformations (Fig. 7, 9). The most significant landforms are large rock landslides mapped on the escarpment. Based on the geomorphological analysis, these landforms are thought to have started as deep seated gravitational slide deformation (D.S.G.S.D.), then evolved as large landslides (Dramis and Sorriso-Valvo, 1994; Dramis et al., 1995).

The distribution of such landforms is linked to the distribution of slope and local relief. In the southern sector of the ridge, the slope and local relief is concentrated in the basal part of the slope, corresponding to the Basal border fault related slope, where the main landslide bodies are located. Poor evidence of D.S.G.S.D. is mostly located in the summit area of the ridge. In the northern sector, however, the distribution of the slope and local relief in two parallel belts seems to have partly prevented the evolution of D.S.G.S.D. into landslides. Evidence for the former is in fact distributed along the lower part of the slope, while landslides are found only on the upper part of the slope, where the gradient becomes steep again.

On the basis of morpho-lithostratigraphic correlations with the relict alluvial fan deposits, these landslides can be dated to the Early?-Mid Pleistocene. The preparatory morphostructural conditions, such as high steep slope on carbonate jointed rocks, and the trigger causes, possibly related to strong seismicity necessary for the occurrence of this type of landslide, could be linked to an important morphotectonic phase during this period. This would have had a great effect on the morphogenesis of the slope. This is confirmed by the intense tectonic activity that took place between the Early Pleistocene and the Mid-Pleistocene, highlighted by various authors in the chain and periadriatic piedmont (Dramis, 1993; Bigi et al., 1996; Centamore and Nisio, 2003). So, possibly a large part of the relief of the slope should have already been formed in the early stages of the slope evolution (Early?-Mid Pleistocene) and would have further growth in later times, as confirmed by the geometry of the foot of the slip surfaces now suspended hundreds of metres above the base of the slope (Fig. 9 b, f, g).

Karst landforms and complex origin landforms on Mt. Morrone are found in the summit areas, as well as on several ridges of the eastern-central Apennines (Montagna Grande, Mt. Godi, Mt. Sirente, Monti Peligni, Maiella). These features have been attributed by many authors to remnants of a summit paleo-landscape and to different periods of shaping from the Late Miocene (Demangeot, 1965) to Late Pliocene-Early Pleistocene (Dramis, 1993; Coltorti and Farabollini, 1995; Centamore and Nisio, 2003). When considering the surface of the mid-slope in the northern sector, it is possible to identify a displacement of the undulated surface brought about by the Schiena d'Asino fault.

In our case, the landform characteristics and the geomorphological correlations with slope forms seem to suggest that the shaping of undulated surfaces and karst depressions may have started before the activity of the landslides between C.le delle Nocelle and Pacentro. This would allow the dating of the first genesis of these forms to a period before the Early?-Mid Pleistocene.

Geomorphological analysis of the alluvial fans has provided a significant contribution to the understanding of the morphostructural evolution of the escarpment and of its base junction with the Sulmona basin. The fans have been useful in defining the morphostratigraphic relationships between the deposits on the slope and in the basin, and also because of the volcanoclastic levels and paleosoil inside them, which have allowed the deposits to be dated (Miccadei et al., 1999). The morphometric analysis of the main fan/catchment systems is summarized in Tab. 3 and Fig. 8. The law which governs the fan area/catchment area relationship (according to Oguchi and Ohmori; 1994; Oguchi, 1997; Allen and Hovius, 1998; Allen and Densmore, 2000) is: $A_f = 0{,}59\ A_b^{0{,}63}$. Note that the constant k (0,63 in this case) has a direct relationship with the erodibility of the materials, as already indicated in Bull (1964), and an inverse relationship with the rate of the movement of the faults at the apex of the fans (Oguchi and Ohmori, 1994).

The analysis of the results that were obtained on the Montagna del Morrone SW escarpment has very clearly demonstrated how the values, and especially the value of the constant k, are among the lowest known in the relevant literature and similar to values measured on fault related slopes with a fault slip rate documented at some mm/yr (Fig. 8e,f; Allen and Hovius, 1998; Allen and Densmore, 2000). This can be only partly due to higher resistance of the bedrock and must therefore also be accounted for by the high slip rate of the slope's basal fault. The relationships between the other morphometric parameters are also governed by a power law, as the graphs of Fig. 8b', c' show. The data are more scattered, but they confirm the morphostructural considerations.

Another important aspect relates to the values for the fan area/catchment relationship, which are markedly far from the gathered data in Fig 8a, as similarly occurs for the other parameters (Fig. 8b, c, d). These values are of fan/catchment systems which have undergone noteworthy perturbation in their geometry (Basin C - Mancini fan; Basin N - Marane fan). In the first case there are several generations of fans that are built up one upon the other. The positioning of the alluvial terraces and the correlation with the terraces of the Sulmona basin demonstrate how the development of the fan itself was affected by external elements, such as the process of regressive erosion from the Gole di Popoli in the Sulmona basin (Ciccacci et al., 1999). This has extended its action headward, leading to a re-cutting of the fan and limiting its growth. The overall catchment geometry and the surface sediment distribution suggest that internal factors such as the existence of sediment storage points in the catchment, which tend to prevent the sediment supply to the fan, have also led to the fan being undersize. In the second case (Basin N) the geometry of the network, of the basin and its hypsometry (Fig. 4,5) show how a large part of the summit area of the basin itself may have been 'captured' during one of the recent phases of the slope's development. This is confirmed by comparing the value of the relationships calculated and illustrated in the graphs of Fig. 8. If the area that is considered the object of capture is excluded from the calculation, the data (triangular dot) clearly approximates to the regression line (cfr. Fig. 8 a, b, c, d, e Fig. 8 a', b', c', d'). Moreover, the anomalous value in the relationships studied shows that the phenomenon must have come about recently, as the re-equilibrium of the fan-basin system has not yet been achieved. Since Allen and Densmore (2000) point to re-equilibrium periods that are in fact rapid (to the order of tens of thousands of years), even considering the presence of resistant lithologies, it seems possible to date the capture to the Late-Pleistocene.

Therefore, it can be stated that the morphometric analysis of fan-basin systems can be exploited in morphostructural contexts such as the central Apennines, whether it be in morphotectonic analysis of fault related slopes or in the assessment of the conditions of equilibrium for single fan-catchment systems, which contributes to the study of local morphostructural evolution.

Finally, the geomorphological evolution of the alluvial fans in the central and southern sectors can be summarized. In the southern sector they are relatively small, with high dip angles, in clear aggradation, and are controlled by structural factors such as the resistant rocks of the catchment bedrock and the high slip rate on the Basal border fault. Considering the relationship between landslide scarps, catchments and alluvial fans, according to Blair (1999), it is possible to argue that the initiation of the catchments was due to the emplacement of the large landslide body.

In the northern sector, the alluvial fans are controlled more by interaction with the geomorphological evolution of the Popoli gorge, the northern outlet of the Sulmona basin, than by these same structural factors (Ciccacci et al., 1999). The regressive erosion due to the incision in the Popoli gorge deeply affected the alluvial fans of this sector, but only just touched those of the central sector, without reaching the southern sector.

4.4 Landscape evolution of the escarpment between the Montagna del Morrone ridge and the Sulmona tectonic basin

The integrated morphotectonic approach to the study of the mountain landscape of the central Apennine chain allows us to outline the main steps of the escarpment between Montagna del Morrone and the Sulmona basin (Fig. 10). The results clearly indicate that it is a high activity fault-generated mountain front according to Bull and McFadden (1977), Bull (1977), Wallace (1978), Bull (1987), Keller and Pinter (1996), and Allen and Densmore (2000). These features include low sinuosity and faceting, high slope and local relief, elongated and out of equilibrium drainage basins, convex and knick pointed stream channel profiles, prevailing areal denudation processes, general aggradation of the alluvial fans at the base of the slope and morphometry of the alluvial fan/catchment system.

This fault-generated mountain front, however, shows a peculiar morphostructural setting, variable both longitudinally and transversally, which led us to define a partition in three distinct sectors: northern, central and southern (Fig. 9). This is closely associated with the morphotectonic evolution of the Montagna del Morrone ridge and the Sulmona basin, which is due to the contrast of local tectonic subsidence on the basin and regional uplift during the Pleistocene (Miccadei et al., 2002).

The geomorphological investigations highlight a complex cyclic evolution in succeeding stages with the dominance either of morphotectonics, linked to the conflicting fault activity and regional uplift, or of erosional processes, particularly during cold stages of Quaternary climate fluctuations (Miccadei et al., 2004).

In a general balance the growth of the escarpment has strongly exceeded and dominated the effect of denudation, due to the local subsidence of the Sulmona basin relative to the Montagna del Morrone blocks along the Basal border fault and the Schiena d'Asino fault and to the general uplift of the area,. This has created relief of up to 1700 m and enabled the maintenance of very steep slopes, on highly resistant rocks, which have been moderately weathered and incised by climate-controlled erosional processes. These processes are mostly due to drainage network linear down-cutting in the mid and lower part of the northern and central sectors, while slope areal denudation is prevailing in the upper part of the northern and central sectors and in the southern sector.

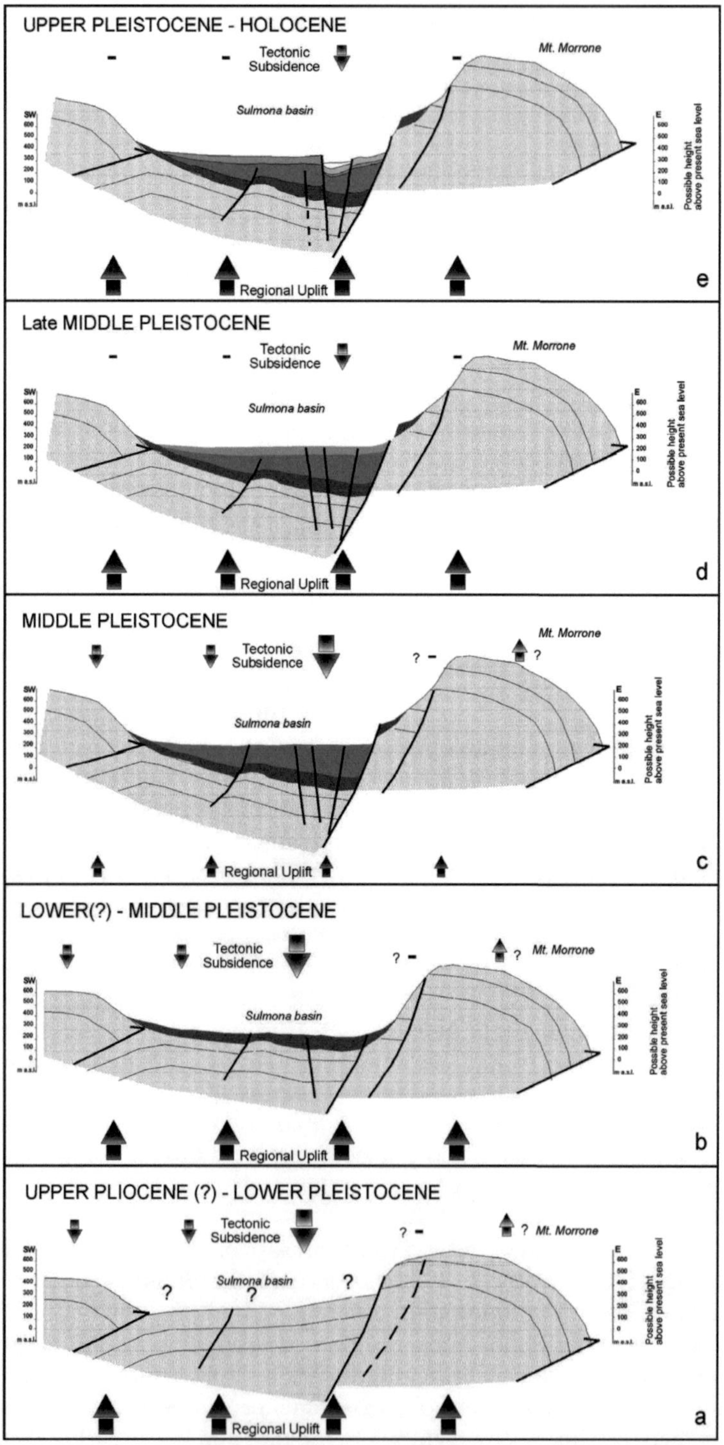

Fig. 10. Evolution of the escarpment between the Montagna del Morrone ridge and the Sulmona basin (Miccadei et al., 2002)

In conclusion, it is possible to define the evolution of the escarpment between the Montagna del Morrone ridge and the Sulmona basin as a growth evolution, rapid in the earlier stages and then continuing in the later phases. We can summarise the main stages of this morphotectonic evolution as follows (Fig. 10):

- Early moderately high relief incised by geomorphic processes, among which possibly karst weathering; remnants of this landscape, though reworked by karst processes and nivation, are preserved on the top of the ridge (Lower? Pleistocene);
- Growth of the slope due to the strong activity of the normal fault; earlier doubling of the ridge in the northern sector; the central sector begins to work as a structural transfer; the occurrence of fractured carbonate rocks, the high local relief and high slope, and eventually, the occurrence of earthquake-triggered landslides, led to the emplacement of large rock slides and to sediment accumulation along the slope (scree slope breccias, alluvial fan conglomerates) (Lower?-Mid Pleistocene);
- Development of drainage basins similar to the present ones (including Basin C, eventually after an early capture of the upper part) and incision of previous alluvial fans (Middle Pleistocene);
- Erosion of the escarpment, mostly due to slope denudation processes in the southern sector (and upper part of the central and northern sector) and to stream incision in the northern sector (Middle-Late Pleistocene);
- Morphotectonics, though possibly less intense than in the earlier stages, led to a progressive renewal with evidence of faulting, along the basal fault scarp of the southern sector and along both fault scarps of the central and northern sector (renewal mostly evident in the upper slope along the Schiena d'Asino fault scarp); a perturbation of some of the alluvial fan/catchment systems, caused in the northern sector by headward regressive erosion on the alluvial fans, controlled by the Sulmona basin outlet evolution and by upstream capture phenomena in the southern sector (Middle-Upper Pleistocene);
- The erosion processes are capable of only partly contrasting the morphotectonic processes (evidence of faulting occurs mostly on Late Pleistocene alluvial fans) and have led to partial reorganization of the drainage basin, still now clearly out of equilibrium, particularly in the southern sector. The present morphotectonic setting is acquired (Upper Pleistocene-Holocene).

5. Case studies - piedmont area: dip stream valley (Sangro river valley)

5.1 Introduction

The Sangro river is, at present, 107 km long and flows on the Adriatic side of central Italy from the inner part of the Apennines to the coast. The direction of the river is variable, from N-S in the upper reach, to WNW-ESE, to SW-NE, to S-N, and, finally, to SW-NE in the lower reach (Fig. 1). The main tributary is the Aventino river, which flows along the eastern side of the Maiella massif and then into the Sangro 20 km away from the coast. The present drainage basin area is about 1560 km^2 and its mean elevation is 970 m a.s.l.; about 70% of the basin lies within the range area; 30% within the piedmont one. The Sangro river's long profile consists of several segments, the highest long valley gradient being in the intermediate sector between the range and the piedmont. Its course and long profile show that the Sangro river can be divided into different reaches based on abrupt bends and/or long gradient variations (Fig. 1). The first part of the Sangro river flows within the range on clayey-arenaceous Miocene foredeep deposits and meso-cenozoic carbonate sequences, and

shows a regular long profile with knick points corresponding to the occurrence of carbonate rocks and thrusts. The intermediate reaches carve into thrusted pre-orogenic clayey and carbonate Oligo-Miocene pelagic sequences, overlain by sinorogenic clayey arenaceous Miocene foredeep deposits. This reach shows a marked convex shape with sharp knick points related to the lithostructural control of alternating clayey and carbonate rocks. The abrupt long gradient decrease corresponds to the front of the range. The lower reach incises with a concave long profile the Plio-Pleistocene clayey-sandy marine sediments of the Adriatic basin (Fig. 11).
The study area, the lower part of the Sangro valley, is located in the Adriatic piedmont, in the south eastern Abruzzi area and lies in a complex geological framework between the central Apennines and the coast (Fig. 11). This area is characterised by a cuesta, mesa and plateau relief at a moderate elevation, sloping from SW to NE, from 500 m a.s.l. to sea level. The Sangro river flows, in this area, from 150 m a.s.l. to sea level.
The geological setting is characterized by late-orogenic Plio-Pleistocene Adriatic foredeep units that, in the SW sector, unconformably overlie pre-orogenic Molise pelagic units (Fig. 1). The Plio-Pleistocene units consist of Middle Pliocene to Early Pleistocene foredeep terrigenous clayey-sandy deposits, up to 2000 m thick, with interbedded conglomerates, coarsening upwards into a sandstone-conglomerate regressive sequence. The structural setting is defined by a regional homocline gently dipping north-east and locally affected by systems of low displacement faults (NW-SE, SW-NE). The Plio-Pleistocene foredeep sequence unconformably overlies folded and thrusted Miocene-Pliocene structures.
In the SW sector (Fig. 12) the pre-orogenic Molise pelagic units are made up of a clayey pelagic Oligocene-Miocene formation (argille varicolori formation) and of a limestone and marly-limestone pelagic Miocene formation, which is followed by a pelitic and arenaceous-pelitic sin-orogenic foredeep Late Miocene sequence. The above mentioned units were affected by fold and thrust Miocene-Pliocene deformations that involved a major NE transport.
Pre-, sin- and late-orogenic sequences are unconformably overlain by Middle-Late Pleistocene and Holocene continental conglomerates and subordinate sands, mainly related to fluvial and alluvial fan deposits.
In the study area, characterized by non-conservative lithology of the bedrock and by widespread superficial deposits, the analysis of geomorphological evidence of tectonics and its correlation with the Quaternary continental deposits and drainage network contribute to defining the role of tectonics in the landscape development (Castiglioni, 1935; Rapisardi, 1982; Ciccacci et al., 1986; Aucelli et al., 1996; Centamore et al., 1996; Del Monte et al., 1996; Bigi et al., 1996; Belisario et al., 1999; Currado and D'Ambrogi, 2002; Molin et al., 2004; Scheidegger, 2004).
The present drainage network and basin of the Sangro river, similarly to those of the main Adriatic rivers, are characterised by general and local geomorphic markers of tectonics and by different types of anomalies: basin asymmetry, irregular long profiles and statistical azimuthal distributions, drainage pattern types, main channel position in the present floodplain, asymmetric distribution of fluvial terraces (Coltorti et al., 1991; Elmi, 1991; Aucelli et al., 1996; Del Monte et al., 1996; Currado and D'Ambrogi, 2002; Molin and Fubelli, 2005; Spagnolo and Pazzaglia, 2005).
The geomorphological study performed in the area allowed us: i) to identify the fluvial and alluvial fan deposits and the transverse and longitudinal geometry of the related terraces; ii) to qualitatively and quantitatively analyse the geometry of the drainage network; and iii) to determine the distribution and geometry of significant morphotectonic evidence such as linear valleys and asymmetric valleys, hanging and beheaded valleys, counterflow streams and river bends.

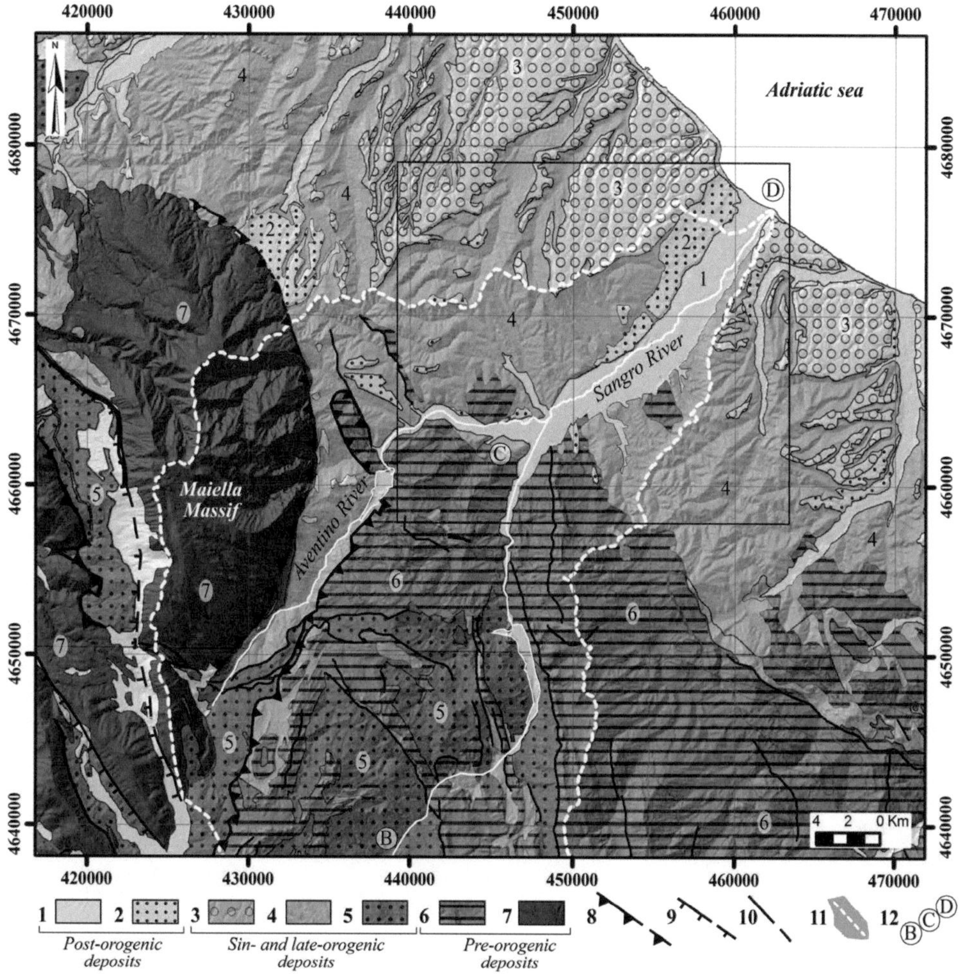

Fig. 11. Geological scheme of south-eastern Abruzzi and location of the study area (black box). Legend: *post-orogenic Quaternary continental deposits,* 1) fluvial deposits (Holocene), 2) terraced fluvial and alluvial fan deposits (Middle-Late Pleistocene); *sin- and late-orogenic terrigenous deposits,* 3) marine to continental transitional sequences (Early Pleistocene), 4) hemipelagic sequences with conglomerate levels (Late Pliocene-Early Pleistocene), 5) turbiditic foredeep sequences (Late Miocene-Early Pliocene); *pre-orogenic carbonate, marly and clayey deposits,* 6) Molise pelagic sequences (Oligocene-Miocene), 7) carbonate platform, slope and pelagic sequences (Jurassic - Miocene); 8) thrust (dashed if buried); 9) normal fault (dashed if buried); 10) fault with strike slip or reverse component (dashed if buried); 11) Sangro river drainage divide; 12) course of the Sangro river (Fig. 1)

5.2 Results

In the lower Sangro valley, five levels of terraced *fluvial and alluvial fan deposits* can be identified, at decreasing heights above the present alluvial plain. The terraces show a heterogeneous plano-altimetric distribution and different sedimentological characteristics, formed in both alluvial fan (T1, T2) and fluvial environments (T3-T5 and alluvial plain) (Fig. 12; Tab. 4).

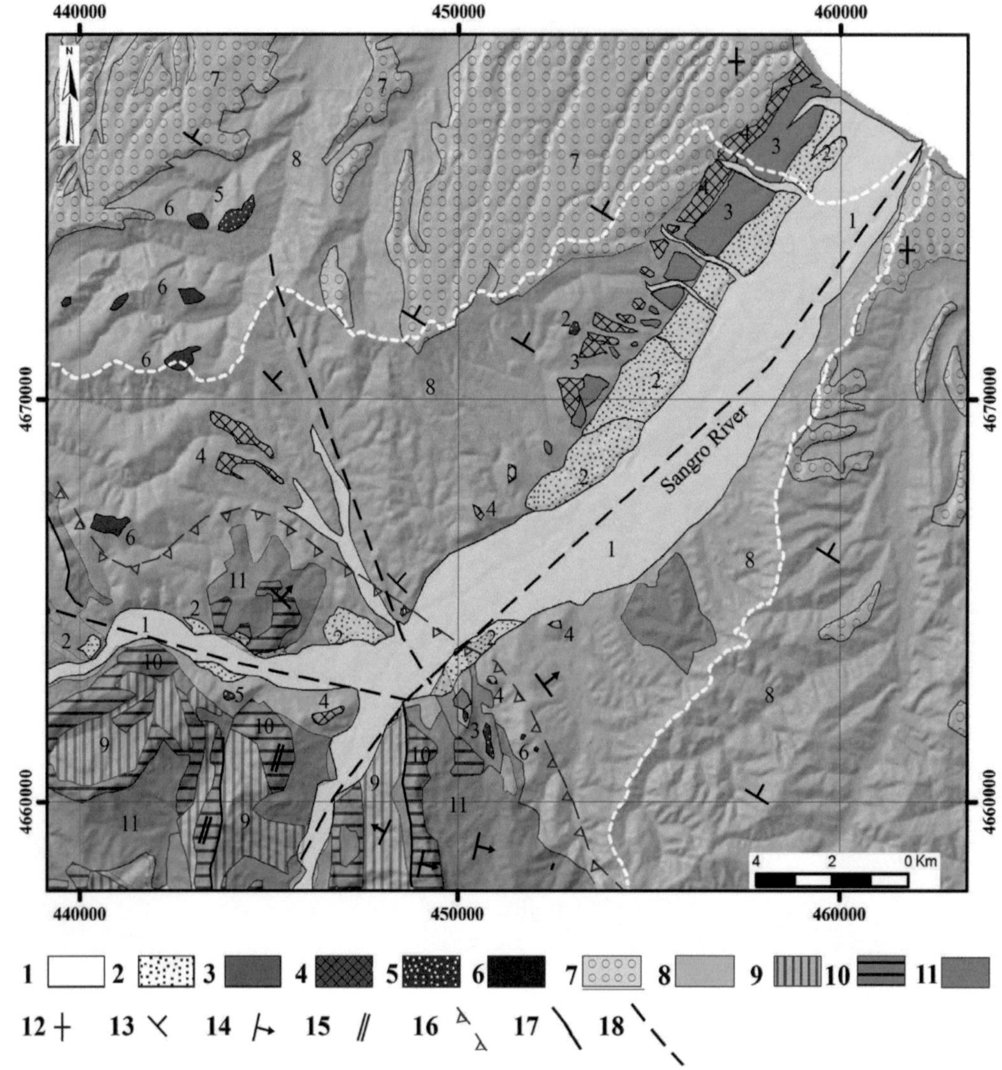

Fig. 12. Geologic and fluvial terraces map of the lower Sangro river valley (D'Alessandro et al., 2008). Legend: *post-orogenic Quaternary continental deposits*, 1) present alluvial plain deposits (Holocene), 2) fluvial terrace T5 deposits (Late Pleistocene), 3) fluvial terrace T4 deposits (late Middle Pleistocene), 4) fluvial terrace T3 deposits (Middle Pleistocene), 5) alluvial fan T2 deposits (Middle Pleistocene), 6) alluvial fan T1 deposits (Middle Pleistocene); *sin- and late-orogenic terrigenous deposits*, 7) conglomerates and sandstone of marine to continental transitional sequences (Early Pleistocene), 8) clays and sands of hemipelagic sequences with conglomerate levels (Late Pliocene-Early Pleistocene), 9) sandstone and siltstone of turbiditic sequences (Late Miocene-Early Pliocene); *pre-orogenic carbonate, marly and clayey deposits*, 10) limestone and marly-limestone of Molise pelagic sequences (Miocene), 11) clays of Molise pelagic sequences (Oligocene-Miocene); 12) 0-10° dipping strata; 13) 10-45° dipping strata; 14) 45-80° dipping strata; 15) 80-90° dipping strata; 16) buried thrust; 17) fault with strike slip or reverse component; 18) inferred neotectonic fault

The alluvial fan deposits (T1 and T2 in Fig. 12) are located at the summit of the hilly relief, at an elevation higher than 300 m a.s.l. and along the drainage divide between the Sangro basin and the surrounding ones. The fan deposits consist of heterometric, poorly sorted and sub-angular conglomerates (ϕ_{max}> 50 cm) with a matrix of fine gravel to sand. They show variable thicknesses up to 20 m and they are often deeply eroded or preserved only as gravel remnants on planar surfaces.
The fluvial deposits are distributed along the lower Sangro valley, more extensively on the NW side. On the SE side the deposits are rare and thinner. They are arranged in four levels, inset in the older alluvial fan terraces, at elevations decreasing from 280 m to the present valley floor (T3, T4, T5 and present alluvial plain, Fig. 12, Tab. 4). Along the NW valley side down to the river mouth, the terrace treads are at various heights above the present valley floor, decreasing from 150-100 m in the case of T3, to 120-60 m (T4), 50-30 m (T5), down to the present alluvial plain (Tab. 4).

Terrace	m *a.s.l.*	m *a.p.c.*	Deposit age
T6 (alluvial plain)	0-130	0	Holocene
T5	30-160	30-50	Late Pleistocene
T4	60-230	60-120	late Middle Pleistocene
T3	100-280	130-150	Middle Pleistocene
T2 (alluvial fan)	220-300	>160	Middle Pleistocene
T1 (alluvial fan)	330-350	>200	Middle Pleistocene

Table 4. The correlation of fluvial terraces, alluvial fan surfaces and related deposits is based on elevation (m a.s.l.) and height above the present channel (m a.p.c.); the age of the deposits is inferred from the correlation with surrounding basins in the Adriatic piedmont (Demangeot, 1965; Calderoni et al., 1991; Coltorti et al., 1991; Nesci et al., 1992, 1995; Fanucci et al., 1996; Di Celma et al., 2000)

The deposits of the four terrace levels are made up of heterometric, moderately-to-well sorted pebble-to-cobble conglomerates; they are generally clast-supported with sandy matrix. The thickness of fluvial deposits is moderate, up to 20 m in the lower part of the valley. The basal erosive unconformity on the Pleistocene clayey and sandy bedrock outcrops in several locations on the valley side, particularly in quarries located in the lower part of the valley.
The transverse profiles show the relative incision of the fluvial terraces in the alluvial fan terraces and of the different terrace levels one into the other. The valley long profile shows a general downstream convergent geometry of the terrace treads.
The age of fluvial and alluvial fan deposits is inferred from the correlation with the surrounding basins in the Adriatic piedmont, as indicated in the previous section. The alluvial fans and the highest fluvial terrace (T1, T2, T3) are ascribed to the Middle Pleistocene, the second fluvial terrace (T4) to the late Middle Pleistocene, the third (T5) is dated to the Late Pleistocene and the alluvial plain to Holocene.
The Sangro river shows a mainly sub-dendritic *drainage pattern* in the piedmont area and a generally angular or trellis one in the mountain zone, related, in general terms, to both lithological and structural control. In the lower Sangro valley (Fig. 13a) the total azimuthal stream analysis of the network confirms the general sub-dendritic pattern, showing a sub-elliptical shape, although E-W and NW-SE main orientations are present (Fig. 13b; Tab. 5).

However, considering only the main streams (3rd order or higher), a general angular pattern can be detected along SW-NE, WSW-ENE and NW-SE orientations, as shown by the azimuthal statistics of the main streams (Fig. 13c; Tab. 5).

Morphotectonic field mapping focused on valley features: linear and asymmetric valleys, hanging and beheaded valleys, river bends and counterflow confluences of streams (Fig. 14a). Planimetric distribution is seemingly non-uniform, however, the analysis of azimuthal distribution can highlight an alignment along preferential orientations (Fig. 14b; Tab. 6).

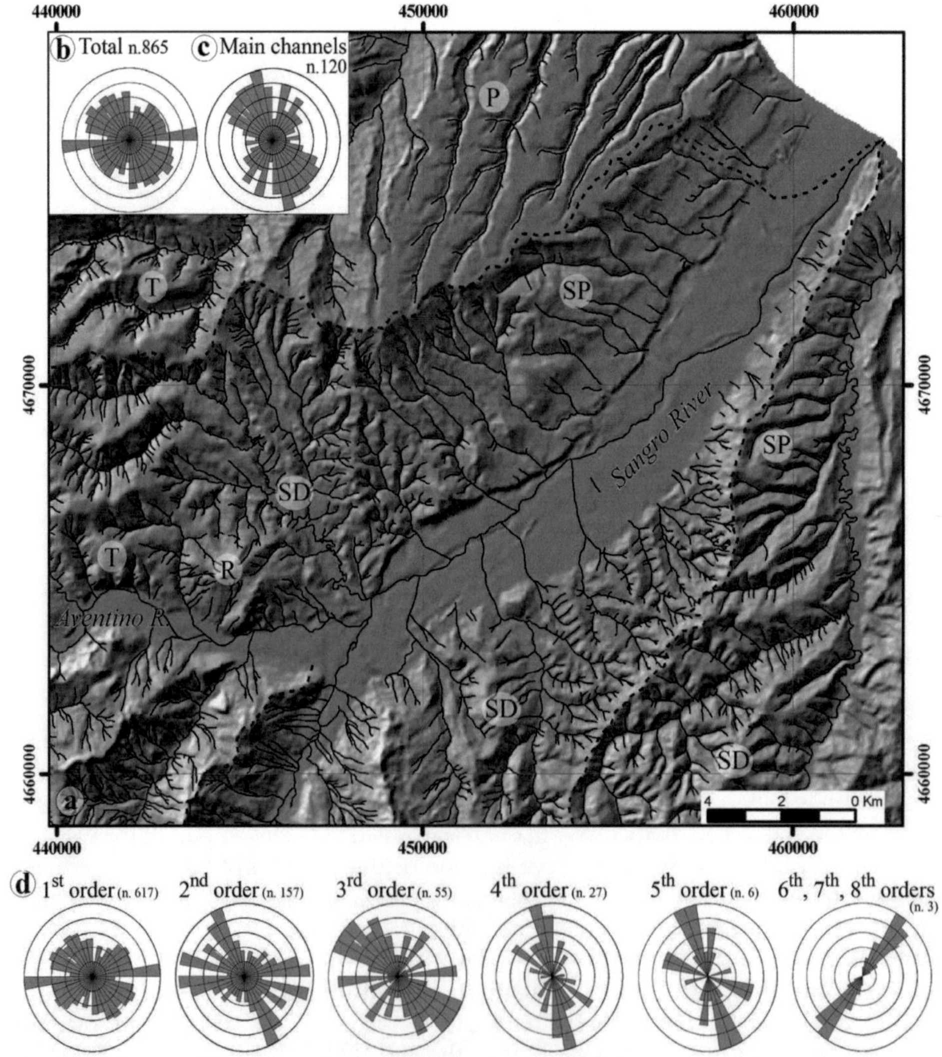

Fig. 13. a) Drainage network of the Sangro basin in the piedmont area and of the surrounding areas; the dashed line marks the drainage divide of the Sangro basin (D'Alessandro et al., 2008). Legend of local drainage patterns: SD) sub-dendritic; P) parallel; SP) sub-parallel; T) trellis; R) radial. b) Total stream azimuth rose diagram for the lower Sangro valley. c) Main stream azimuth rose diagram. d) Azimuth rose diagram of the streams ordered according to Strahler (1957)

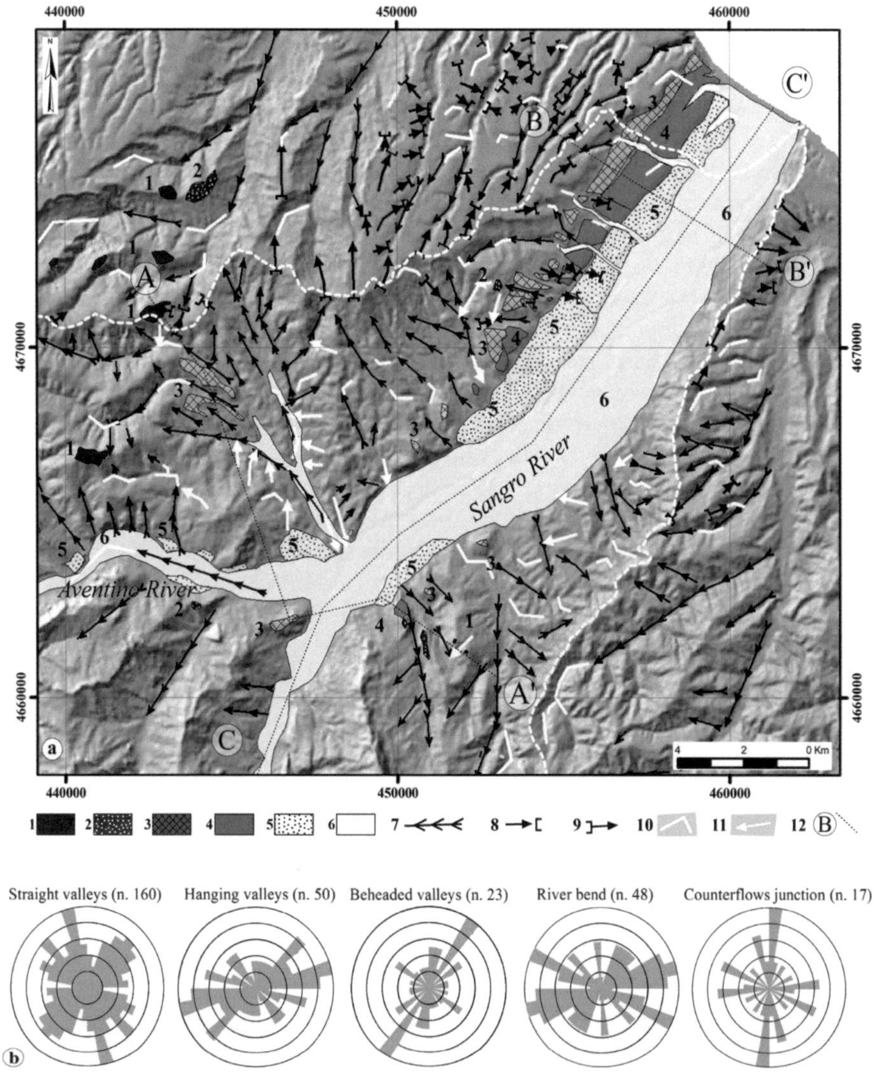

Fig. 14. a) Map of the morphological field evidence of tectonics (D'Alessandro et al., 2008). Legend: 1) alluvial fan surface T1; 2) alluvial fan surface T2; 3) fluvial terrace T3; 4) fluvial terrace T4; 5) fluvial terrace T5; 6) present alluvial plain; 7) linear and asymmetric valley; 8) hanging valley; 9) beheaded valley; 10) river bend; 11) counterflow confluence of streams; 7) profile locations (see Fig. 15). b) Azimuth rose diagram of morphological field evidence of tectonics

Linear valleys and asymmetric valleys show a main NNW-SSE orientation and secondary NW-SE and SW-NE orientations. These features - which are linked to the Late Pleistocene terrace (T5) - incise the Middle Pleistocene alluvial fan surfaces and the later fluvial terraces (T1 to T4) on the NW side of the Sangro valley, whereas they incise the Early Pleistocene clayey bedrock on the SE side.

The hanging valleys are located particularly on the northern and eastern sectors of the area, with WSW-ENE and SW-NE dominant orientations. River bends are frequent all over the

study area and particularly on the NW side of the Sangro valley (Fig. 14). The azimuthal analysis shows two prevailing orientations: WSW-ENE and NW-SE (Fig. 14b).
The planimetric distribution of the beheaded valleys indicates a dominant SW-NE orientation (Fig. 14). In particular, they are located along the present northern and southern drainage divide of the Sangro basin, and drain into the adjacent basins (Fig. 14).
Finally, the counterflow confluences are mainly on the NW side of the Sangro valley (Fig. 14), they show a prevailing N-S direction and are connected to NNW-SSE linear valleys. These elements are located in an area characterised by beheaded valleys.
The morphological evidence of tectonics allows us to detect morphotectonic lineaments, mainly SW-NE, along the main valley, on its right side and in the northern area outside the Sangro basin. These lineament are intersected by NNW-SSE and WNW-ESE ones, particularly in the SW sector of the study area (Fig. 14, Tab. 6). On the NW side of the valley the correlation between terraces and morphotectonic elements indicates that the SW-NE elongated beheaded valleys are intersected by NNW-SSE and WNW-ESE linear valleys and river bends. The beheaded SW-NE valleys lie on, or slightly incise, the sandy-conglomeratic regressive sequence on top of the marine Pleistocene succession; locally, they are correlated with T1 terraces. The linear valleys incise the Middle Pleistocene fluvial and alluvial fan terraces (T1-T4) and are correlated with the Late Pleistocene terrace (T5). On the right (SE) valley side, it is possible to detect the intersection of NE-SW minor elements (counterflow confluences, linear valleys) with major NNW-SSE and WNW-ESE linear valleys, as already suggested by the drainage pattern.

order	orientation
1st	E-W NW-SE
2nd	NNW-SSE E-W
3rd	NW-SE
4th	NNW-SSE
5th	NNW-SSE
6th 7th 8th	SW-NE

Table 5. Dominant azimuthal distribution of the ordered streams (Strahler, 1957)

elements	orientation
Linear valleys	NNW-SSE SW-NE
Asymmetric valleys	NNW-SSE E-W
Hanging valleys	SW-NW WSW-ENE
Beheaded valleys	SW-NE
River bends	WSW-ENE WNW-ENE
Counterflow confluences	N-S

Table 6. Dominant azimuthal distribution of the morphotectonic evidence

5.3 Discussion

The analysis and the correlation of fluvial terraces and alluvial fan surfaces, drainage patterns and morphotectonic evidence provide several indications concerning drainage development in the piedmont area of the Sangro valley since the Middle Pleistocene. Moreover, they allow us to evaluate the role and timing of tectonics in the development of the piedmont area of the Sangro valley.

The remnants of alluvial fan surfaces (T1 and T2; Fig. 12, 15), on top of the hilly relief and along the present drainage divide, outline a landscape completely different from the present one: a Middle Pleistocene wide piedmont plain on which - after the emergence and, more specifically, during early continental morphogenesis - large alluvial fans formed, as already observed in the northern sector of the Adriatic piedmont (Nesci and Savelli, 2003).

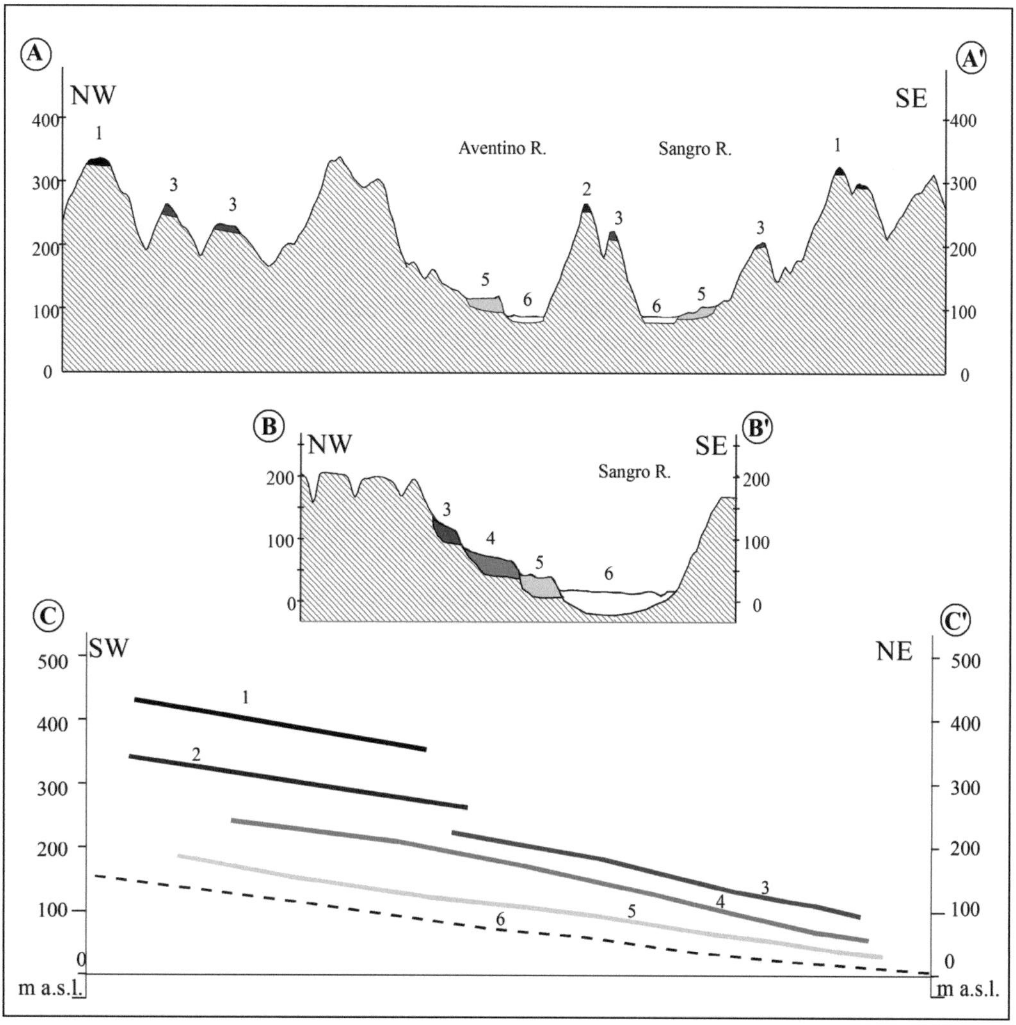

Fig. 15. Cross-valley and long profiles showing the relation of the terrace deposits with the local valley geometry (A-A′, B-B′) and the relations among the terrace tread levels (C-C′) (D′Alessandro et al., 2008). Location and numbers legend are referred to Fig. 14

The plano-altimetric distribution of T3 and T4 fluvial terraces (Fig. 14, 15), elongated in the SW-NE direction and entrenched into the ancient fan surfaces and along both margins of the present valley, indicates the development of a subsequent SW-NE consequent drainage starting in the Middle Pleistocene (Coltorti et al., 1991; Nesci et al., 1992; Fanucci et al., 1996). This is confirmed by the SW-NE parallel pattern with marked SW-NE linear valleys, preserved in the northern area outside the drainage divide of the Sangro basin (Fig. 13).
On the NW side of the Sangro valley, a sub-parallel drainage is present along the NNW-SSE direction. Moreover, the sub-dendritic drainage shows NW-SE to NNW-SSE elongated main streams on both valley sides (3rd, 4th, 5th order, Fig. 13). These streams are marked by linear valleys that cut Middle Pleistocene fluvial terraces (T3 and T4) and are connected to the Late Pleistocene terrace (T5) (Fig. 14). This configuration indicates the control of local tectonics on drainage development along NNW-SSE faults and fractures, and suggests that the age of these elements can be ascribed to the late Middle Pleistocene (Fig. 12). The analysed counterflow junction of streams confirms tectonic control along the NNW-SSE orientation (Fig. 14). In the lower part of the valley, along the coastal area, the occurrence of SW-NE, NW-SE river bends and the anomaly in the path of the drainage divide can be related to the role of gravity deformation parallel to the coast.
On the SE side of the Sangro valley, SW-NE counterflow confluences and linear valleys carved on the clayey bedrock are aligned to the straight flank of the main valley and intersect NNW-SSE and WSW-ENE linear valleys. In this case, local tectonic control on the drainage is exerted also along SW-NE lineaments (Fig. 14).
Hence, a rectangular drainage network (*sensu* Zernitz, 1932) developed during the late Middle Pleistocene, as a result of the junction of previous SW-NE streams' directions - controlled by differential uplift and tilting - and new NNW-SSE, WNW-ESE and SW-NE streams' directions, controlled by local tectonics along faults and fracture systems (Figs. 12, 14). The development of the NNW-SSE streams, on the NW side of the Sangro valley, beheaded the previous SW-NE drainage, as highlighted along the present divide. A major rearrangement of the drainage widened the basin towards the left (NW) side. In this scenario local tectonics could be an additional explanation of the drainage basin asymmetry in the Adriatic piedmont.
Finally, the development of a sub-dendritic drainage pattern on the clayey Plio-Pleistocene bedrock in the peripheral parts of the basin, together with a local radial pattern on the clayey-calcareous Miocene sequence, indicates a prevailing lithostructural control in the last phase of drainage incision and in the definition of the present fluvial landscape.

5.4 Sangro dip-stream valley landscape evolution

The integrated morphotectonic approach to the study of the fluvial landscape of the Adriatic piedmont allows us to outline the main steps of the lower Sangro river valley evolution:

- alluvial fan development in a wide piedmont plain (Middle Pleistocene), following the emergence of the Adriatic basin, which progressively occurred during the Early Pleistocene due to regional uplift, as already observed in the northern Abruzzi and Marche area (Nesci and Savelli, 2003; Cantalamessa and Di Celma 2004);
- consequent SW-NE parallel drainage controlled by a regional topographic gradient, due to regional differential uplift with NE tilting and by SW-NE tectonic structures (Middle Pleistocene);
- superimposition of the SW-NE drainage on the uplifting piedmont and development of fluvial terraces (Middle Pleistocene – late Middle Pleistocene);

- development of NNW-SSE, WNW-ESE and SW-NE faults with low displacements and fractures (late Middle Pleistocene);
- adaptation of the drainage network to fault and fracture systems, and development of a rectangular pattern (late Middle Pleistocene - Late Pleistocene);
- rearrangement of the drainage network due to lithology-controlled morphoselective processes (Late Pleistocene - Holocene).

The role of regional uplift with NE tilting in the development of the piedmont consequent valleys is confirmed in the case of the Sangro river valley. Local tectonics, mainly along NNW-SSE, WNW-ESE and SW-NE faults and fractures (Fig. 10), played a crucial role in the rearrangement and configuration of the drainage network during the late Middle Pleistocene.

6. Conclusion

This chapter provides examples of morphotectonic studies undertaken in Abruzzi (central Italy) within chain and piedmont areas:
- chain area - escarpment between the Montagna del Morrone ridge and the Sulmona tectonic basin (central Abruzzi);
- piedmont area - dip stream valley (Sangro river valley, south-eastern Abruzzi)

These areas are characterized by different geomorphological features: 1) bedrock lithologies (conservative carbonate bedrock in the montane area, not conservative clayey-arenaceous-conglomeratic bedrock in the piedmont area), 2) surface deposits (slope and alluvial fan deposits in the montane area, fluvial and alluvial fan deposits in the piedmont area), 3) landforms (slope landforms and fluvial, and water erosion landforms in the montane area, fluvial and water erosion landforms in the piedmont area), and 4) tectonic framework (strong Pleistocene-Holocene extensional tectonics and uplift in the montane area, Pleistocene-Holocene uplift with poor local tectonics in the piedmont area).

The studies suggest that different features of morphostructural domains require similar approaches and methods with suitable adaptation, based on (a) terrain analysis, (b) morphostructural analysis of the relief, (c) analysis of geomorphic markers such as certain landforms (geomorphological evidence of tectonics) and deposits (developed in a continental environment), (d) analysis and morphometry of drainage basins, (e) dating of deposits and landforms.

These studies are focused on deciphering the role of morphotectonics and selective erosion in the landscape's evolution, incorporating regional morphostructural analysis (based on DEM analysis), Quaternary fluvial deposits mapping, local morphostructural analysis (based on field mapping and aerial photo interpretation), drainage network analysis and orography and hydrography morphometry. The key point is the integration of field geomorphological methods and modern morphotectonic analysis (including drainage network and terrain analysis). Only the correlation of these methods of analysis at drainage basin scale allow us to find out and define geomorphic markers of tectonics and landscape evolution, suggesting also its timing.

The results allow us to outline the main morphostructural features of central Italy (chain area and piedmont area) and to suggest the use of a similar methodological approach, but focused also on different geomorphological landscapes. In addition, discussion and conclusions of the studies show how integrative morphotectonic studies allow for the description of drainage network and landscape evolution driven by the alternating action of tectonic forces and geomorphic processes due to orography and Pleistocene climate fluctuations. It is possible also to define the role of morphotectonics and selective erosion in the landscape evolution and suggest the relative timing of this evolution, while only incorporating specific geochronology studies enables us to define the absolute timing of tectonics and landscape evolution.

7. Acknowledgments

The authors wish to thank the Struttura Speciale di Supporto Sistema Informativo Regionale of Abruzzo Region (http://www.regione.abruzzo.it/xcartografia/), for providing the topographic data and aerial photos used for the geomorphological investigations and in the figures of this work.

8. References

Allen P.A., Densmore A.L. (2000). Sediment flux from an uplifting fault block. Basin Res., 12, 367-380.

Allen P.A., Hovius N. (1998). Sediment supply from landslide-dominated catchments: implications for basin-margin fans. Basin Res., 10, 19-35.

Ambrosetti, P., Bonadonna, F.P., Bosi, C., Carraro, F., Cita, B.M., Giglia, G., Manetti, P., Martinis, B., Merlo, C., Panizza, M., Papani, G., Rampoldi, R., (1976). Proposta di un progetto operativo per l'elaborazione della carta neotettonica d'Italia. Progetto finalizzato geodinamica, pp. 1-49.

APAT (2006). Carta Geologica d'Italia alla scala 1:50.000, Foglio 369 "Sulmona".

APAT (2007). Carta Geomorfologica d'Italia 1:50.000 - Guida alla rappresentazione cartografica. Presidenza del Consiglio dei Ministri, Dip. Servizi Tecnici Nazionali, Servizio Geologico, Quaderno serie III, 10, 59 pp,

Ascione A., Cinque A. (1997). Le scarpate su faglia dell'Appennino Meridionale: genesi, età e significato tettonico. Il Quaternario, 10(2), 285-292, Verona.

Ascione A., Cinque A. (1999). Tectonics and erosion in the long term relief history of the southern Apennines (Italy). Zeit. Geomorph. N.F., 118, 1-16.

Ascione A., Cinque A., Miccadei E., Villani F., Berti C. (2008). The Plio-Quaternary uplift of the Apennine chain: new data from the analysis of topography and river valleys in central Italy. Geomorphology, 102, 105-118.

Aucelli P.P.C., Cavinato G.P., Cinque A., (1996). Indizi geomorfologici di tettonica plio-quaternaria sul piedimonte adriatico dell' Appennino abruzzese. Il Quaternario 9, 299-302.

Avena G.C. and Lupia Palmieri E. (1969). Analisi geomorfica quantitativa. In: Idrogeologia dell'alto bacino del Liri (Appennino centrale). Geol. Romana, 8, 319-378.

Avena G.C., Giuliano G., Lupia Palmieri E. (1967). Sulla valutazione quantitativa della gerarchizzazione ed evoluzione dei reticoli fluviali. Boll. Soc. Geol. It., 86, 781-796.

Bartolini, C., Peccerillo, C. (2002). I fattori geologici delle forme del rilievo. Lezioni di Geomorfologia strutturale, Pitagora Editrice, Bologna.

Belisario F., Del Monte M., Fredi P., Funiciello R., Lupia Palmieri E., Salvini F., (1999). Azimuthal analysis of stream orientations to define regional tectonic lines. Zeitshrift fur Geomorphologie N.F. Suppl. Bd. 118, 41-63.

Bigi S., Cantalamessa G. Centamore E. Didaskalu P. Dramis F. Farabollini P. Gentili B. Invernizzi C. Micarelli A. Nisio S. Pambianchi G., Potetti M. (1996). La fascia periadriatica marchigiano-abruzzese dal Pliocene medio ai tempi attuali: evoluzione tettonico-sedimentaria e geomorfologica. Studi Geol. Camerti, Vol. Spec. 1995/1, 37-49.

Blair T.C. (1999). Alluvial fan and catchment initiation by rock avalanching, Owens Valley, California. Geomorphology, 28, 201-221.

Blair T.C., McPherson J.G. (1994). Alluvial fans and their natural distinction from rivers based on morphology, hydraulic processes, sedimentary processes and facies assemblages. Journ. Sediment. Res., A64(3), 450-489.

Blumetti A.M., Dramis F., Michetti A.M. (1993) - Fault-generated mountain front in the central Apennines (central Italy): geomorphological features and seismotectonic implications. Earth Surf. Proc. and Landf., 18, 203-223.

Bosi C., Galadini F., Messina P. (1993). Neotectonic significance of bedrock fault scarps: case studies from the Lazio-Abruzzi Apennines (central Italy). Zeit. Geomorph. N.E., Suppl.-Bd. 94, 187-206.

Brancaccio L., Cinque A., Sgrosso I. (1978). L'analisi morfologica dei versanti come strumento per la ricostruzione degli eventi neotettonici. Mem. Soc. Geol. It., 19, 621-626

Bull W. B., (2007), Tectonic geomorphology of mountains, Blackwell Publishing, Malden, MA. 316 p.

Bull W.B. (1964). Relations of alluvial-fan size and slope to drainage-basin size and lithology in western Fresino Country, California. Abstract, article 19, B-51/B-53, 1 fig.

Bull W.B. (1977). The alluvial fan environment: Progress in Phys. Geogr., 1, 222-270.

Bull W.B. (1987). Relative rates of long term uplift of mountain fronts. In Crone A.G., Omdahl E.L. eds Directions in paleoseismology, U.S. Geol. Surv. Open-File Rep. 87-693, 192-202.

Bull W.B., McFadden L.D. (1977). Tectonic geomorphology north and south of the Garlock fault, California. In: Doehring D.D. (ed.): Geomorphology in arid regions. 3rd Geomorphology Symposium, State Univ. New York, 115-138.

Bull, W. B. (1991). Geomorphic response to climatic change. Oxford University Press, New York.

Burbank, D.W., Anderson, R.A. (2001). Tectonic Geomorphology, Blackwell Science, Malden, MA, USA.

Burbank, D.W., Pinter, N. (1999). Landscape evolution: the interaction of tectonics and surface processes. Basin Research 11 (1), pp. 1-6.

Cailleux, A., Tricart, J. (1956). Les problème de la classification des fait geomorphologiques. Annales de Geographie 65, pp. 162-186.

Calderoni, G., Nesci, O., Savelli, D., (1991). Terrace fluvial deposits from the middle basin of Cesano river (Northern Marche, Apennines): reconnaissance study and radiometric constraints on their age. Geografia Fisica e Dinamica Quaternaria 14, 201-207.

Cantalamessa, G. ,Di Celma, C., (2004). Sequence response to syndepositional regional uplift: insights from high-resolution sequence stratigraphy of the late Early Pleistocene strata, Periadriatic Basin, central Italy. Sedimentary Geology 164, 283-309.

Carrara C. (1998). I travertini della Valle del Pescara tra Popoli e Torre dè Passeri. Il Quaternario, 11(2), 163-179.

Castiglioni, B., (1935). Ricerche geomorfologiche nei terreni pliocenici dell'Italia centrale. Pubbl. Ist. Geogr. R. Univ. Roma, s. A, 4.

Cavallin A., Crescenti U., Dramis F., Prestininzi A., Sorriso-Valvo M. (1987). Tipologia e diffusione delle deformazioni gravitative profonde di versante in Italia: prime valutazioni. Mem. Soc. Geol. It., 37, 241-252.

Centamore E., Nisio S. (2003). Effects of uplift and tilting in the central-northern Apennines (Italy). In: Bartolini C. (ed.): Uplift and erosion: driving processes and resulting landforms, international workshop, Siena, September 20 - 21, 2001. Quaternary International, 101-102C, 93-101.

Centamore, E., Ciccacci, S., Del Monte, M., Fredi, P., Lupia Palmieri, E., (1996). Morphological and morphometric approach to the study of the structural arrangement of the north-eastern Abruzzo (central Italy). Geomorphology 16, 127-137.

Ciccacci S., D'Alessandro L., Dramis F., Miccadei E. (1999). Geomorphological evolution and neotectonics of the Sulmona intramontane basin (Abruzzi, Apennine, central Italy). Zeit. Geomorph., 118, 27-40.

Ciccacci S., D'Alessandro L., Fredi P., Lupia Palmieri E. (1992). Relation between morphometric characteristics and denudational processes in some drainage basins of Italy. Zeit. Geomorph. N.F., 36, 1, 53-67.

Ciccacci S., Del Monte M., Fredi P., Lupia Palmieri E. (1995). Plano altimetric configuration, denudational processes and morphodynamics of drainage basins. Geol. Romana, 31, 1-13.

Ciccacci, S., Fredi, P., Lupia Palmieri, E., Salvini, F., (1986). An approach to the quantitative analysis of the relations between drainage pattern and fracture trend. International Geomorphology. John Wiley & Sons, Chichester, pp. 49-68.

Coltorti M., Farabollini P. (1995). Quaternary evolution of the "Castelluccio di Norcia" basin (Umbro-Marchean Apennines, central Italy). Il Quaternario, 8(1), 149-166.

Coltorti, M., Consoli, M., Dramis, F., Gentili, B., Pambianchi, G., (1991). Evoluzione geomorfologica delle piane alluvionali delle Marche centro-meridionali. Geografia Fisica e Dinamica Quaternaria 14, 87-100.

Crescenti U., Dramis F., Gentili B., Pambianchi G. (1989). Deformazioni gravitative profonde di versante e grandi frane nell'area a sud di Monte Porrara (Appennino centrale, Abruzzo). Mem. Soc. Geol. It., 39, 477-486.

Currado, C., D'Ambrogi, C., (2002). Plio-Pleistocene morphostructural evolution of Chieti sector in the Periadriatic Basin: an example of integrated analysis. Memorie della Società Geologica Italiana 57, 501-508.

D'Agostino, N., Jackson, J.A., Dramis, F., Funiciello, R. (2001). Interactions between mantle upwelling, drainage evolution and active normal faulting: an example from central Apennines (Italy). Geophysical Journal International 141, pp. 475-497.

D'Alessandro, L., Miccadei, E., Piacentini, T. (2008). Morphotectonic study of the lower Sangro river valley (Abruzzi, central Italy). Geomorphology 102, pp. 145–158.

D'Alessandro, L., Miccadei, E., Piacentini, T. (2003). Morphostructural elements of central–eastern Abruzzi: contributions to the study of the role of tectonics on the morphogenesis of the Apennine chain. Quaternary International 101-102C, pp. 115-124.

Del Monte, M., Di Bucci, D., Trigari, A., (1996). Assetto morfotettonico della regione compresa tra la Majella e il Mare adriatico (Appennino Abruzzese). Memorie della Società Geologica Italiana 51, 419-430.

Della Seta, M., Del Monte, M., Fredi P., Miccadei, E., Nesci, O., Pambianchi, G., Piacentini T., Troiani, F. (2008). Morphotectonic evolution of the Adriatic piedmont of the Apennines: an advancement in the knowledge of the Marche-Abruzzo border area. Geomorphology 102, pp. 119-129.

Demangeot J. (1965). Geomorphologie des Abruzzes Adriatiques, C. Rech. et Doc. Cart. Mem. Doc., 1-403, Paris

Di Celma, C., Farabollini, P., Moscatelli, U., (2000). Landscape, settlement and roman cadastres in the lower Sangro valley (Italy). Proceedings - Geoarchaeology of the landscape of classical antiquity, International Colloquium Gent, 23-24 October 1998, Babesch Supplement, pp. 5-14.

Doglioni C., D'Agostino N., Mariotti G. (1998). Normal faulting vs. regional subsidence and sedimentation rate. Marine and Petroleum Geol., 15, 737-750.

Dramis F. (1993). Il ruolo dei sollevamenti tettonici a largo raggio nella genesi del rilievo appenninico. In: Farabollini P., Invernizzi C., Pizzi A., Cavinato G.P., Miccadei E. (eds.): Evoluzione geomorfologica e tettonica quaternaria dell'Appennino centro-meridionale. Studi Geol. Camerti, vol. spec. 1992/1, 9-15.

Dramis F., Farabollini P., Gentili B., Pambianchi G. (1995). Neotectonics and large-scale gravitational phenomena in the Umbria-Marche Apennines, Italy. In Slaymaker O. (ed.), Steepland geomorphology. J. Wiley, Sons, New York, 199-217.

Dramis F., Sorriso-Valvo M. (1994). Deep-seated gravitational slope deformations, related landslides and tectonics. In: N. Oyagy, M., Sorriso-Valvo and B. Voight (eds.): Deep-seated landslides and large-scale rock avalanches. Engineering Geol., 38 (3-4), 231-243.

Dramis, F., (1993). Il ruolo dei sollevamenti tettonici a largo raggio nella genesi del rilievo appenninico. In: Farabollini, P., Invernizzi, C., Pizzi, A., Cavinato, G.P., Miccadei, E. (Eds.), Evoluzione geomorfologica e tettonica quaternaria dell'Appennino centro-meridionale. Studi Geologici Camerti, 1992/1, 9-15.

Elmi, C., (1991). Anomalie del reticolo idrografico nell'Appennino centro-settentrionale: evoluzione geomorfologica e neotettonica. Giornale di Geologia, ser. 3a, 53/2, 81-92.

ENEL (1981). Elementi di neotettonica del territorio italiano, 3, 1-94.

Fanucci, F., Moretti, E., Nesci, O., Savelli, D., Veneri, F., (1996). Tipologia dei terrazzi vallivi ed evoluzione del rilievo nel versante adriatico dell'Appennino centro-settentrionale. Il Quaternario 9, 255-258.

Frankel, K.L., Pazzaglia, F.J., (2005). Tectonic geomomorphology, drainage basin metrics and active mountain fronts. Geografia Fisica e Dinamica Quanternaria 28, 7-21.

Gerasimov, I.P. (1946). Experience with geomorphological interpretation of the general scheme of geological structure of USSR, Probleme Fizische Geographie 12, pp. 89-115.

GNGFG, Gruppo Nazionale Geografia Fisica e Geomorfologia (1994). Proposta di legenda geomorfologica a indirizzo applicativo. Geogr. Fis. Dinam. Quat., 16(2), 129-152.

Horton R.E. (1945). Erosional development of streams and their drainage basin; hydrophysical approach to quantitative morphology. In: Schumm S.A. (Ed.), Drainage Basin Morphology, Geol. Soc. America Bull, 56, 275-370.

ISPRA (2009). Carta Geologica d'Italia 1:50.000 - Aggiornamento ed integrazioni delle linee guida della Carta Geologica d'Italia alla scala 1:50.000. Presidenza del Consiglio dei Ministri, Dip. Servizi Tecnici Nazionali, Servizio Geologico, Quaderno serie III, 12(1,2,3).

Keller E.A., Pinter N. (1996). Active tectonics, Prentice Hall, Upper Saddle River, New Jersey, 338 pp.

Kühni, A., Pfiffner, O.A., (2001). Drainage patterns and tectonic forcing: a model study for the Swiss Alps. Basin Research 13, 169-197.

Leeder M.R., Jackson J.A. (1993). The interaction between normal faulting and drainage in active extensional basins, with examples from the western United States and central Greece. Basin Res., 5, 79-102.

Lombardo M., Calderoni G., D'Alessandro L., Miccadei E. (2001). The travertine deposits of the upper Pescara valley (Central Abruzzi, Italy): a clue for the reconstruction of the late Quaternary Palaeoenvironmental evolution of the area. In: Visconti G., Beniston M., Iannorelli E. D., Barba D (eds.): Global Changes, Protected Areas. Advances in global change research, 9, 459-464.

Lupia Palmieri, E., Ciccacci, S., Civitelli, G., Corda, L., D'Alessandro, L., Del Monte, M., Fredi, P., Pugliese, F., (1996). Geomorfologia quantitativa e morfodinamica del territorio abruzzese. I. Il Bacino del Fiume Sinello. Geografia Fisica e Dinamica Quaternaria 18, 31-46.

Mayer L. (1986). Tectonic geomorphology of escarpments and mountain fronts. In: Wallace R.E., Allen C.R.(eds.): Active tectonics. National Academy Press, Washington D.C., 125-135.

Merrits, D.J., Vincent, K.R., Wohl, E.E. (1994). Long river profiles, tectonism, and eustasy: a guide to interpreting fluvial terraces. Journal of Geophysical Research 99, pp. 14031-14050.

Mescerjakov, J.P. (1968). Les concept de morphostructure et de morphosculpture: un nouvel instrument de l'analyse géomorphologique. Annales de Géographie, 423, pp. 539-552.

Miccadei E., Barberi R., Cavinato G.P. (1999). La geologia quaternaria della Conca di Sulmona (Abruzzo, Italia centrale). Geol. Romana, 34, 58-86.

Miccadei E., Mascioli F., Piacentini T. (2011). Quaternary geomorphological evolution of the Tremiti Islands. Quaternary International, 233, 3-15.

Miccadei E., Paron P., Piacentini T. (2004). The SW escarpment of the Montagna del Morrone (Abruzzi, central Italy): geomorphology of a faulted-generated mountain front. Geografia Fisica e Dinamica Quaternaria, 27, pp. 55-87.

Miccadei E., Piacentini T., Barberi R. (2002). Uplift and local tectonic subsidence in the evolution of intramontane basins: the example of the Sulmona basin (central Apennines, Italy). In: Dramis F., Farabollini P., Molin P. (eds.): Large-scale vertical movements and related gravitational processes, International Workshop Camerino-Rome, 21th-26th June, 1999. Studi Geologici Camerti, Numero Speciale 2002, 119-134.

Miller V.C. (1953). A quantitative geomorphology study of drainage basin characteristic in the Clinch Mountain Area, Virginia and Tennessee. Dept. of Geology, 3, 30.

Molin, P., Fubelli, G., (2005). Morphometric evidence of the topographic growth of central Apennines. Geografia Fisica e Dinamica Quanternaria 28, 47-61.

Molin, P., Pazzaglia, F.J., Dramis, F., (2004). Geomorphic expression of active tectonics in a rapidly-deforming arc, Sila Massif, Calabria, southern Italy. American Journal of Sciences 304, 559-589.

Morisawa, M., Hack, T. (1985). Tectonic Geomorphology. Allen and Unwin, Boston & London.

Nesci, O., Savelli, D., (2003). Diverging drainage in the Marche Apennines (central Italy). Quaternary International 101-102, 203-209.

Nesci, O., Savelli, D., Calderoni, G., Elmi, C., Veneri, F., (1995). Le antiche piane di fondovalle nell'Appennino nord-marchigiano. In: Assetto fisico e problemi ambientali delle pianure italiane. Memorie Società Geografica Italiana 53, 293-312.

Nesci, O., Savelli, D., Veneri, F., (1992). Terrazzi vallivi e superfici di spianamento nell'evoluzione del rilievo appenninico nord-marchigiano. Studi Geologici Camerti, spec. vol. 1992/1, pp. 175-180.

Oguchi T. (1997). Late Quaternary sediment budget in alluvial-fan-source-basin systems in Japan. Journ. Quat. Science, 12(5), 381-390.

Oguchi T., Ohmori H. (1994). Analysis of relationships among alluvial fan area, source basin area, basin slope and sediment yield. Zeit. Geomorph. N.F., 38(4), 405-420.

Ollier C.D. (1999). Geomorphology and mountain building. Geogr. Fis. Dinam. Quat., 22, 49-60.

Ollier, C.D. (1981). Tectonics and landforms. Longman, London.

Panizza, M., Castaldini, D. (1987). Neotectonic research in applied geomorphologic studies. Zeitschrift für Geomorphologie Suppl. Bd. 63, pp. 173 - 211.

Patacca E., Scandone P., (2007), Geology of the southern Apennines, Boll. Soc. Geol. It. Spec. Issue 7, 75 -119.

Pazzaglia F.J. (in press). Fluvial terraces, in Wohl, E., ed., Treatise of Geomorphology. New York, NY: Elsevier.

Pazzaglia, F.J., Brandon, M.T., (2001). A fluvial record of long term steady-state uplift and erosion across the Cascadia forearc high, Western Washington State. American Journal of Science 301, 385–431.

Peulvast J.P., Vanney J.R. (2001). Géomorphologie structurale. Tome 1, Relief et structure. Gordon and Breach Science Publisher, 505 pp.

Picotti V., Ponza A., Pazzaglia F. J., (2009), Topographic expression of active faults in the foothills of the northern Apennines, Tectonophysics 474, 285-294.

Rapisardi, L., (1982). Tratti di neotattonica al confine molisano-abruzzese. CNR - Progetto finalizzato Geodinamica, Roma, pp. 223-232.

S.G.N. (1994). Carta Geomorfologica d'Italia 1:50.000 - Guida al rilevamento. Presidenza del Consiglio dei Ministri, Dip. Servizi Tecnici Nazionali, Servizio Geologico, Quaderno serie III, 4, 42 pp,

Saito K. (1982). Classification of alluvial fans in Japan by topographical and geological data of drainage basins. Geogr. Rev. Japan, 55, 334-349.

Scheidegger, A. (2004). Morphotectonics. Springer, Amsterdam

Schumm S.A. (1956). Evolution of drainage system and slopes in bad-lands at Perth Amboy, New Jersey. In: Schumm S.A. (ed.): Drainage Basin Morphology. Geol. Soc. America Bull., 67, 597-598.

Schumm, S.A. (1969). River metamorphosis: proceedings of the American Society of Civil Engineers. Journal of the Hydraulics Division 95, pp. 255–273.

Spagnolo, M., Pazzaglia, F.J., (2005). Testing the geological influences on the evolution of river profiles: a case from northern Apennines (Italy). Geografia Fisica e Dinamica Quanternaria 28, 103-113.

Stewart I.S., Hancock P.L. (1994). Neotectonics. In: Hancock P.L. (ed.): «Continental deformation». Pergamon Press, 370-409.

Strahler A.N. (1952). Hypsometric (area-altitude) analysis of erosional topography. Geol. Soc. America Bull., 63, 1117-1142.

Strahler A.N. (1957). Quantitative Analysis of Watershed Geomorphology. Am. Geophys. Union Trans., 38(6), 913-920.

Sylos Labini S., Bagnaia R., D'Epifanio A (1993). Il Quaternario del Bacino di Sulmona (Italia Centrale). Quaternaria Nova, 3, 343-360.

Twidale, C.R. (2004). River patterns and their meaning. Earth-Science Reviews 67, pp. 159–218.

Vittori E., Cavinato G.P., Miccadei E. (1995). Active faulting along the north-eastern edge of the Sulmona basin (central Apennines), Special Issue Bull. Am. Ass. Eng. Geol., 6, 115-126.

Wallace R.E. (1977). Profiles and ages of young fault scarps in north-central Nevada. Geol. Soc. Am. Bull., 88, 107-172.

Wallace R.E. (1978). Geometry and rate of changes of fault-generated range fronts, north-central Nevada. Geol. Surv. Journ. Res., 6, 637-650.

Zernitz, E.R., (1932). Drainage patterns and their significance. The Journal of Geology 40, 498- 521.

8

Neogene Tectonics in Croatian Part of the Pannonian Basin and Reflectance in Hydrocarbon Accumulations

Tomislav Malvić[1] and Josipa Velić[2]
[1]INA-Industry of Oil Plc. & University of Zagreb,
Faculty of Min., Geol. and Petr. Eng.,
[2]University of Zagreb, Faculty of Mining, Geology and Petroleum Engineering, Zagreb,
Croatia

1. Introduction

The structural complex Croatian part (CPBS) of the Pannonian Basin System (PBS; Fig. 1) is regional known example of Neogene and Quaternary transtensional and transpressional tectonics (Royden, 1988; Velić, 2007). Tectonical and sedimentation features in this area are very well described in numerous publications and can be used as examples for interpretation of Upper Cenozoic geology in similar geological provinces in the world. Also, PBS is characterised with many hydrocarbon discoveries, making this basin system still one of the most important hydrocarbon province in Europe, including significant undiscovered reserves in subtle, satellite or simply smaller traps. CPBS covers entire SW/S part of PBS and is characterised with many features that are recognizable in other areas of basin. But as it is marginal southern part of that basinal system there are several unique tectonic characteristics. Also, CPBS is characterised with numerous hydrocarbon discoveries in rocks from Palaeozoic to Lower Pontian ages (Velić, 2007; Velić et al., 2002, 2008, 2010; Malvić, 2003). Structurally, hydrocarbon fields include compartmentalization controlled by fault system and lithological variations. These properties make some of field's structures in CPBS as classical examples of Neogene and Quaternary tectonics and hydrocarbon reservoirs in PBS as a whole.

Also, geology of SW margin of CPBS is characterised by contact with Dinarides, as the next regional geological province. This remarkable contact is very deep fault zone called Periadriatic-Vardar lineament (Fig. 1). During Neogene it was the margin of Pannonian (as part of Paratethys) Sea and later Sava Lake. This sea or lake, comprised several Neogene depositional areas in CPBS, which is surrounded with many inland and highland areas (Malvić, 2003; Velić, 2007; Vrbanac, 2002; Vrbanac et al., 2010). As it is seen CPBS has relatively isolated position in PBS, thanks to numerous mountains which existed as uplifted palaeorelief through entire Neogene (like Medvednica, Kalnik, Moslavačka gora, Papuk, Psunj, Krndija and Dilj Mts.). The elongated shape of depressions, which clearly follows the SW margin of PBS, indicated on relatively shallow environment with irregular palaeo-cost line. All these facts asked for many detail studies of Neogene and Quaternary evolution of CPBS and its reflectance in hydrocarbon genesis, migration and accumulation, what had been done through several decades and still are object of scientific interest. Analysis presented here is some kind of review of and contribution in our present knowledge of that area.

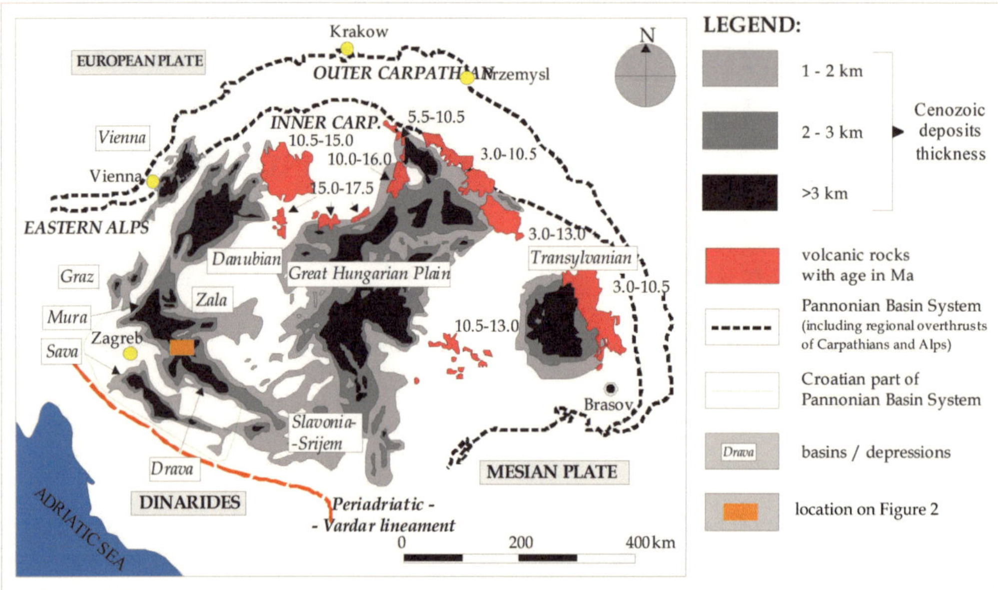

Fig. 1. Geotectonic units of the Pannonian Basin System (modified after Royden, 1988)

2. Neogene tectonics in Croatian part of the Pannonian Basin system

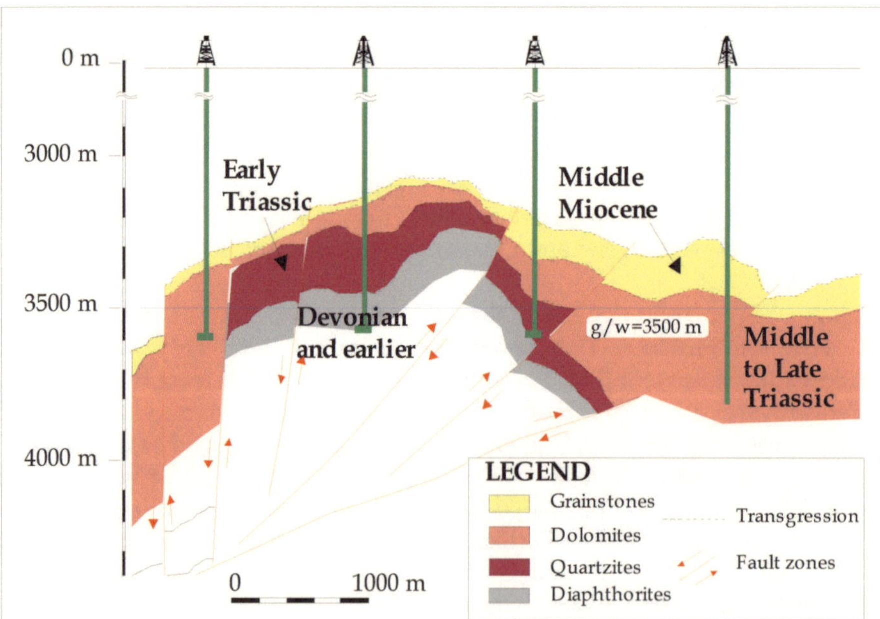

Fig. 2. Typical large, faulted anticline in the Drava Depression

The oldest, still recognisable, tectonic displacements are the results of Palaeozoic and Mesozoic orogenesis. Many Mesozoic deposits were eroded during the Palaeogene. Palaeozoic and Mesozoic rocks are separated by regional unconformities from Neogene sediments, but also inside Neogene sequence can be found some local unconformities. However, the chronostratigraphic units mostly have retained their original superpositional relations.

The present-day structures are developed in the Miocene, starting with extensional tectonics in opening of the PBS and uplifting of the Apennines and the Dinarides (Royden, 1988; Yilmaz et al., 1996). Middle Miocene was generally reflected in numerous extensional intra-basin strike-slip structures.

Such strike-slip structures regular started as negative flower structures in Middle (rarely Lower) and Late Miocene, but later in Pliocene and Quaternary were inverted in positive (Malvić and Velić, 2010). It assumed reactivation and changing character of many faults, especially main bordering structure faults, from normal to reverse (Fig. 2). Eventually, many such uplifted anticlines had been connected through mutual spill points in larger anticlinorium. The best example is the largest anticlinorium in CPBS, the large Molve-Kalinovac-Stari Gradac structure.

2.1 Transtensional and transpressional phases in Neogene and Quaternary

Royden (1988) and Rögl (1996, 1998) documented the ages of different Miocene episodes in the Pannonian area, establishing a scale used to categorize tectonic and sedimentation cycles. Tectonic extension in the PBS (probably sporadically also in CPBS) started in the *Ottnangian* (19.0 to 17.2 Ma; Haq & Eysinga, 1998), locally accompanied by a marine transgression, often paralic environments and strike-slip displacements. Extension continued in the *Karpatian* (17.2 to 16.4 Ma; Haq & Eysinga, 1998), when the some parts of CPBS possible were covered by brackish or marine sediments.

1st transtensional phase in CPBS maximal extension reached *in the Badenian* (16.4 to 13.0 Ma; Haq & Eysinga, 1998), when strike-slip tectonics was main mechanism of structural development. Vrbanac (2002) described an extensive marine environment as dominant in CPBS during the Late Badenian, with several large mountains as islands. During extension, clastics sources were (a) siliciclastic basement rocks or (b) carbonates eroded by weathering of *corallinaceaen* sp. and *briozoan* sp. reefs (Malvić, 1998, 2003). This material was deposited in alluvial-fan environments. Coarse-grained sediments were deposited in the proximal part, and medium- and fine-grained sandstones in the middle or distal part such fans (Fig. 3). This explains the gradual decrease of porosity values in Middle Miocene facies in many field's structures, mostly toward SE (Fig. 5).

Middle Badenian sediments are coarse-grained breccia and breccia-conglomerates, representing important reservoir lithofacies for hydrocarbon accumulations. Sometimes these reservoirs can contain significant part of non-hydrocarbon components (mercury, N_2, CO_2, H_2S, chlorides and mercaptans) like reservoir fluid in the Molve Field. It can be explained by geological complexity, reservoir thermodynamics and generation in carbonate rocks (Barić et al., 1991, 1998). Upper Badenian sediments are mostly fine-grained sandstones, siltstones and, at the top, limestones and calcitic marlstones. It is very hard to determined border between Badenian and Sarmatian due to lithological similarity. The Late Badenian generally represented a transition between extensional (transtensional) and post-extensional phases in Northern Croatia (Pavelić, 2002). Generally, evolution of the Badenian environment was mostly controlled by palaeorelief of the pre-Neogene basement, erosion rate, stream power of alluvial fans and local tectonics.

Sarmatian was period of when **1st transpression in CPBS** started. After series of transgressive-regressive cycles in the Badenian, it was followed by a overall regression during the Sarmatian (13.0 to 11.5 Ma; Haq & Eysinga, 1998), typical for the Central Paratethys (e.g. Rögl & Steininger, 1984; Kovač et al., 1997; Pavelić, 2001; Vrsaljko et al., 2006). This post-extensional phase was characterized by local thermal subsidence of the base of the Pannonian Basin. Extensional tectonics (dominantly strike-slip negative structures and normal faulting) were replaced by a compressional style (mostly reverse faulting) over almost the entire PBS. Tectonics still controlled sedimentation locally (Royden, 1988; Rögl & Steininger, 1984; Pavelić, 2001, 2002), but the importance of alluvial fans significantly decreased. Sea level was lowered, salinity was reduced (existence of brackish environment) and, most importantly, the deep-water turbidites that originated in the Alps had been activated in some parts of PBS.

These turbidites were a much more abundant source of sediments in CPBS from the Early Pannonian to the Late Pontian, when almost completely replaced mechanism of relatively small alluvial fans. Sedimentation of organic-rich material (later often matured source rocks) continues from Late Badenian. These limestones, marly limestones and calcitic marls today can be found on the depths between 2000-4000 m. Such source rocks also are commonly seals, causing long (10-15 km; Malvić, 2003) lateral migration as a favourable mechanism of hydrocarbon accumulating in CPBS.

In the Early *Pannonian* (11.5 to 9.3 Ma; Haq & Eysinga, 1998) the **2nd transtensional** localized strike-slip tectonics took place in the PBS (Royden, 1988). In a large lacustrine, brackish and eventually fresh-water environment, characterised by depths up to several hundreds metres (Vrbanac et al., 2010), salinity was continuously reduced owing to fresh-water inflow and a lack of connection with other open-sea environments.

Turbidite currents of high-density were especially active in Late *Pannonian* (9.3 to 7.1 Ma; Haq & Eysinga, 1998) and Early *Pontian* (7.1 to 6.3 Ma; Haq & Eysinga, 1998) time. Significant quantities of clastics were transported from the Eastern Alps in CPBS through several turbidite events initiated by ramp-fault activity. Each depositional episode moved clastics over tens of kilometres, deposited them next to the tectonic ramp, and pushing sediments to a final depositional centres in the Mura, Sava, Drava and Slavonia-Srijem depressions to structurally determined areas. Successive turbidite events represented the dominant Late Miocene sedimentation mechanism in the Croatian depressions (Vrbanac, 2002). Turbiditic sequences were source of medium to fine-grained sandstones that represents the main reservoir lithofacies in CPBS. This mechanism was active periodically, interrupted typical hemipelagic sedimentation in lacustrine environment in CPBS. That resulted in monotonous alteration of sandstone and marlstone lithofacies through most of Pannonian and Pontian stages.

The Late *Pontian* (6.3-5.6 Ma), *Pliocene* and *Quaternary* (2.6 to 0.0 Ma) were period of **2nd transpressional phase**, when negative flower structures and faulted anticlines had been uplifted and hydrocarbons migrated in recent reservoirs. It was characterized by sporadically lacustrine, and mostly marsh, river and continental sediments (loess). Sedimentation took place in fresh lacustrine waters. This sedimentation represented periods of overall regression and reduction of the depositional area in final continental phase. The main described regional tectonical and depositional events in chronostratigraphical time-scale are schematically given in Table 1.

Million of years in the past (Ma)

19 18 17 16 15 14 13 12 11 10 9 8 7 6 5 4 3 2 1 0

Ottnangian	Karpatian	Badenian	Sarmatian	Pannonian Early	Pannonian Late	Pontian Early	Pontian Late	Pliocene	Quaternary
Maybe sporadical extensions in CPBS?		1st transtensional phase	1st transpression	2nd transtensional phase			2nd transpressional phase		
		Local inland weathering and alluvial fans mechanism	Shallowing and salinity reducing	Distant clastic source and turbidite mechanism			Final structural forming, migration of hydrocarbons, eventually continental environment		

Table 1. Time-scale of the main tectonical and depositional events in Neogene and Quaternary in CPBS

2.2 Neogene and Quaternary sedimentation through megacycles as genetic units

Neogene and Quaternary sediments are divided in 3 megacycles (Velić et al. 2002, Velić 2007). Each is described by depth and age of sediments, trap types and petrophysical properties. These sediments can be also found as outcrops on hills of surrounding mountains, but only with approx. 3% of total volume, but the largest part is covered with Quaternary deposits (Velić, 2007). In marginal parts of CPBS depressions Neogene thickness ranges 500-1500 m, and in central areas it could reach up to 3500 m in Slavonia-Srijem Depression, 5500 m in Sava and Mura Depressions and near 7000 m in Drava Depression.

Neogene sediments are often cyclic. Such regularities and lithological characteristics were reason for selection 3 megacycles, and their division to lithostratigraphic formations and members. Each megacycles included rocks deposited in one super cycle of relative sea level changing, which is firstly characterised by gradually increasing, and secondly by relative abruptly decreasing of water depth (Mitchum, 1977). All megacycles are lithologically different, because they resulted from different transtensions and transpressions phases, different basins palaeorelief, as well as from eustatic and climate changes (Šimon, 1980).

2.2.1 Lower and Middle Miocene (1st) megacycle

The oldest megacycle lasted in Lower and (mostly) Middle Miocene, i.e. approx. 6.8 Ma (geochronological data after Rögl, 1996). It is characterized by heterogenetic clastic lithologies. Lithostratigraphically, this megacycle includes Prečec Formation in the Sava Depression, Vukovar Formation in the eastern Drava and Slavonia-Srijem Depression, Moslavačka gora Formation in the western Drava Depression and Murska Sobota Formation (except uppermost part) in the Mura Depression.

These sediments are characterised by coarse-grained clastics (breccia, conglomerates or sandstones), fine-grained and pelitic clastics (clays, marls, calcitic marls, sandy and clayey marls with tuff intercalations) and carbonates (limestones, often *Lithotamnium*, and calcarenite sandstones). Some authors origin of tuffs and effusives in sediments of this age placed in Egerian and Egenburg (Pamić, 1997), and related sediments in Early Miocene paralic and transition environments. But Badenian is generally described as period when marine sedimentation took place everywhere in CPBS (e.g. Lučić et al., 2001, Vrbanac, 2006). Some new results located begging of all Neogene sedimentation in CPBS just in Badenian (Ćorić et al., 2009), what is today mostly accepted.

Heterogeneity of rocks is accompanied with large (and somewhere rapid) changes in thicknesses. For example, in the Sava and Drava Depressions thickness of these sediments can differ more than 2000 m on distance of 3-4 km. Generally, psamitic sediments of this megacycle are mostly deposited on margins as results of alluvial fans activity, and pelitic in central parts of depressions where normal basin plain sedimentation took place. As it was mentioned, pelitic sediments of this age are often source rocks (e.g. Troskot-Čorbić et al., 2009), including mature rocks from Upper Badenian to Lower Pannonian (2nd megacycle).

2.2.2 Late Miocene (2nd) megacycle

The second megacycle had Late Miocene age (Pannonian and Pontian). It includes sedimentary association of the Sava Group (Ivanić-Grad, Kloštar-Ivanić and Široko Polje Formations) in the Sava and western Drava Depressions, then Vinkovci and Vera Formations in the eastern Drava and Slavonia-Srijem Depressions and Lendava Formation and lower part of Mura Formation in the Mura Depression.

Deposition lasted approx. 5.9 Ma (Rögl, 1996). The oldest part is located deeper than 5000 m in Virovitica area. Generally, it is represented by uniform sequences (or homogeneous alteration) of sandstones (subgrauwacke and calcarenite subgrauwacke), siltites and marls with transitional lithotypes. Sandstones are mostly grey coloured, dominantly with quartz, and rock clasts (mostly limestones), mica and feldspat. The Pannonian sediments are deposited in brackish and Pontian in fresh water lake environment. The thickness varies from approx. 2000 m (in small pull-aparts of Hrvatsko Zagorje Subdepression) to more than 4000 m in the deepest parts of the Drava Depression (Velić et al., 2002).

The maximal total thicknesses are proven in wells located in the central part of depressions, where is also proven the maximum of sandstone thicknesses. These sandstones gradually pinch out toward margins, where are laterally gradually changed in marls. Deposition had been active in lake environment mostly deeper than 200 m (Vrbanac et al., 2010), where low energy basin sedimentation had been interrupted with periodical turbidites, which direction had been determined by sub-water, basement palaeorelief.

The main sources of silt and sand particles were Eastern Alps. These clastics which had been several times re-deposited, and moved to tectonically instable slope or ramp and, after tectonical displacement, transported by turbidites in different parts of depressions (e.g. Velić et al., 2002; Vrbanac et al., 2010). Homogenous alteration of sandstones and marls in deeper parts is characterized by regional e-log markers (or key-beds) that are also chronostratigraphic border. These markers are regularly used in regional correlation of lithostratigraphic units in CPBS.

2.2.3 Pliocene and Quaternary (3rd) megacycle

The youngest is 1st megacycle of Pliocene and Quaternary periods, which lasted approx. 5.6 Ma (Rögl, 1996.). There are described Lonja Formation in the Sava and western Drava Depressions, Vuka Formation in the eastern Drava and Slavonia-Srijem Depressions, and Mura Formation in the Mura Depression. This megacycle is lithologically represented by alteration of poorly consolidated sandstones and soft clay with rare lignite. Only in the deepest parts (below 500 m) compaction can be enough strong for clayey limestones, marls and fine-grained sandstones. In the shallowest parts sediments are of gravels, loess and silts. Thickness can reach more than 1500 m in the Sava and 2200 m in the Drava Depression (Velić et al., 2002). All these sediments are product of mostly continental environments.

2.3 Main depositional mechanisms for reservoir rocks in Neogene

Depositional systems through Neogene of CPBS can be considered through two main cycles when large part of clastic and reservoir lithofacies deposited, with different transport mechanisms and sources of clastic materials. The first one is period of Badenian that tectonically belongs to period of 1st transtension in CPBS, and 3rd megacycle. The second type of depositional mechanism had been active from Late Pannonian to Early Pontian, i.e. through period of 2nd transtension and 2nd megacycle.

2.3.1 Depositional mechanism during Badenian (1st transtension and 1st megacycle)

Sediments of Middle (maybe locally also Lower) Miocene in CPBS are common reservoir, seal and source rocks. These are the earliest sediments deposited during Neogene transgression covering the entire CPBS (Ćorić et al., 2009). These sequences embracing mostly the Badenian, Sarmatian and Lower Pannonian ages are mainly clastic rocks. Badenian started by coarse-grained alluvial sediments and Lower Pannonian finished by lacustric marls.

Marine environments (Vrbanac, 1996; Rögl, 1996, 1998; Ćorić et al., 2009) covered entire northern Croatia. Irregular, previously faulted and weathered, palaeorelief of Palaeozoic and Mesozoic basement resulted in significant depth differences and played important geomorphological role on the sea bottom. It is considered that large areas of present-day mountains in northern Croatia, like Medvednica, Kalnik, Moslavačka gora, Psunj and Papuk Mts., remained more or less isolated islands above the sea-level.

Palaeozoic rocks and basement gave siliciclastic and Mesozoic carbonate detritus for coarse-grained sediments in Badenian. Moreover, lithoral algae reefs were abundant in shallow marine environments, where such reefs (mostly *Coralinacea* and *Briozoa* origin) were eroded by sea currents, representing important source of carbonate clasts later re-deposited in potential reservoir rocks.

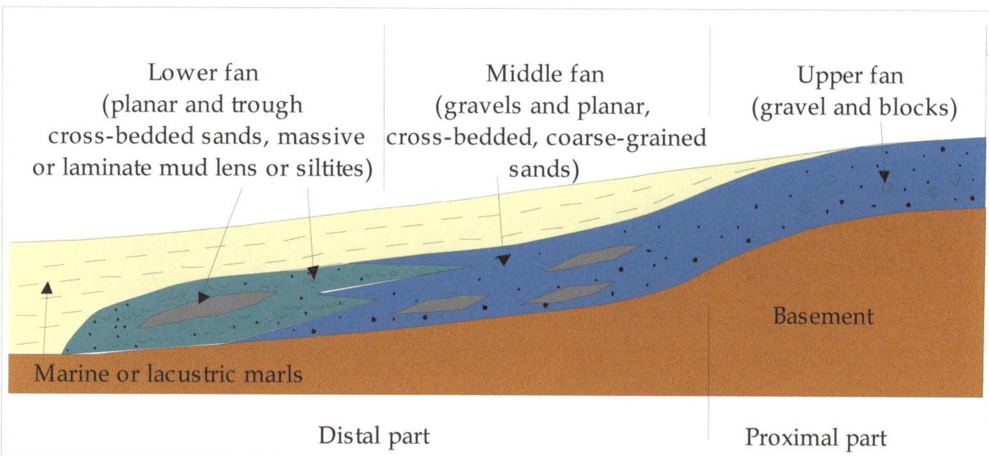

Fig. 3. Schematic review of typical alluvial fan and lithofacies distribution

The main mechanism that transported clastics in shallow sea encompassed alluvial fans, divided in proximal and distal parts (Malvić, 1998). Coarse-grained sediments, as promising reservoir lithofacies, were deposited in the proximal part (Fig. 3). Sometimes in such fans, especially in proximal part, can not be found carbonate clasts originated form reef organisms, what is indicator of strong erosion and high energy of fan. Mostly medium-grained sands were deposited in the middle part of alluvial fan (Fig. 3). Sometimes their green colour, due to mica and chlorite minerals, is the indication of weak reductive environment in sea with normal salinity (Odin and Matter, 1981; Tišljar, 1993). Fine-grained sands and silts had been deposited in the lower part of alluvial fan (Fig. 3) and this sequence characterise a fan migration in time and space. Also, such vertical transition described change in brackish, shallow and stagnant environment of Upper Badenian. Badenian top is represented with marlstones and marly limestones deposited in stagnant and shallow (mostly up to 100 m deep) sea.

The typical Badenian erosional slope had been mapped in many hydrocarbon fields in the CPBS (Tišljar, 1993; Malvić, 2003, 2006). It comprises Middle Badenian reservoir rocks connected in one hydrodynamic unit by catalysed basement rocks along unconformities. Upper Badenian included seal rocks, which are sometimes also source rocks, and such pelitic sequence is often continued in Sarmatian and Early Pannonian.

One typical field structures, with Badenian reservoirs, located between major extensional faults and connected with basement buried hills, is shown on Fig. 4. The structural position and shape determined distribution of lithofacies and their thickness during Middle Miocene sedimentation. It is possible using maps of some geological reservoir variables like porosity (Fig. 5), thickness or depth for defining transition and spatial positions of different reservoir lithofacies (e.g. Malvić, 2006) as well as direction and energy of alluvial fan that had been active as mechanism for sedimentation of potential reservoir rocks.

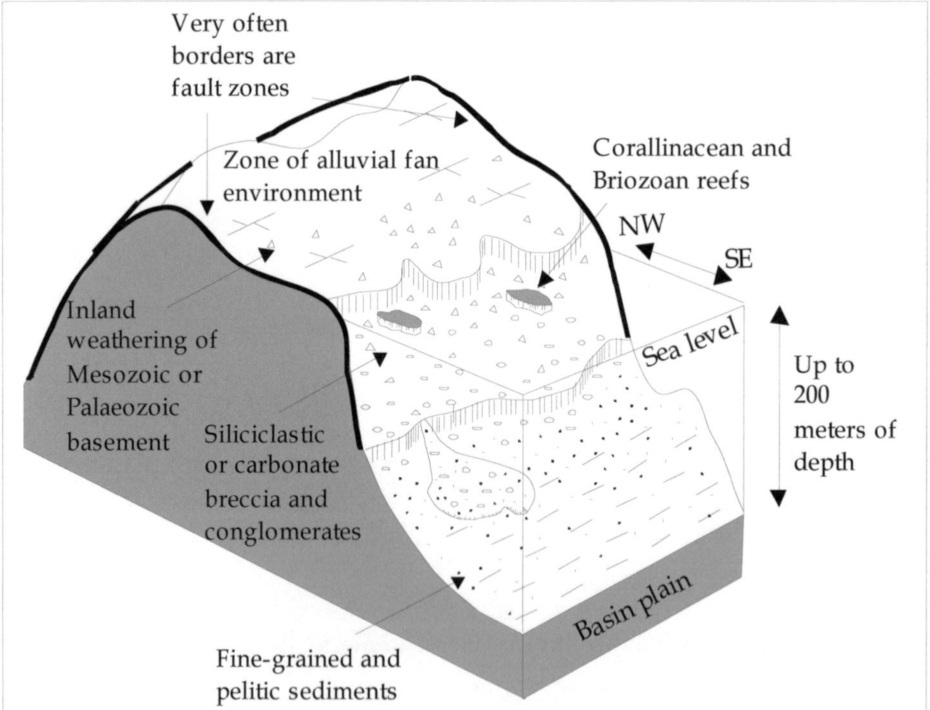

Fig. 4. Schematic review of typical depositional environment in Early Badenian (modified after Malvić, 2006)

The coarse-grained sediments of alluvial fans have the great hydrocarbon potential, because they are, after few million years, often consolidated into good reservoir rocks, with significant primary porosity (more than 10%). Such rocks are mostly overlain by pelitic seal deposits (resulted from distal fan or basin plain sedimentation) sometimes including organic rich source facies (kerogene type II and III, Barić et al., 1991, 1993). It means that Badenian sequences can often be described as complete petroleum systems, what is confirmed by large number of oil and gas discoveries in such sediments in entire CPBS (Velić, 2007).

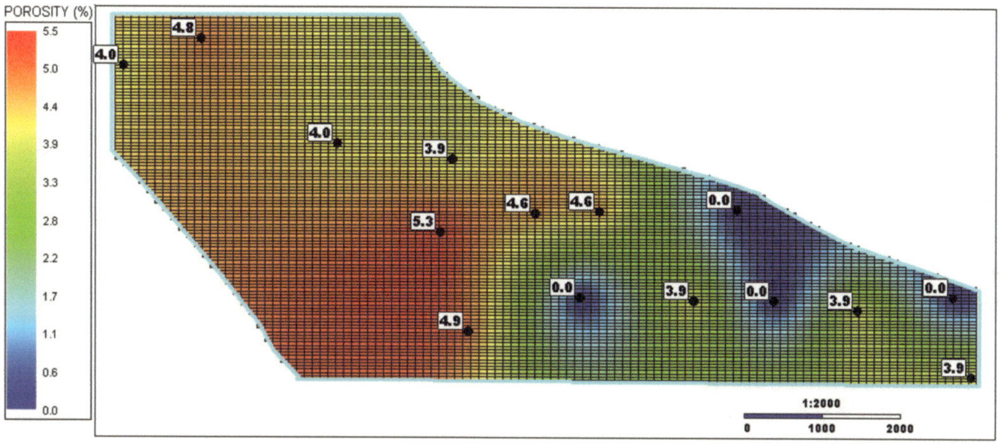

Fig. 5. Distribution of porosity in alluvial fan environment in Stari Gradac-Barcs Nyugat Field (from Malvić, 2006)

Alluvial environments are relatively short distribution area of few kilometres distance (Malvić, 1998), but they are also characterised by frequent changes of petrophysical properties due to local material source and depositional area. It means that Badenian lithofacies are characterised by heterogeneity of thickness and porosity already in small scale (within the range of several hundreds of meters).

It could be observed in Fig. 5, where palaeo-activity of proximal part of alluvial fan is observed on the NW of the Stari Gradac-Barcs Nyugat structure. That area contains coarse-grained sandstones with primary porosity. Those sandstones originated from uplifted Mesozoic basement located close to the NW margin of the structure (Fig. 6). It was weathered and eroded close to strike-slip depression opened just on SE along of the main bordering faults (strike NW-SE and WNW-ESE; Fig. 6).

The region of smaller (mostly lower than 3%, what was cut-off for HC reserve calculation) porosities (Fig. 5) and larger thicknesses belong to distal part of alluvial fan, which through Badenian existed on the SW part of the structure shown on Fig. 6. Distal part was tectonically downlifted along the main fault of the NNE-SSW striking, located in the middle of structure (Fig. 6). Lithologically, this area is characterised by fine-grained planar and trough cross-bedded sands or basin-plain marls (Malvić, 2006).

Later, in the 1st and especially 2nd transpressional phase, entire structure was uplifted (changing of fault character) along the same main fault lines (NW-SE strike) that bordered the structure, which also changed fault character from normal to reverse. Described mechanism of alluvial fan lithofacies distribution and tectonic inversion is the most often geological mechanism of Middle Miocene reservoir forming in CPBS.

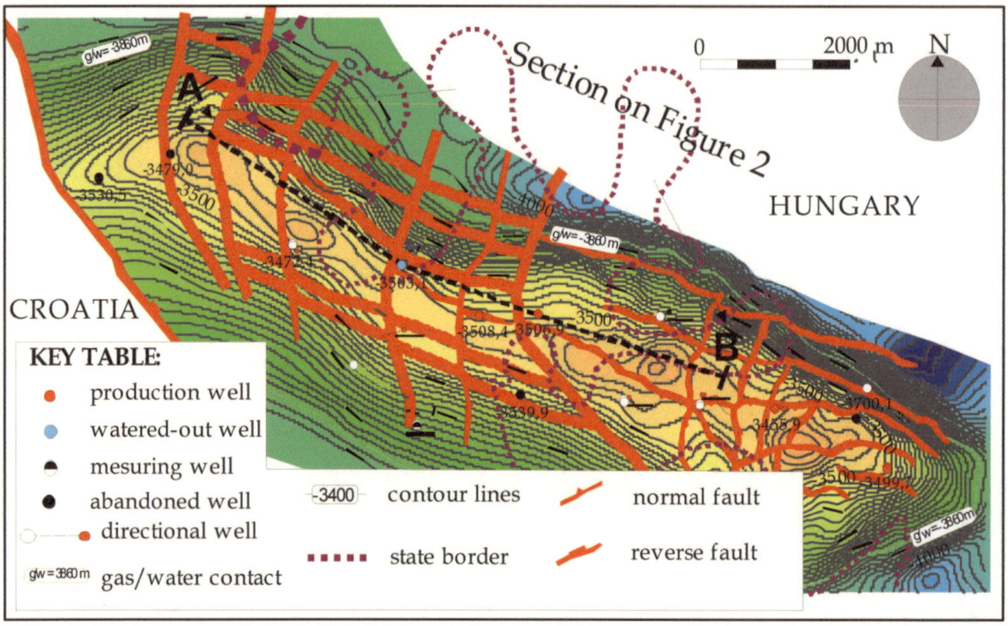

Fig. 6. Structural map of Badenian top (from Malvić, 2006; modified after Gaćeša, S., Futivić, I., Gyorgy, F. and Horvath, Z. (2001): Barcs Nyugat - Stari Gradac field study. INA-Naftaplin & MOL KTD, company archive, INA-Naftaplin, Zagreb)

2.3.2 Deposition mechanism during Late Pannonian and Early Pontian (2nd transtension or 2nd megacycle)

In the Late Pannonian and Early Pontian (9.3-5.6 Ma), CPBS depressions were lacustric parts of the Pannonian Basin System, divided in partially independent elongated areas of brackish environments (e.g. Sava Depression had width of 25 km and a length of 100 km). In this period PBS was an open lake system, composed of several connected basins with active inflows and outflows (rivers) that have been described in detail by Bérczi et al. (1988). The chemical composition of water was determined by continuous inflows of fresh water from rivers, what resulted in the development of slightly brackish and eventually fresh lake environment.

Croatian depressions had a very complex palaeorelief, especially the Sava and Drava Depressions, which represented typical tectonic grabens, with deep, elongated central part, and marginally uplifted parts of shallow water or continental palaeorelief (Fig. 7). Consequently, the largest thickness and lithofacies types had been described for central depression parts, simply because there was enough large accommodation space for the most sediment quantity. Mechanism of sediments transportation included turbidite flows. The major part of the siliciclastic material originates from one regional source (Eastern Alps). The last slope before CPBS, which contained detrital material that was accumulated with varying intensity, was probably located into the Mura depression (north of Varaždin). From this point, due to tectonic events, sediment was relocated and transported to the Drava, Sava and Slavonia-Srijem depressions. Such relocation was highly subdued by depression's palaeorelief. In the central parts depressions were surrounded by basement highs (somewhere even above sea-level), where water depths was mostly deeper than 200 m. The erosion level of turbidites had been the strongest just in these central, deeper channels. But, such central parts are also place of large and thick sandstone bodies. However, it is necessary to outline that in the central part of PBS (e.g. Békés Basin in Hungary) the depositional mechanism had been partially different then on the margins of PBS. For example, Juhász and Magyar (1992) proved two main delta systems were dominated, which were Gilber-type deep water deltas. The sediment accumulation is reconstructed from two directions, from NW and NE, which encountered in Békés Basin.

But, observing again CPBS turbiditic flows (or currents) had been determined by positions of Medvednica, Kalnik and Moslavačka gora Mts., i.e. location of the Mura Depression at very NW margin of CPBS. This depression was the first regional depositional centre for turbidites. As it was relatively shallow, the large quantities of clastics had been accumulated on eastern margin (tectonic ramp) of the Mura Depression, and later tectonically moved on the S/SE. Southern from uplifted palaeomountains re-activated turbidites transported material toward south in the Sava and northern in the Drava and Slavonia-Srijem Depressions. The strongest indicator of turbidite direction is the geometry of the sandstone bodies. The sandy detritus was transported into the depression exclusively by turbidites, and the primary flow direction had an arching (curved) shape from the NW, and later N, and somewhere even NE (influenced by palaeorelief). During a period of inactivity, permanent basin sedimentation resulted in deposition of the finest detritus throughout the entire CPBS. Structurally, main channels carried along the main part of the sandy material, while the minor part was deposited in the inter-channel areas. In addition, the channel locations were changed through space and time, continually migrating, thereby defining the inter-channel area pattern.

In such transport directions the crucial role had Moslavačka gora Mt. (Fig. 7) that separated Sava and Drava Depressions, and regional dipping of the depressions toward SE and E. Moreover it is proven that turbidites in the Sava Depression had been transported through two main channels - northern and southern, and during the Early Pontian deposition in the NW part of the Sava Depression was locally inclined firstly toward S (Vrbanac et al., 2010) and locally even toward the SW (Hernitz & Jurak, 1973). This was likely the result of the uplifting of Moslavčka gora Mt., stronger subsidence of the Donja Jelenska and Ilova sags in the NE and existence of Martinska Ves horst in the SW margin of the Sava Depression. Also, this depression had been also S margin of the CPBS, which was closed by mostly underwater palaeorelief located between the Psunj and Prosara Mts. (Blašković, 1982), what also was pinchout zone of the turbidites.

The Drava Depression had more irregular shape due to SW "extension" of the rhomboidal Bjelovar Subdepression (Fig. 7). However sedimentation was similar to the Sava Depression. It means monotonous alteration of sandstones and marls in the central part, while the Bjelovar Subdepression as relatively closed basin of moderate palaeorelief included significantly more hemipelagic marls, especially in Late Pannonian. The final forming of the Drava Depression was took part during Pliocene and Quaternary when northern margin of the Bjelovar Subdepression, the Bilogora Mountain was uplifted (2-3 Ma ago) and separated these depositional areas in practically different depressions.

Reconstruction of the palaeorelief can indicate the palaeodepth of the CPBS Neogene lakes. The most reliable indicator of water depth and the depositional environment are the remains of animals and plants (biotop) that characterise particular environments. Unfortunately, the Upper Pannonian sediments contain very few fossils. However, some macrofossils, microfossils and palinological analysis showed at the point at which the transition from brackish to fresh water occurred just in Late Pannonian. Some other indicates the presence of shallow water in that period (e.g. Pletikapić, 1965, Lučić et al., 1995). Relatively rare samples of sporomorphs would indicate that land was relatively far, but the presence of kerogen type III (terrestrial) supports conclusion (Lučić et al., 1995) that the shore was close.

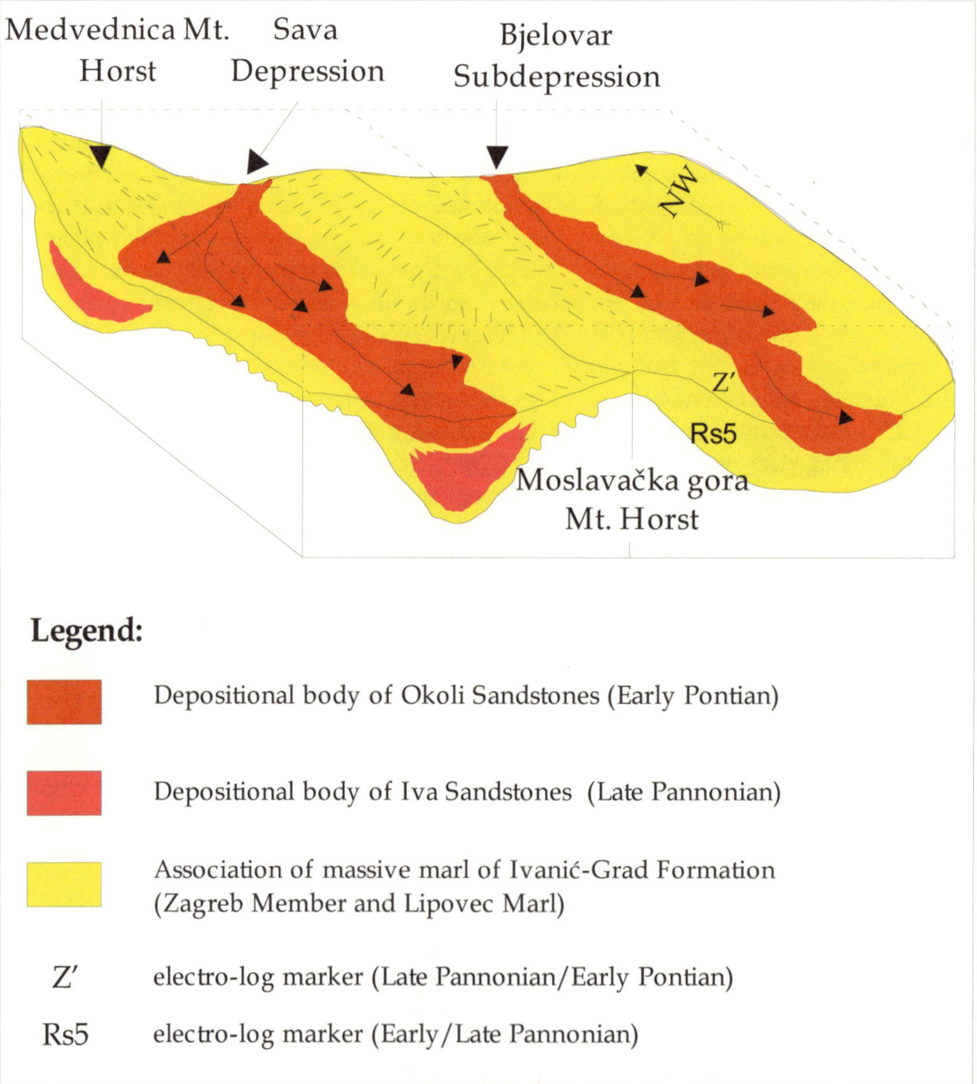

Fig. 7. Schematic palaeographic situation on the border between Sava and Drava Depressions at the end of Late Pannonian (from Vrbanac et al., 2010)

Regarding the relatively narrow depositional depressions it is not clear where the beaches and shoals could be located and where the transported material was eroded only from the close land. However, if marls are defined as sediments of swamps and lagoons, but which do not include in present cores any characteristics that indicate the presence of close land areas, it can hardly be accept. Even if they are reworked some fossils could be found in collection of numerous well cores.

Also it is illogical assume that throughout the entire Late Pannonian (which was about 2 Ma long), in a tectonically active depressions, shallow marsh environment existed all the time with constant water level. Indeed, this would means that no parts of the shallow-water (marsh, lagoons) were exposed to the air, due to continuously deepening and compensating sedimentation rate. Also, there are no evidences for delta activities and erosion of on-shore and near-shore material (like mud and marl) from exposed land. All these facts (assumptions of far land or near shore) can be explained by such mechanism that allowed clastic materials could be derives from distant near-shore environment, and then delivered by turbidity currents to the deeper parts of the basins, also several times reworked (like also skeletal and plant remains) and redeposited.

So, today for CPBS is accepted deeply water (more than 200 m in the deepest parts; Fig. 8), turbidite mechanism as the main source of Upper Miocene sandstones in CPBS. Also complete sequence of massive, hemipelagic marls deposited over the basin plain has been described both in the marginal and central parts of depressions, indicating on calm basin environment. Such lake environment can explain all of the rapid changes in lithologies (especially vertically) and the nearly identical depositional conditions that were observed over a geological ages of Late Pannonian and Early Pontian.

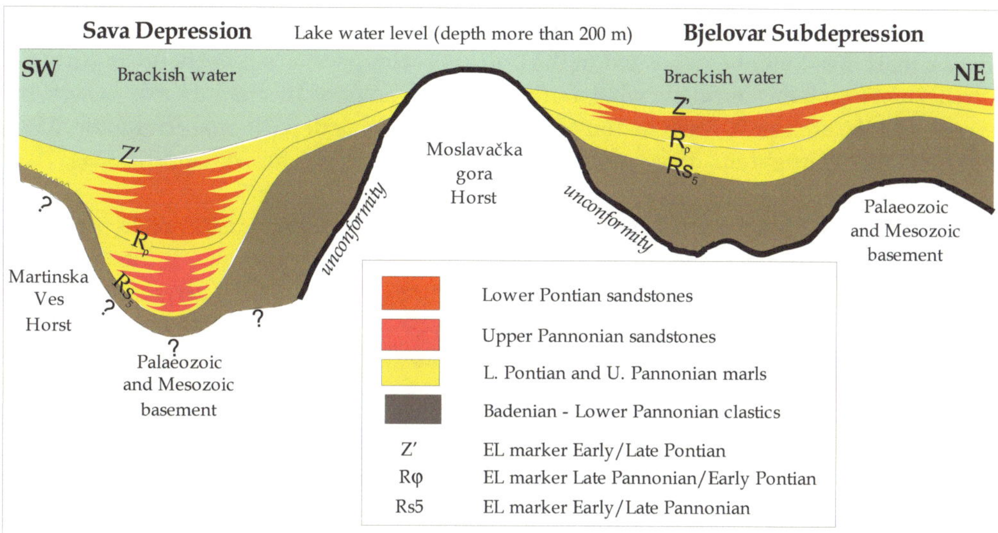

Fig. 8. Schematic palaeographic reconstruction in the beginning of Late Pontian at margin between the Sava and Drava Depressions (modified after Vrbanac et al., 2010)

Sedimentary, mineralogical and petrographic characteristics of the Upper Pannonian and Lower Pontian sediments in CPBS are very similar. The majority of the detritus originates from the Alpine mineral association (Šćavničar, 1979). Transport from Alps has taken a very long time, and the material was re-deposited several times. This explains why the sandstones are mostly medium- to fine-grained, with relatively rounded grains. Conversely, the influence of local horst as a source of detritus has been limited as they probably were under water at that time. According to large distance from Eastern Alps it can be assumed that margins of CPBS (especially eastern Drava and Slavonia-Srijem Depression) were not only filled by turbidites, but also with clastics originated from much closer palaeo-mountains, like Mecsek in southern Hungary. However, it is still not undoubtedly proven.

2.3.3 Main facies associations in Upper Miocene turbiditic environments

Turbiditic clastic environments are the most complex depositional facies through entire Neogene in CPBS, also characterised with the largest thickness and extension of reservoir lithofacies. It is why here will be described such lithofacies, using the most characteristics basin facies association for period from Upper Pannonian to Lower Pontian. There can be distinguished several facies association in all depression. Recently it was very detailed described for the Sava Depression and Bjelovar Subdepression (Vrbanac et al., 2010), but can be almost in the same way accepted for entire CPBS.

The facies with the main reservoir potential is **turbidite channel fill facies association** consists of thick-bedded sandstone facies and thin-bedded sandstone facies. The channel axis is usually characterised by thick sandstone beds with rare marl layers. The number of marl layers is greater in the channel margins, while the basin plain sandstones are thinner and contain more fine-grained sediments.

Usually this association in the lower part is composed of thick bedded sandstones, which strongly eroded marls in bottom of turbiditic flow. In fact, the energy of currents was strongest inside the channel, which resulted in strong bottom erosion. Therefore, the missing marl that could be deposited in the previous calm period due to energy of turbidite currents was probably eroded, what is the result of activity in channel axis where the current was the strongest. Such feature describes channels as depositional (Hamilton, 1967) and erosional (Laughton, 1968) environments in which the main current mass was transported.

Moreover, channel facies association is composed of thin bedded sandstones in the upper parts (when the channel began to be less active and eventually inactive). If the channel is abandoned rapidly, there is no observed upward fining and thinning sequences what is often event in many CPBS hydrocarbon fields. Then SP log is characterised by a cylindrical shape (Pirson, 1970). Erosion of channel bottom was strongest in the central parts of the channels and weaker in the margins, which indicates that the positions of the channel in distributary system can be determined based on the number of altered marl layers. Specifically, the positions can be deduced based on the direction of the main current during different periods. In the proximal and bottom parts of the current the particle concentration was highest, while the majority of suspended detritus was deposited in distal areas due to the decreasing energy. This resulted in the maximum sandstone thicknesses being developed in channels that were formed in according to basin morphology and synsedimentary tectonics.

Turbidite overbank-levee facies association is composed of facies of laminated sandstones, siltstones and marls in the lower part and facies of sandstone layers in the upper part. Sediments of this facies association were deposited outside the distributary channels, where the turbidity currents were significantly weaker, which created overbank-levee, fringe and channel bank deposits. These characteristics resulted in the erosional effects being smaller than in the channel facies. As a result, amalgamations are very rare and the thickness of eroded marls is probably not higher than several to several dozen centimetres. However, the total thickness of the sandstone body in this association can be more than 30 m and shows upward thickening. This association is characteristic at the marginal parts of channels and the channel's banks. This indicates that the distribution was not restricted to the channels, but also occurred in the wider channel area, which resulted in thick sandstones occasionally being deposited in bank (levee) areas.

Lateral or distal turbidite facies association consists of a monotonous alternation of very thin (laminated), thin and medium thick sandstone beds passing into siltstones and marls. In this association, erosion on the bedding plane is almost completely absent. The sub-layers of marls and sandstones are so thin that they are beyond the resolution of well-log curves. Sediments of this association are recognized in distal areas, where the maximal range of turbidites occurred. Moreover, such sediments are discovered on morphologically uplifted parts between channels or further away from the channels for great distances. Consequently, the portion of siltstones and marls in the facies is increased and the portion of marly sandstones and sandstones is decreased. The total thickness of this association can reach several dozens of meters, and can cover several hundreds of meters laterally. This distal turbidite association can be compared with classical turbidite sequences that are deposited (Walker, 1978) at the end of the distal part of a lower fan in the basin plain, which is the area in which the maximal extensions of the turbidity current exist. Such distal turbidite associations can also be found on the banks of the upper fan (Walker, 1978). Generally, they are the result of the deposition of sandy and fine-grained detritus in the marginal, transitional parts of the basin (or in CPBS areas between strike-slip depressions), between the main current flow and morphologically uplifted areas. These results indicate that deposition was active in areas where turbidity current is still active, but only from a thin portion of the current cloud. This part of current is characterised with low energy, which resulted in only small quantities of sandy detritus being transported.

Massive marls facies association is primarily represented by massive marl facies with rare intercalations of thin siltstone or sandstone laminae. Sediments in this association are continuously deposited in marginal areas or areas outside the range of turbidites. However, deposition of these marls was also active in the central parts of the depressions during periods of calm basin sedimentation in which there were no turbidite activities. This association can be compared to hemipelagic sediments of open seas.

2.4 Hydrocarbon reserves and reservoir characteristics in CPBS

Approximately 104 x 10^6 m^3 of oil from 39 fields, 6.93 x 10^6 m^3 of condensate (11 fields) and 64.92 x 10^9 m^3 of gas (52 fields) were recovered in the Croatian part of the Pannonian basin during 63 years of exploitation. Detail calculation is performed for period 1941-2004 in Velić et al. (2010). The total production peak was attained in 1980–1989, when exploration resulted in discoveries of 12 oil (Fig. 9) and 20 gas (Fig. 10) fields. Generally 30 years (1960–1989) can be described as a highly successful exploration period (Fig. 9 and Fig. 10).

The total (geological) reserves are about 740 x 10^6 m^3 of equivalent oil, with recoverable reserves of 112 x 10^6 m^3 of oil, 10.74 x 10^6 m^3 of condensate and 100.67 x 10^9 m^3 of gas (Belošić, 2003; Velić et al., 2008). Up to 2004 a total of 175.89 x 10^6 m^3 of equivalent oil had been recovered from Croatian fields, including 104.05 x 10^6 m^3 of oil from 39 fields, 6.93 x 10^6 m^3 of condensate from 11 fields and 64.91 x 10^9 m^3 of gas from 52 fields (Velić et al., 2010; Fig. 11).

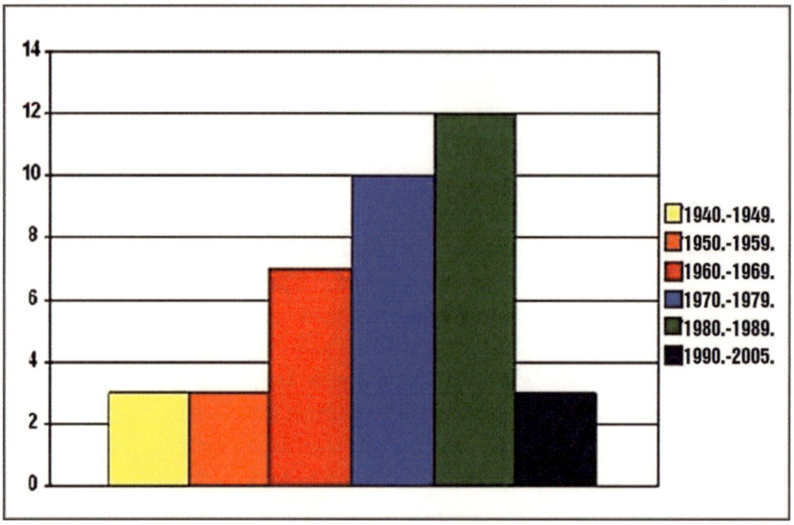

Fig. 9. Number of discovered oil fields during 1945-2005 (from Velić et al., 2010)

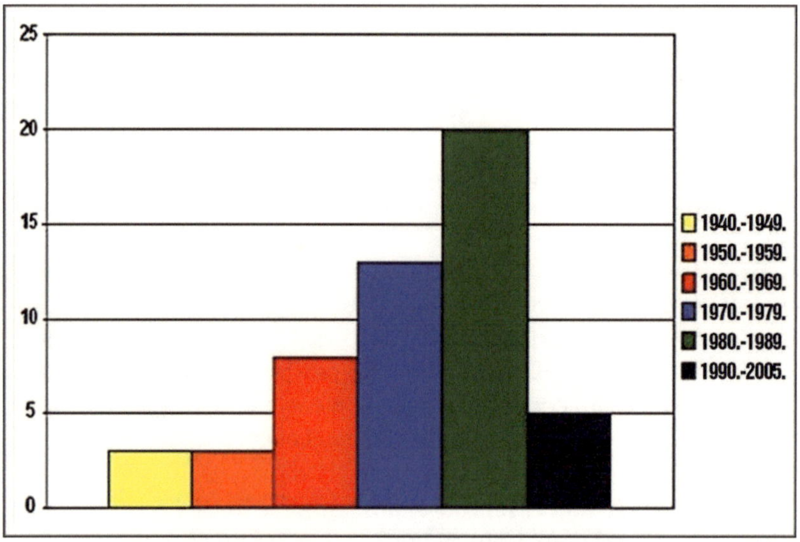

Fig. 10. Number of discovered gas fields during 1945-2005 (from Velić et al., 2010)

Production is decreased after 1989 caused by depletion (maturation) of older fields, and substantial decrease in exploration activities. Dependence between these two factors it is not favourable condition, because reserves decreasing can be only ceased by discovering of new proven reserves as result of new exploration drilling. But total recovery of proven reserves is continuously increasing due to application of additional method for improving recovery. The most applied secondary recovery method for many fields is water injection. However, the fields of all sizes are often revitalized with mechanisms of artificial (gas) lift. Also, it is proven, especially in the largest oil fields, that sandstone reservoirs of Pannonian and Pontian age have the most favourable values of petrophysical properties and consequently the highest recoveries and the longest production period (Velić et al., 2002).

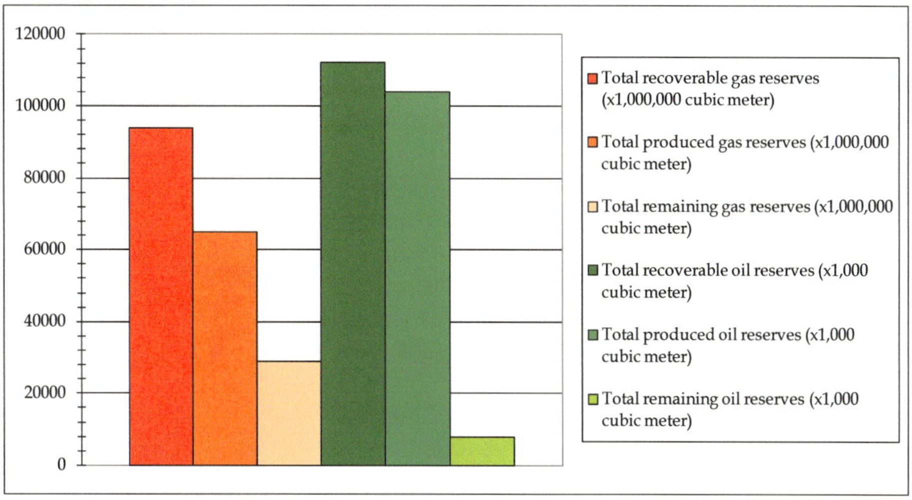

Fig. 11. Total hydrocarbon recoverable, produced and remaining reserves in CPBS (from Velić et al., 2010)

2.4.1 Hydrocarbon reserves in numbers

Regarding reserves, there are many sources and estimations. But, recently there are published two estimations in internationally reliable source (Dobrova et al., 2003; Velić et al., 2010) for CPBS. Lower estimation originated from 2010 (Velić et al., 2010), and higher from 2003 (Dobrova et al., 2003). Also, the complete deterministical calculation procedure for reserve estimation in CPBS through geological categories that define petroleum system is presented in Malvić (2003) and later upgraded in Malvić & Rusan (2009).

According to earlier estimation (Dobrova et al.; 2003) Croatia is ranked in third place according to the remaining recoverable liquid reserves in the central and eastern European states regarding reserves at the end of 2002 (Romania first, 39%; Ukraine second, 25%; and Croatia, Austria and Belarus third, 6%). But later Velić et al. (2010) included newer production data from CPBS are stated that recoverable oil reserves are lower - 112.06×10^6 m^3 of oil (+ 10.73×10^6 m^3 of condensates), versus 141.18×10^6 m^3 of oil in Dobrova et al. (2003). For gas ratio is 100.67×10^6 m^3 (Velić et al., 2010) vs. 108.11×10^9 m^3 (Dobrova et al., 2003). But decreasing published for remaining reserves is more dramatic and for oil it is 8.01×10^6 m^3 + 3.80×10^6 m^3 of condensate (Velić et al., 2010) vs. 31.48×10^6 m^3 (Dobrova et al., 2003), and for gas 35.76×10^9 m^3 (Velić et al., 2010) vs. 81.95×10^9 m^3 (Dobrova et al., 2003).

In any case, significant remaining hydrocarbons could be recovered from reservoirs with proven reserves in CPBS. It is sure that there are some subtle, potentially medium or small reservoirs of about 200,000 m^3 recoverable oil equivalent (Malvić, 2003; Malvić & Rusan, 2009), where can be located remaining reserves (Fig. 11). There are two stratigraphic plays as targets, logically based on two previously described depositional environments, which could be considered for future explorations. The first ones are smaller structures in Badenian sediments. Reservoirs of such age are generally coarse-grained sediments, deposited in smaller (generally up to 5 km in longer axis) strike-slip depressions. The largest such fields are already discovered, but there is for sure smaller structures of such type that were not recognized earlier on regional seismic sections, subsurface geological maps or include large uncertainty about possible accumulated hydrocarbons (e.g. in Malvić, 2003).

The second ones targets are Upper Miocene sandstones, where hydrocarbon recovery can be increased by producing from marginal, thin or marly (inter-channel) sandstone layers. Also, the largest sandstone reservoirs in CPBS are very probable accompanied with several satellite or smaller reservoirs as separate hydrodynamic units.

2.4.2 Hydrocarbon reservoirs characteristics

Hydrocarbon reservoirs absolute depths in CPBS range approx. from 400 to 4000 m. Initial reservoir pressure is hydrostatic or higher. Average geothermal gradient is about 4.5°C/100 m, but locally can vary 30-40%. Gas and condensate reservoirs are always over-pressured, with average gradient 0.16 bars/m. Temperature of reservoirs are sometimes high (180-220°C) with large part of non-hydrocarbon gases CO_2 (10-25%), H_2S (60-200 ppm) and mercury.

The oldest reservoir rocks are proven in the Šumećani (Palaeozoic schists and granites), Bunjani (Palaeozoic granites and gneisses), Molve, Kalinovac and Stari Gradac-Barcs Nyugat Fields (highly tectonized Palaeozoic schists and quartzites of Permian and Triassic systems). On the east, the oldest reservoirs are discovered in Ilača, Privlaka and Đeletovci Fields (pre-Cenozoic). There are significant reserves found in Mesozoic (Triassic) dolomites in Molve, Kalinovac and Stari Gradac Fields.

The most important Badenian reservoirs in CPBS are located in the Sava and Drava Depressions. In the eastern part of Drava Depression the largest is Beničanci Field, from where is recoverable about $18x10^6$ m^3 of oil. Smaller, but economical, reservoirs of that age are located in surrounding Obod, Kučanci-Kapelna, Bokšić-Klokočevci and Števkovica Fields. Badenian reservoirs are proven also in Mura Depression (Legrad Field), Sava Depression (Vezišće, Stružec and Kloštar Fields) and elsewhere. Generally, the most hydrocarbons from reservoirs of 1st megacycle are recovered from eight fields: Beničanci, Molve, Kalinovac, Žutica, Stružec, Ivanić, Šandrovac and Bokšić.

The largest hydrocarbon reserves (recovered and remained) in CPBS belongs to sandstones of Upper Pannonian and Lower Pontian. These reservoirs are deposited during 2nd megacycle, and the best properties are proven in central parts of depressions for thick and lithologically clean, medium and fine-grained sandstones. Average primary (inter-granular) porosity is often higher than 20% and permeability is about $60x10^{-3}$ μm^2. The largest sandstone reservoirs are discovered in the Sava Depression on Stružec Field ($16x10^6$ recovered oil tons), Žutica Field (14 $x10^6$ recovered oil tons), Ivanić Field ($7x10^6$ recovered oil tons) and Šandrovac Field (6.5 $x10^6$ recovered oil tons) in the Drava Depression. Three largest fields in CPBS (Beničanci, Molve and Kalinovac) contain hydrocarbon reserves of approx. 200 $x10^6$ tons of oil equivalent, and are classified in 1000 the world's largest hydrocarbon fields (Velić, 2007). The main gas reservoirs (also rich in condensate) are Molve, Kalinovac and Stari Gradac-Barcs Nyugat in the western Drava Depression, with 70% gas reserves in CPBS.

3. Conclusion

Neogene and Quaternary tectonics in the Croatian basin is very complex, due to two phases of transtension (Badenian and Late Pannonian-Early Pontian) and two of transpression (Sarmatian-Early Pannonian and Late Pontian-recent). Transtensions were periods of main sediment accumulation and transpression of uplifting and structural forming. Consequently, there are lithological heterogeneities laterally and vertically as well as occurrence of numerous fault zones. Such faults separate regional tectonic blocks and very often bordering hydrocarbon field structures, acting as partial or complete seals. Pressure and production anomalies can be a useful indicator of fault sealing, observing at wells located on opposite blocks.

Middle Miocene was period when a numerous strike-slip negative flowers had been formed inside regional depressions of CPBS. Such places were depositional centres of alluvial fans, where coarse-grained sediments from local sources of clastics, with good reservoir properties, were deposited. Moreover, numerous intra-reservoir micro-zones of secondary porosity are present in basement of such reservoirs owing to the complex Middle Miocene transtensional tectonic reflected in Palaeozoic and Mesozoic basement rocks. Results are single hydrodynamic units of heterogeneous reservoirs of Palaeozoic, Mesozoic and Middle Miocene (rarely Lower Miocene), where unconformities play the most important role. Stratigraphycally, such unconformities are loci for lengthy exposure at the surface before Badenian transgression. Heterogeneity and the ages of reservoir rocks, which span several geological erathems, strongly emphasise diagenesis as a highly important process in Mesozoic basement rocks. Rocks of these sediments today represent the 2nd reservoir unit according with importance of remaining and potential hydrocarbon reserves. The modern resolution of petroleum exploration techniques today make possible to detect also smaller, but economically interesting reservoirs as the new discoveries.

Upper Miocene encompasses turbiditic depositional mechanism strongly characterised by single source area located at western margins of PBS (Eastern Alps). Reservoir rocks are sandstones, which reached the highest thicknesses in central parts of CPBS depressions. The major influence on sandstone reservoir quality had increasing of silty and marly components in marginal depression parts and decreasing of thickness. Also, mechanical diagenesis, as a process of compaction, caused a decrease in porosity for depth difference more than 400 meters in the same lithostratigraphic member (Malvić et al., 2005). It happened especially in deeper parts, but has minor influence compared with increasing of silty and marly components in marginal parts. The third process, also of minor influence, had been chemical diagenesis including several processes caused by dissolved ions, pH value, pressure and temperature. in any case, sandstone reservoirs contain the largest proven hydrocarbon reserves. Also, there are projected the highest possible (undiscovered) quantities of oil and gas, but for difference of Badenian coarse-grained reservoirs, in sandstones such reserves would be located close to existing fields in the so called subtle traps, i.e. as satellite reservoirs. Such targets were not explored in the past, due to smaller areal extension and often not so favourable lithological composition (transitional lithofacies). But, as total recovery increase thanks to new technologies, such reservoirs start to be important hydrocarbon source.

Total remaining hydrocarbon potential in CPBS makes this area interesting for future exploration. Remaining reserves are probably at least 8×10^6 m^3 of oil, 3.80×10^6 m^3 of condensate and 36×10^9 m^3 of gas (Velić et al., 2010). Middle Miocene sediments probably hide some undiscovered smaller structures on depressions margins, and in Upper Miocene subtle traps are probably remained as satellite structures around existing larger sandstone reservoirs. The majority of remaining hydrocarbons is assumed in Upper Miocene sandstones.

The strike-slip tectonics played crucial role in shaping CPBS with contemporary structures and hydrocarbon fields. These fault systems play two important roles. In transtensional phases they formed negative flower structures where sediments had been accumulated (both in Middle and Upper Miocene). On contrary, in transpressional phases they were changed in positive flower structures, especially in 2nd transpressional period, forming traps for hydrocarbon accumulations as well as migration pathways (e.g. in Malvić, 2003 or Velić, 2007). The main fault displacement happened along the bordering strike-slip faults, between which structure had been formed.

Also, tectonic and sedimentation in CPBS, which occurred through 2 transtensional and 2 transpressional phase, can be analysis additionally through three depositional megacycles. Properties of such megacycles can be followed on seismic sections, well cores, logs, and outcrops on surface, even in large scale (Blašković et al., 1984). The 1st megacycle corresponds with 1st transtensional phase and mostly included coarse-grained reservoir sediments in older, and pelites (often source rocks) in younger part. The 2nd megacycle corresponds to 2nd transtensional phase of Late Miocene, and included the sandstones that are the main reservoirs regarding volumes and recoverable hydrocarbons in CPBS. These rocks resulted from periodical, strong turbiditic currents that generally moved from NW/N toward SE/S, and which direction had been strongly determined by marginal depression's faults, local strike-slip structures and uplifted, subaqueous palaeoreliefs remained from 1st transtensional phase. The 3rd megacycles is connected with 2nd transpressional phase, which took place the most of time in continental environment. This phase has importance for final structural evolution of CPBS but also in the last decade some researching results opened possibilities for economical biogenic methane accumulation in Pliocene and Lower Quaternary sediments. Some projections indicate that such accumulations are often located above existing reservoirs as represent mixture of thermogenic and biogenic gases.

Croatian part of Pannonian Basin System is large and well geologically described Neogene and Quaternary regional basin system. Many analyses offered enough data and results for describing geological evolution of this area and transfer methods and conclusion in other similar geological provinces. This basin system is also well known area of numerous hydrocarbon reservoirs, where some of them are classified as very large in world scale. Although this province is today considered as mature petroleum basin, there is still enough remaining reserves for production in next several decades, what is here described numerically for CPBS. But, technological improvements also make possible increasing of recovery from discovered reservoirs as well as discovering some smaller and subtle traps. Presented analysis for CPBS makes easier to understand in which stratigraphical units and tectonical environments such traps can be found in CPBS.

4. Acknowledgment

This work represents part of a multidisciplinary research carried out in 2010 and 2011 within the larger project entitled "Stratigraphical and geomathematical research of petroleum geological systems in Croatia" (project no. 195-1951293-0237), financed by the Ministry of Science, Education and Sports of the Republic of Croatia.

5. References

Barić, G., Mesić, I., Jungwirth, M. & Španić, D. 1991. Gas and gas condensate-field in the north-west of the Drava depression, Yugoslavia, In: *Generation, Accumulation, and Production of Europe's Hydrocarbons.* A.M. Spencer (Ed.), Special Publication of the

European Association of Petroleum Geoscientists, 508 p., Oxford University Press , No. 1, pp. 323–339, ISBN-10: 0198542828, Oxford, UK.

Barić, G., Mesić, I. & Jungwirth, M. (1998). Petroleum geochemistry of the deep part of the Drava Depression, Croatia. *Organic Geochemistry*, Vol. 29, (December 1998 on-line), No. 1–3, pp. 571–582, ISSN 0146-6380.

Belošić, Ž. (2003). Istraživanje i proizvodnja nafte i plina [Exploration and production of oil and gas]. *Proceedings of 2nd International scientific-professional congress on petroleum industry*, Zadar, September-October 2003.

Bérczi I., Hamor G., Jambor A. & Szentgyörgyi K. (1988). Neogene Sedimentation in Hungary. In: *The Pannonian Basin*, L.H. Royden and F. Horváth (Eds.), AAPG Memoir, Vol. 45, pp. 57-67, ISBN 0891813225, Tulsa, USA.

Blašković I. (1982). The Neogene of the Ilova River depression (Northern Croatia). *Jugosl. akad. znan. i umjetnosti, Acta geologica (Prirodoslovna istraživanja 46)*, Vol. 12, No. 2, pp. 23-67, ISSN 0448-0155.

Blašković, I., Tišljar, J., Dragičević, I. & Velić, J. (1984). Razvoj sedimentacijskih okoliša miocenskih naslaga na zapadnim obroncima Psunja (sjeverna Hrvatska) [Application Evolution of depositional environments of Miocene sediments on western hills of Psunj Mt. (northern Croatia)] (in Croatian). *Geol. vjesnik*, Vol. 37, pp. 11-32, ISSN 0016-7924.

Ćorić, S., Pavelić, D., Rögl, F., Mandić, O., Vrabac, S., Avanić, R., Jerković, L. & Vranjković, A. (2009). Revised Middle Miocene datum for initial marine flooding of North Croatian Basins (Pannonian Basin System, Central Paratethys) The Pannonian Basin System (PBS) originated during the Early Miocene as a result of extensional processes between the Alpine-Carp. *Geologia Croatica*, Vol. 62, No. 1, pp. 31-34, ISSN 1330-030X.

Dobrova, H., Kolly, E. & Schmitz, U. (2003). E&P Ventures in the Eastern-Central Europe Transformation States after 1989 - a Review of Expectations and Results. *Oil Gas European Magazine*, Vol. 30, No. 4., pp. 172-182, ISSN 0342-5622.

Hamilton E.L. (1967). Marine geology of abyssal plains in the Gulf of Alaska. *Journal of Geophysical Research*, Vol. 72, No. 16, pp. 4189-4213, ISSN 0148-0227.

Haq, B. U. & Eysinga, F. W. B. (Eds.) (1998). Geological Time Table, Fifth Edition (Wall Chart). ISBN 0444828656, Elsevier Science, Amsterdam, The Netherlands.

Hernitz Z. & Jurak V. (1973). Primjena paleostrukturne i statističke analize naslaga mlađeg tercijara u području Ivanić-Grada (sjeverna Hrvatska) [Application of Palaeostructural Analysis of Late Tertiary sediments in the area of Ivanić-Grad (northern Croatia)] (in Croatian). *Nafta*, Vol. 24, No. 7-8, pp. 343-367, ISSN 0027-755X.

Juhász, G. & Magyar, I. (1992): Review and correlation of the Late Neogene (Pannonian s.l.) lithofacies and mollusc biofacies in the Great Plain, eastern Hungary. *Foldtani Kozlony*, Vol. 122, No. 2-4, pp. 167-194, ISSN 0015-542X.

Kovač, M., Barath, I. & Nagymarosy, A. (1997). The Miocene collapse of the Alpine-Carpathian-Pannonian junction - an overview. *Acta Geol. Hungarica*, Vol. 40, No. 3, pp. 241-264, ISSN 1788-2281.

Laughton A.S. (1968). New evidence of erosion on the deep ocean floor. *Deep Sea Research (continued as Deep Sea research and Oceanographic Abstracts)*, Vol. 15, No. 1, pp. 21-29, ISSN 0146-6291.

Lučić D., Krizmanić K. & Dalić N. (1995). Lithofacies and Sequence Stratigraphy of Upper Miocene Siliciclastic Deposits in Okoli Region (Sava Depression, Croatia). *Abstract book of the 1st Croatian Geological Congress*, ISBN 953-96520-4-9, Opatija, October 1995.

Lučić, D., Saftić, B., Krizmanić, K., Prelogović, E., Britvić, V., Mesić, I. & Tadej, J. (2001). The neogene evolution and hydrocarbon potential of the Pannonian Basin in Croatia. *Marine and petroleum Geology*, Vol. 18, No. 1, pp. 133-147, ISSN 0264-8172.

Malvić, T. (1998). *Strukturni i tektonski odnosi, te značajke ugljikovodika širega područja naftnoga polja Galovac-Pavljani* [Structural and Tectonic Relations in the Wider Area of Galovac-Pavljani Oil Field]. MSc. thesis, Faculty of Mining, Geology and Petroleum Engineering, University of Zagreb.

Malvić, T. (2003). *Naftnogeološki odnosi i vjerojatnost pronalaska novih zaliha ugljikovodika u bjelovarskoj uleknini [Oil-Geological Relations and Probability of Discovering New Hydrocarbon Reserves in the Bjelovar Sag].* PhD. thesis, Faculty of Mining, Geology and Petroleum Engineering, University of Zagreb.

Malvić, T. (2006). Middle Miocene Depositional Model in the Drava Depression Described by Geostatistical Porosity and Thickness Maps (Case study: Stari Gradac-Barcs Nyugat Field). *Rudarsko-geološko-naftni zbornik*, Vol. 18, pp. 63-70, ISSN 0353-4529.

Malvić, T., Velić, J. & Peh, Z. (2005): Qualitative-Quantitative Analyses of the influence of Depth and lithological Composition on Lower Pontian Sandstone Porosity in the Central Part of Bjelovar Sag (Croatia). *Geologia Croatica*, Vol. 58, No. 1, pp. 73-85, ISSN 1330-030X.

Malvić, T. & Rusan, I. (2009). Investment Risk Assessment of Potential Hydrocarbon Discoveries in a Mature Basin. *Oil Gas European Magazine*, Vol. 35, No. 2, pp. 67-72, ISSN 0342-5622.

Malvić, T. & Velić, J. (2010). Relation between Effective Thickness, Gas Production and Porosity in Heterogeneous Reservoirs, an Example from the Molve Field, Croatian Pannonian Basin. *Petroleum Geoscience*, Vol. 16, No. 1, pp. 41-51, ISSN 1354-0793.

Middleton G.V. (1967). Experiments on density and turbidity currents: III. Deposition of sediment. *Canadian Jour. Earth Sci.*, Vol. 4, No. 3, pp. 475-505, ISSN 0008-4077.

Mitchum, R.M. Jr. (1977). Glossary of terms used in seismic stratigraphy, In: *Seismic Stratigraphy-Applications to Hydrocarbon Exploration*, U. Payton (Ed.), AAPG Memoir, Vol. 26, pp. 205-212, ISBN 0891813020, Tulsa, USA.

Odin, S.G. & Matter, A. (1981). De glauconiarum origine. Sedimentology. Vol. 28, No. 5, pp. 611-641, ISSN 0037-0746.

Pamić, J. (1997). Vulkanske stijene Savsko-dravskog međurječja i Baranje (Hrvatska) [Volcanic rocks in Sava-Drava inter-river area and Baranja (Hrvatska)] (in Croatian). *Nafta*, special issue, 192 p., ISSN 0027-755X.

Pavelić, D. (2001). Tectonostratigraphic model for the North Croatian and North Bosnian sector of the Miocene Pannonian Basin System. *Basin Research*, Vol. 13, (September 2001), No. 3, pp. 359–376.ISSN 0950-091X.

Pavelić, D. (2002). The south-western boundary of Central Paratethys. *Geologia Croatica*, Vol. 55, No. 1, pp. 83-92, ISSN 1330-030X.

Pletikapić Ž. (1965). *Stratigrafija, paleogeografija i naftoplinonosnost Ivanić - Grad formacije na obodu Moslavačkog masiva* [Stratigraphy, Palaeogeography and Hydrocarbon Potential of Ivanić-Grad formation on the margin of the Moslavačka gora Mt.] (in

Croatian). PhD Thesis, Faculty of Mining, Geology and Petroleum Engineering, University of Zagreb.

Pirson S.J. (1970). *Geologic well log analysis* (1st edition). Gulf Publ. Co., ISBN 0-87201-901-2, Houston.

Rögl, F. (1996). Stratigraphic correlation of the Paratethys Oligocene and Miocene. *Mitteilungen der Gesellschaft der Geologie- und Bergbaustudenten in Wien*, Vol. 41, (annually), pp. 65–73, 1 tab., ISSN 0435-8104.

Rögl, F. (1998). Palaeographic consideration for Mediterranean and Paratethys Seaways (Oligocene to Miocene). *Annalen des Naturhistorischen Museums in Wien*, Vol. 99 (Serie A), (annually), pp. 279–310, ISSN 0255-0091.

Rögl, F. & Steininger, F. (1984). Neogene Paratethys, Mediterranean and Indo-Pacific seaways, In: *Fossils and Climate*, P. J. Brenchey (Ed.), 352 p., Geological Journal special issue, no. 11, Wiley and Sons, pp. 171–200, ISBN 047190418X, Chichester, UK.

Royden, L. H. (1988). Late Cenozoic tectonics of the Pannonian Basin System, In: *The Pannonian Basin*, L.H. Royden and F. Horváth (Eds.), AAPG Memoir, Vol. 45, pp. 27–48, ISBN 0891813225, Tulsa, USA.

Šćavničar B. (1979). Pješčenjaci pliocena i miocena Savske potoline [Pliocene and Miocene sandstones of Sava depression] (in Croatian). *3rd annual scientifc meeting ZSN JAZU*, Novi Sad, March 1977.

Šimon, J. (1980). *Prilog stratigrafiji u taložnom sustavu pješčanih rezervoara Sava-grupe .naslaga mlađeg tercijara u Panonskom bazenu sjeverne Hrvatske [Contribution to stratigraphy in depositional system of sandstone reservoirs of the Sava Group of Late Tertiary sediments in Pannonian basin of Northern Croatia]* (in Croatian). PhD Thesis, Faculty of Mining, Geology and Petroleum Engineering, University of Zagreb.

Tišljar, J. (1993). Sedimentary bodies and depositional models for the Miocene oil-producing areas of Ladislavci, Beničanci and Obod. *Nafta*, Vol. 44, No. 10, pp. 531-542, ISSN 0027-755X.

Troskot-Čorbić, T., Velić, J. & Malvić, T. (2009). Comparison of the Middle Miocene and the Upper Miocene source rock formations in the Sava Depression (Pannonian Basin, Croatia). *Geologia Croatica*, Vol. 62, No. 2, pp. 123-133, ISSN 1330-030X.

Velić, J. (2007). *Geologija ležišta nafte i plina [Geology of oil and gas reservoirs] (in Croatian)*. ISBN 978-953-6923-10-6, Faculty of Mining, Geology and Petroleum Engineering, University of Zagreb.

Velić, J., Malvić, T., Cvetković, M. & Vrbanac, B. (2010). Characteristics of Hydrocarbon Fields in the Croatian Part of the Pannonian Basin. *Oil Gas European Magazine*, Vol. 36, No. 3, pp. 146-147, ISSN 0342-5622.

Velić, J., Malvić, T., Cvetković, M., Weisser, M. & Curi, V. (2008). Observation useful for increase of hydrocarbon fields exploitation period within Croatian part of the Pannonian basin. *Proceedings of SGEM conference "Modern Management of Mine Producing Geology and Environmental Protection"*. ISBN 954918181-2, Albena, Jun 2008.

Velić, J., Weisser, M., Saftić, B., Vrbanac, B. & Ivković, Ž. (2002). Petroleum-geological characteristics and exploration level of the three Neogene depositional megacycles in the Croatian part of the Pannonian basin (in Croatian). *Nafta*, Vol. 53, No. 6-7, pp. 239-249, ISSN 0027-755X.

Vrbanac, B. (1996). *Paleostrukturne i sedimentološke analize gornjopanonskih naslaga formacije Ivanić grad u Savskoj depresiji* [*Palaeostructural and sedimentological analysis of the Upper Pannonian deposits of the Ivanić Grad Formation in the Sava depression*] (in Croatian). PhD Thesis, Faculty of Natural Sciences, University of Zagreb.

Vrbanac, B. (2002). Facies and facies architecture of the Ivanic Grad Formation (late Pannonian)—Sava Depression, NW Croatia. *Geologia Croatica,* Vol. *55,* No. 1, pp. 57–77, ISSN 1330-030X.

Vrbanac, B., Velić J. & Malvić T. (2010). Sedimentation of deep-water turbidites in main and marginal basins in the SW part of the Pannonian Basin. *Geologica Carpathica,* Vol. 61, No. 1, pp. 55-69, ISSN 1335-0552.

Vrsaljko, D., Pavelić, D., Miknić, M., Brkić, M., Kovačić, M., Hećimović, I., Hajek-Tadesse, V., Avanić, R. & Kurtanjek, N. (2006). Middle Miocene (upper Badenian/Sarmatian) palaeoecology and evolution of the environments in the area of Medvednica Mt. (North Croatia). *Geologia Croatica,* Vol. 59, (December 2006), No. 1, pp. 51–63, ISSN 1330-030X.

Walker R. G. (1978). Deep-water sandstone facies and ancient submarine fans; model for exploration for stratigraphic traps. *Bulletin of Am. Assoc. Petroleum Geologists,* Vol. 62, No. 6, pp. 932-966, ISSN 0149-1423.

Yilmaz, P. O., Norton, I. O., Chuchla, R. J. & Leary, D. A. (1996). Tectonic evolution and paleogeography of Europe, In: *Peri-Tethys memoir 2: structure and Prospects of Alpine Basins and Forelands,* P.A. Ziegler and F. Horváth (Eds.), Mémoires du Muséum National d'Histoire Naturelle, Vol. 170, pp. 47-60, ISBN 2-85653-507-0, Paris, France.

Part 3

Dynamics of the Eastern Mediterranean

Architecture of Kinematics and Deformation History of the Tertiary Supradetachment Thrace Basin: Rhodope Province (NE Greece)

Adamantios Kilias, George Falalakis, Aristides Sfeikos, Eleftheria Papadimitriou, Agni Vamvaka and Chara Gkarlaouni
Aristotle University of Thessaloniki, School of Geology
Greece

1. Introduction

The Rhodope province in Northern Aegean region where the Thrace basin, one of the largest Tertiary basins in this region, is located (Fig. 1) belongs to a highly extended continental terrain during Tertiary time. Recent studies indicate that extensional deformation began in Eocene (Kilias et al., 1999; Burchfiel et al., 2003, Brun & Sokutis, 2007) or earlier in Early Paleocene (Bonev et al., 2006). Dinter & Royden (1993), Dinter et al. (1995) and Wawrenitz & Krohe (1998) suggest that the onset of extension can be set in the Miocene. Extension took place simultaneously with compression at the more external parts of Hellenides (Fig. 1; Schermer et al. 1990; Kilias et al. 1991, 1999; Jolivet & Brun 2010).

Extensional basins and crustal subsidence are often temporally and spatially associated with uplift/exhumation of metamorphic rocks (Friedmann & Burbank, 1995; McClaughry & Gaylord, 2005). Two end-members of extensional basins can be recorded during a continuous extensional tectonism, which are: rift basins and supradetachment basins (Friedmann & Burbank, 1995). The differences in magnitude and rate of extension, volcanism, heat flow and structural architecture define the basin style and suggest the tectonic setting related to the basin evolution (Friedmann & Burbank, 1995; McClaughroy & Gaylord, 2005).

The Thrace basin formed on the metamorphic rocks of the Rhodope massif in Northern Greece and the Strandja and Sankarya massifs in NW-ern Turkey (Fig. 1, 2). In Southern Bulgaria and Western Former Yugoslavian Republic of Macedonia (FYROM) a series of equivalent basins with similar deposits infilling are also developed between the metamorphic rocks of the Bulgarian Rhodope massif (Fig.1, 2; Burchfiel et al., 2003; Dumurtzanov et al., 2005). The basins evolution is related to an intensive calc-alkaline and locally shoshonitic magmatism forming the NW-SE trending Eocene-Oligocene volcanic arc of the Hellenides. Its SW-ward progressive migration until the present day active Hellenic volcanic arc has been recorded by several works (Fig. 1; Innocenti et al., 1982; Fytikas et al., 1984). The position of the Thrace basin on the highly extended Rhodope metamorphic sequence, behind of the SW-ward contemporaneous nappes stack of Hellenides, give the Thrace basin an important role to understand better the geodynamic context of the Hellenides.

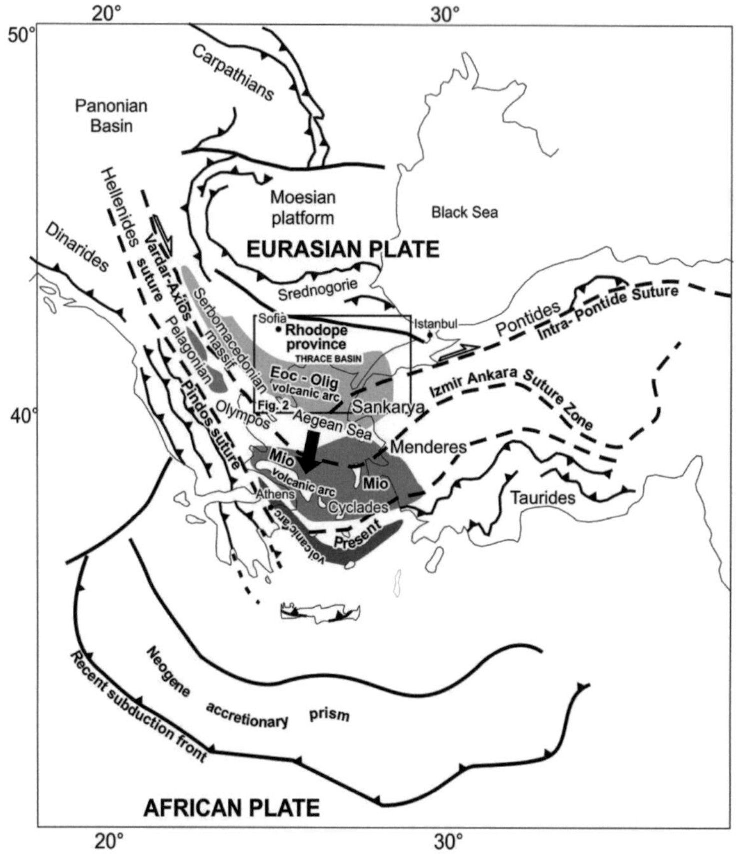

Fig. 1. Position of the Rhodope province and Thrace basin with respect to the main tectonic units of southeastern Europe and eastern Mediterranean region. The SSW ward migration of the Tertiary magmatic activity in the Hellenic arc is also shown

Our study is focused on the Greek part of the Thrace basin stretched from the Pangaion mountainous range until the Greek borders with Turkey along the Evros river. Metamorphic units of the Rhodope massif constitute the Northern tectonic boundary of the Greek Thrace basin, while its Southern parts are lost under the Aegean Sea and are exhumed again as small remnants on some islands of the Northern Aegean (e.g. Limnos, Samothraki; Fig. 1, 2). The Thrace basin in NW-ern Turkey has been intensely studied because of its importance as a natural hydrocarbon reservoir (Okay et al., 1990; Goeruer & Okay, 1996; Coskun, 1997; Turgut & Eseller, 2000; Siyako & Huvaz, 2007).
For the Greek Thrace basin published results are also available (e.g. Christodoulou, 1950; Kopp, 1965; Lescuyer et al., 2003; Kilias et al., 2006; Tranos, 2009). However, details on the basin evolution and its tectonic position in the geodynamic context of North Aegean region, as well as a relationship to the exhumation history of the Rhodope massif are lacking.
The aim of the current study is to present detailed information on the structural evolution of the Greek Thrace basin and its kinematics of deformation, as well as to clarify its geodynamic setting in the frame of the Hellenic orogenic belt. Furthermore, the calculation of the paleostress tensor for each deformational event recorded in the overall tectonic history of the basin is performed. We interpret the Tertiary Thrace basin as a

supradetachment basin related to the extensional exhumation history of the footwall Rhodope metamorphic rocks.

2. Geological setting

2.1 The Rhodope metamorphic basement

The Rhodope metamorphic province consists of high to moderate grade metamorphic units forming a complicated metamorphic nappe pile complex. It records a series of tectonic events of pre-Alpine and Alpine age and a different Tertiary exhumation age of the several metamorphic units. Alpine nappe stacking and crustal thickening were resulted from the Cretaceous-Paleocene SW-ward accretion of a number of structural units of mainly continental origin, sandwiched between Moesian continental block (Eurasia) in the North and Adriatic-Apulia continental block (Africa) in the South (Burg et al., 1996; Zagorchev, 1998; Bonev et al., 2006; Georgiev et al., 2010; Jahn-Awe et al., 2010).

According to their composition and tectonometamorphic setting the several metamorphic units are distinguished into a Lower and an Upper structural plate (Fig. 2; Krohe & Mposkos, 2002; Marchev et al., 2005; Bonev et al., 2006; Georgiev et al., 2010). The Upper plate contains the Kimi unit and the Lower plate includes the Sidironero, Kardamos-Kesebir and Kechros units (Fig. 2). At the westernmost Rhodope massif, the Sidironero unit overthrusts the lowermost Rhodope unit, the Pangaion unit (Fig. 2; Papanikolaou & Panagopoulos, 1981; Kilias & Mountrakis, 1990; Dinter et al., 1995).

Lower and Upper units are divided by low angle normal detachment fault systems, resulting in a progressively extensional exhumation of the several tectonic units from top to the bottom and the building of several metamorphic core complexes and tectonic windows (Fig. 2; e.g. Kessebir-Kardamos, Kechros and Pangaion domes). The structurally uppermost Kimi unit was the earliest exhumed Rhodope part (cooling age 65-60 Ma), followed by the exhumation of the Sidironero, Kessebir-Kardamos and Kechros units of the Lower plate (cooling age 42-30 Ma). Last to be exhumed was the lowermost Pangaion unit (cooling age 26-10 Ma), forming a well defined metamorphic core complex (Kilias et al., 1999; Krohe & Mposkos, 2002; Mposkos & Krohe, 2006; Georgiev et al., 2010).

Three main metamorphic events of alpine age from Cretaceous to Oligocene-Miocene time are identified: An earlier ultra-high to high pressure eclogite facies event, a second medium pressure amphibolite facies event and a late greenschist facies retrogressive event. The lowermost Pangaion unit was affected by lower P-T metamorphic conditions in the greenschist facies (Kilias & Mountrakis, 1990; Wawrenitz et al., 1998; Liati & Gebauer, 1999; Krohe & Mposkos, 2002; Mposkos & Krohe, 2006).

Acid to basic magmatism of Paleogene to Neogene age, associated with syn- to late-orogenic extension and thinning of the former over-thickened continental crust during Cretaceous-Paleocene time, characterizes the Rhodope metamorphic nappe pile (Koukouvelas & Pe-Piper, 1991; Dinter et al., 1995; Burg et al., 1996; Kilias & Mountrakis, 1998; Christophides et al., 2001, 2004; Marchev et al., 2005; Bonev et al., 2006).

To the West the Rhodope massif is separated from the Serbomacedonian massif by a mylonitic, low angle, normal detachment fault zone defining very well the Rhodopian Pangaion metamorphic core complex (Fig. 2; Dinter & Royden, 1993; Sokoutis et al., 1993; Kilias et al., 1999). The Serbomacedonian massif forms also a metamorphic nappe pile complex of Paleozoic or older age's rocks, including the Upper Vertiskos and the Lower Kerdylia metamorphic units (Fig. 2; Kockel et al., 1971; Papadopoulos & Kilias, 1985; Kilias et al.,

1999). Moreover, recent works distinguish a normal detachment fault zone also along the tectonic contact between the two Serbomacedonian units (i.e. Kerdylia and Vertiskos units), regarding the lower Kerdylia unit as a part of the Rhodope Massif (Brun & Sokoutis, 2007).

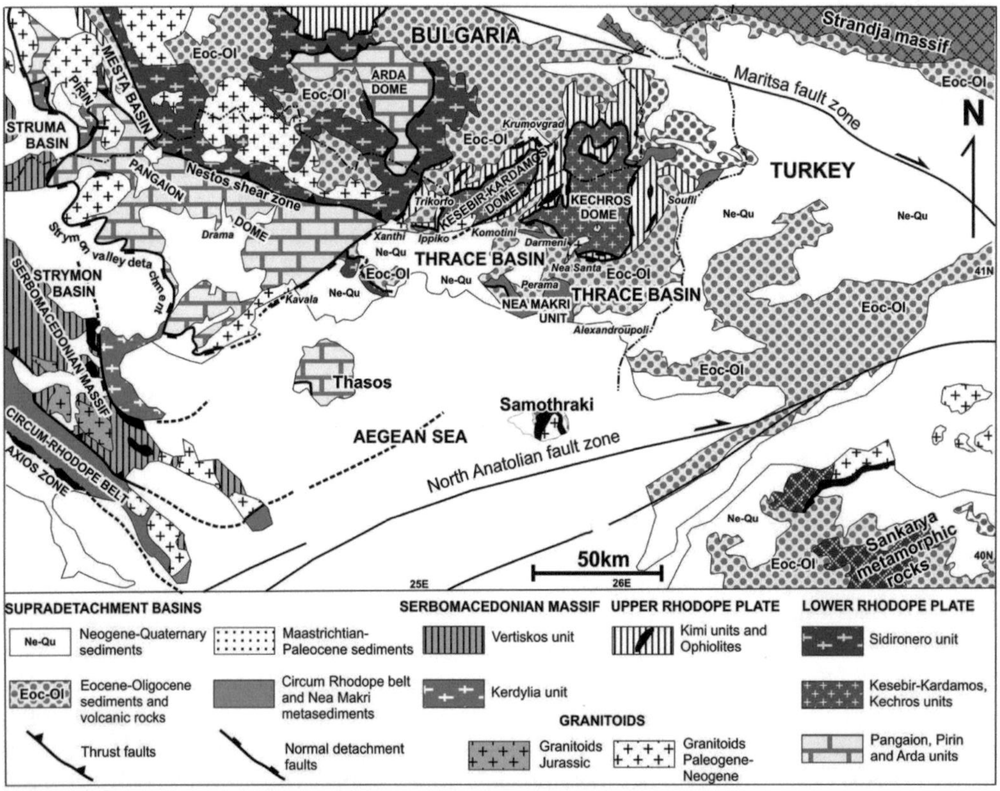

Fig. 2. Simplified geological map of the Rhodope province. The Thrace basin and the main structural units of the Rhodope and Serbomacedonian massifs with their tectonic relationships, as well as the Sankarya and Strandja metamorphic rocks in northwestern Turkey (Pontides-Anatolides belt) are shown (modified after Kilias et al., 1999; Turgut & Eseller, 2000; Krohe & Mposkos, 2002; Bonev et al., 2006)

A low grade metamorphic volcano-sedimentary unit of Upper Paleozoic-Mesozoic age, named Nea Makri unit, forming possibly the E-ward continuation of the Circum Rhodope Belt (Fig. 2; Kockel et al., 1971; Kaufmann et al., 1976), occurs in some places as small rests over the high grade Rhodope metamorphic rocks. The contact between the Nea Makri metasediments and the underlying Rhodope metamorphic units usually develops as a normal detachment fault zone with sense of shear towards S to SW (Fig. 2), related to the Tertiary extensional collapse history of the Rhodope basement rocks (Kilias et al., 1999; Bonev et al., 2006; Georgiev et al., 2010).

2.2 The deposits of the Thrace basin

The Tertiary sediments in the Greek part of the Thrace Basin are overlaid with a total well as in a few cases over the Nea Makri metasediments bounded usually with the underlying units with big fault zones. In the Turkish part, a thickness up to 9 km is reported (Fig. 2, 3; 4; Huvaz et al., 2007; Siyako & Huvaz, 2007; Mainhold & BouDagher-Fadel, 2010; Jolivet & Brun, 2010).

The deposits of the Thrace basin consist of Paleogene molassic type sedimentary rocks, as well as Neogene and Quaternary sediments (Christodoulou, 1950; Kopp, 1965; Lalechos, 1986; Mainhold & BouDagher-Fadel, 2010). The Paleogene molassic sediments are intercalated with a lot of calc-alkaline and partly shoshonitic type, volcanic products (Fig. 3). The latter form lava flows, debris-flows, hydroclastites, domes, dykes and numerous pyroclastics. Their chemical composition ranges from basaltic pyroxene andesite to biotite rhyolith through pyroxene or biotite-hornblende andesite, pyroxene trachyandesite, biotite-hornblende dacite and pyroxene-biotite trachydacite (Innocenti et al., 1982; Christofides et al., 2004). K/Ar data show ages from 33.4 to 20 Ma, establishing an Early Oligocene to Early Miocene volcanic activity (Christofides et al., 2004). However, intercalations of volcanic products with the earliest Middle-Late Eocene clastic sediments of the Thrace basin indicate that volcanic activity started earlier than Early Oligocene, in Middle-Late Eocene times. Furthermore, the Oligocene-Miocene Xanthi granite (Koukouvelas & Pe-Piper, 1991) intrudes in some places into the Paleogene strata of the basin (Fig. 2, 4), forming a contact metamorphic aureole of a few meters thickness.

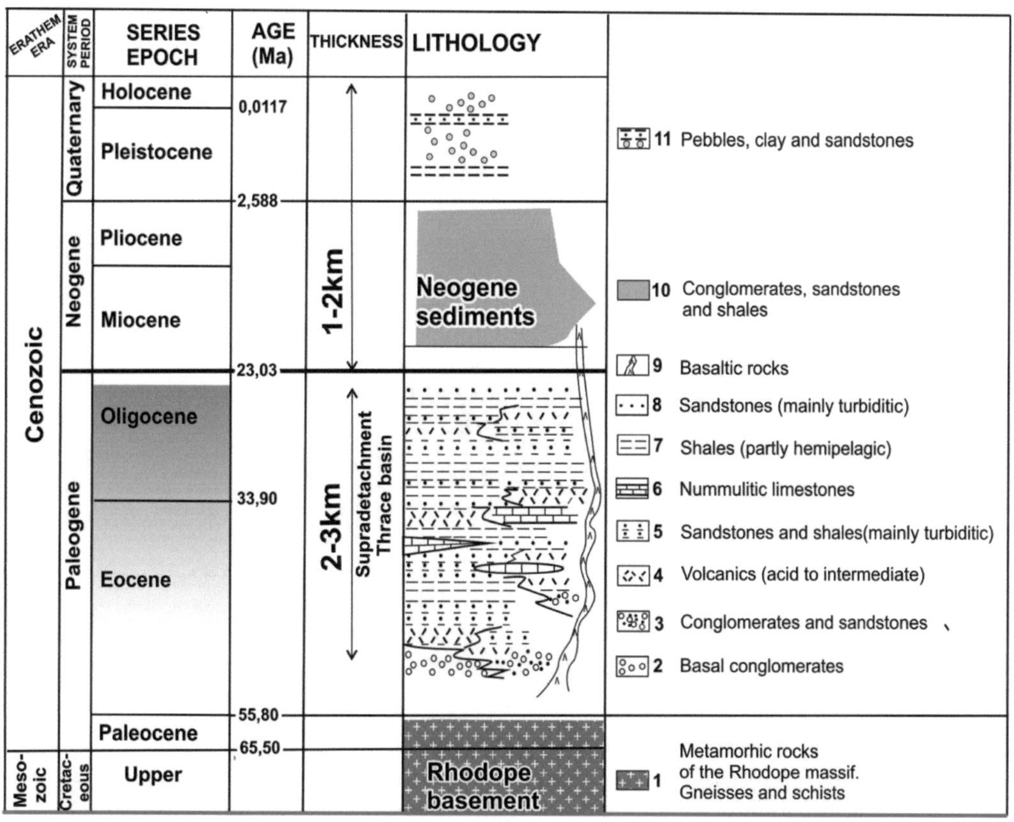

Fig. 3. Schematic stratigraphic column of the Thrace basin compiled after data in Christodoulou, 1950; Kopp, 1965; Lescuyer et al., 2003; Burchfiel et al., 2003

The Paleogene sediments of the Thrace basin show an age from Middle-Late Eocene to Oligocene (Christodoulou, 1950; Dragomanov et al., 1986; Zagorcev, 1998; Burchfiel et al., 2003; Meinhold & BouDagher - Fadel, 2010). They constitute a complicated stratigraphic sequence composed by intercalations of bedded conglomerates, breccia conglomerates, sandstones, numulitic limestones, turbiditic layers, and shales (Fig. 3). Sedimentation starts during Lutetian- Bartonian time with initial deposition of continental sediments (mainly breccias conglomerates and sandstones), followed during Late Eocene-Oligocene by marine turbiditic type deposits and limestones interbedded by the volcanogene products of the Late Eocene-Oligocene volcanic activity in the area (Fig. 3, 4; Dragomanov et al., 1986; Zagorcev; 1998; Burchfiel et al., 2003).

Fig. 4. a. Looking E the detachment zone between Rhodope basement rocks and molassic sediments near Nea Santa village. b & c. ductile fabrics at the footwall of the detachment fault basement rocks, σ-clasts and shear bands, showing a top to SSW sense of shear. d. basal breccia conglo-merate of the basin deposits. e. molassic sediments interbedded with volcanics with high angle dipping (>65 dip angle). f. Oligocene granite intrusion into the molassic sediments

A thick sedimentary sequence of Neogene age overlies unconformably the Paleogene strata of the basin containing conglomerates, sandstones and siltstones. Pebbles (breccio-conglomerates), clay and sandstones form the youngest Quaternary sediments of the whole stratigraphic column (Fig. 2, 3).

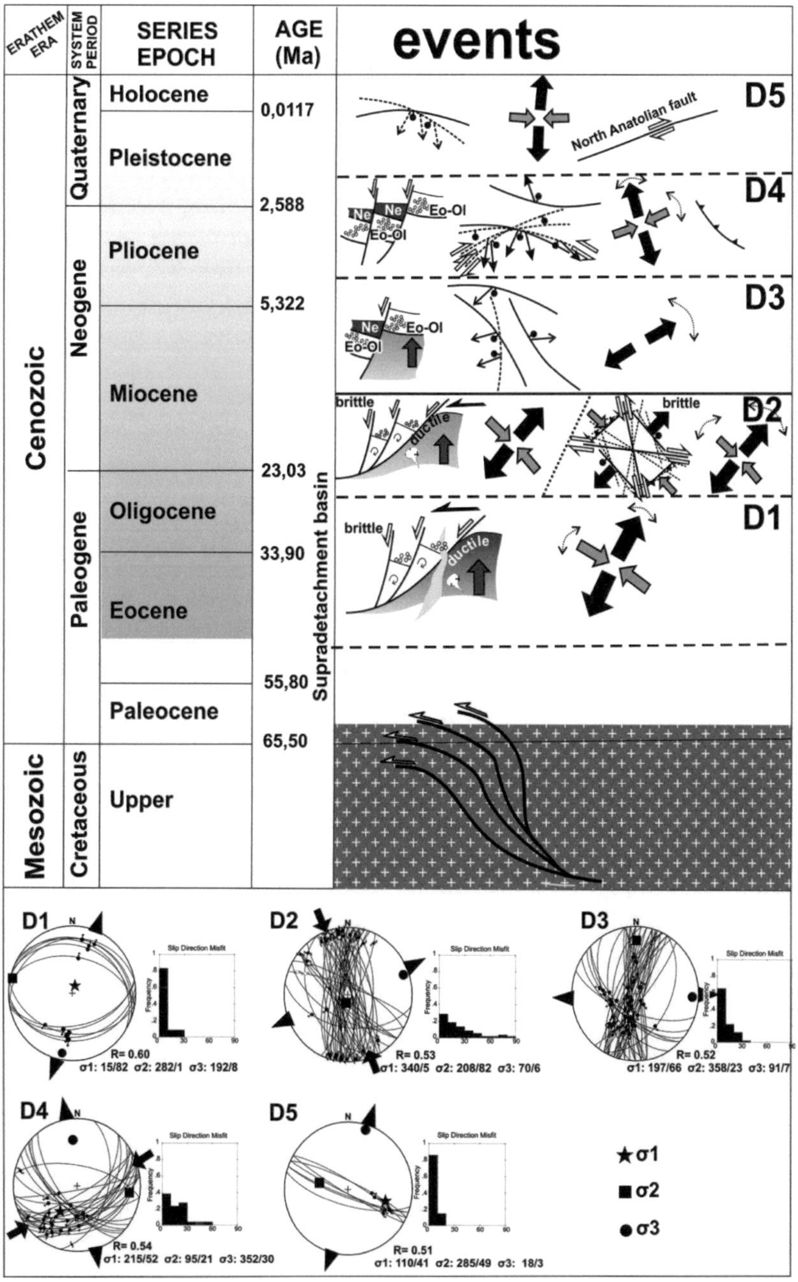

Fig. 5. Distinctive features and kinematics of the five main deformational processes (D1 to D5) related to the evolution of the Thrace basin from Eocene to recent. Mesozoic-Early Paleogene contractional tectonic and nappe stacking in Rhodope massif (Burg et al., 1996; Jahn-Awe et al., 2010) were replaced during Paleogene to Neogene by extensional tectonic, exhumation of deep crustal metamorphic sequences and coeval magmatism, as well as subsidence of the Thrace basin. Active tectonic dominates during Quaternary time. Paleostress analysis diagrams (σ1>σ2>σ3) for each tectonic event are also shown

3. Data and observations. Structural analysis

The Thrace basin shows a complicated structural evolution from its initial stages until present. Several successive deformational events interacted making their distinction difficult in several cases. The deformation progression documents semi-ductile to brittle conditions. Subhorizontal extension characterizes the overall deformation history.
The geometry and kinematics of deformation, cross cutting relationships, kinematic indicators and slickenslides overprinting criteria on the fault planes record the progressive activity of five (D1 to D5) tectonic events. They took place from Middle-Late Eocene up to present and are related to the basin evolution and the unroofing of the Rhodope metamorphic units (Fig. 5). Fault-slip data were used in order to calculate the paleostress tensor, for each tectonic event following the direct stress inversion method (Angelier, 1979, 1990). The solution is considered satisfactory if more than 80% of the fault-slip data from a site show a misfit angle less than 30o between the theoretic and real slip vector. The program My fault, Version 1.03 (Pangea scientific, 2005) and StereoNett (Duyster, 2000) were used for the graphical presentation of the tectonic data.

3.1 D1 event

In places where the contact between Paleogene sediments of the Thrace and the metamorphic basement rocks of the Rhodope massif or the low grade metasediments of the Nea Makri belt has not been reworked by more recent events, this is developed as a low angle, normal detachment fault (Fig. 5, 6, 7).
In many places, the entire Nea Makri belt is omitted or exists as small remains in between Paleogene sediments and underlying Rhodope metamorphic sequence, indicating the extensional regime of the tectonic contact and the basin evolution (Fig. 5, 6). The omission of several tectonic units of the whole Rhodope metamorphic nappe stack and the extensional unroofing and exhumation of the underlying deep crustal units during the Tertiary deformation of the Rhodope province have been described in detail by several authors (Kilias & Mountrakis, 1990; Sokoutis et al., 1993; Kilias et al., 1999; Bonev et al., 2006; Brun & Sokoutis, 2007; Georgiev et al., 2010).
The detachment fault zone indicates brittle to semi-ductile conditions and consists commonly of a cataclastic domain of a few meters thickness (ca. 2-5m; Fig. 7, 8b). Furthermore, pseudotachylites between the cataclastic fault structures are observed in some cases. On the fault plane, a well-developed striation, some times with an important strike-slip component, is recorded. Kinematic indicators (Hancock, 1985) show a main top to SSW-S sense of movement (Fig. 8b). In some places (e.g., Perama area), the tectonic contact is evolved with opposite top to the N-NNE sense of movement. It is related to secondary, conjugate faults with opposite N-NNE-ward dip direction to the main fault zones. These faults are evolved on the hangingwall fault segment of the main detachment fault zone but rooted into it. (Fig. 6, 9).
The strike of the fault zone is not constant. It is usually developed corrugated or folded with a SW to SSW dipping fold axis (Fig. 6), following the dipping of the fault plane. The hangingwall Paleogene strata dips in some cases opposite to the main dipping of the detachment fault plane or with a low angle conformably to the fault dipping (Fig. 7).
At the footwall of the detachment fault zone, the metamorphic basement rocks contain a S-SW-ward dipping mylonitic fabric with the same kinematic symmetry (top to the S-SSW, through ductile kinematic indicators, Simpson & Schmid, 1983) to the fault zone (Fig. 4).

Antithetic top to the N-NNE sense of movements are interpreted as secondary, of local important movements, due to strain inhomogenity or as a result of an important coaxial component of deformation (Kilias at al., 1999; Bonev et al., 2006). A NE-SW to NNE-SSW micro folding of the mylonitic fabric parallel to the stretching lineation (X- axis of the finite strain ellipsoid in the field) is also mapped, following the described folding geometry of the detachment fault plane.

Paleostress analysis based on fault-slip data from the detachment zone indicates NE-SW to NNE-SSW trending subhorizontal minimum σ3- and subvertical maximum σ1-axes of the stress tensor (Fig. 5, 10).

3.2 D2 event

The D2 event is related to the further opening and reconstruction of the Thrace basin and is associated with intense magmatic activity guided by the D2 crustal structures. D2 is characterized by a conjugate pair of major strike-slip faults forming very important structures in the Paleogene strata of Thrace basin. The first set is developed with NNE-SSW to NNW-SSE strike with a sinistral component. The second set strikes WNW-ESE to NW-SE with a dextral horizontal component (Fig. 5, 6, 11, 8a).

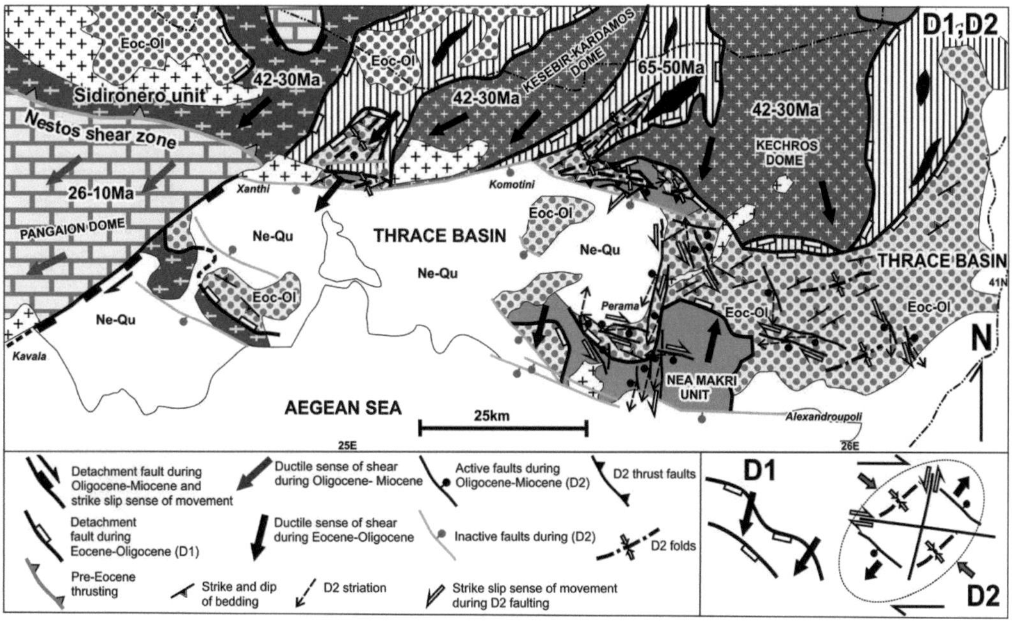

Fig. 6. Main structures and kinematics of D1 and D2 events. Ductile and brittle conditions take place simultaneously during D1 and D2 events at several tectonic crustal levels; ductile at deeper and brittle at the higher crustal levels. Legend as in Fig. 2

Reverse faults and open to tight folds in between the Paleogene deposits of the Thrace basin (Fig. 12) were interpreted from us as simultaneous to the D2 strike-slip structures as concluded from their geometry of kinematics, as well as structural and stratigraphic relationships. They are developed parallel to D2 extension and are also related to horizontal shortening perpendicular to D2 stretching (Fig. 5, 6). Overthrusting of the metamorphic basement rocks (Rhodope basement or Nea Makri metasediments) on the Paleogene strata of the Thrace basin along NE-SW to E-W trending thrust faults associates usually the D2 compressional structures (Fig. 5, 6).

Vertical or high angle dipping Paleogene layers of the basin should be regarded as a result of D2 compressional overprinting (Fig. 6, 4). Thus, a series of NE-SW trending rest basins are formed as a result of the NW-SE to N-S directed compressional component (Fig. 2, 6).

Furthermore, NW-SE directed veins and normal faults with an important downward dip-slip movement in between the Palaogene sediments develop simultaneously with the others described D2 structures. The kinematic and dynamic compatibility of all these brittle structures show their contemporaneous origin. Moreover the infilling of the NW-SE directed veins with products of the Oligocene magmatism, some of which are also gold bearing, indicates clearly that at least D2 extension was contemporaneous with magmatism (Kilias & Mountrakis, 1998; Christofides et al., 2001). The geometry and kinematics of the described complicated network of the D2 structures suggest that during the D2 event NE-SW to ENE-WSW stretching took place simultaneously with NW-SE to NNW-SSE shortening (Fig. 5, 6).

The paleostress analysis of D2 fault-slip data estimated the orientation of the lower most σ3-stress axis to a NE-SW to ENE-WSW strike, dipping with low angle and the maximum σ1-axis subhorizontal, acting perpendicularly (Fig. 5, 13).

Fig. 7. a. The detachment fault zone between Rhodope basement rocks and molassic sediments with the development of a strong cataclastic domain at the footwall basement rocks (inset photo). Road to Trikorfo. b.Looking W, the low angle normal fault zone between Rhodope basement rocks and molassic sediments northern of Darmeni village. Shear bands in the footwall basement rocks showing a top to SSW sense of shear (inset photo)

3.3 D3 event

D3 is recognized from the occurrence of a second younger, oblique to ca. dip-slip striation on the planes of the D2 strike-slip faults (Fig. 5,11, 8a), showing a reactivation of the D2 structures during D3. On the other hand, D3 is characterized by large WNW-ESE to NNW-SSE trending, oblique to dip-slip, normal faults that dismember the Paleogene basin and are related to the evolution of the Neogene deposits of the Thrace basin (Fig. 11, 14). The D3 normal faults were developed subparallel to the D2 dip-slip normal faults and veins but are distinguished from that because D3 faults bound the Neogene subbasins forming usually steep slopes in between Neogene sediments and the Paleogene strata or the metamorphic basement rocks. (Fig. 11, 14).

Fault plane analyses yielded subvertical σ1- and subhorizontal ENE-WSW to E-W trending σ3-axes (Fig. 5, 15). In some places, D3 kinematics coincides with the kinematics of the D2 event. Nevertheless, D3 and D2 can be easily distinguished because the D2 strike-slip faults exhibit a progressively oblique to about dip-slip D3 component of slide, as well as by the fact that D3 normal faults border the Neogene basins (Fig. 5, 11, 14).

3.4 D4 event

The D4 event is related to large WNW-ESE to NE-SW fault zones some of which are older and were reactivated during D4. This is indicated by the existence of at least two striations generations on their fault planes (Fig. 5, 11, 14), with the younger one to be compatible with the D4-kinematics. The Kavala-Xanthi, Xanthi-Komotini, Komotini-Nea Santa, Aisymi-Soufli and Maronia-Alexandroupoli fault zones belong to the main fault structures of the D4 event (Fig. 5, 11, 14). On their fault planes the older D2 or D3 striations were clearly overprinted by the D4 kinematics. The D4 striation is mainly developed as oblique lineation with an important dextral or sinistral strike-slip component. Pitch angle was measured equal to 50° -75° to the E or W, being dependent on the dip direction of the fault plane (Fig. 5, 11, 14).

Fig. 8. a. D2 strike-slip and D3 oblique to dip-slip slickenlines on the same fault plane. (Egnatia road from Komotini to Alexandroupoli). b. D4 high angle normal faults cut the cataclastic domain of the low angle normal detachment fault zone related to initial opening of the Thrace basin. Inset photo: the cataclastic domain of the detachment fault zone. Well developed shear bands and kinematic indicators show a downward top to SSW sense of movement (near Ippiko village)

In many places, the D4 normal faults cut with a high angle the cataclastic, pseudotachylite bearing domain related to the D1 low angle normal detachment fault (Fig. 8b). In these cases, due to the uplift of the footwall fault segment and the intensive erosion of the overlying strata, the related detachment fault zone cataclastites are exhumed. The latter form a cataclastic gouge-bearing cap, below of which the top to the SW-SSW D1 mylonitic, downward shearing of the metamorphic basement rocks dominates (Fig. 4, 9,14; Kilias & Mountrakis, 1990; Kilias et al., 1999; Krohe & Mposkos, 2002). NW-SE trending oblique thrust faults, as well as NW-SE trending, open, commonly angular folds, younger than the D2 contractional structures, are interpreted as structures related to the D4 kinematic setting. They are linked with a ca. ENE-WSW compressional component that coincides with the D4 dynamics.

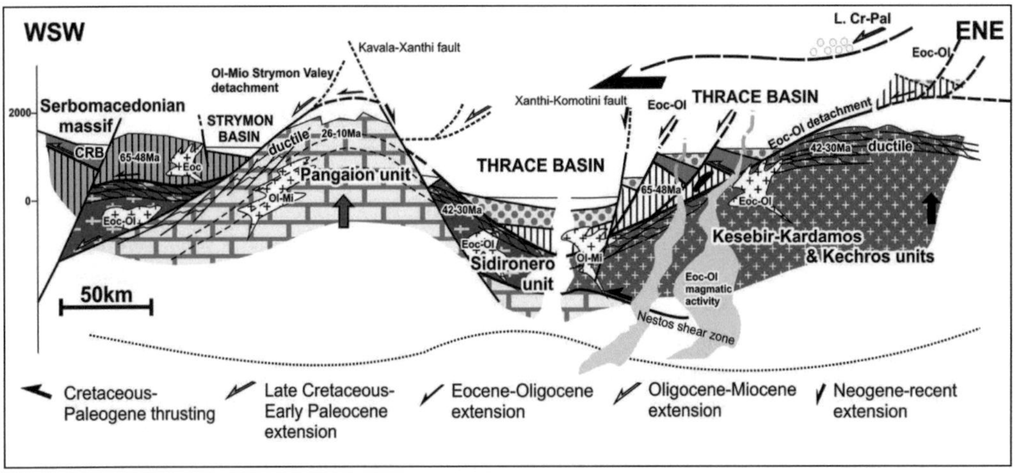

Fig. 9. Geometry of kinematics during Paleogene-Neogene extension of Rhodope Massif and formation of the Paleogene volcano-sedimentary Thrace basin on top of the Paleogene extensional detachment fault system, simultaneously with uplift of deep crustal metamorphic Rhodope units. Extension migrates towards SW-SSW.
Legend as in Fig. 2

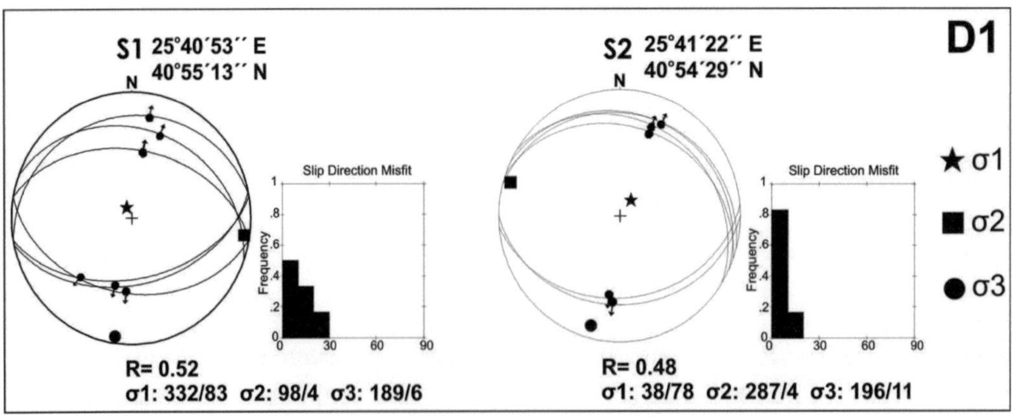

Fig. 10. Faults–striae data and deduced paleostress field (σ1>σ2>σ3) orientation during D1 (equal area, lower hemisphere). Fluctuation histograms of deviation angles (angle between the calculated slip vector and the measured slickenside) and stress ratio R=σ2-σ3/σ1-σ3 are shown. The geographic coordinates for each station are also indicated

Paleostress analysis from the D4 fault-striae data revealed the minimum σ3-axis with NNW-SSE to NNE-SSW strike and the maximum σ1-axis with ENE-WSW to ESE-WNW strike both with a low to moderate dip angle, respectively (Fig. 5, 16).

3.5 D5 event

Some of the D4 fault zones remain active until recent time as it is shown by the international accepted geological and seismotectonic criteria (Hancoock, 1985; Kilias et al., 2006; Mountrakis et al., 2006; Gkarlaouni et al., 2007). They form large active faults reactivated during the present stress field of the area (Fig. 5, 11, 14). On their fault planes, the old generations of slickensides together with the younger ones of the D5 striations are usually recognized. The latter dip mainly to the SE with a high to moderate angle of dipping (i.e., pitch 50°-80° to the E on fault planes dipping to the S-SSW; Fig. 5, 11, 14).

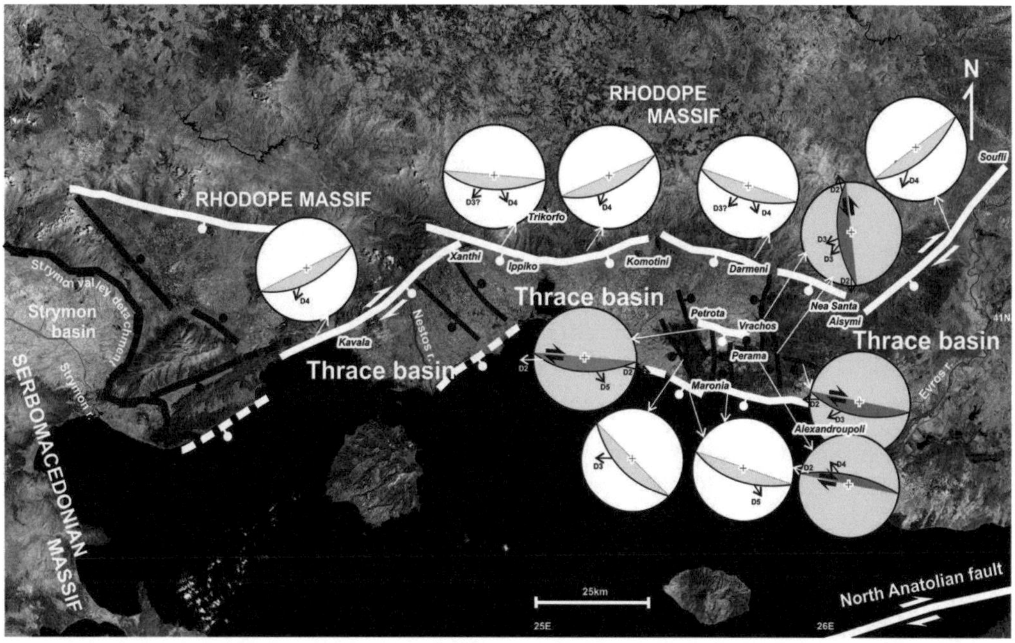

Fig. 11. Shaded relief map with the surface traces of the main fault zones mapped in the study area; black lines, faults related to the D2 event with their possible reactivations, white lines, faults related to the D3, D4 and D5 events. The main geometry of kinematics for each fault zone is shown in the stereographic diagrams (equal area, lower hemisphere). Arrows in the diagrams indicate the main development of the slickensides on the fault planes and their relative dating

Representative D5 active fault zones form the Maronia - Alexandroupoli fault zone, consisting of two main subparallel ca. E-W trending fault segments, and the Petrota-Vrachos fault with the same geometry of kinematics with the Maronia-Alexandroupoli fault (Fig. 11). According to their visible length and taking into account the relationship between fault length and earthquake magnitude (Papazachos et al., 2004), the most probable earthquake magnitude of the reactivation of the active fault zones, either as continuous structures or separated into smaller segments, as they are shown on the geological maps of Fig. 11, 14, is ranged between M=3,9 and M=5,3.

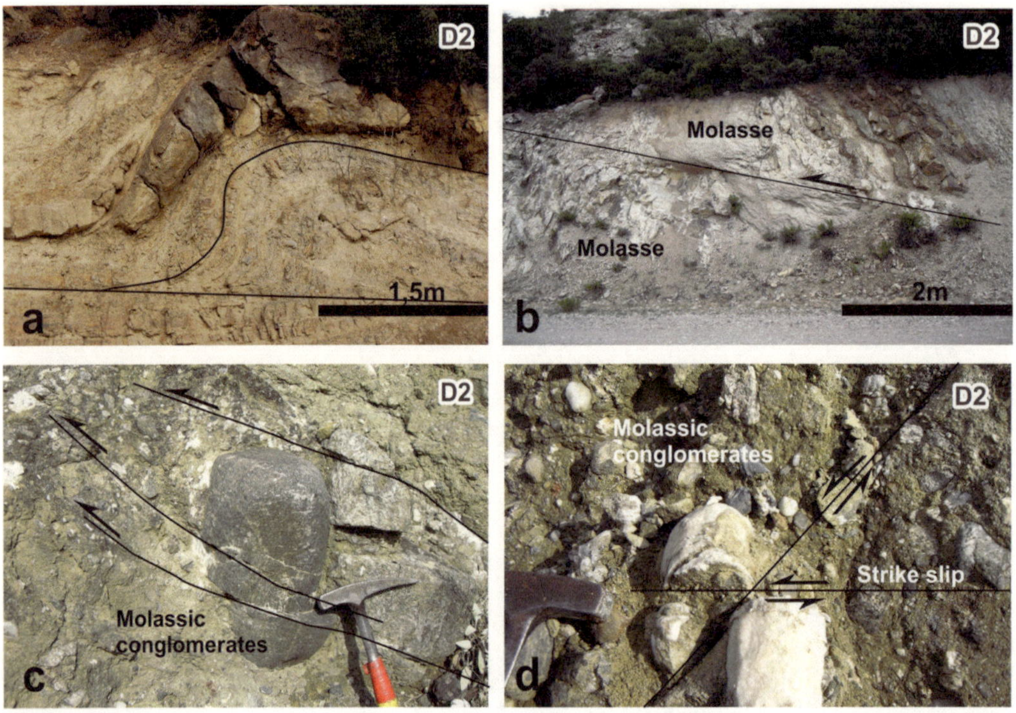

Fig. 12. During D2 kinematics; contractional structures (a,b & c) and strike-slip movements (d)

The paleostress analysis of the D5 faults revealed a minimum σ3-stress axis subhorizontal with NNE-SSW strike (i.e., 10° to 30° strike; Fig. 5,17) which coincides exactly with the active extension in the broader area, as it is concluded from the focal mechanisms of the strong earthquakes (Papazachos et al., 1998).

4. Age of deformation

Aiming to determine the time of deformation and its progression, the stratigraphic succession of the basin's strata and overprinting structural criteria, as well as the record of the structures evolution in certain layers in the whole stratigraphic column of the basin, have been considered. Additionally, the existing geochronological data for the ductile deformation recognized on the footwall segment of the detachment fault zone were also taken into account (Liati & Gebauer, 1999; Dinter et al., 1995; Wawrenitz & Krohe, 1998).

4.1 D1 event

For the D1 event related to the initial opening of the Thrace basin and the orogenic collapse along normal detachment faults (Kilias et al., 1999; Krohe & Mposkos, 2002; Bonev et al., 2006; Brun & Sokoutis, 2007) an age of Middle-Late Eocene is concluded (Fig. 5, 9). This age is indicated by the syndetachment deposition of the basal Middle-Late Eocene clastic sediments of the basin, supporting indeed the initiation of the extension during the Middle-Late

Eocene time (Kilias et al., 1999; Burchfiel et al., 2003; Brun & Sokoutis, 2007). Ductile fabrics of the metamorphic basement rocks at the footwall segment of the detachment fault with the same kinematic sense to the hangingwall have also been dated of Eocene age (Wawrenitz & Krohe, 1998; Liati & Gebauer, 1999; Krohe & Mposkos, 2002).

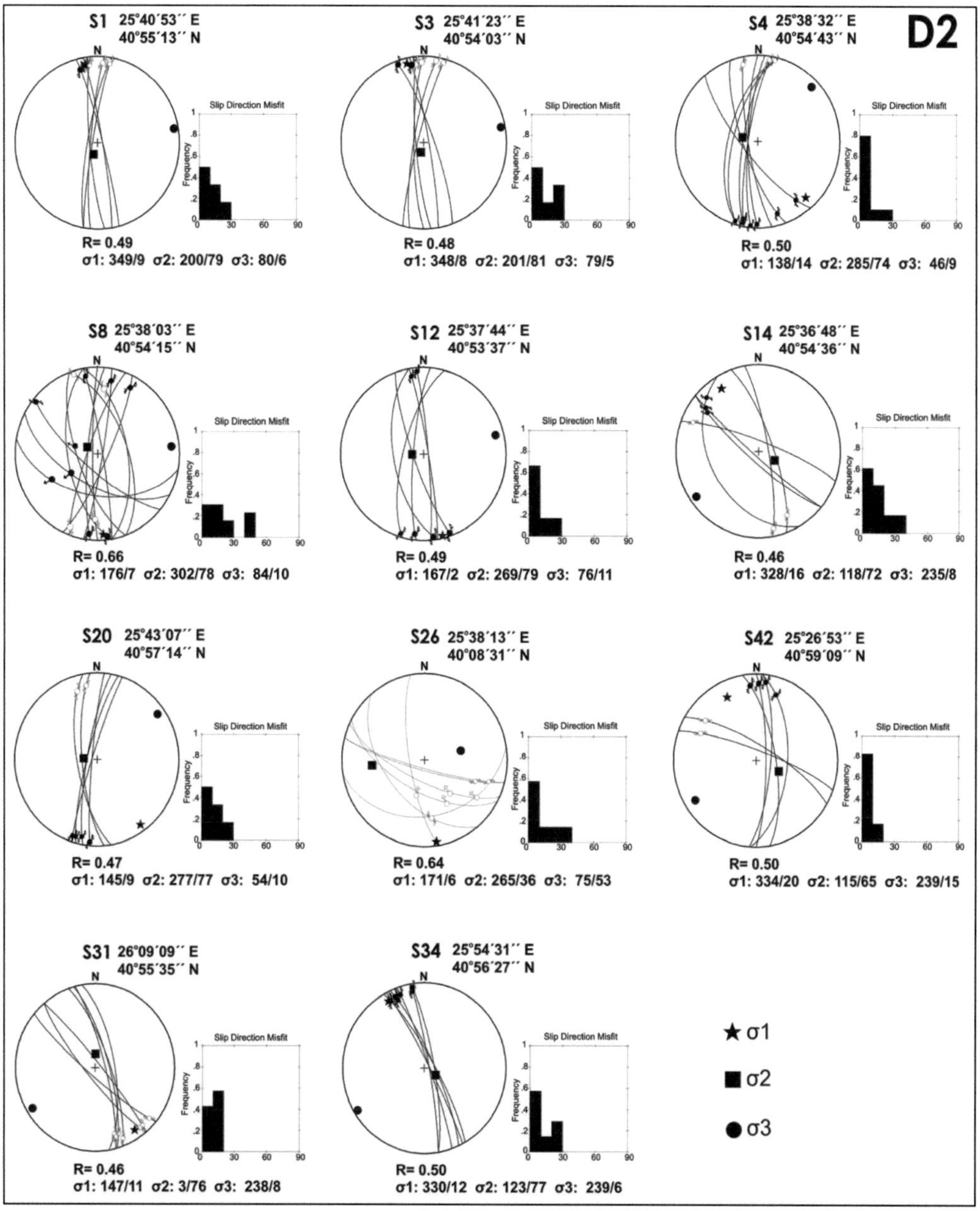

Fig. 13. Faults–striae data and deduced paleostress field (σ1>σ2>σ3) orientation during D2 (equal area, lower hemisphere). Fluctuation histograms of deviation angles (angle between the calculated slip vector and the measured slickenside) and stress ratio R=σ2-σ3/σ1-σ3 are shown. The geographic coordinates for each station are also indicated

However, at the same time other parts of the Rhodope nappe pile (e.g. Kimi and Nea Makri units) were already exhumed (Mposkos & Krohe, 2006; Bonev et al., 2006) and behaved as brittle segments on which the sedimentary sequence of the Thrace basin initially deposited (Fig. 5, 9).

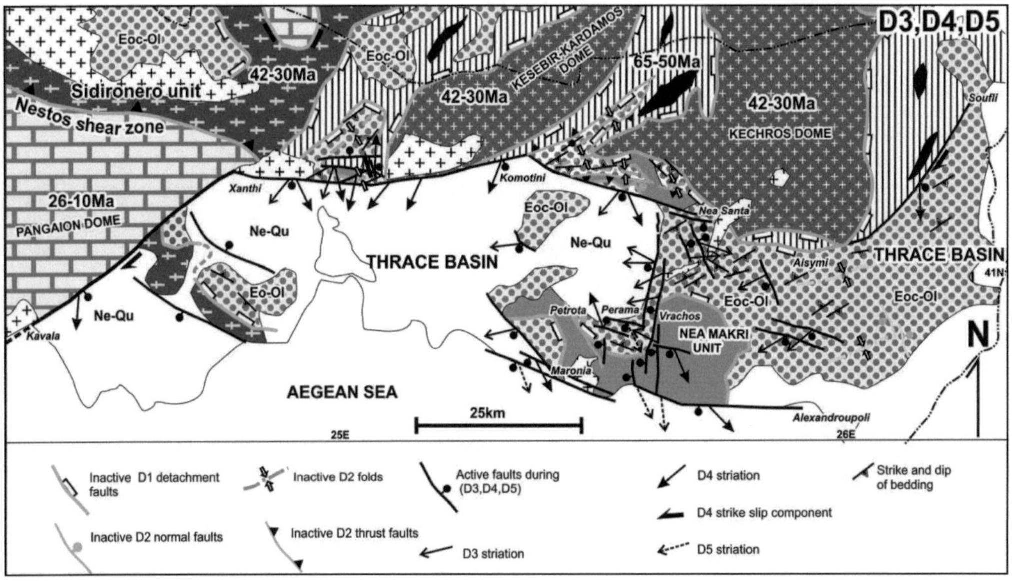

Fig. 14. Main structures and kinematics of D3, D4 and D5 events. Brittle conditions dominate in whole Rhodope region (including metamorphic basement and volcano-sedimentary series of the Thrace Basin). Legend as in Fig. 2

4.2 D2-event

D2-structures cut or overprint clearly the Eocene-Oligocene strata of Thrace basin but do not affect the Neogene sediments of the basin that lie unconformably on the Paleogene suite. In this case, a Late Oligocene-Early Miocene age could be supposed for D2 deformation having taken place under a brittle conditions regime. Moreover, ductile mylonitic fabric of the same Oligocene-Miocene age is referred for the lowermost Pangaion unit at the western edge of the Rhodope province (Kilias & Mountrakis, 1990; Dinter et al., 1995; Wawrenitz & Krohe, 1998; Kilias et al., 1999). This is related to the Oligocene-Miocene SW-ward detachment of the Serbomacedonian massif and the exhumation of the Pangaion metamorphic core complex (Dinter & Royden, 1993; Sokoutis et al., 1993; Dinter et al., 1995; Kilias et al., 1999).

4.3 D3-event

D3 structures dismember the Paleogene volcanosedimentary sequence of the Thrace basin into NW-SE to WNW-ESE trending Neogene basins (Fig. 5, 9, 14). D3 normal faults evolved as synsedimentary structures in relation to the Neogene basins deposition, as it is shown by the tilting of the older Miocene strata towards the fault planes and the overlying horizontal to subhorizontal younger Pliocene sediments which occur in direct contact with the fault planes. Thus, the age of D3 structures should range in between Middle Miocene and

Pliocene time and certainly before D4-structures as the latter clearly overprint the D3-structures.

4.4 D4 event

The complicated network of the D4 structures is recognized to affect the deposits up to the Pliocene and is related to different geometry of kinematics from the former D3 deformation. Therefore, at least a Pliocene age can be concluded for the activation of D4 structures and clearly just subsequent to the D3 event (Fig. 5, 14).

4.5 D5-event

This event concerns the last up to the present active structures, some of which are related to historical earthquakes in the broader Thrace area (Papazachos et al., 1998) and thus its age is ranged between Pleistocene up to present, postdating the D4 deformation.

5. Geodynamic implications-basin evolution

Our detailed tectonic analysis and geological mapping, on both meso- and micro-scale, of the Thrace basin in Greek mainland allowed to constrain the geometry and kinematic of deformation and the tectonic position of the basin, as well as its relationships to the Cenozoic unroofing and exhumation history of the Rhodope metamorphic deep crustal rocks.

The stratigraphic and structural data indicate initially, during Middle-Late Eocene time clastic deposition above a detachment fault zone, continued until Late Oligocene time by molassic type sedimentation. It is linked with the Tertiary NE-SW extension and collapse, which affected the whole Rhodope province (Dinter & Royden, 1993; Kilias et al., 1999; Krohe & Mposkos, 2002; Brunn & Sokoutis, 2007; Bonev et al., 2006; Georgiev et al., 2010) after Cretaceous-Paleocene nappe stacking and crustal thickening (Fig. 5, 9; Burg et al., 1996; Liati & Gebauer, 1999; Jahn-Awe et al., 2010).

The position of the Thrace basin deposits above the Eocene detachment fault zone of the collapsed Rhodope nappe stack allows us to define the Thrace basin as a Paleogene supradetachment basin or as basin above an asymmetrical collapsed volcanic arc, taking in to account the abundant magmatic products in between the basin deposits (Fig. 5, 9; Einselie, 2000; Frisch & Meschede, 2007). As a supradetachment basin, it could be also seen as a type of "piggy-back" basin formed above the primary bounding detachment fault. According to our data the interpretation of the Thrace basin as a forearc basin (Görür & Okay 1996; Tranos, 2009) should be challenged. The interpretation of the Thrace basin as a forearc basin was mainly based on the occurrence along its southern margin, of a sequence of chaotic deposits interpreted as a tectonic melange developed in an accretionary wedge. However, this tectonic melange may represent olistoliths in the Paleogene sediments of the basin.

Basin subsidence during the Paleogene took place simultaneously with the exhumation of the metamorphic core complexes of the Eastern Rhodope massif (e.g., the Kesebir-Kardamos and Kechros domes; Fig. 5, 9; Krohe & Mposkos, 2002; Bonev et al., 2006; Georgiev et al., 2010). A series of smaller Paleogene basins that extended in the broader Rhodope-Serbomacedonian province in Bulgaria FYROM or Turkey (Fig. 2) should be interpreted as rift basins related to the same Paleogene kinematic setting and normal detachment faulting (e.g., Mesta half graben in Bulgaria; Buchfiel et al., 2003) or other basins of Paleogene age in FYROM, Bulgaria and Turkey (Okay et al., 1990; Turgut & Eselle 2000; Dumurdzanov et al., 2005; Bonev et al., 2006).

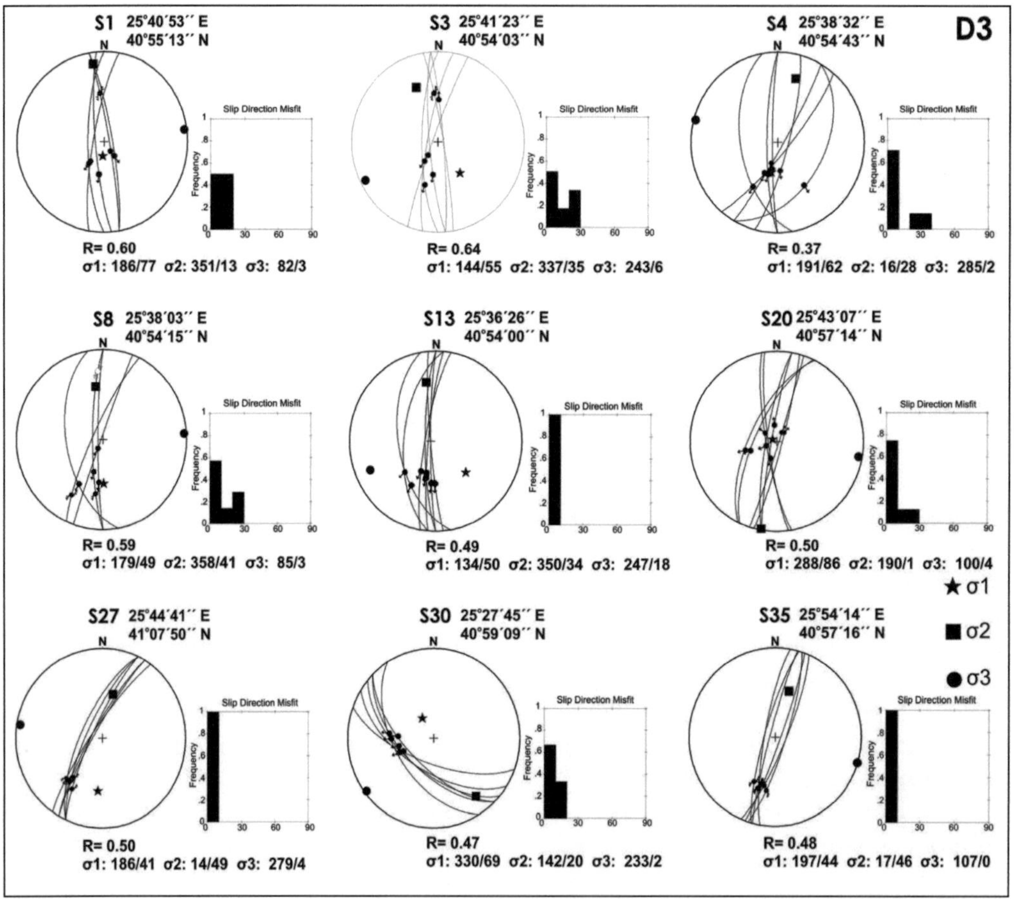

Fig. 15. Faults–striae data and deduced paleostress field (σ1>σ2>σ3) orientation during D3 (equal area, lower hemisphere). Fluctuation histograms of deviation angles (angle between the calculated slip vector and the measured slickenside) and stress ratio R=σ2-σ3/σ1-σ3 are shown. The geographic coordinates for each station are also indicated

The second D2 deformation affected the Paleogene strata of the Thrace basin, dismembering the basin during Oligocene-Miocene time. Contraction perpendicular to the NE-SW to NNE-SSW trending main extension also continued during that time. Compressional structures parallel to the extension (X-axis of the finite strain ellipsoid) associated with strike-slip fault zones and tension fractures are interpreted as contemporaneous formed structures. Compression can act parallel to the Y-axis of the finite strain ellipsoid simultaneously with a main subhorizontal extension. In this case, the 3-D deformation is thought to be a constrictional type. On the other hand, the described geometrical relationships between reverse, normal and strike-slip faults of the same age could be explained by a transpressional type of deformation in accordance with which a strike-slip movement acts simultaneously with a pure shear component or simultaneous action of a non-coaxial and coaxial component of deformation (Sanderson & Marchini, 1984).

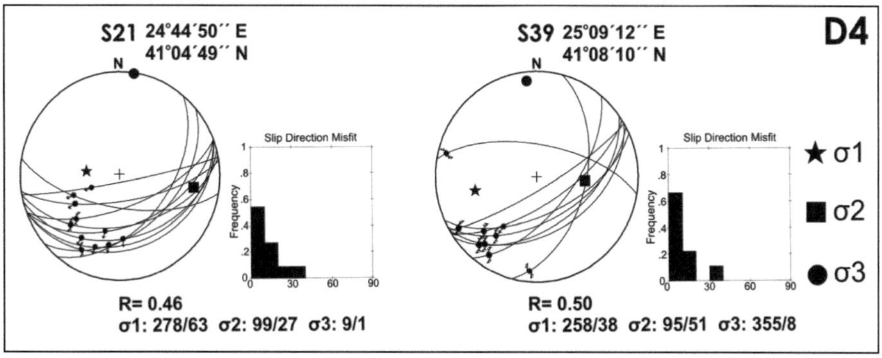

Fig. 16. Faults–striae data and deduced paleostress field (σ1>σ2>σ3) orientation during D4 (equal area, lower hemisphere). Fluctuation histograms of deviation angles (angle between the calculated slip vector and the measured slickenside) and stress ratio R=σ2-σ3/σ1-σ3 are shown. The geographic coordinates for each station are also indicated

The Oligocene-Miocene D2 brittle tectonics of the upper crustal parts in Eastern Rhodopes took place simultaneously with ductile deformation in the lower- most Pangaion metamorphic core complex in Western Rhodopes related to the collapse and tectonic denudation of the upper Rhodope tectonic nappes along the Strymon-Valey normal detachment fault zone (Dinter & Royden, 1993; Kilias et al., 1999). Oligocene-Miocene ductile deformation indicates the same kinematic symmetry with D2 brittle event at the Eastern structurally higher Rhodope nappes (Kilias & Mountrakis, 1990; Sokoutis et al., 1993; Dinter & Royden, 1993; Dinter et al., 1995; Kilias et al., 1999; Brun & Sokoutis 2007).The Strymon basin at the western edge of the Pangaion metamorphic core complex evolved as a syndetachment rift basin (Dinter & Royden, 1993; Dinter et al., 1995; Sokoutis et al., 1993) or as an Oligocene-Miocene supradetachment basin, above the Strymon-Valey detachment fault zone (Fig. 2, 9; Friedmann & Burbank, 1995).

According to our descriptions, a progressive migration of the exhumation towards W-SW through time can be concluded during Tertiary time due to successive Eocene and Oligocene-Miocene stages of extension and metamorphic core complexes formation. Firstly, the structurally higher Rhodope metamorphic units were exhumed during the Eocene, and then the structurally lower units during Oligocene-Miocene time. Therefore, a multi stage extensional unroofing is suggested for the several Rhodope metamorphic units.

The same conclusions for a progressive formation and subsidence of the supradetachment basins in space and time are extracted. Initially, the Paleogene Thrace basin at the Eastern Rhodopes was opened, and then followed the opening of the Neogene Strymon basin at the Western margin of the Rhodope province. This gradual evolution is fairly consistent with the main W-SW-wards sense of movement of the tectonic hangingwall during the Tertiary orogenic collapse of the Rhodope nappe stack, at least for the parts in the Greek mainland (Kilias & Mountrakis, 1990; Sokoutis et al., 1993; Dinter & Royden, 1993; Kilias et al., 1999; Brun & Sokoutis, 2007).

The next D3 deformational stage, during the Miocene-Pliocene involved intense dismemberment of the Thrace basin into Neogene basins, filled with terrigene sediments, covered unconformably the Paleogene strata. NE-SW to E-W extension, under brittle conditions, common for all Rhodope province, was related to the Neogene basins subsidence, as well as the finally exhumation of the lowermost Pangaion metamorphic core complex.

During the Pliocene the extension continued with NNW-SSE to NNE-SSW strike (D4 event) with simultaneous strike-slip and dip-slip movements taking place. Shortening structures in the Pliocene strata, related to the D4 kinematics, are of local importance. The same complicated kinematic geometry and faults network is also revealed for the older D2 deformation, indicating that an important strike-slip component was active during the overall Tertiary structural evolution of the Rhodope province. Co-existence of strike-slip faults with normal and reverse fault zones is common for many areas (Hancock, 1985; Ratschbacher et al., 1989; Kilias et al., 1999).

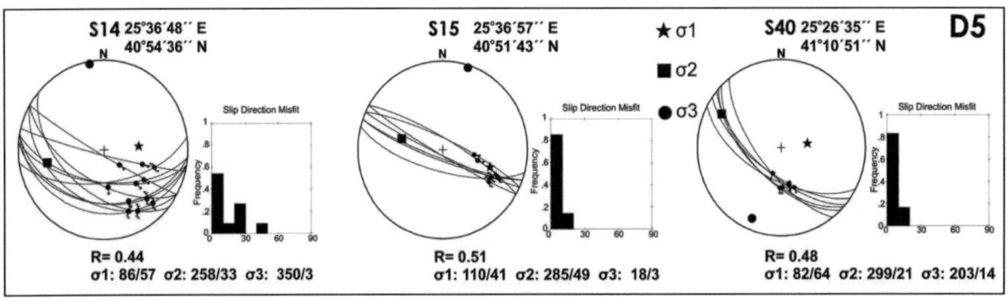

Fig. 17. Faults–striae data and deduced paleostress field (σ1>σ2>σ3) orientation during D5 (equal area, lower hemisphere). Fluctuation histograms of deviation angles (angle between the calculated slip vector and the measured slickenside) and stress ratio R=σ2-σ3/σ1-σ3 are shown. The geographic coordinates for each station are also indicated

Extension remains active with a little change of its trend until present, now with N-S to NNE-SSW trend (D5 event), related to the active faults of the study area. D5 dynamic is well coincided with the dynamic of the large active dextral strike-slip Anatolian fault zone (Papazachos et al., 1988).

Basin subsidence and deposition during the Paleogene were associated with intense magmatic activity (Kilias & Mountrakis, 1998; Marchev et al., 2005; Bonev et al., 2006). The origin of the Late Eocene-Oligocene (34-26 Ma) syndepositional magmatic activity (Christofides et al., 2001; Kilias et al., 2006) could be attributed to the subduction processes evolved during Paleogene time further to the W-SW, in the Olympos-Ossa and Cyclades areas, due to subduction of a thin continental crust or parts of the Pindos oceanic lithosphere beneath the Pelagonian continent (Fig. 1, 18; Schermer et al., 1990; Kilias et al., 1991; Robertson et al., 1996; Lips et al., 1998). However, the remnant of the subducted slab of the Eastern margin of the opened, until the Tertiary time, Vardar-Axios ocean (Brown & Robertson, 2004; Sharp & Robertson, 2006) under the Internal Hellenides could also be the driving mechanism for the Tertiary magmatic activity in the Rhodope province, simultaneously with the opening and sedimentation of the Thrace basin (Fig. 1, 18).

However, Marchev et al., (2005) explain the origin of the Paleogene magmatism and the simultaneous extension and crustal thinning of the Rhodope continental crust as having been due to convective removal of the lithosphere and mantle diapirism. A close analogy was found to the magmatic activity and orogenic collapse taken place in the Menderes massif in SW-ern Turkey during Eocene-Miocene time that resulted in a series of exhumation events (Lips et al., 2001). Furthermore, Bonev et al., (2006) suggest that extension in the Rhodope metamorphic belt initiated a little earlier than the Eocene, during Early Paleogene time, with syndetachment deposition of Paleocene-Early Eocene clastic sediments in the Krumovgrad half graben basin at the SW Bulgaria (Fig. 2, 9).

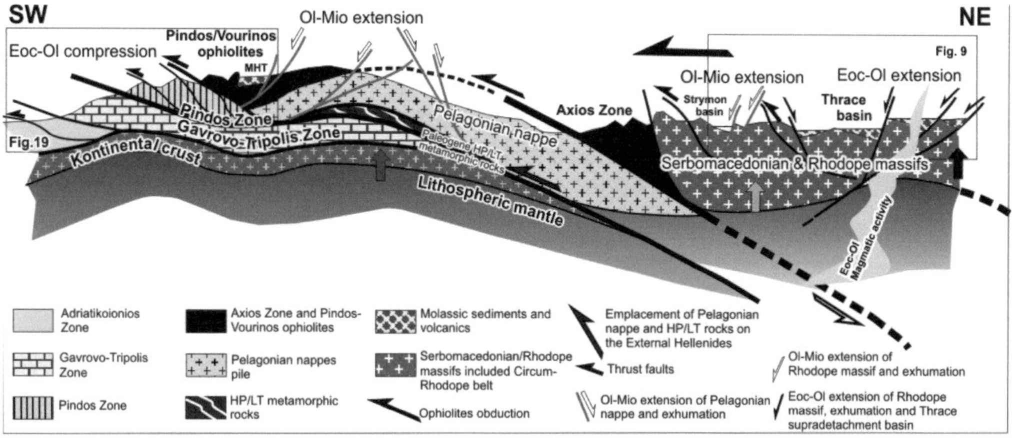

Fig. 18. Schematic lithospheric-scale cross-section through the Hellenides during Paleogene. The future extensional structures leaded to the finally Oligocene-Miocene exhumation of the Hellenides units and basins subsidence are respectively shown. Contraction tectonic and nappe stacking associated with HP/LT metamorphism dominate during Paleogene in External Hellenides and Pelagonian nappes pile whereas extension and orogens collapse take place simultaneously in Internal Hellenides (Serbomacedonian and Rhodope massifs)

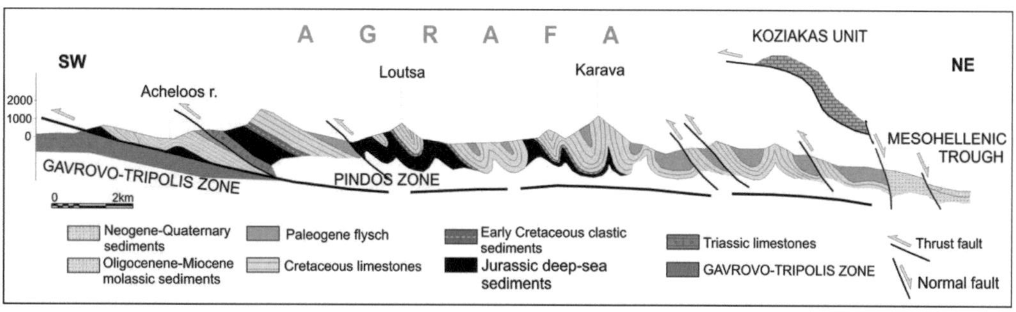

Fig. 19. SW ward thrusting during Paleogene thin skinned tectonic of External Hellenides thrust belt takes place simultaneously with extension and orogenic collapse at the inner parts of Hellenides

Contraction, nappe stacking and crustal thickening associated with HP/LT metamorphism take place simultaneously with the described Rhodope Tertiary extension at the suture zone between External Hellenides and Pelagonian nappe pile more W-wards (Fig. 18; Godfriaux, 1968; Schermer et al., 1990; Kilias et al., 1991). This structural setting at the specific orogen part of Hellenides was further followed by extensional orogenic collapse and crustal thinning during the Oligocene-Miocene, while compression migrated towards the SW to the more external parts of Hellenides at the Hellenic foreland (Fig. 18, 19; Schermer et al., 1990; Kilias et al., 2002, 1999).

In conclusion, in the Hellenic orogen during Tertiary time, a progressive migration towards W to SW, of compression vs. extension under a continuous plate convergence regime, is well established. As a result, the deep crustal levels were successively exhumed during extensional tectonics (e.g., Rhodope metamorphic core complexes, Olympos-Ossa unit, Pelion, Cyclades and Crete complexes), simultaneously with burial of lithospheric material and nappe stacking at the deformed orogenic front towards SW (Fig. 18, 19; Kilias et al., 1999, 2002; Jolivet & Brun, 2010; Ring et al., 2010). Supradetachment basins were successively formed at the internal Hellenides above the extensional detachment fault zones and the footwall segment of the collapsed nappe stack.

The dynamic migrating system extension vs. compression in the Hellenides during the Tertiary time, could be ascribed to, (I) successive subduction processes towards the SW-SSW under oblique plate convergence conditions (Karfakis & Doutsos, 1995; Kilias et al., 1999, 2002; Vamvaka et al., 2006; Schmid et al., 2008), (II) periodic changes in the convergence rate between Africa and European plates, combined with a retreat of the subduction zone towards SW (Royden, 1993, Kilias & Mountrakis, 1998; Dumurdzanov et al., 2005; Papanikolaou, 2009; Jolivet & Brun, 2010; Ring et al., 2010), as well as (III) dynamic instabilities and gravitational collapse of the over-thickened Hellenic accretionary prism (Platt, 1993) or the clockwise rotation of Western External Hellenides away from more Eastern parts of the Internal Hellenides (Kontopoulou, 1986; Kissel & Lai, 1998).

6. Conclusions

The structure and sedimentation of the Thrace basin in Northwestern Greek mainland indicate that NE-SW extension initiated during Middle-Late Eocene time. The Thrace basin evolved during the Eocene-Oligocene as a supradetachment basin above a normal detachment zone with a main W-SW direction of movement of the hangingwall (D1).

The Paleogene Thrace basin was affected during Oligocene-Miocene time by contemporaneous activity of large strike-slip, reverse and normal faults, as well as folds (D2). Extension was trended NE-SW to NNE-SSW almost subhorizontally while compression NW-SE to WNW-ESE, also subhorizontally. During Miocene-Pliocene, Neogene sub-basins developed unconformably over the Paleogene strata bounded by high angle dipping normal faults linked to NE-SW to E-W extension (D3). The next deformational stage (D4) produced simultaneously during the Pliocene local oblique shortening structures and normal faults related to NNW-SSE to NNE-SSW extension and ca. ENE-WSW compression. Some of the D4 normal faults remain active until present (D5) coinciding with the active tectonic dynamics of the broader Rhodope province and the North Aegean region. Tertiary basin subsidence and structural evolution was associated by unroofing and successively, multistage extensional exhumation of the deep crustal metamorphic Rhodope levels, as well as intense calc-alkaline and locally shoshonitic magmatic activity. Exhumation and basin subsidence ages are recorded younger towards the W-WSW and they were linked to a general SW-ward migration of syn-orogenic extension related to an oblique plate convergence regime. Finally, from our descriptions about the structural evolution and geotectonic position of the Thrace basin is clearly concluded, apart from the industrial interesting of the basin (gold bearing deposits, hydrocarbon reservoir), its important contribution to the general geodynamic knowledge.

7. Acknowledgments

This research was supported by the project "Pythagoras" funded by the EPEAEK II. We thank Vasilio Karakosta and Alexandra Gemitzi for various helpful comments and the improvement of the English. Editorial handling by Uri Schattner and Aleksander Lacinica is greatly appreciated.

8. References

Angelier, J. (1979). Determination of the mean principal directions of stresses for a given fault population. *Tectonophysics*, Vol.56, No.3-4, pp. T17-T26

Angelier, J. (1990). Inversion of field data in fault tectonics to obtain the regional stress - III. A new rapid direct inversion method by analytical means. *Geophysical Journal International*, Vol.103, No.2, pp. 363-376.

Bonev, N., Burg, J.-P. & Ivanov, Z. (2006). Mesozoic-Tertiary structural evolution of an extensional gneiss dome - The Kesebir-Kardamos dome, eastern Rhodope (Bulgaria-Greece). *International Journal of Earth Sciences*, Vol.95 No.2, pp. 318-340.

Brown, S.A.M. & Robertson, A.H.F. (2004). Evidence for the Neotethys ocean rooted in the Vardar zone : evidence from the Voras Mountains, NW Greece. *Tectonophysics*, Vol. 381, pp. 143-173.

Brun J.P. & Sokoutis, D. (2007). Kinematics of the Southern Rhodope Core Complex (North Greece). *International Journal of Earth Sciences*, Vol. 99, No.1, pp. 109-138.

Burchfiel, B.C., Nakov, R. & Tzankov, T. (2003). Evidence from the Mesta half-graben, SW Bulgaria, for the Late Eocene beginning of Aegean extension in the Central Balkan Peninsula. *Tectonophysics*, Vol.375, No.1-4, pp. 61-76.

Burg, J.P., Ricou, L.E., Ivanov, Z., Godfriaux, I., Dimov, D. & Klain, L. (1996). Syn-metamorphic nappe complex in the rhodope massif. Structure and kinematics. *Terra Nova*, Vol. 8, No.1, pp. 6-15.

Christodoulou, G. (1958). Ueber das Alter einiger Formationen von Samothraki (in Greek with German summary). *Bulletin of Geological Society of Greece*, Vol. III/1, pp. 40-45.

Christofides, G., Koroneos, A., Soldatos, T., Eleftheriadis, G., & Kilias, A. (2001). Eocene magmatism (Sithonia and Elatia plutons) in the Internal Hellenides and implications for Eocene-Miocene geological evolution of the Rhodope massif (Northern Greece). *Acta Vulcanologica*, Vol. 13, pp. 73-89.

Christofides, G., Pécskay, Z., Eleftheriadis, G., Soldatos, T. & Koroneos, A. (2004). The Tertiary Evros volcanic rocks (Thrace, Northeastern Greece): Petrology and K/Ar geochronology. *Geologica Carpathica*, Vol. 55, No.5, pp 397-409.

Coskun, B. (1997). Oil and gas fields - Transfer zone relationships, Thrace basin, NW Turkey. *Marine and Petroleum Geology*, Vol.14, No.4, pp. 401-416.

Dinter, D.A. & Royden, L. (1993). Late Cenozoic extension in northeastern Greece: Strymon Valley detachment system and Rhodope metamorphic core complex. *Geology*, Vol.21, No.1, pp. 45-48.

Dinter, D.A., Macfarlane, A.M., Hames, W., Isachsen, C., Bowring, S. & Royden, L. (1995). U-Pb and 40Ar/39Ar geochronology of Symvolon granodiorite: implications for the thermal and structural evolution of the Rhodope metamorphic core complex, Northeastern Greece. *Tectonics* Vol.14, pp.886–908

Dragomanov, L. K., Grigorov, V., Ioncher, A., Jelev, A. & Darakchieva, St. (1986). Lithological indications of the presence of Middle Eocene in the Eastern Rhodopes. *Ann. Higher Inst. Mining Geol.*, 32, 37-41 (In Bulgarian with an English abstract).

Dumurdzanov, N., Serafimovski, T. & Burchfiel, B.C. (2005). Cenozoic tectonics of Macedonia and its relation to the South Balkan extensional regime. *Geosphere*, Vol.1, No.1, pp. 1-22.

Duyester, J.D. (1999). StereoNett version 2.4. http://homepage.ruhr-uni-bochum.de/Johannes.P.Duyester/Stereo/Stereo1.htm.

Einsele, G. (2000). Sedimentary basins. 2. Auflage, Springer Verlag, Berlin, pp. 792.

Friedmann, S.J. & Burbank, D.W. (1995). Rift basins and supradetachment basins: Intracontinental extensional end-members. *Basin Research*, Vol.7, No.2, pp. 109-127.

Frisch, W. & Meschede, M. (2007). Plattentektonik, Kontinentverschiebung und Gebirgsbildung 2. überarb. Auflage, Primus Verlag u. Wiss. Buchges,. Darmstadt,196 s.

Fytikas, M., Innocenti, F., Manetti, P., Mazzuoli, R., Peccerillo, A. & Villari, L (1984). Tertiary to Quaternary evolution of volcanism in the Aegean region The geological evolution of the Eastern Mediterranean. *Geol. Soc. Spec. Pb.*, Vol.17, pp. 687-699 In: Dixon EJ, Robertson AHF (eds), Geological Society, London.

Garlaouni C., Papadimitriou E., Kilias A., Falalakis G. & Gemitzi, A. (2007). The evolution of the stress field in Eastern Macedonia and Thrace. *Bulletin of the Geological Society of Greece*, Vol. 1, pp. 32-40.

Georgiev, N., Pleuger, J., Froitzheim, N., Sarov, S., Jahn-Awe, S. & Nagel, T.J. (2010). Separate Eocene–Early Oligocene and Miocene stages of extension and core complex formation in the Western Rhodopes, Mesta Basin, and Pirin Mountains (Bulgaria) *Tectonophysics*, Vol. 487, pp. 59-84.

Georgiev, N.; Pleuger, J., Froitzheim, N., Sarov, S., Jahn-Awe, S. & Nagel, T.J. (2010). Separate Eocene-Early Oligocene and Miocene stages of extension and core complex formation in the Western Rhodopes, Mesta Basin, and Pirin Mountains (Bulgaria). *Tectonophysics*, Vol.487, No.1-4, pp. 59-84

Godfriaux, I. (1968). Etude géologique de la région de l' Olympe (Grèce). *Ann Géol Pays Héll.*, Vol.19, pp. 1-280.

Goeruer, N. & Okay, A.I. (1996). A fore-arc origin for the Thrace Basin, NW Turkey. *Geologische Rundschau*, Vol. 85, pp. 662-668.

Hancock, P.L. (1985). Brittle microtectonics: principles and practice. *Journal of Structural Geology*, Vol. 7, No.3-4, pp. 437-457.

Huvaz O., Karahanoglu, N. & Ediger, V. (2007). The thermal gradient history of the Thrace Basin, NW Turkey: correlation with basin evolution processes. *Journal Petrol. Geol.*, Vol.30, No.3–24.

Innocenti, F., Kolios, N., Manetti, P., Rita, F. & Villari, L. (1982). Acid and basic late Neogene volcanism in central Aegean Sea: its nature and geotectonic significance. *Bulletin Volcanologique*, Vol.45,No.2, pp.87-97.

Jahn-Awe, S., Froitzheim, N., Nagel, T.J., Frei, D., Georgiev, N. & Pleuger, J. (2010). Structural and geochronological evidence for Paleogene thrusting in the western

Rhodopes, SW Bulgaria: Elements for a new tectonic model of the Rhodope Metamorphic Province. *Tectonics*, Vol.29, No. 3, TC3008

Jolivet L. & Brun J-P (2010). Cenozoic geodynamic evolution of the Aegean. *International Journal of Earth Sciences*, Vol.99, pp.109–138

Karfakis, I. & Doutsos, T. (1995). Late orogenic evolution of the Circum-Rhodope Belt, Greece. *Neues Jahrb. Geol. Palaeontol. Monatsh.*, pp. 305-319.

Kauffman, G., Kockel, F. & Mollat, H. (1976). Notes on the stratigraphic and paleogeographic position of the Svoula formation in the innermost zone of the Hellenides (northern Greece). *Bull Soc Geol France*, Vol.18, pp. 225-230.

Kilias A., Falalakis G., Gemitzi A. & Christaras V. (2006). Faulting and paleostress evolution in the Maronia-Petrota basin. Evidence of active faults. *Bulletin of the Tethys Geological Society*, Vol.1, pp. 7-16.

Kilias, A. & Mountrakis, D. (1998). Tertiary extension of the Rhodope massif associated with granite emplacement (Northern Greece). *Acta Vulcanologica*, Vol.10, No.2, pp. 331-337.

Kilias, A. & Mountrakis, D. (1990). Kinematics of the crystalline sequences in the western Rhodope massif. *Geologica Rhodopica*, Vol.2, pp. 100-116.

Kilias, A., Falalakis, G. & Mountrakis, D. (1999). Cretaceous-Tertiary structures and kinematics of the Serbomacedonian metamorphic rocks and their relation to the exhumation of the Hellenic hinterland (Macedonia, Greece). *International Journal of Earth Sciences*, Vol. 88, No.3, pp. 513-531

Kilias, A., Frisch, W., Ratschbacher, L. & Sfeikos, A. (1991). Structural evolution and metamorphism of blueschists, Ampelakia nappe, eastern Thessaly, Greece. *Bulletin of the Geological Society of Greece*, Vol.25, No.1, pp. 81-89.

Kilias, A.A., Tranos, M.D., Orozco, M., Alonso-Chaves, F.M. & Soto, G.I. (2002). Extensional collapse of the Hellenides: A review. *Revista de la Sociedad Geoló Gica de España*, Vol.15, No.3-4, pp. 129-139.

Kissel, C. & Laj, C. (1988). The Tertiary geodynamical evolution of the Aegean arc: a palaeomagnetic reconstruction. *Tectonophysics*, Vol.146, pp. 183-201.

Kockel, F., Mollat, H. & Walther, H. (1971). Geologie des Serbomazedonischen massivs und seines mesozoischen Rahmes (Nordgriechenland). *Geol Jahrb*, Vol.89, pp. 529-551.

Kondopoulou, D. (1986). Tertiary rotational deformation in the Greek Serbomacedonia Massif. *Bulgarian Geophysical Journal*, Vol.12, No.4, pp. 71-80.

Kopp, K.O. (1965). Geologie Thrakiens III : Das Tertiaer zwischen Rhodope und Evros. *Annales geologiques des pays Helleniques*, Vol. 16, pp. 315-362

Koukouvelas, I. & Pe-Piper, G. (1991). The Oligocene Xanti pluton, northern Greece: A granodiorite emplaced during regional extension. *J. Geol. Soc. London*, Vol.148, pp. 749-758.

Krohe, A. & Mposkos, E. (2002). Multiple generations of extensional detachments in the Rhodope Mountains (northern Greece): Evidence of episodic exhumation of high-pressure rocks. *Geol. Soc. Spec. Publ.*, The Timing and Location of Major Ore Deposits in an Evolving Orogen, D. J. Blundell, F. Neubauer, and A. Von Quadt (eds), Vol.204, pp. 151-178.

Lescuyer, J.L., Bailly, L., Cassard, D., Lips, A.L.W., Piantone, P. & McAlister, M. (2003). Sedimenthosted gold in south-eastern Europe: the epithermal deposit of Perama,

Thrace, Greece In: *Mineral exploration and sustainable development*, Eliopoulos D.G. et al. (eds),. Millpress, Rotterdam pp.499-502.

Liati, A. & Gebauer, D. (1999). Constraining the prograde and retrograde P-T-t path of Eocene HP rocks by SHRIMP dating of different zircon domains: Inferred rates of heating, burial, cooling and exhumation for central Rhodope, northern Greece. *Contributions to Mineralogy and Petrology*, Vol.135, No.4, pp. 340-354.

Lips, A.L.W., Cassard, D., Soezbilir, H., Yilmaz, H. & Wijbrans, J.R. (2001). Multistage exhumation of the Menderes Massif, Western Anatolia (Turkey). *International Journal of Earth Sciences*, Vol. 89, pp. 781-792.

Lips, A.L.W.; White, S.H. & Wijbrans, J.R. (1998). 40Ar/39Ar laserprobe direct dating of discrete deformational events: A continuous record of early Alpine tectonics in the Pelagonian Zone, NW Aegean area, Greece. *Tectonophysics*, Vol.298, No.1-3, pp. 133-153.

Marchev, P.; Kaiser-Rohrmeier, M.; Heinrich, C.; Ovtcharova, M.; von Quadt, A.& Raicheva, R. (2005). 2: Hydrothermal ore deposits related to post-orogenic extensional magmatism and core complex formation: The Rhodope Massif of Bulgaria and Greece. *Ore Geology Reviews*, Vol. 27, No.1-4, pp. 53-89.

McClaughry, J.D. & Gaylord, D.R. (2005). Middle Eocene sedimentary and volcanic infilling of an evolving supradetachment basin: White Lake Basin, south-central British Columbia. *Canadian Journal of Earth Sciences*, Vol.42, No.1, pp.49-66

Meinhold, G. & BonDagher-Fadel, M. (2010). Geochemistry and biostratigraphy of Eocene sediments from Samothraki islald, NE Greece. *Neues Jahrbuch fuer Geologie und Palaeontologie Abhandlungen*, Vol. 256(1), pp. 17-38.

Mountrakis, D.; Tranos, M.; Papazachos, C.; Thomaidou, E.; Karagianni, E. & Vamvakaris, D (2006). Neotectonic and seismological data concerning major active faults, and the stress regimes of Northern Greece. *Geological Society Special Publication*, Vol.260, pp. 649-670.

Mposkos, E. & Krohe, A. (2006). Pressure-temperature-deformation paths of closely associated ultra-high-pressure (diamond-bearing) crustal and mantle rocks of the Kimi complex: Implications for the tectonic history of the Rhodope Mountains, northern Greece. *Canadian Journal of Earth Sciences*, Vol.43, No.12, pp. 1755-1776.

Okay, A., Siyako, M. & Burkan, K.A. (1990). Geology and Tectonic Evolution of the Biga Peninsula. *Bull. TPJD-C*, Vol.2, No.1, pp. 83-121.

Pangaea Software Scientific (2007). MyFault, version 1.03, http://www. pangaeasci.com

Papadopoulos, C. & Kilias, A. (1985). Altersbeziehungen zwischen Metamorphose und Deformation im zentralen Teil des Serbomazedonischen Massivs (Vertiskos Gebirge, Nord-Griechenland). *Geologische Rundschau*, Vol.74, No.1, pp. 77-85.

Papanikolaou, D. & Panagopoulos, A. (1981). On the structural style of southern Rhodope, Greece. *Geologica Balkanica*, Vol.11, pp. 13-22.

Papanikolaou, D. (2009). Timing of tectonic emplacement of the ophiolites and terrane paleogeography in the Hellenides. *Lithos*, 108, pp. 262-280

Papazachos, B.C., Scordilis, E.M., Panagiotopoulos, D.G., Papazachos, C.B. & Karakaisis, G. (2004). Global relation between seismic fault parameters and moment magnitude of earthquakes. Bulletin Geological Society Greece, Vol. 25, pp. 1-8.

Papazachos, B.C., Papadimitriou, E.E., Kiratzi, A.A., Papazachos, C.B. & Louvari, E.K. (1998). Fault plane solutions in the Aegean Sea and the surrounding area and their tectonic implications. *Boll. Geof. Teor. Appl.*, Vol.39, pp. 199-218.

Platt, J.P. (1993). Exhumation of high-pressure rocks: a review of concept and processes. *Terra Nova*, Vol.5, pp.119–133.

Ratschbacher, L., Frisch, W., Neubauer, F., Schmid, S.M. & Neugebauer, J. (1989). Extension in compressional orogenic belts: The eastern Alps. *Geology*, Vol.17, pp. 404-407.

Ring, U., Glodny, J., Will, T. & Thomson, S. (2010). The Hellenic subduction System: High-pressure Metamorphism, Exhumation Normal Faulting and Large scale extension. *Annual Review of Earth and Planetary Sciences*, Vol.38, pp. 45-76.

Robertson, A.H.F., Dixon, J.E., Brown, S., Collins, A., Morris, A., Pickett, E., Sharp, I. & Ustaömer, T. (1996). Alternative tectonic models for the Late Palaeozoic-Early Tertiary development of Tethys in the Eastern Mediterranean region. *Geological Society Special Publication*, Vol.105, pp. 239-263.

Royden, L.H. (1993). Evolution of retreating subduction boundaries formed during continental collision. *Tectonics*, Vol. 12(3), pp. 629-638.

Sanderson, D.J. & Marchini, W.R.D. (1984). Transpression. *Journal of Structural Geology*, Vol.6, No.5, pp. 449-458.

Schermer, E.R., Lux, D.R. & Burchfiel, B.C.(1990). Temperature-time history of subducted continental crust, Mount Olympos region, Greece. *Tectonics* Vol.9, No.5, pp. 1165-1195.

Schmid, S.M., Bernoulli, D., Fuegenschuh, B., Matenco, L., Schefer, S., Schuster, R., Tischler, M. & Ustaszewski, K. (2008). The Alpine-Carpathian-Dinaric orogenic system: correlation and evolution of tectonic units. *Swiss Journal of Geosciences*, Vol. 101, pp. 139-183.

Sharp, I.R. & Robertson, A.H.F. (2006). Tectonic sedimentary evolution of the western margin of the Mesozoic Vardar Ocean: evidence from the Pelagonian and Almopias zones, northern Greece. In: Robertson A.H.F. and Mountrakis D. (eds), *Geological Society of London, Special Publication*, Vol. 260, pp. 373-412.

Simpson, C. & Schmid, S.M. (1983). An evaluation of criteria to deduce the sense of movement in sheared rocks. *Geological Society American Bulletin*, Vol. 94, pp. 1281-1288.

Siyako M. & Huvaz O. (2007). Eocene stratigraphic evolution of the Thrace Basin, Turkey. *Sedimentary Geology*, Vol. 198, pp.75–91

Sokoutis, D., Brun, J.P., Van Den Driessche, J. & Pavlides, S. (1993). A major Oligo-Miocene detachment in southern Rhodope controlling north Aegean extension. *Journal Geological Society London*, Vol.150, No.2, pp. 243-246.

Tranos, M.D. (2009). Faulting of Lemnos Island: a mirror of the North Aegean Trough (Northern Greece). *Tectonophysics*, Vol. 467, pp. 72-88.

Turgut, S. & Eseller, G. (2000). Sequence stratigraphy, tectonics and depositional history in eastern Thrace Basin, NW Turkey. *Marine and Petroleum Geology*, Vol.17, No.1, pp.61-100

Vamvaka, A., KIlias, A., Mountrakis, D. & Papaoikonomou J.(2006). Geometry and structural evolution of the Mesohellenic Trough (Greece): a new approach. In: Robertson A.H.F. and Mountrakis D. (eds*), Geological Society of London, Special Publication*, Vol. 260, pp.132-146.

Wawrzenitz, N. & Krohe, (1998). Exhumation and doming of the Thasos metamorphic core complex (S Rhodope, Greece): Structural and geochronological constraints. *Tectonophysics,* Vol.285, No.3-4, pp.301-332

Zagorchev, I.S. (1998). Pre-Priabonian Paleogene formations in southwest Bulgaria and northern Greece: stratigraphy and tectonic implications. *Geological Magazine,* Vol.135, pp.101–119.

10

The North Anatolian Fault Zone: an Evaluation of Earthquake Hazard Parameters

Yusuf Bayrak, Hakan Çınar and Erdem Bayrak
Karadeniz Technical University / Geophysics Department, Trabzon
Turkey

1. Introduction

Turkey is located in one of the most actively deforming regions in the world. The tectonic in and around Turkey depends on relative motions between the African, the Aegean, the Arabian, the Anatolian, the Black Sea and the Eurasian plates (Kasapoğlu & Toksöz, 1983). The neotectonics of Turkey is governed by three major elements (Bozkurt, 2001): (1) the Aegean–Cyprean Arc, a convergent plate boundary where the African Plate to the south is subducting beneath the Anatolian Plate to the north; (2) the dextral North Anatolian Fault Zone; and (3) the sinistral East Anatolian Fault Zone. Also, the sinistral Dead Sea Fault Zone has an important role. The North Anatolian Fault Zone (NAFZ) is one of the best-known dextral strike-slip faults in the world because of its remarkable seismic activity, extremely well developed surface expression and importance for the tectonics of eastern Mediterranean region. The NAFZ is a dextral strike-slip fault system that extends from eastern Turkey in the east, to the north Aegean in the west. Along much of its length, this fault zone consists of a few shorter sub-parallel fault strands that sometimes display an anastomosing pattern (Bozkurt, 2001). To the east, the NAFZ forms a typical triple-junction and joins with the sinistral East Anatolian Fault Zone at Karlıova. The NAFZ does not terminate at the Karlıova triple junction, but continues towards south east.

The development of earthquake hazard assessment in Turkey has a substantial history and it has been produced considerable progress and innovation because Turkey has frequently suffered from major damaging earthquakes since the year 2000 BC. It has a long history of devastating earthquakes that have killed many thousands of people and caused economic loss, such as the İzmit (M_S=7.8) and Düzce (M_S=7.2) earthquakes of 17 August and 12 November 1999,respectively. As a result, a great number of local or regional earthquake hazard studies have been applied in recent years in order to present quantitative results on earthquake hazard of Turkey using the instrumental earthquake data. Several authors have used different statistical models to estimate the size of earthquake occurrences such as expected magnitude, intensity, ground acceleration, velocity or displacement. Amongst these studies are involving maximum seismic intensity, maximum magnitude and modal values (Bayrak et al., 2005, 2008a), peak ground acceleration (Erdik et al., 1999; Kayabalı & Akın 2003), ω-upper bound magnitude of GIII distribution (Bayrak et al., 2008b), the index K-value defined as relative earthquake hazard scale (Bayrak et al., 2009).

Although many reports are known that Turkish instrumental and historical records are far from being incomplete for probabilistic approach of seismic hazard, a serious effort is undertaken for such analysis in the present work. For this reason, a method developed by Kijko & Sellevoll (1989, 1992) has been carried out. The proposed approach is very flexible and provides several attractive properties. It accommodates the "gaps" in both historical and complete parts of the catalogue. It makes possible to estimate the maximum regional magnitude from the largest historical known earthquake, which occurred before the catalogues began. It allows the combination between the earthquakes of historical epoch and those which are extracted from short time periods of instrumental data. The complete part of the catalog can be divided into time intervals of different level of completeness. An illustration of the quality of the data, which can be used to obtain the seismic parameters through this approach, can be seen in Kijko & Sellevoll (1992).

The earthquake hazard assessment requires the knowledge of the earthquake potential in a region. Qualitative techniques (epicenter maps, etc) as well as quantitative ones have been applied for presentation of the geographical distribution of earthquake hazard in several regions of the Earth and Turkey (e.g. Makropoulos, 1978; Tsapanos & Papazachos, 1998; Hamdache et al., 1998; Tsapanos, 2001; Tsapanos, 2003; Bayrak et al., 2005). Common quantities considered as measures of seismicity are the maximum observed magnitude M_{max} (e.g. Yegulalp & Kuo, 1974; Koravos et al., 2003; Ryall & van Wormer, 1980), the annual number $N(M)$ or the mean return period T_m of earthquakes (Comninakis, 1975) with magnitudes greater or equal to a given value M, etc. The knowledge of the return period is of great importance in studying and analyzing earthquake hazard and/or seismicity. It contributes with great importance to the determination of the national seismic code, according to which building if different categories (normal strategic, monumental) must be constructed; it conditions the priority of interventions on existing buildings (Hamdache et al., 1998). However, estimation of earthquake hazard involves the computation of long-term probabilities for the occurrence of earthquakes of a specified size in a given area during a given time interval (Mäntyniemi et al., 2004).

In this study, in order to estimate the maximum regional magnitude and the other related parameters such as the magnitude-frequency relationship, and the mean seismic activity rate a method introduced by Kijko & Sellevoll (1989) is used. For this purpose, the technique of the maximum likelihood estimation is applied in the different regions of NAFZ on a basis of a procedure which utilizes data from both historical and instrumental files. The computations of the method are based on the assumptions of the Poisson occurrence of the earthquakes in time with a mean activity rate and the doubly truncated frequency-magnitude relation of Gutenberg-Richter. The standard deviations of these parameters are also estimated. Moreover, the mean return periods of the earthquakes with a certain lower magnitude, the most probable maximum magnitude of earthquakes in a given time interval, and the probability for a large earthquake occurrence are determined.

1.1 The tectonics and seismicity of the North Anatolian Fault zone

The North Anatolian Fault zone (NAFZ), one of the most seismoactive faults in the world, is a 1200-km-long and broad arc-shaped (Bozkurt, 2001) dextral strike-slip fault zone (Figure 1). NAFZ takes up the relative motion between the westward moving Anatolian Block, due mainly to the collision of the Arabian and African plates against the Eurasian and Anatolian ones (Sengör et al., 1985; Sengör et al., 2005), and Black Sea Plate (Şengör, 1979; Barka &

Kadinsky-Cade, 1988), and it forms a border between Eurasian and Anatolian plates. NAFZ was formed by progressive strain localization in a generally westerly widening right-lateral keirogen in northern Turkey mostly along an interface juxtaposing subduction-accretion material to its south and older and stiffer continental basements to its North (Şengör et al., 2005). Morphologically distinct and seismically active this fault extends from the Gulf of Saros in the northern Aegean Sea to the town of Karlıova (39°18N, 41°01E) in Eastern Turkey for 1200 km, paralleling roughly the southern Black Sea shores and keeping a fairly regular distance of some 100 km to the coast, connecting the Aegean taphrogen (Şengör et al. 2005) with the East Anatolian high plateau (Koçyiğit et al. 2001; Şengör et al. 2003). This fault formed approximately 13 to 11 Ma ago in the east and propagated westward. It reached the Sea of Marmara no earlier than 200 ka ago, although shear related deformation in a broad zone there had already commenced in the late Miocene (Şengör et al., 2005). Along much of its lengths, NAFZ has well developed surface expressions and geological features such as a few shorter subparallel fault strands that sometimes display an anastomosing pattern, second order faults that splay from it into the Anatolian Plate, pull-apart basins, where there is an over-step along the fault trace (e.g.Niksar Basin, Taşova-Erbaa Basin, Erzincan Basin (Aydın&Nur, 1982)), fault-wedge basins (e.g. Suşehri Basin, Gölova Basin, Vezirköprü Basin), complex pull-apart basins, bounded by an active strike-slip along one margin and a thrust fault along the other (e.g. Geyve Basin), composite pull-aparts, formed by coalescence of smaller pull-apart basins (e.g. Merzifon Basin, Suluova Basin, Erzincan Basin), and negative flower structure, bounded by strike-slip faults which show considerable amount of extension (e.g. Kazova basin along the Almus Fault Zone) (Bozkurt, 2001). In the East the NAFZ starts around Karlıova triple junction, and it runs NW to Vezirköprü where it makes a left bend and continues westward. Around Kargı, it makes another left bend and then runs in a SWdirection (Bozkurt, 2001). Total offset of NAFZ is 85 ±5 km (Bozkurt, 2001; Şengör et al. 2003). Along most of the NAFZ, the right-lateral slip has a rate of 24±1 mm/yr (McClusky et al., 2000; Flerit et al., 2004; Reilinger et al., 2006). This estimate is made by assuming that all motion of Anatolia is accommodated by slip on the North Anatolian fault, which serves as the primary boundary between Anatolia and Eurasia (Reilinger et. al., 1997; Reilinger et al., 2006). Focal mechanisms of the earthquakes along this fault zone give consistently pure right-lateral strike-slip solutions (Canıtez&Üçer, 1967; McKenzie, 1972; Jackson & McKenzie, 1984; Eyidoğan et al., 1991; Barka, 1996).

Just east of the Marmara Sea, NAFZ splays into two major strands controlling and steering the tectonic regime of the Marmara region. These are the Northern Marmara strand and the Southern Marmara strand (Parsons, 2004). In the Marmara Sea, the NAFZ begins to lose its single fault line character and splays into a complex fault system (Erdik et al., 2004). The complexity of the NAFZ in this region may actually be related to the transition between an intermediate stress regime to the east and the extension regime which characterizes the Aegean Sea (Gürbüz et al., 2000). Northern Marmara strand consists of İzmit segment, Prince's Islands segment, and Ganos segment, from east to west respectively. The southern strand splits up again into a middle and southern branch, the former passing south of Iznik and the latter south of Bursa, by the Lakes of Apolyont and Manyas into the North Aegean extensional province (Ambraseys, 2002). Several marine-based geophysical surveys have been carried out in Marmara Sea in order to delineate fault systems of this region. The results obtained from these surveys are presented in the literature by Le Pichon et al. (1999, 2001, 2003), Imren et al. (2001), Demirbağ et al. (2003), Carton (2005), Carton et al. (2007),

Laigle et al. (2008) and Becel et al. (2009). The Sea of Marmara is a large pull-apart that appears to have been a geometrical/mechanical obstacle encountered by the NAFZ during its propagation (Armijo et al., 2002, 2005). The pull-aparts of Marmara are Çınarcık, Central Marmara and Tekirdağ basins, from east to west respectively. These basins consist of several strike-slip and normal faults. GPS velocities suggest that the Anatolia/Eurasia motion is accommodated across the Marmara region by 18–20 mm/yr of right-lateral slip and 8 mm/yr of extension (Flerit et al., 2003, 2004).

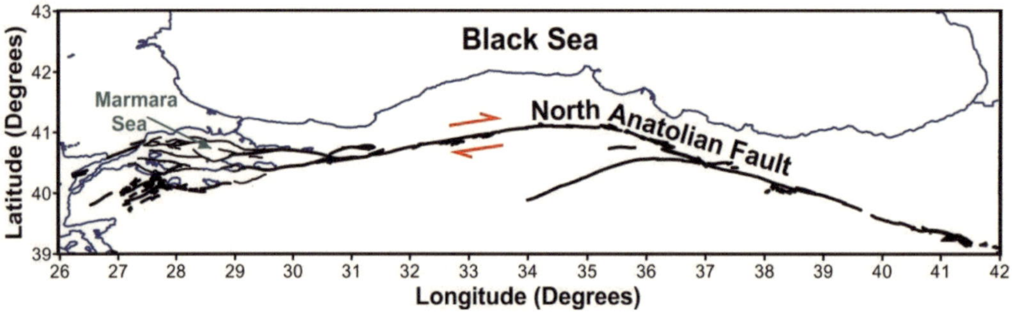

Fig. 1. Tectonic map of the North Anatolian Fault zone (modified from Şaroğlu, 1992; Le Pichon et al. 2003)

NAFZ has produced devastating earthquakes along its different sections in both historical and instrumental periods. The earthquakes of $M_S \geq 7.0$, used in this study, are listed in Table 1 and their epicentral distribution is shown in Figure 2. In instrumental period, the earthquakes in the NAFZ migrated from East to West. Beginning with 1939 Erzincan earthquake (M_S=7.8), which produced about 350 km of ground rupture, the NAFZ ruptured, and formed more than 1000 km surface rupture along the fault in nine moderate large earthquakes (M_S>6.7) (Bozkurt, 2001).

1.2 Data, source zonation and completeness analysis

The database and the seismicity data in this work were compiled from different sources and catalogues such as TURKNET, International Seismological Centre (ISC), Incorporated Research Institutions for Seismology (IRIS) and The Scientific and Technological Research Council of Turkey (TUBITAK) and provided in different magnitude scales. The catalogues contain the origin time, different magnitudes scales (m_b-body wave magnitude, M_S-surface wave magnitude, M_L-local magnitude, M_D-duration magnitude, and M_W-moment magnitude), epicenter and depth information of earthquakes. Turkey earthquake catalogue, obtained from the Boğaziçi University, Kandilli Observatory and Earthquake Research Institute (KOERI), starts from 1974 until 2010. The earthquakes from 1900 to 1974 come from the International Seismological Centre (ISC) and instrumental catalogue of KOERI. The historical earthquake catalogue used in this study is taken from database of GSHAP (Global Seismic Hazard Assessment Program) being compiled by Erdik et al. (1999).

An earthquake data set used in seismicity or seismic hazard studies must certainly be homogenous, in other words it is necessary to use the same magnitude scale. But, the earthquake data obtained from different catalogues have been reported in different magnitude scales. So, all earthquakes must be defined in the same magnitude scale. Bayrak et al. (2009) developed some relationships between different magnitude scales (m_b-body wave magnitude, M_S-surface wave magnitude, M_L-local magnitude, M_D-duration magnitude, and M_W-moment magnitude) in order to prepare a homogenous earthquake catalogue from different data sets. We prepared a homogenous earthquake data catalogue for M_S magnitude using these relationships. The time interval considered for the present work changed between BC 1010 and AD 2010.

Region No	*Region Location*	*Date*	*Longitude*	*Latitude*	M_S
1	Saroz Gulf	1010	27.00	40.60	7.4
		01.03.1354	27.00	40.70	7.3
		05.08.1766	27.11	40.74	7.6
		09.08.1912	27.20	40.60	7.3
2	Marmara Sea	18.10.1343	28.30	40.80	7.5
		22.05.1766	29.10	40.80	7.3
		10.07.1894	28.70	40.60	7.0
3	İzmit-Düzce	25.05.1719	29.58	40.66	7.0
		02.09.1754	30.00	40.60	7.2
		22.07.1967	30.69	40.67	7.1
		17.08.1999	29.99	40.70	7.8
		12.11.1999	31.21	40.74	7.2
4	The Southern Branch of NAF				
5	The Southern of Marmara	10.05.1556	28.00	40.30	7.0
		06.03.1737	26.60	39.80	7.2
		28.02.1855	29.10	40.20	7.4
		11.04.1855	29.20	40.20	7.0
		18.03.1953	29.49	40.01	7.2
6	Düzce-Tosya	26.11.1943	33.22	40.97	7.3
		01.02.1944	33.20	41.10	7.3
		16.05.1957	31.00	40.58	7.0
7	Tosya-Erbaa	1510	35.20	40.90	7.0
		20.12.1942	36.35	40.66	7.1
8	Tokat-Erzincan	24.01.1916	36.83	40.27	7.1
		26.12.1939	39.38	39.80	7.9
9	The Eastern of Erzincan				

Table 1. The earthquakes with $M_S \geq 7.0$ in the North Anatolian Fault zone

A complete understanding of the historical and instrumental seismicity, tectonics, geology, paleoseismology, and other neotectonic properties of the considered region are necessary for an ideal delineation of seismic source zones. Several authors used different seismic source zones to study seismic hazard of Turkey (e.g., Alptekin, 1978; Erdik et al., 1999; Kayabalı, 2002; Bayrak et al., 2005; Bayrak et al., 2009). Bayrak et al. (2009) used different 24 source regions considering the different previous zonation studies for modeling of seismic hazard in Turkey and 3 seismic source zones in these 24 regions are related to NAFZ. In this study, we divided NAFZ into nine seismic zones for detailed study as seen in Figure 2. Also, these regions are listed in Table 1. The epicentral distributions of the historical and instrumental earthquakes are shown in Figure 2 on different seismic source zones in NAFZ.

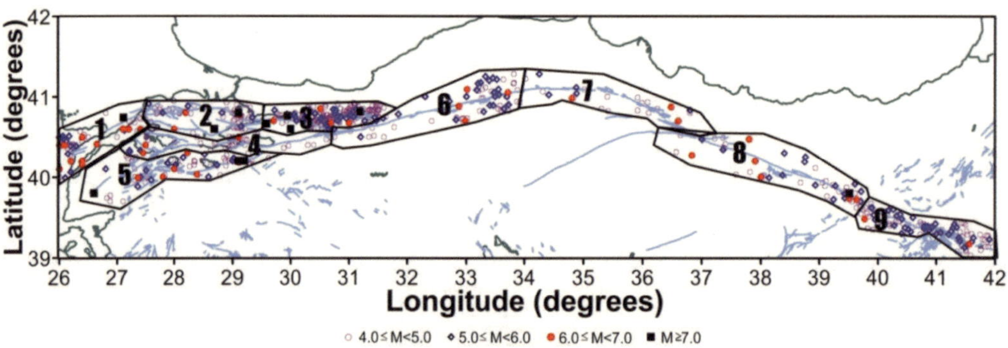

Fig. 2. Earthquake epicenter distribution and different 9 seismic regions in the North Anatolian Fault zone

It is frequently necessary to use a great number of events available for high-quality results in seismicity studies. The fact that magnitude completeness changes with time in most catalogues and usually decreases is well known. So, the minimum magnitude of completeness is an important parameter for seismicity studies. The catalogue used in this study encompasses the time period between BC 1010 and AD 2010 and this is the historical and instrumental parts of the catalogue. For the historical and instrumental periods, results of completeness analysis are made in this study are shown in Table 2. The method used to assess the completeness of the data of this catalogue has been described in the literature (e.g., Tsapanos, 1990; Tsapanos & Papazachos, 1998). The completeness was assessed on the basis of the cumulative frequency distribution of the magnitudes, and of the cumulative frequency distribution of the number of earthquakes with magnitudes larger than a certain value.

Region	Period	Cut off Magnitude
1	1984	$M_S \geq 2.1$
	1976	$M_S \geq 3.2$
	1912	$M_S \geq 4.1$
	1010	$M_S \geq 6.2$

2	1990	$M_S \geq 2.0$
	1975	$M_S \geq 3.0$
	1942	$M_S \geq 4.3$
	1296	$M_S \geq 6.3$
3	1988	$M_S \geq 2.2$
	1974	$M_S \geq 3.0$
	1907	$M_S \geq 4.2$
	1719	$M_S \geq 6.6$
4	1993	$M_S \geq 2.4$
	1976	$M_S \geq 3.5$
	1935	$M_S \geq 4.4$
	1863	$M_S \geq 6.4$
5	1991	$M_S \geq 2.3$
	1977	$M_S \geq 33$
	1907	$M_S \geq 4.3$
	1327	$M_S \geq 6.0$
6	1999	$M_S \geq 2.0$
	1975	$M_S \geq 3.2$
	1918	$M_S \geq 4.0$
7	1996	$M_S \geq 2.0$
	1978	$M_S \geq 3.1$
	1910	$M_S \geq 4.0$
	1510	$M_S \geq 6.5$
8	1995	$M_S \geq 2.1$
	1975	$M_S \geq 3.1$
	1904	$M_S \geq 4.0$
	1165	$M_S \geq 6.5$
9	1995	$M_S \geq 2.2$
	1974	$M_S \geq 3.3$
	1907	$M_S \geq 4.0$

Table 2. The results of completeness analysis for the different seismic regions in the North Anatolian Fault zone

1.3 Theoretical descriptions for the estimation of the earthquake hazard parameters

Estimation of earthquake hazard parameters (maximum regional magnitude, $\hat{M}_{\max}$, earthquake activity rate $\hat{\lambda}$, and b parameter in the Gutenberg-Richter equation) is extended to the case of mixed data containing large historical and recent instrumental events. The method accepts variable quality of complete data in different parts of a catalogue with different threshold magnitude values. The available earthquake catalogues usually contain two types of information: historical observations including major seismic events that occurred over a period of a few hundred years, and instrumental data for relatively short periods of time. The most suitable methods for analyzing the historical part of the catalogue are the extreme distributions, extended to allow varying time intervals from which maximum magnitudes are selected. Assuming that this part of catalogue contains only the largest seismic events, and having the possibility of dividing the catalogue into time intervals of different lengths, it can be in practice to analyze all the historical data. This method of incorporating the incomplete part of the catalogue into the analysis is very far from being optimum, as a great deal of information contained in small shocks is wasted (Kijko & Sellevoll (1989). Another method for estimating the earthquake hazard parameters is to reject the historical parts that are incomplete and to use any standard method for the data from the other complete part of the catalogue. It is obvious that this procedure is also highly ineffective, as the quantitative assessment of recurrence of strong seismic events based on observations over a short period of time is burdened with large errors (Knopoff & Kagan, 1977). This work presents a different approach, making it possible to combine the information contained in the historical part of the catalogue with that contained in the instrumental part of the catalogue.

The area-specific parameters that have to be determined are the area-specific mean seismic activity rate, the Gutenberg-Richter parameter and the maximum regional earthquake magnitude. In the present study, the maximum regional earthquake magnitude, $\hat{M}_{\max}$, is defined as the upper limit of the magnitude for the given seismic tectonic source (Reiter, 1990). The procedure for the evaluation of the maximum regional magnitude $\hat{M}_{\max}$ is based on the equation that compares the maximum *observed* magnitude $M_{\max}^{obs}$ and the maximum *expected* magnitude $E(\hat{M}_{\max} / T)$ of the catalogue in the time interval T (Kijko, 1988, 2004).

For the Gutenberg-Richter relation, a frequency-magnitude relation, the respective CDF (cumulative distribution function) of earthquake magnitudes which are bounded above by $M_{\max}^{obs}$ is given as:

$$F_M(M|M_{\min}, M_{\max}^{obs}) = \begin{cases} 0 & \text{for } M < M_{\min}, \\ \dfrac{1-\exp\left[-\beta(M-M_{\min})\right]}{1-\exp\left[-\beta\left(M_{\max}^{obs}-M_{\min}\right)\right]}, & \text{for } M_{\min} \le M \le M_{\max}^{obs}, \\ 1, & \text{for } M > M_{\max}^{obs}. \end{cases} \quad (1)$$

where $\beta = b\ln(10)$, and b is the parameter of the Gutenberg-Richter relation.

From the condition that compares the largest observed magnitude $M_{\max}^{obs}$ and the maximum expected magnitude during a specified time interval T, we obtain, after integration by parts and simple transformations, the maximum regional magnitude $\hat{M}_{\max}$ (Kijko & Graham, 1998):

$$\hat{M}_{\max} = M_{\max}^{obs} + \int_{M_{\min}}^{M_{\max}^{obs}} \left[F_M(M|M_{\min}, M_{\max}^{obs}) \right]^n dM \qquad (2)$$

Further modifications of estimator (2) are straightforward. For example, following the assumption that the number of earthquakes occurring in unit time within a specified area obeys the Poisson distribution with parameter $\hat{\lambda}$, after replacing n by $\hat{\lambda}T$, estimator (2) becomes:

$$\hat{M}_{\max} = M_{\max}^{obs} + \int_{M_{\min}}^{M_{\max}^{obs}} \left[F_M(M|M_{\min}, M_{\max}^{obs}) \right]^{\hat{\lambda}T} dM \qquad (3)$$

It is not difficult to show that the Gutenberg-Richter-based magnitude CDF (1), the estimator (2) takes the form (Kijko, 1988):

$$\hat{M}_{\max} = M_{\max}^{obs} + \frac{E_1(Tz_2) - E_1(Tz_1)}{\beta \exp(-Tz_2)} + M_{\min} \exp\left(-\hat{\lambda}T\right) \qquad (4)$$

The above estimator of $\hat{M}_{\max}$ for doubly truncated Gutenberg-Richter relation was first obtained by Kijko (1988). The quantities in equation (4) are computed as: $Z_1 = \hat{\lambda}A_1 / (A_1 - A_2)$, $Z_2 = \hat{\lambda}A_2 / (A_1 - A_2)$, $A_1 = \exp(-\beta M_{\min})$, $A_2 = \exp(-\beta M_{\max}^{obs})$ and $E_1(.)$ denotes an exponential integral function (Abramowitz & Stegum, 1970):

$$E_1(z) = \int_z^{\infty} \exp(-\zeta) / \zeta d\zeta \qquad (5)$$

From equations (3) and (4), the approximate variance of the maximum regional magnitude $\hat{M}_{\max}$, is equal to that derived by Kijko (2004):

$$Var\left(\hat{M}_{\max}\right) = \sigma_M^2 + \left[\frac{E_1(Tz_2) - E_1(Tz_1)}{\beta \exp(-Tz_2)} + M_{\min} \exp\left(-\hat{\lambda}T\right) \right]^2 \qquad (6)$$

where σ_M^2 is the variance in the determination of the largest observed magnitude $M_{\max}^{obs}$. It is assumed that the observed (apparent) magnitude is distorted by an observational error, which is distributed normally with a known standard deviation $\hat{\sigma}_M$ (Kijko & Dessokey, 1987).

The parameters β and $\hat{\lambda}$ for a given area are estimated by the maximum likelihood procedure described by Kijko & Sellevoll (1989, 1992). This method allows for the utilization of all available seismicity information, as it makes use of an earthquake catalogue containing both incomplete historical observations and more congruous and complete instrumental data. Periods with gaps in the catalogue can also be taken into account. Equation (4) is applicable even in the cases where the considered magnitude interval, $M_{\max} - M_{\min}$, is short and the number of events small. Further details may be found in Kijko & Sellevoll (1989).

The probability of a particular magnitude being exceeded in the area at least once during a specified time interval t, $H(M|t)$, and its associated mean return period RP can give an indication of the level of seismic hazard in the area.

Under the assumption that earthquakes with magnitude $M \geq M_{min}$, follow a Poisson process with mean activity rate $\hat{\lambda}(M) = \hat{\lambda}_A[1 - F_M(M)]$, where λ_A is the area-specific mean seismic activity rate of earthquakes exceeding M_{min}, we obtain:

$$H(M \mid t) = 1 - \exp\left\{-\hat{\lambda}_A t\left[1 - F_M(M)\right]\right\} \quad (7)$$

and,

$$RP(M) = \frac{1}{\hat{\lambda}_A\left[1 - F_M(M)\right]} \quad (8)$$

In (7), substituting t=1, we obtain the typical seismic hazard curve. Substituting for other values of t (eg. 25, 50, 100, 500), we obtain the probabilities of exceeding within the respective t.

The procedure is applied for the evaluation of the earthquake hazard potential in the NAFZ. For this purpose, we calculated the expected time interval for the occurrence of an earthquake, the most probable maximum magnitude of earthquakes in a given time period and the probability for an earthquake occurrence.

2. Discussion and results

In order to evaluate the earthquake hazard potential in the NAFZ using the historical and instrumental data, NAFZ is divided into different 9 source regions. The earthquake catalogue includes the time period between BC 1010 and AD 2010. The earthquake hazard has been assessed in terms of the maximum regional magnitude $\hat{M}_{max}$, the mean seismic activity rate $\hat{\lambda}$, the mean return period RP, probability for an earthquake occurrence Pr and the b parameter of the magnitude-frequency relationship. The magnitude of the historical period earthquakes usually suffers from large errors which cause many problems in seismic hazard evaluation. In order to overcome this inconsistency Kijko & Sellevoll (1989, 1992) introduced an alternative maximum likelihood approach, utilizing the information derived by any part of a data file including both historical and/or instrumental data. So, the earthquake hazard parameters, namely the maximum regional magnitude, $\hat{M}_{max}$, activity rate of seismic events, $\hat{\lambda}$ and the b-values are obtained through the application of the Kijko & Sellevoll (1989, 1992) approach. These earthquake hazard parameters estimated 9 different regions of NAFZ , listed in Table 3.

Regional variability of the maximum expected magnitudes for different 9 regions in NAFZ is shown in Figure 3. The estimated $\hat{M}_{max}$ values are between 6.09 and 8.03. These values were distributed into four groups, smaller than 6.50, 6.50-7.00, 7.00-7.50 and greater than 7.50. These three groups of $\hat{M}_{max}$ values are shown with different color scales, as shown in Figure 3. The values greater than 7.50 are found in regions 1, 2, 3, 5 and 8. The largest $\hat{M}_{max}$ value is calculated in and around Erzincan (region 8 with $\hat{M}_{max} = 8.03$), where the largest earthquake occurred in the instrumental period in 1939, with a maximum observed magnitude $M_{max}^{obs} = 7.90$. The other largest values of $\hat{M}_{max}$ are calculated in the region between Kocaeli-Düzce (region 3 with $\hat{M}_{max} = 7.97$), where the largest event recently occurred in 1999 with $M_{max}^{obs} = 7.80$; in the west of the Marmara (region 1 with $\hat{M}_{max} = 7.72$), where the largest earthquake of the present century occurred in 1766 with a maximum observed magnitude of $M_{\max}^{obs} = 7.60$; in the Marmara (region 2 with $\hat{M}_{\max} = 7.69$), where the largest earthquakes occurred in 1343 with $M_{\max}^{obs} = 7.50$; in the south of Marmara (region 5 with $\hat{M}_{\max} = 7.52$), where the largest earthquake occurred in 1855, with a maximum observed magnitude $M_{\max}^{obs} = 7.40$. The $\hat{M}_{\max}$ values changing between 7.00 and 7.49 are calculated in the 6 and 7 related to the central part of NAFZ located in Anatolia. The largest earthquakes in these regions are occurred in 1943 ($M_{\max}^{obs} = 7.30$) and 1942 ($M_{\max}^{obs} = 7.10$), respectively. The $\hat{M}_{\max}$ lower than 7.00 are related Region 4 and 9. The largest earthquakes in these regions are lower than 6.40 and occurred in the instrumental period.

Region	$M_{\max}^{obs}$	$\hat{b}$	$\sigma\hat{b}$	$\hat{\lambda}$	$\sigma\hat{\lambda}$	$\hat{M}_{\max}$	$\sigma\hat{M}_{\max}$
1	7.60	0.74	0.02	20.53	3.37	7.72	0.23
2	7.50	1.06	0.02	73.10	13.60	7.69	0.28
3	7.80	0.79	0.01	52.30	8.60	7.97	0.26
4	6.40	0.89	0.02	40.19	7.60	6.54	0.24
5	7.40	0.78	0.02	33.41	5.57	7.52	0.23
6	7.30	0.64	0.02	23.62	3.57	7.48	0.27
7	7.10	0.69	0.03	10.68	1.62	7.22	0.23
8	7.90	0.66	0.02	14.37	1.83	8.03	0.24
9	6.00	0.65	0.02	27.321	4.06	6.09	0.13

Table 3. Earhquake hazard parameters computed from Kijko method for the different seismic regions in the North Anatolian Fault zone. $M_{\max}^{obs}$ is observed maximum magnitude

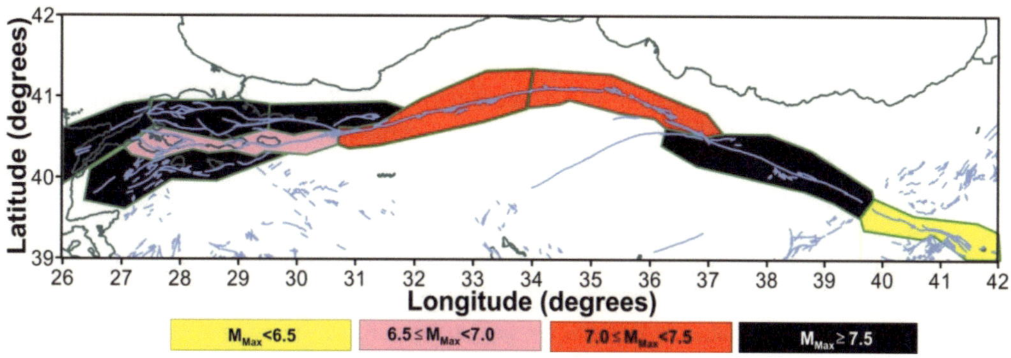

Fig. 3. M_{max} values computed Kijko method for the different 9 seismic regions in the North Anatolian Fault zone

The *b*-values for different 9 regions in NAFZ change between 0.64 and 1.06. Computed *b*-values were distributed into four groups varying lower than 0.70, 0.70-0.79, 0.80-0.89 and larger than 0.90. Figure 4 shows these four groups which is plotted with different color scale. The *b*-values greater than 0.70 are found in regions 1, 2, 3, 4 and 5 which are in and around Marmara Sea. The highest *b*-value in all regions is equal to 1.06 and computed in region 2 covering Marmara Sea. The b-values smaller than 0.70 are calculated in regions 6, 7, 8 and 9. These regions are located to Anatolian part of NAFZ. The lowest *b*-value in all regions is equal to 0.64 and computed in region 6 between Düzce and Tosya.

The *b* parameter is considered to be closely related to tectonic characteristics of a region (Hatzidimitriou et al., 1985; Wang, 1988; Tsapanos, 1990). It seems to be in close connection with the geological age of an area (Allen et al., 1965). Tsapanos (1990) found significantly different *b*-values in east and west Pacific and suggested that this is related to the difference in the mechanical structure of the material in each area, as well as to their tectonic evolution. Manakou &Tsapanos (2000) suggested that low *b*-values are related to low degree of heterogeneity, large strain rate, large velocity of deformation and therefore large fault. The *b* parameter is considered to be closely related to tectonic characteristics of a region (Hatzidimitriou et al., 1985; Wang, 1988; Tsapanos, 1990). Scholz (1968) stated that low *b*-values correspond to great stress and strain in a given region. This can be interpreted that this region is a promised one for an earthquake generation.

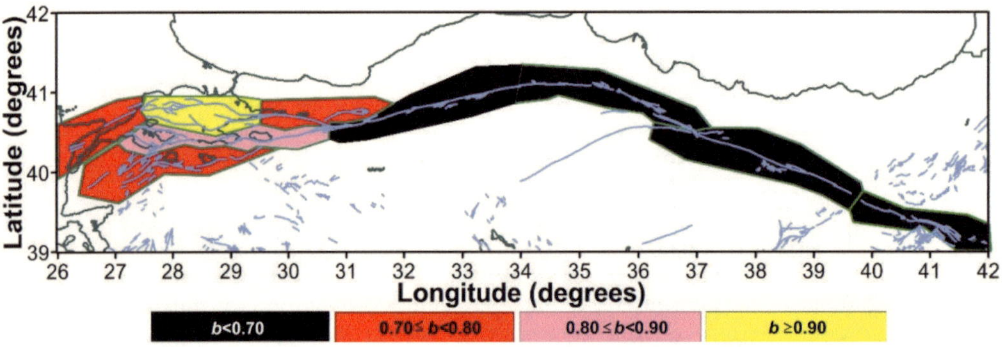

Fig. 4. *b*-values computed Kijko method for the different 9 seismic regions in the North Anatolian Fault zone

It is interesting that the *b*-values greater than 0.70 are related to Marmara region and the *b*-values lower than 0.70 are observed in Anatolian part of NAFZ. The *b* values computed for different 9 regions of NAFZ reflect tectonic characteristics of the different parts of the fault. Along part of Anatolia of the NAFZ, the right-lateral slip has a rate of 24±1 mm/yr (McClusky et al., 2000; Flerit et al., 2004; Reilinger et al., 2006). The regimes of right-lateral slip which has a rate of 18–20 mm/yr and extension which has a rate of 8 mm/yr dominate in the Marmara region (Flerit et al., 2003, 2004). The highest *b*-value in 9 regions of NAFZ is observed in the Marmara Sea named region 2 in this study. In the Marmara Sea, the NAFZ begins to lose its single fault line character and splays into a heterogeneous fault system because of large pull-apart. It is resulted that high *b*-values in regions 1, 2, 3, 4, and 5 located in and around Marmara Sea are related to pull-apart system, high degree of heterogeneity, low strain rate and low velocity of deformation.

The mean return periods (*RP*) of earthquakes, with a certain magnitude, will not be exceeded in any year, are listed in Table 4. Also, the earthquake hazard curves expressed in terms of the mean return period of earthquakes that are expected for the maximum observed magnitudes are shown in Figure 5. The lowest value of the mean return period for $M_S \geq 7.0$ is revealed in region 6, having a value of 32.9 years. The second lower value is founded in region 3 as 35.0 years. The other regions with mean return periods which not exceeded 100 years for magnitude larger than or equal to 7.0 are the seismic regions: 1, 5 and 8. The mean return periods in the other regions 2 and 7 are greater than 100 years and changes between 148.0 and 235.0 years. Since 1999 İzmit and Düzce earhquakes, it has intensively been debated that in the near future an earthquake greater than 7.0 would occur in the Marmara sea covering region 2. The last earthquake with magnitude of 7.0 occurred in this region in 1894. The main return period such an earthquake is found as 148.0 years in this study. According to our results, we would expect that an earthquake greater than 7.0 may take place between 2030 and 2050 in this region taking into account uncertainties. Although the debates are concentrated on the Marmara Sea, the seismic risk of occurrence for an earthquake with $M_S \geq 7.0$ in the region 6 is higher than that of Marmara Sea. We observed that the main return period such an earthquake is only 32.9 years and an earthquake with $M_S \geq 7.0$ in this region has not been occurred such an earthquake since 1957.

	Magnitüd					
Region No	5.0	5.5	6.0	6.5	7.0	7.5
1	3.7	7.2	14.0	28.2	68.3	303.0
2	4.4	9.9	22.5	53.3	148.0	844.0
3	2.0	4.0	7.9	15.9	35.0	101.0
4	5.6	12.4	32.1	83.9		
5	2.9	5.8	11.8	26.0	72.0	2550.0
6	1.7	3.1	5.9	12.1	32.9	
7	5.7	11.0	22.5	53.2	235.0	
8	3.5	6.3	11.4	21.6	44.2	115.0
9	2.1	5.35	47.0			

Table 4. Earthquake return periods for magnitudes between 5.0-7.5 for the different seismic regions in the North Anatolian Fault zone

The probabilities (P_r) are computed for a certain magnitude which will not be exceeded in 50 and 100 years are listed in Table 5. Also, the earthquake hazard curves expressed by the probability expected for earthquakes with the maximum observed magnitudes are plotted and shown in Figure 6.

As seen Table 5 and Figure 6, the probability of occurrence for the earthquakes with $M \geq 7.0$ is highest in region 6 covering area between Düzce and Tosya. In this region the probability of occurrence such an earthquake is 84% in the next 50 years.

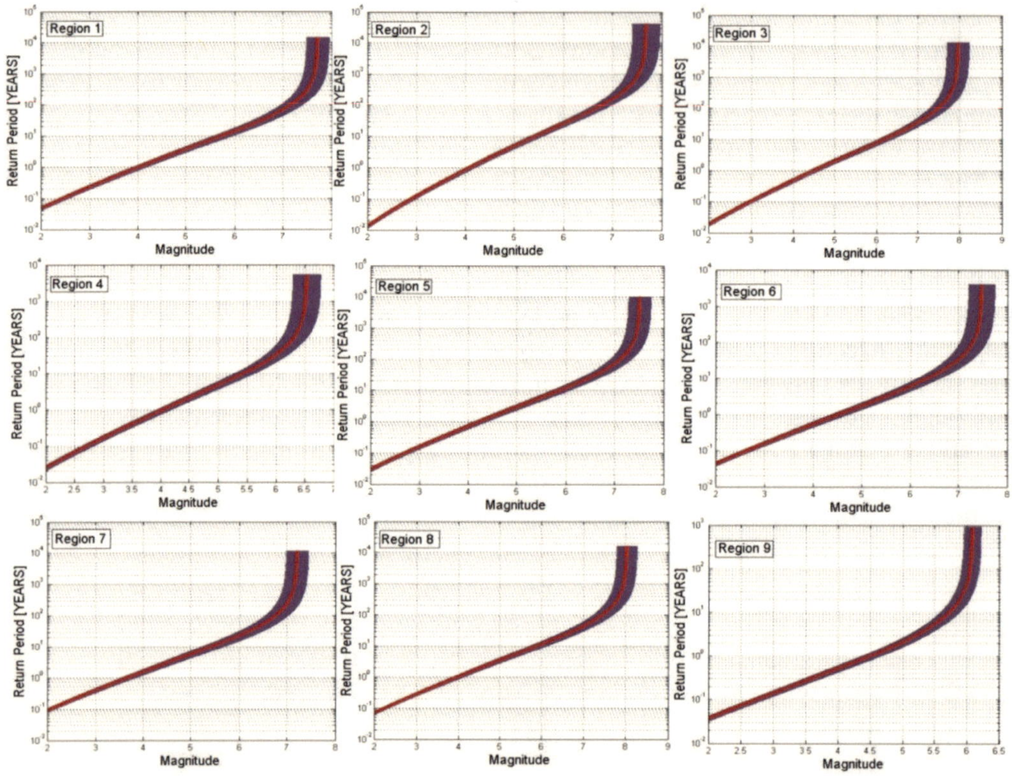

Fig. 5. Return period versus magnitudes for the different 9 seismic regions in the North Anatolian Fault zone

Region No	6.0		6.5		7.0		7.5	
	50	100	50	100	50	100	50	100
1	0.96	1.00	0.81	0.96	0.51	0.75	0.15	0.28
2	0.85	0.97	0.63	0.85	0.27	0.46	0.06	0.11
3	1.00	1.00	0.96	1.00	0.74	0.93	0.46	0.71
4	0.77	0.94	0.08	0.15				
5	0.98	1.00	0.84	0.97	0.49	0.74	0.02	0.04
6	0.99	1.00	0.97	1.00	0.84	0.97		
7	0.88	0.98	0.60	0.83	0.19	0.34		
8	0.98	1.00	0.85	0.97	0.66	0.88	0.35	0.57
9	0.64	0.87						

Table 5. Earthquake probabilities versus magnitudes between 6.0-7.5 for 50 and 100 years for the different seismic regions in the North Anatolian Fault zone

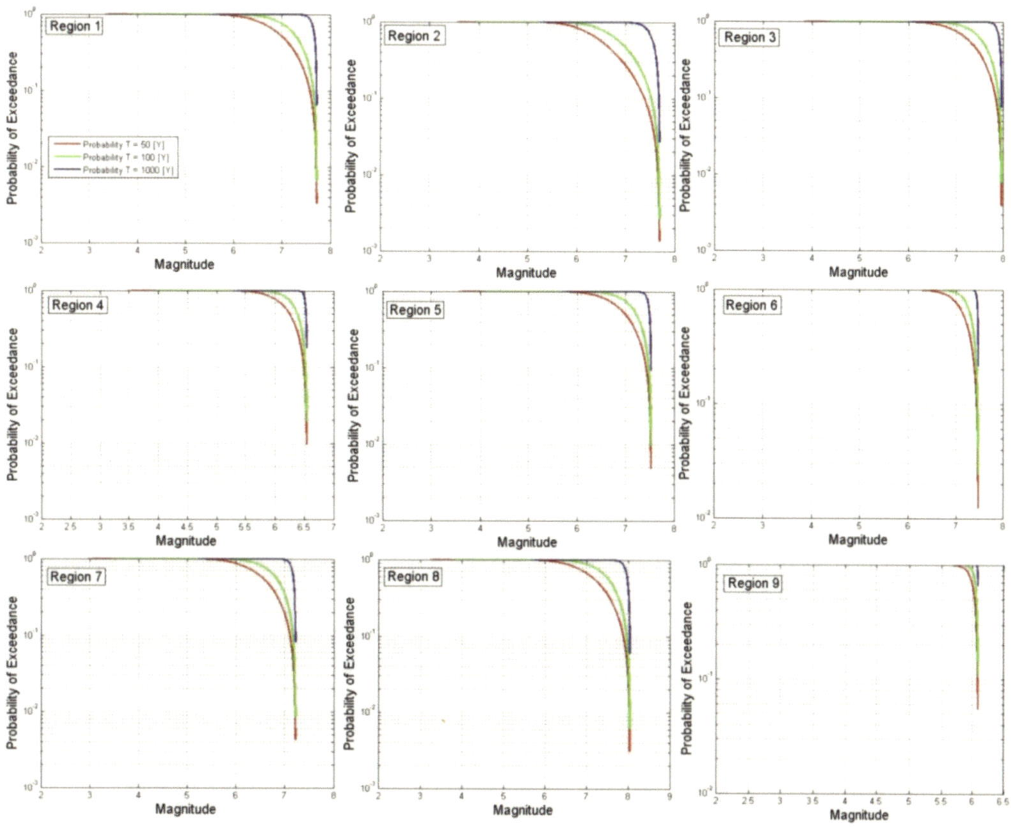

Fig. 6. Earthquake probabilities versus magnitudes for 50, 100 and 1000 years for the different 9 seismic regions in the North Anatolian Fault zone

3. Conclusions

A combination of historical and instrumental earthquake catalogs was used to evaluate earthquake hazard potential for 9 different source regions of NAFZ. For this purpose, the maximum regional magnitude $\hat{M}_{max}$, the mean seismic activity rate $\hat{\lambda}$, the mean return period *RP*, probability for an earthquake occurrence *Pr* and the *b* parameter of the magnitude-frequency relationship are computed.
According to computed $\hat{M}_{max}$ values, the largest earthquake occurrence potential on NAFZ is expected in and around Erzincan. The *b*-values exhibit two different clusters. While high *b*-values are found in regions 1, 2, 3, 4 and 5 which are in and around Marmara Sea due to pull-apart system and high degree of heterogenic faulting, low *b*-values are obtained in regions 6, 7, 8 and 9 which are located in Anatolian part of NAFZ showing homogenous faulting. The lowest *b* value is observed between Düzce and Tosya where the mean return period is lowest and probability of occurrence for the earthquakes with $M_S \geq 7.0$ is highest. Although such a great earthquake is expected in the Marmara Sea in the near future, our results show that a great earthquake occurrence potential between Düzce and Tosya is higher than that of Marmara Sea. So, it should certainly be paid attention to this region. This

study concludes that an earthquake greater than 7.0 may take place the dates between 2030 and 2050 in the Marmara Sea region which has intensively been debated by scientists.

4. Acknowledgment

The authors would like to express their sincere thanks to Prof. Dr. A. Kijko for providing computer program written in Matlab and the suggestions made in order for the paper to be improved. Also, they thank to Prof. Dr. T. M. Tsapanos for training the method used in this research.

5. References

Abramovitz M. & Stegum I. R. (1970). Handbook of Mathematical Functions, 9th Edition. Dover Publ., New York, pp. 1046.

Allen, C. R.; P. S. & Richter, C. F. & Nordquist, J. M. (1965). Relation between seismicity and geological structure in southern California region, *Bull. Seismol. Soc. Am.*, 55, 752-797, 1965.

Ambraseys, N. (2002). The seismic activity of the Marmara Sea region over the last 2000 years. *Bull. Seism. Soc. Am.* 92, 1-18.

Armijo, R.; Meyer, B. & Navarro, S. & King, G. & Barka, A. (2002). Asymmetric slip partitioning in the Sea of Marmara pull-apart: a clue to propagation processes of the North Anatolian Fault? *Terranova* 14, 80–86.

Armijo, R.; Pondard, N. & Meyer, B. & Uçarkus, G. & Mercier de Lépinay, B. & Malavieille, J. & Dominguez, S. & Gustcher, M.A. & Schmidt, S. & Beck, C. & Catagay, N. & Cakir, Z. & Imren, C. & Eris, K. & Natalin, B. & Ozalaybey, S. & Tolun, L. & Lefèvre, I. & Seeber, L. & Gasperini, L. & Rangin, C. & Emre, O. & Sarikavak, K. (2005). Submarine fault scarps in the Sea of Marmara pullapart (North Anatolian Fault): Implications for seismic hazard in Istanbul. *Gcube,*Vol. 6, No.6, Q06009, doi:10.1029/2004GC000896.

Aydın, A. & Nur, A. (1982). Evolution of pull-apart basins and their scale independence, *Tectonics*, 1, 91–105.

Alptekin, Ö. (1978). Magnitude-frequency relationships and deformation release for the earthquakes in and around Turkey. Thesis for Promoting to Associate Professor Level. Karadeniz Technical University, 107 pp. (in Turkish).

Barka, A. & Kadinsky-Cade, K.(1988). Strike-slip fault geometry in Turkey and its influence on earthquake activitiy. *Tectonics*, 7, 663-684.

Barka, A. (1996). Slip distribution along the North Anatolian Fault associated with the large earthquakes of the period 1939 to 1967. *Bull. Seismol. Soc. Am.* 86:1238–1254.

Bayrak, Y.; Yılmaztürk, A. & Öztürk, S. (2005). Relationships between fundamental seismic hazard parameters for the different source regions in Turkey. *Natural Hazards* 36: 445-462.

Bayrak, Y.; Öztürk, S. & Koravos, G.Ch. & Leventakis, G.A. & Tsapanos, T.M. (2008a). Seismicity assessment for the different regions in and around Turkey based on instrumental data: Gumbel first asymptotic distribution and Gutenberg-Richter cumulative frequency law. *Natural Hazards and Earth System Sciences* 8: 109-122.

Bayrak, Y.; Öztürk, S. & Çınar, H.& Koravos, G.Ch. & Tsapanos, T.M. (2008b). Regional variation of the ω-upper bound magnitude of GIII distribution in and around

Turkey: Tectonic implications for earthquake hazards. *Pure and Applied Geophysics* 165: 1367-1390.

Bayrak, Y.; Öztürk, S. & Çınar, H. & Kalafat, D. & Tsapanos, T.M. & Koravos, G.Ch. & Leventakis, G.A. (2009). Estimating earthquake hazard parameters from instrumental data for different regions in and around Turkey. *Engineering Geology* 105: 200-210.

Becel, A.; Laigle,M. & de Vood, B. & Hirn, A. & Taymaz, T.& Galve, A.& Shimamura, H.& Murai, Y. & Lepine, J-C. & Sapine, M. & Özalaybey, S. (2009). Moho, crustal architecture and deep deformation under the North Marmara trough, from the SEISMARMARA Leg1 offshore-onshore reflection-refraction survey, *Tectonophysics,* 467, 1–21.

Bozkurt, E. (2001). Neotectonics of Turkey- a Synthesis. *Geodinamica Acta,* 14, 3-30.,

Canıtez, N. & Üçer, B. (1967). Computer determinations for the fault plane solutions in and near Anatolia. *Tectonophysics,* 4, 235-244.

Carton, H. (2005). Three-dimensional seismic images of an active-pull apart basin: the Cinarcik Basin along the North Anatolian Fault. *Lithos Sci* Rep 7:63–72

Carton, H.; Singh, S.C. & Hirn, A. & Bazin, S. & de Voogd, B. & Vigner, A. & RIcolleau, A. & Cetin, S. & Oçakoglu, N. & Karakoç, F.& Sevilgen, V.(2007). Seismic imaging of the three-dimensional architecture of the Cinarcik Basin along the North Anatolian Fault. *J.Geophys. Res.* 112, 1–17. doi:10.1029/2006JB004548 B06101.

Comninakis, PE. (1975). Contribution to the study of the seismicity of Greece. PhD Thesis, University of Athens, 110pp.

Demirbağ, E.; Rangin, C. & Le Pichon, X. & Sengor, A.M.C. (2003). Investigation of the tectonics of the Main Marmara Fault by means of deep-towed seismic data. *Tectonophysics* 361, 1–19.

Erdik, M. ; Alpay, BY. & Onur, T. & Sesetyan, K. & Birgoren, G. (1999). Assessment of earthquake hazard in Turkey and neighboring regions. *Annali di Geofisica* 42: 1125-1138.

Erdik, M.; Demircioğlu, M. & Sesetyan, K. & Durukal, E. & Siyahi, B. (2004). Earthquake hazard in Marmara region, Turkey. *Soil Dynamics and Earthquake Engineering,* 24, 605–631.

Eyidoğan, H.; Utku, Z. & Güçlü, U. & Değirmenci, E. (1991). Türkiye Büyük Depremleri Makro-Sismik Rehberi (1900–1988). İstanbul Tek.Univ. Maden Fak., Jeofizik Mühendisliği Bölümü, [II]+198 pp.

Flerit, F.; Armijo, R. & King, G.C.P. & Meyer, B. & Barka, A. (2003). Slip partitioning in the Sea of Marmara pull-apart determined from GPS velocity vectors. *Geophys. J. Int.* 154, 1–7.

Flerit,F.; Armijo, R. & King, G. & Meyer, B. (2004). The Mechanical interaction between the propagating North Anatolian Fault and the back-arc extension in the Aegean, *Earth Planet. Sci. Lett.,* 224, 347–362.

Gürbüz, C.; Aktar, M. & Eyidoğan, H. & Cisternas, A. & Haessler, H. & Barka, A. & Ergin, M. & Türkelli, N. & Polat, O. & Üçer, S.B. & Kuleli, S. & Barış, S. & Kaypak, B. & Bekler, T. & Zor, E. & Biçmen, F. & Yörük, A. (2000). The seismotectonics of the Marmara region (Turkey): results from a microseismic experiment. *Tectonophysics* 316, 1 - 17.

Hamdache, M.; Bezzeghoud, M. & Mokrane, A. (1998). Estimation of seismic hazard parameters in the northern part of Algeria. *Pure and Applied Geophysics* 151: 101-117.

Hatzidimitriou, P. M.; Papadimitriou, E. E. & Mountrakis, D. M. & Papazachos, B. C. (1985). The seismic parameter b of the frequency-magnitude relation and its association with geological zones in the area of Greece, *Tectonophysics*, 120, 141-151, 1985.

İmren, C.; Le Pichon, X. & Rangin, C. & Demirbağ, E. & Ecevitoğlu, B. & Gorur, N. (2001). The North Anatolian fault within the Sea of Marmara: A new interpretation based on multi-channel seismic and multi-beam bathymetry data, *Earth Planet. Sci. Lett.*, 186, 143- 158,

Jackson, J. & McKenzie, D. (1984). Active Tectonics of the Alpine-Himalayan Belt between western Turkey and Pakistan. *Geophys. J.R. Astr.Soc.*, 77,185- 264.

Koravos, G.; Main, I. & Tsapanos, TM. & Musson, RMW. (2003). Maximum earthquake magnitudes in the Aegean area constrained by tectonic moment release rates. *Geophysical Journal International* 152: 94-112.

Kasapoglu, EK. & Toksöz, M.N. (1983). Tectonic consequences of the collision of the Arabian and Eurasian plates: finite element models. *Tectonophysics* 100: 71–95.

Kayabalı, K. (2002). Modeling of seismic hazard for Turkey using the recent neotectonic data. *Engineering Geology* 63, 221-232.

Kayabali, K. & Akin, M. (2003). Seismic hazard map of Turkey using the deterministic approach. *Engineering Geology* 69: 127-137.

Koçyiğit, A.; Rojay, B. & Cihan, M. & Özacar, A. (2001). The June 6, 2000, Orta (Çankırı, Turkey) earthquake: sourced from a new antithetic sinistral strike-slip structure of the North Anatolian Fault System, the Dodurga Fault Zone. *Turkish Journal of Earth Sciences*, 10, 69–82.

Kijko, A. & Dessokey, M.M. (1987). Application of extreme magnitude distributions to incomplete earthquake files. *Bulletin of the Seismological Society of America* 77: 1429-1436.

Kijko, A. (1988). Maximum likelihood estimation of Gutenberg-Richter b parameter for uncertain magnitudes values. *Pure and Applied Geophysics* 127: 573-579.

Kijko, A. & Sellevoll, M.A. (1989). Estimation of earthquake hazard parameters from incomplete data files. Part I. Utilization of extreme and complete catalogs with different threshold magnitudes. *Bulletin of the Seismological Society of America* 79: 645-654.

Kijko, A. & Sellevoll, M.A. (1992). Estimation of earthquake hazard parameters from incomplete data files, Part II: Incorporation of magnitude heterogeneity. *Bulletin of the Seismological Society of America* 82: 120-134.

Kijko, A. & Graham, G. (1998). Parametric-historic procedure for probabilistic seismic hazard analysis, Part I: Estimation of maximum regional magnitude M_{max}. *Pure and Applied Geophysics* 152, 413-442.

Kijko, A. (2004). Estimation of the maximum earthquake magnitude M_{max}. *Pure and Applied Geophysics* 161: 1-27.

Knopoff, L. & Kagan, Y. (1977). Analysis of the theory of extremes as applied to earthquake problems. *Journal of Geophysical Research* 82: 5647–5657.

Laigle, M.; Bécel, A. & de Voogd, B. & Hirn, A. & Taymaz, T. & Ozalaybey, S & the Members of the SEISMARMARA Leg1.(2008). A first deep seismic survey in the Sea of Marmara:whole crust and deep basins. *Earth Planet. Sci. Lett.* 270, 168–179.

Le Pichon, X.; Taymaz, T. & Şengör, A.M.C. (1999). The Marmara Fault and the future Istanbul earthquake, in International Conference on the Kocaeli Earthquake, 17 August 1999, edited by M. Karaca and D. N. Ural, pp. 41–54, Istanbul Tech. Univ. Press House, Istanbul, Turkey.

Le Pichon, X.; Sengor, A.M. C. & Demirbag, E. & Rangin, C. & Imren, C. & Armijo, R. & Görür, N. & Çağatay, N. & Mercier de Le´pinay, B. & Meyer, B. & Saatcilar, R. & Tok, B. (2001). The active Main Marmara fault, *Earth Planet.Sci. Lett.*, 192, 595–616.

Le Pichon, X.; Chamot-Rooke, N. & Rangin, C. & Sengör, A.M.C. (2003). The North Anatolian fault in the Sea of Marmara. *J. Geophys. Res.* 108 (B4), 2179.

Makropouos, K.C. (1978). The statistics of large earthquake magnitude and an evaluation of Greek seismicity, PhD Thesis, University of Edinburgh, Edinburgh, 193 pp.

Manakou, MV. & Tsapanos, T.M. (2000). Seismicity and seismic hazard parameters evaluation in the island of Crete and the surrounding area inferred from mixed files. *Tectonophysics* 321: 157-178.

Mäntyniemi, P.; Tsapanos, T.M. & Kijko, A. (2004). An estimate of probabilistic seismic hazard for five cities in Greece by using the parametric-historic procedure. *Engineering Geology* 72: 217-231.

Mc Clusky, S.; Balassanian, S. & Barka, A. & Demir, C. & Ergintav, S. & Georgiev, I. & Gurkan, O. & Hamburger, M. & Hurst, K. & Kahle, H. & Kastens, K. & Kekelidze, G. & King, R. & Kotzev, V. & Lenk, O. & Mahmoud, S. & Mishin, A. & Nadariya, M. & Ouzounis, A. & Paradissis, D. & Peter, Y. & Prilepin, M. & Reilinger, R. & Sanli, I. & Seeger, H. & Tealeb, A. & Toksoz, M.N. & Veis, G. (2000). Global Positioning system constraints on plate kinematics and dynamics in the eastern Mediterranean and Caucasus. *Journal of Geophysical Research*, 105, B3, 5695-5719.

McKenzie, D.P. (1972). Active tectonics of the Mediterranean region, *Geophys. J.R.Astron. Soc.*, 30, 109-185.

Parsons, T. (2004). Recalculated probability of M ≥7 earthquakes beneath the Sea of Marmara, Turkey. *Journal of Geophysical Research*, vol. 109, B05304, doi: 10. 1029 / 2003 JB 002667.

Reilinger, R.E.; McCLusky, S. & Oral, M.B. & King, R.W. & Toksoz, M.N. (1997). Global Positioning System measurements of present-day crustal movements in the Arabia-Africa- Eurasia plate collision zone. *Journal of Geophysical Research*,102:9983–9999.

Reilinger,R.; McClusky,S. & Vernant,P. & Lawrence,S. & Ergintav,S. & Cakmak,R. & Özener,H. & Kadirov,F. & Guliev,I. & Stepanyan,R. & Nadariya,M.& Hahubia,G. & Mahmoud,S. & Sakr, K. & ArRajehi,A. & Paradissis, D. & Al-Aydrus, A. & Prilepin, M. & Guseva,T. & Evren, E. & Dmitrotsa, A. & Filikov, S.V. & Gomez, F. & Al-Ghazzi, R. & Karam,G. (2006). GPS constraints on continental deformation in the Africa-Arabia- Eurasia continental collision zone and implications for the dynamics of plate interactions. *Journal of Geophysical Researchs.* VOL. 111, B0541, doi:10.1029/2005JB004051.

Ryall, A.S. & Van Wormer, JD. (1980). Estimation of maximum magnitude and recommended seismic zone changes in the western great basin. *Bulletin of the Seismological Society of America* 70: 1573-1581.

Scholz, C.H. (1968). The frequency-magnitude relation of microfracturing in rock and its relation to earthquakes, *Bull. Seism. Soc. Am.*, 58, 399-415, 1968.

Şaroğlu, F.; Emre, O. & Kuşçu, I. (1992). Active fault map of Turkey. General Directorate of Mineral Research and Exploration, Ankara, Turkey

Şengör, A.M.C. (1979). The North Anatolian transform fault: its age, offset and tectonic significance. *Journal of the Geological Society of London* 136, 269-282.

Şengör, A.M.C.; Görür, N. & Şaroğlu, F. (1985). Strike-slip faulting and related basin formation in zones of tectonic escape: Turkey as a case study, *The Society of Economic Paleontologists and Mineralogists*, Special Publication, 37, 227-264.

Şengör, A.M.C.; Özeren, S. & Genc¸ T. & Zor E. (2003). East Anatolian high plateau as a mantle-supported, north-south shortened domal structure. *Geophys. Res. Lett.* 30: 8045.

Şengör, A.M.C.; Tüysüz, O. & İmren, C. & Sakınç, M. & Eyidoğan, H. & Görür, N. & Le Pichon, X. & Rangin, C. (2005). The North Anatolian Fault: A New Look. *Annual Review of Earth and Planetary Sciences*, 33, 37-112.

Tsapanos, T.M. (1990). b-values of two tectonic parts in the circum-Pacific belt. *Pure and Applied Geophysics* 134: 229-242.

Tsapanos, T.M. & Papazachos, BC. (1998). Geographical and vertical variation of the earth's seismicity. *Journal of Seismology* 2: 183-1998.

Tsapanos, T.M. (2001). Evaluation of seismic hazard parameters for selected regions of the world: the maximum regional magnitude. *Annali di Geofisica* 44: 69-79.

Tsapanos, T.M. (2003). Appraisal of seismic hazard parameters for the seismic regions of the east Circum-Pasific belt inferred from a Bayesian approach. *Natural Hazards* 30: 59-78.

Wang, J. H. (1988). b-values of shallow earthquakes in Taiwan, *Bull. Seismol. Soc. Am.*, 78, 1243-1254, 1988.

Yegulalp, TM. & Kuo, JT. (1974). Statistical prediction of occurrence of maximum magnitude earthquakes. *Bulletin of the Seismological Society of America* 64: 393-414.

11

Tectonic Escape Mechanism in the Crustal Evolution of Eastern Anatolian Region (Turkey)

Ömer Elitok[1] and M. Nuri Dolmaz[2]
[1]*Süleyman Demirel University, Department of Geological Engineering, Isparta,*
[2]*Süleyman Demirel University, Department of Geophysical Engineering, Isparta*
Turkey

1. Introduction

The Neotectonic evolution of Turkey has been dominated by the collision of the African and Arabian plates with the Eurasian plate along Hellenic arc to the west and the Bitlis-Zagros suture zone to the east. The Eastern Anatolian Contractional Province (EACP) including the Eastern Anatolian High Plateau (EAHP) and the Bitlis-Pötürge Thrust Zone (BPTZ) consists of an amalgamation of fragments of oceanic and continental crusts that squezzed and shortened between the Arabian and Eurasian plates. This collisional and contractional zone is being accompanied by the tectonic escape of most of the Anatolian plate to the west by major strike-slip faulting on the right-lateral North Anatolian Transform Fault Zone (NATFZ) and left-lateral East Anatolian Transform Fault Zone (EATFZ) which meet at Karlıova forming an east-pointing cusp (Figure 1). The present-day crust in the area between the easternmost part of the Anatolian plate and the Arabian Foreland gets thinner from north (ca 44 km) to south (ca 36 km) relative to its eastern (EAHP) and western sides (central Anatolian region). The contraction and thickening of the crust to ca 50-52 km (Dewey et al., 1986; Pearce et al., 1990) in the collisional zone has been accompanied by the tectonic escape of most of the Anatolian crustal block (Anatolian wedge in Dewey et al., 1986; Anatolian platelet in Koçyiğit et al., 2001; Anatolian plate in Yılmaz et al., 1998; Anatolian block in Lyberis et al., 1992) to the west-southwest towards the Aegean-Cyprean arc system by major strike-slip faulting on the right-lateral North Anatolian Transform Fault Zone (NATFZ) and left-lateral East Anatolian Transform Fault Zone (EATFZ) (Dewey et al., 1986; Şengör, 1979; Şengör and Yılmaz, 1981 and references therein; Şengör et al., 1985; Dilek and Moores, 1990; Tatar et al., 1996; Hubert-Ferrari et al., 2003; Allen et al., 2004; Barazangi et al., 2006) at a slip rate of 24±1 and 9±1 mm/yr, respectively (McClusky et al., 2000).
Major tectonic structures such as the Aegean-Cyprean arc system, NATFZ, EATFZ, and Dead Sea Fault Zone (DSFZ) have been played a key role on the neotectonic evolution of the eastern Mediterranean region especially in Late Cenozoic. The westward escape of the Anatolian plate has been accompanied by counterclockwise rotation as well as lateral translation deduced from GPS and paleomagnetic data (Tatar et al., 1996; Reilinger et al., 1997; Platzman et al., 1998; Gürer and Gürer, 1999; McClusky et al., 2000; Özçep and Orbay, 2000; Bozkurt, 2001; Büyüksaraç, 2007). Therefore, the timing of collision-induced westward extrusion of the Anatolian plate has been widely accepted as the commencement of the Neotectonic history of Turkey (Şengör and Yılmaz, 1981; Koçyiğit et al., 2001). Koçyiğit et al.

(2001) investigated the eastern Anatolian compressional tectonic regime in two main stages: the compressional-contractional and compressional-extensional tectonic regimes corresponding to the pre-tectonic escape period and the syn-tectonic escape period, respectively. However, the driving force causing the westward escape of the Anatolian plate is still under debate, e.g. orogenic collapse or gravity spreading following orogenic crustal shortening and thickening within the Anatolian plate (Seyitoğlu and Scott, 1996; Gautier et al., 1999; Dhont et al., 2006), mantle flow driven lithospheric extrusion (Faccenna et al., 2006; Çoban, 2007).

Eastern Anatolia is dissected into three main lithospheric portions (the Eastern Anatolian High Plateau-EAHP, the Anatolian block, and the Arabian Foreland) by the major tectonic structures (Bitlis-Pötürge Suture Zone, NATFZ and EATFZ). The crust in the easternmost part of the Anatolian plate bounded by the NATFZ and the EATFZ, which corresponds to inner east Anatolia in Büyüksaraç (2007), mainly gets thinner from north (ca 43-44 km) to south (ca 36-37 km) relative to its eastern (EAHP) and western sides (central Anatolian region) (Figure 2), as proposed by Gök et al. (2007) for the east-southeast Anatolia. On the other hand, this thinner crustal terrain is characterized by very low Pn velocities (< 7.8 km/s) and high Sn attenuation which indicate existence of partially molten to eroded mantle lid or occurrence of asthenospheric mantle beneath the crust (Al-Lazki et al., 2003; Gök et al., 2003). This chapter is based on mainly the study of Elitok and Dolmaz (2008) who investigated the geodynamic evolution of crustal thinning in the area between the easternmost part of the Anatolian plate and the Arabian Foreland using the geological and geophysical data. They outlined Curie Point Depth (CPD) map of an area including the eastern Anatolia and the Arabian Foreland, and redrawn the crustal thickness map of the same area on the basis of the estimations of Gök et al. (2007) which are based on receiver functions from recordings of 29 broadband seismological stations.

2. Geological setting

On the basis of the nature of underlying basement, the eastern Anatolian contractional zone can be separated into four main crustal terrains: from north to south i) the Pontide Belt, ii) the Eastern Anatolian Accretionary Complex (EAAC), iii) the Bitlis-Pötürge Thrust Zone (BPTZ), and iv) the Arabian Foreland. Most of the eastern Anatolian region and Arabian Foreland are covered by compositionally different and young volcanic units that erupted during the Neogene and Quaternary times (Yazgan, 1981; Pearce et al., 1990; Yılmaz et al., 1998; Keskin, 2003; Keskin et al., 2006).

The Arabian Foreland in the southeastern Turkey is made up of an almost continuous stratigraphic sequence of mainly shelf sediments in Early Paleozoic to Miocene age resting on a Precambrian basement (Hall, 1976; Yazgan, 1981 and references therein; Pearce et al., 1990; Yılmaz, 1993; Yiğitbaş et al., 1993; Yiğitbaş and Yılmaz, 1996 and references therein). The foreland volcanism is dominated by basaltic shield and fissure eruptions of transitional tholeiitic-alkaline composition (Pearce et al., 1990). Karacadağ volcano, a low shield volcano situated on the Arabian Foreland, erupted since the Pliocene along a N-S trending set of fissures and craters, spatially associated with the nearby Akçakale graben (Figure 1) (Şengör and Yılmaz, 1981; Şengör et al., 1985 and references therein; Pearce et al., 1990). Pearce et al. (1990) interpreted that small degrees of streching might have caused melting of metasomatized lithosphere by perturbation of the geotherm by heat from upwelling of hot asthenosphere. They also explained that Akçakale graben, a small rift structure with N-S normal faults and located to the SW of Karacadağ, is another indicator of E-W extension, although this rift appears to act, at least in part, as a transfer structure between some of the outer thrust of the Arabian Foreland sedimentary sequence.

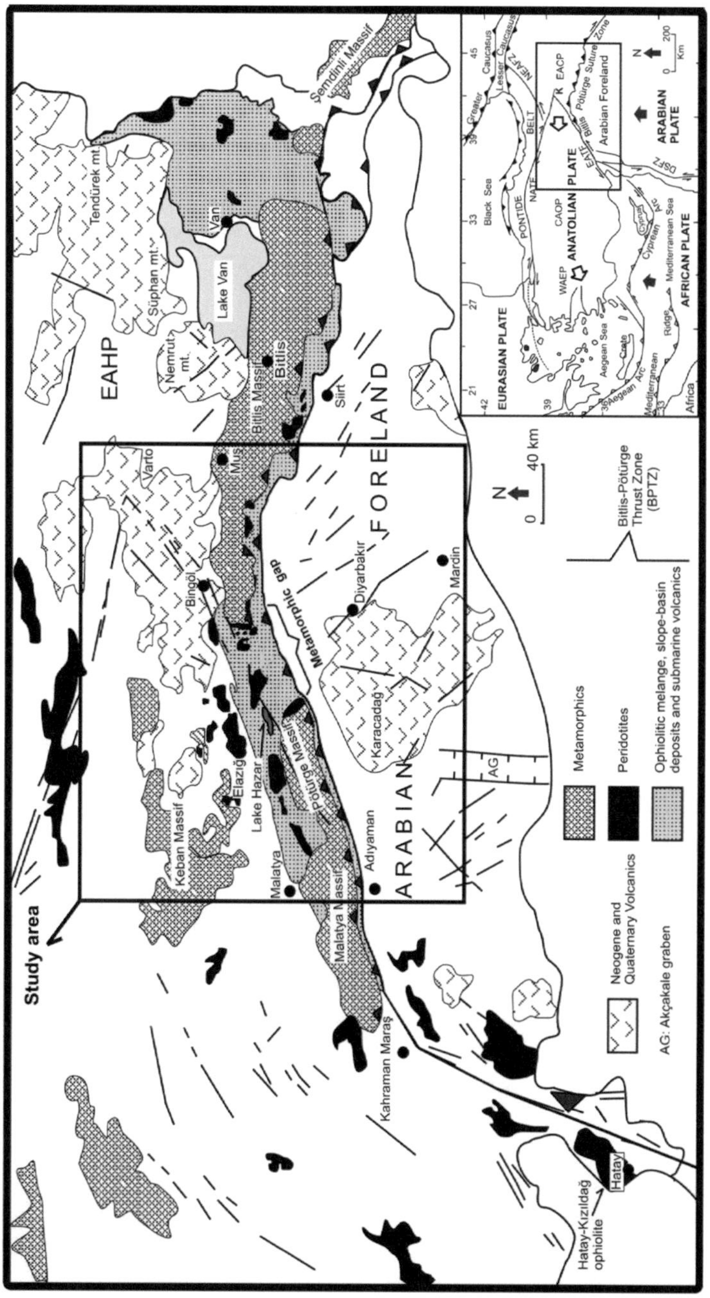

Fig. 1. Simplified geological map of the eastern Anatolian region (from Elitok and Dolmaz, 2008) and the inset location map from Bozkurt (2001), EACP: Eastern Anatolian Contractional Province, EAHP: Eastern Anatolian High Plateau, CAOP: Central Anatolian Ova Province, WAEP: Western Anatolian Extensional Province

The autochthonous Arabian Foreland is tectonically overlain in the north generally by a nappe package consisting of metamorphic massifs at the top and imbricated ophiolitic rock associations laying below the massifs (Hall, 1976; Yazgan, 1981; Yılmaz, 1993; Yiğitbaş et al., 1993; Yılmaz and Yıldırım, 1996; Yiğitbaş and Yılmaz, 1996). Yılmaz (1993) separated the nappe successions running along the BPTZ into two main groups: i) the nappe zone consisting of two nappe stacks (the upper nappe represented mainly by Bitlis and Pötürge metamorphics; the lower nappe is characterized by the slices of polyphase metamorphic ophiolitic assemblage and the Maden Group), ii) the imbricated zone sandwiched between the Arabian Platform and the nappe zone. In this study we used "the Bitlis-Pötürge Thrust Zone-BPTZ" for the nappe successions including both the nappe zone and the imbrication zone of Yılmaz (1993). In the BPTZ, the metamorphic massifs display an arc-shaped geometry in parallel to Arabian promontory (Figure 1). The Bitlis and Pötürge metamorphic massifs form the highest tectonic units of the Southeast Anatolian orogenic belt (Yiğitbaş and Yılmaz, 1996). However, there is no direct link through a gateway between these metamorphic massifs. This gateway covered by accretionary materials and Tertiary sediments is termed "metamorphic gap" in this study (Figure 1). Consistently, Hall (1976) interpreted that the Bitlis Massif is thrust southward over the ophiolite-flysch complex, whereas in the east and west they are structurally below the same complex. Presumably, the Bitlis and Pötürge massifs might have been dismembered during the indentation of the Arabian plate to the Eurasian plate since the Miocene time. In the metamorphic gap area, even if the metamorphic rocks take place within the accretionary materials and Tertiary sediments, most likely they might be the blocks broken off from the pre-collisional unique Bitlis-Pötürge metamorphic body. The Keban metamorphic rocks and Upper Triassic-Upper Cretaceous platform carbonates of Munzur Mountains are located to the north of the Pötürge Massif.

3. Crustal structure

Based on the Eastern Turkey Seismic Experiment (ETSE) project (Sandvol et al., 2003), crustal thickness of east Turkey has been estimated by Zor et al. (2003) and Gök et al. (2007) using receiver functions obtained from teleseismic recordings of 29 broadband seismological stations. Zor et al. (2003) and Şengör et al. (2003) also contoured the crustal thickness of east Anatolia. Elitok and Dolmaz (2008) produced the crustal thickness map of the eastern Anatolian region contouring the moho depth estimates from Figure 7 in Gök et al. (2007) using the standard gridding routine (Figure 2). In the map, the crustal thickness in the easternmost part of the Anatolian plate gets thinner from the NAFZ in the north (ca 44 km) to the Arabian Foreland in the south (ca 36 km) through the metamorphic gap between the Bitlis Massif and the Pötürge Massif relative to its eastern (EAHP) and western sides (central Anatolian region) (Figure 2). However, the thickness of the crust reaches to ca 45-46 km in the north of the Bitlis Massif and the Pötürge Massif. The crust sharply thickens in a short interval from 40 to 46 km in the northern side of the Bitlis Massif (Figure 2). Gök et al. (2007) estimated the average crustal thickness as 36 km in the northern margin of the Arabian Plate, 44 km in the Anatolian Block and 48 km in the Anatolian Plateau. Although Pn velocities are high in the south of the Bitlis Massif implying a stable mantle lid region beneath the northernmost part of the Arabian plate, it sharply stops along the Bitlis Suture Zone (Al-Lazki et al., 2003).

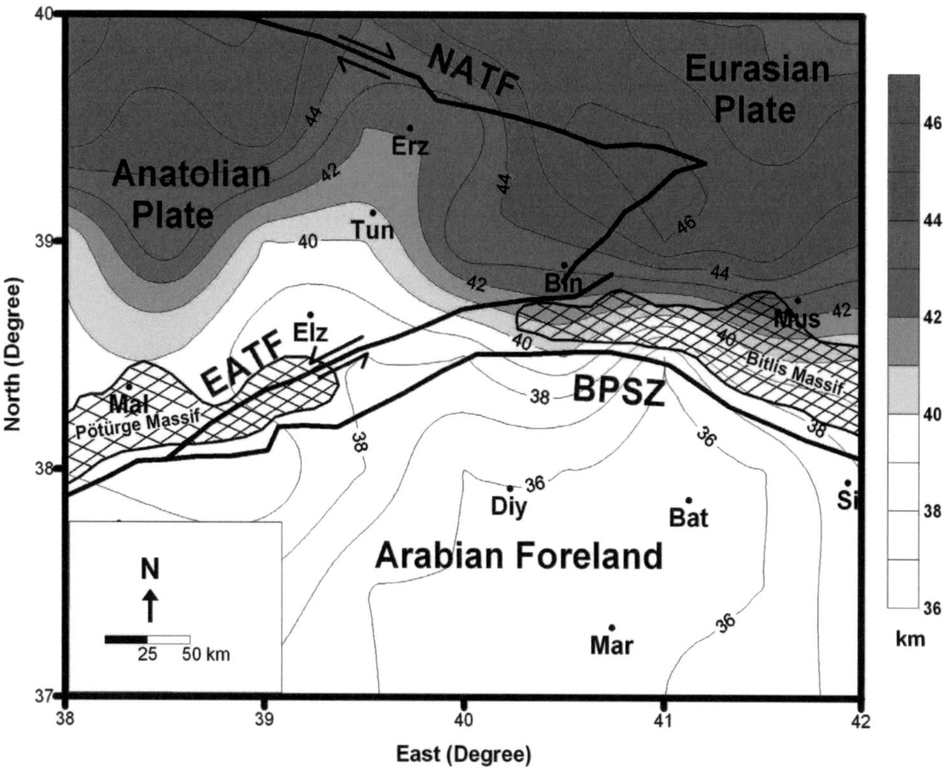

Fig. 2. Map showing the moho depth variation in eastern Turkey (from Elitok and Dolmaz, 2008). Contour interval: 1 km. NATF: North Anatolian Transform Fault; EATF: East Anatolian Transform Fault; BPSZ: Bitlis Pötürge Suture Zone; Mal: Malatya; Ady: Adıyaman; Elz: Elazığ; Tun: Tunceli; Erz: Erzincan; Bin: Bingöl; Diy: Diyarbakır; Mar: Mardin; Bat: Batman; Si: Siirt

4. Curie point depth estimates

The Curie Point Depth (CPD) is known as the depth at which the dominant magnetic mineral (ca 580 °C for magnetite) in the crust passes from a ferromagnetic state to a paramagnetic state under the effect of increasing temperature (Nagata, 1961). Beneath the Curie point depth the lithosphere shows virtually non-magnetic properties. Therefore, the basal depth of a magnetic source from aeromagnetic data is considered to be the CPD. Elitok and Dolmaz (2008) implied that the CPD is in close relationship with the crustal thickness, crust-hot asthenospheric mantle interactions, magmatic events, thermal structure of the crust and hence brittle upper crust-ductile lower crust transition zone. In capturing the estimates of the bottom of the deepest magnetic layer (Elitok and Dolmaz, 2008), the centroid method (Bhattacharyya and Leu, 1977; Okubo et al., 1985; Tanaka et al., 1999) was used. Moreover, the method of Okubo et al. (1985) has an advantage since large grid sizes of data are not necessary, whereas other studies require large grid sizes. The CPD may reflect the broad average temperature and regional crustal thermal structure (Tanaka and Ishikawa, 2005). It is important that the relative relations of estimated depths in a region are important rather than their numerical values.

Elitok and Dolmaz (2008) performed two dimensional spectral analyses using divided subregions of size 90 x 90 km overlapped 50% with the adjacent subregion. The average depth to the bottom of the deepest magnetic layer, the CPD (z_b), is derived for an interested subregion by means of the Equation $z_b = 2z_0 - z_t$. The top of the magnetic sources was found to be ca 2.31 km b.s.l. for the 90 x 90 km subregions. On the other hand, the CPD vary between 11.05 and 23.57 km b.s.l. Figure 3 shows the CPD map of the study area (Elitok and Dolmaz, 2008) which is constructed from the CPD estimates by using the standard gridding routine. Three cross sections (Profiles 1, 2 and 3) are produced from the CPD to show variation in depth of the 580 °C isotherm across the region and to compare the thermal structure of the crust with the topography and geological data of the region. Figures 6a, 6b, and 6c show cross-sections including the topography, CPD and moho dephts.

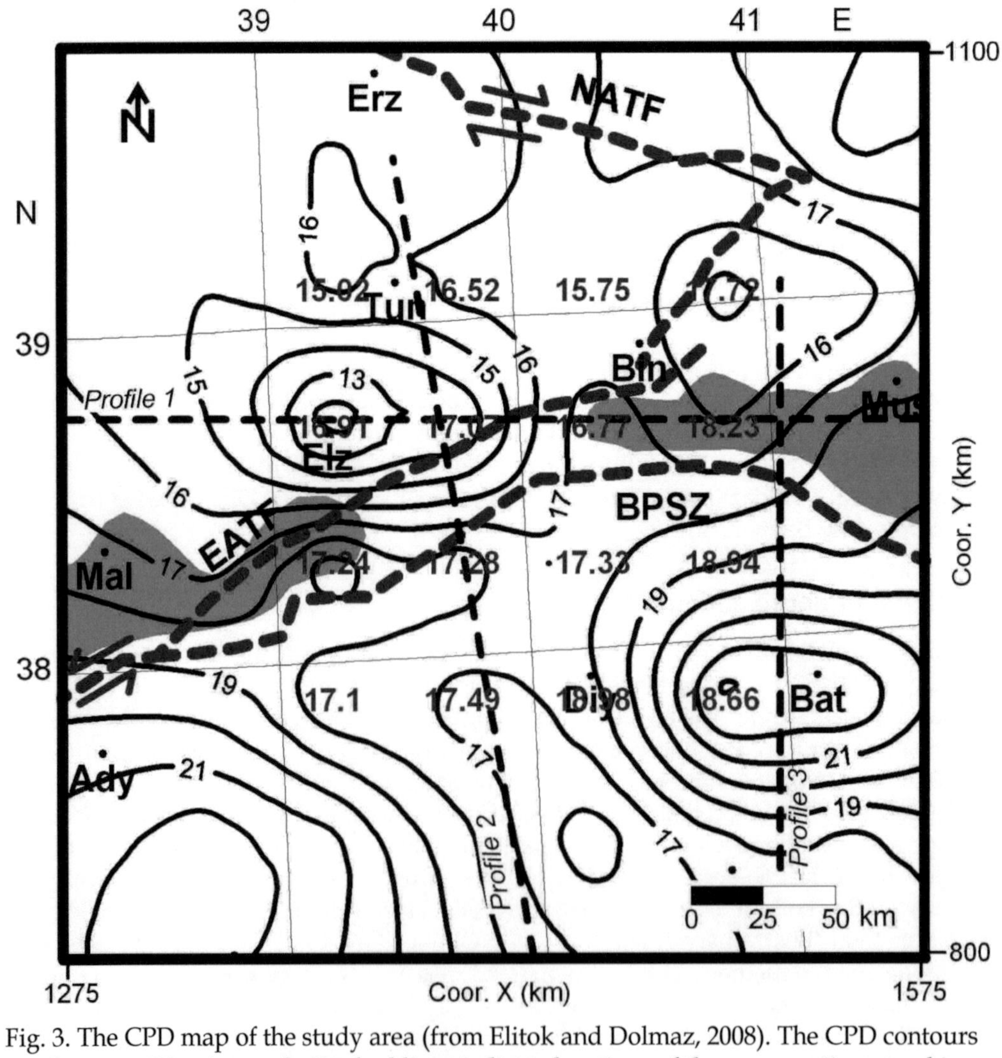

Fig. 3. The CPD map of the study area (from Elitok and Dolmaz, 2008). The CPD contours are drawn at 2 km intervals. Dashed lines indicate locations of the cross-sections used in Figure 4. City and major tectonic features names are same with Figure 2

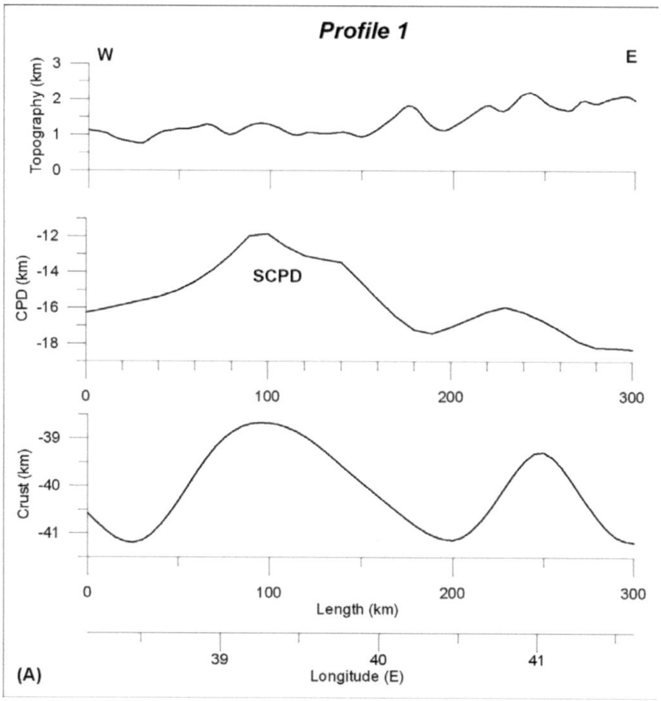
Profile 1
W
E
Topography (km)
3
2
1
0
CPD (km)
-12
-14
-16
-18
SCPD
0
100
200
300
Crust (km)
-39
-40
-41
0
100
200
300
Length (km)
39
40
41
Longitude (E)
(A)

a)

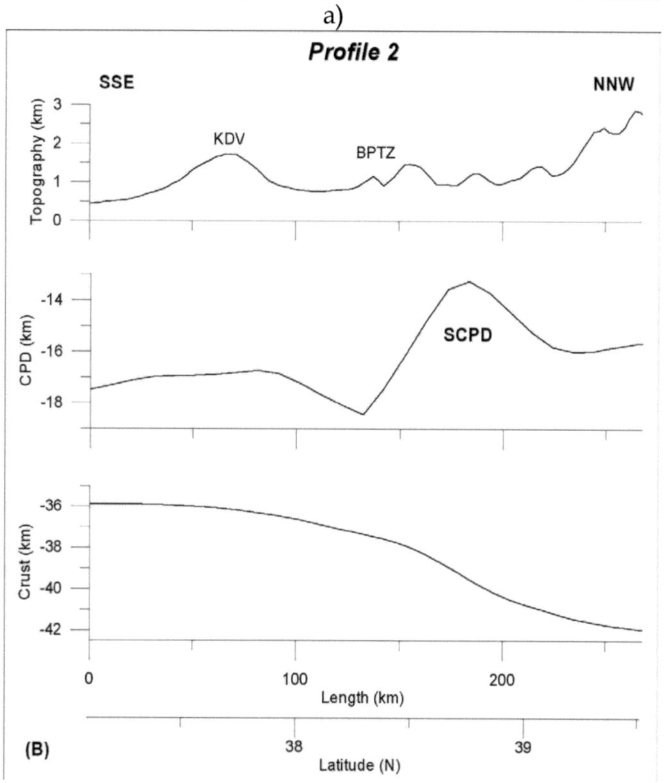
Profile 2
SSE
NNW
Topography (km)
3
2
1
0
KDV
BPTZ
CPD (km)
-14
-16
-18
SCPD
Crust (km)
-36
-38
-40
-42
0
100
200
Length (km)
38
39
Latitude (N)
(B)

b)

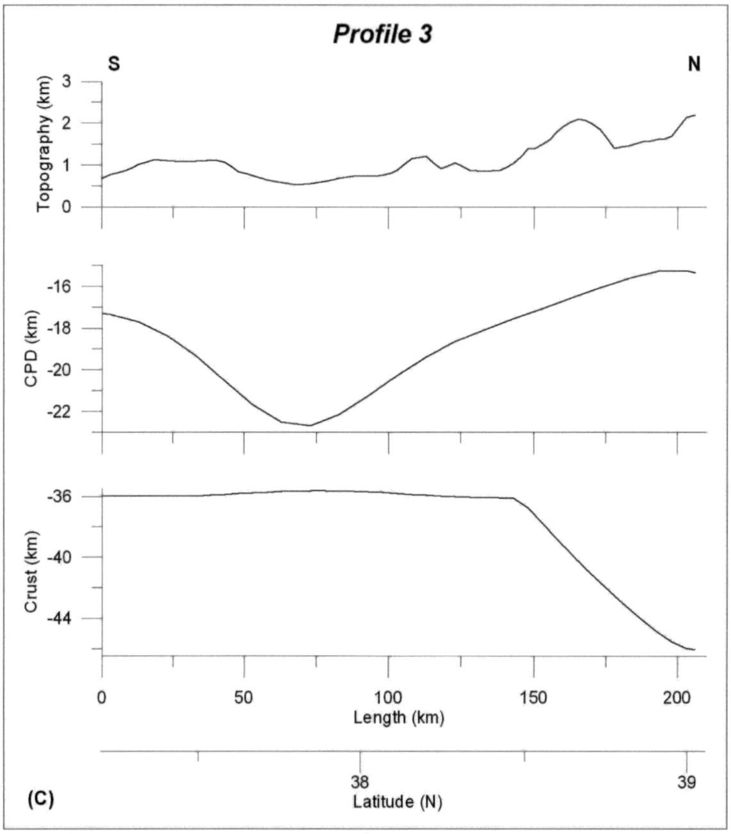

c)

Fig. 4. The cross-sections taken along the profiles in Figure 3 (from Elitok and Dolmaz, 2008), including topography, Curie isotherm depth (CPD), and moho depth (crust) a) Profile 1, b) Profile 2, and c) Profile 3. SCPD, shallow CPD; KDV, Karacadağ volcano, BPTZ: Bitlis-Pötürge Thrust Zone. Length of profile is in km and also in degree sheet format

Profile 1: Profile 1 involves the E-W directed cross-sections with 300 km between latitudes 38°-39° N (Figure 4a). The cross-section indicating CPD clearly illustrates the shallow Curie Point Depth (SCPD) between longitudes 39° and 40° E. This area is characterized by thinner crust of ca 38-39 km and very low Pn velocity zone (< 7.8 km/s; Figure 2a from Al-Lazki et al., 2003), corresponding to the metamorphic gap between the Bitlis Massif and the Pötürge Massif. The Curie point depth estimates in between 200 and 300 kms of the profile tend to get shallow reaching a depth of ca 16 km. This area with a 2200 m topographic height is represented by 39.5 km crust and low Pn velocities (< 8 km/s; Figure 2a from Al-Lazki et al., 2003).

Profile 2: Profile 2 indicates the NNW-SSE directed cross-sections with 270 km between longitudes 39°-40° E (Figure 4b). The cross-section displays a shallow CPD (SCPD) in between 170-220 kms of the profile which corresponds to the location of shallow CPD of Profile 1. The Curie point depth estimates in the Karacadağ volcanic area (KDV) in between 30 and 80 kms of the profile are characterized by ca 17 km deep. The Karacadağ volcanic area in the Arabian Foreland has an average crustal thickness of 36 km and low Pn velocity values (ca 8.0 km/s). In between latitudes of 38° and 38.5° N on the profile, moderate deep CPD reaching a maximum depth of ca 18.5 km is located under the BPTZ, forming a plateau at a height of 1.65 km rugged topography. The negative spike on the CPD is consistent with the suture.

Profile 3: Profile 3 involves the N-S directed cross-sections with 206 km long around the longitude of 41° E (Figure 4c). The cross-section generally shows the deep Curie point depth estimates reaching to a maximum depth of ca 22.5 km. When compared the Profile 2 with the Profile 3, it is seen that the CPD on the Profile 2 is shallower (ca 17 km) than the Profile 3 (ca 22.5 km) in the south of the suture zone (in the Arabian Foreland) between the latitudes of 37.5°-38°, suggesting most probably variations in the thickness of mantle lithosphere beneath the Arabian Foreland. High Al-Lazki et al. (2003) interpreted that Pn velocities beneath the Arabian Foreland sharply stop along the Bitlis Suture Zone.

5. Conclusions

There is a general consensus that the Neotectonic evolution of Turkey has been dominated by the collision of the African and Arabian plates with the Eurasian plate along the Hellenic arc to the west and the Bitlis-Zagros suture zone to the east (Wilson and Bianchini, 1999 and references therein). The EACP bounded by the PB in the north and the arc-shaped BPSZ in the south consists mainly of remnants of Tethyan oceanic lithosphere and Gondwana-derived continental blocks or microplates. This contractional province is still being squezzed between the Arabian and Eurasian plates, and experienced a compressional-extensional tectonic regime in connection with westward extrusion of the Anatolian plate. The eastern Anatolian contractional zone is dissected into three main lithospheric portions (EAHP, the Anatolian block, and the Arabian Foreland) by the major tectonic structures (BPSZ, NATFZ and EATFZ). Translation of the collision-related compressional stresses to the north (the Anatolian region) and to the south (the Arabian Foreland), mantle flow-driven extrusion tectonics and hot asthenospheric-lithospheric mantle interactions have been played a key role on the lithospheric evolution of the collisional zone. The present-day crust in the area between easternmost part of the Anatolian plate and the Arabian Foreland gets thinner from the NATFZ in the north (ca 43-44 km) to the Arabian Foreland in the south (ca 36-37 km) through the metamorphic gap between the Bitlis and Pötürge massifs. This thinner crustal area is characterized by shallow CPD (12-16 km), very low Pn velocities (< 7.8 km/s) and high Sn attenuation which indicate partially molten to eroded mantle lid or occurrence of asthenospheric mantle beneath the crust. Northernmost margin of the Arabian Foreland in the south of the Bitlis-Pötürge metamorphic gap area is represented by moderate CPD (16-18 km) relative to its eastern and western sides, and low Pn velocities (8 km/s). Elitok and Dolmaz (2008) infer from the geophysical data that the lithospheric mantle gets thinner towards the Bitlis-Pötürge metamorphic gap area in the northern margin of the Arabian Foreland which has been most probably caused by mechanical removal of the lithospheric mantle during mantle invasion to the north following the slab breakoff beneath the BPSZ. Mantle flow-driven rapid extrusion and counterclockwise rotation of the Anatolian plate gave rise to stretching and hence crustal thinning in the area between the easternmost part of the Anatolian plate and the Arabian Foreland which is currently dominated by wrench-tectonics (Figure 5).

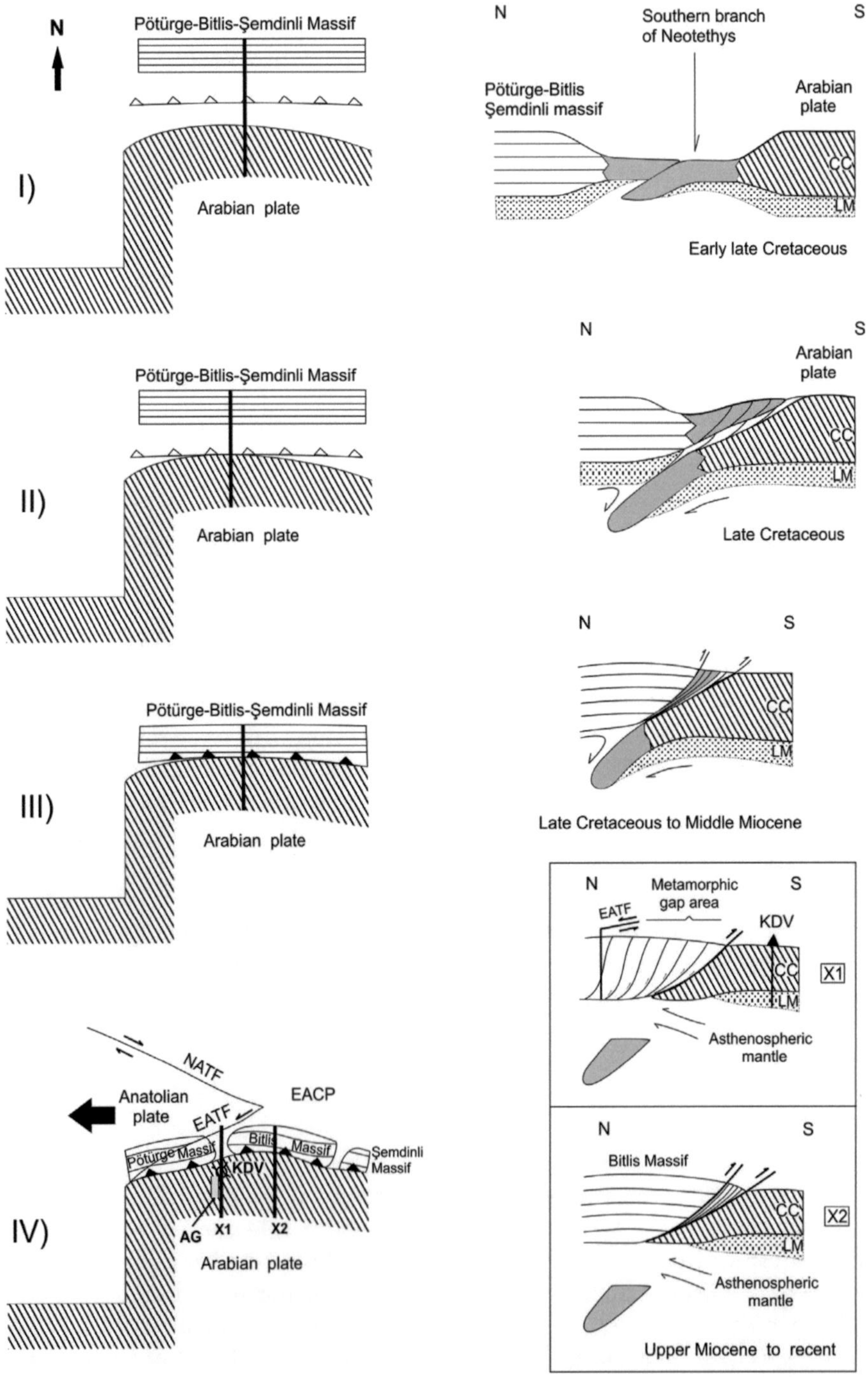

Fig. 5. Schematic tectonic model with cross-sections indicating Arabia-Eurasia collision and lithospheric evolution from late Cretaceous to recent (from Elitok and Dolmaz, 2008). (I) The northerly motion of Africa during Late Cretaceous resulted in a rapid convergence between Eurasia and Africa, and hence initiation of intraoceanic subduction zones in the Neotethys

(Dilek and Moores, 1990; Dilek and Whitney, 2000 and references therein). (II) The continued convergence of the Africa-Arabia plate resulted in the emplacement of southern Neotethyan oceanic lithosphere on the Arabian platform during the Late Cretaceous (Dilek and Moores, 1990; Yılmaz et al., 1993). (III) During the initial continental collision, the northern margin of the Arabian plate subducted following the subducting oceanic lithosphere into the trench due to slab pull effect. (IV) The nappe stacks together with the accreted Gondwana-derived continental blocks were emplaced on the Arabian platform mainly in three phases (late Cretaceous, late-early Middle Eocene, Early Middle Miocene; Yılmaz et al., 1993; Bozkurt, 2001). The subducting oceanic lithosphere generated a large downward force and hence the oceanic lithosphere commenced to detach from the Arabian continental plate as proposed by Şengör et al. (2003), Keskin (2003 and 2007), Lei and Zhao (2007). The slab detachment most likely began 11 Ma ago and lasted untill 8 Ma (Şengör et al., 2003). The two continental masses (Bitlis-Pötürge metamorphic block and Arabian plate) came in contact with each other in the Upper Miocene (Innocenti et al., 1982). Northward motion of the Arabian plate continued after the slab breakoff and hence caused the breaking apart of the the Bitlis-Pötürge metamorphic block (Innocenti et al., 1982 and references therein). (IV) Compressional-contractional tectonic regime replaced by the compressional-extensional tectonic regime as proposed by Koçyiğit et al. (2001) following the slab breakoff, which corresponds to commencement of the westward extrusion of Anatolian plate. AG: Akçakale graben, CC: continental crust, EACP: Eastern Anatolian Contractinal Province, NATF: North Anatolian Transform Fault, EATF: East Anatolian Transform Fault, KDV: Karacadağ volcano, LM: lithospheric mantle.

6. References

Al-Lazki, A., Sandvol, E., Seber, D., Türkelli, N., Mohamad, R., Barazangi, M., 2003. Tomographic Pn velocity and anisotropy structure beneath the Anatolian plateau (eastern Turkey) and the surrounding regions. *Geophysical Research Letters* 30, 8043, doi:10.1020/2003GL018912.

Allen, M., Jackson, J., Walker, R., 2004. Late Cenozoic reorganization of the Arabia-Eurasia collision and the comparison of short-therm and long-therm deformation rates. *Tectonics* 23, TC2008, doi:10.1029/2003TC001530.

Barazangi, M., Sandvol, E., Seber, D., 2006. Structure and tectonic evolution of the Anatolian plateau in eastern Turkey. In: Dilek, Y., and Pavlides, S. (Eds.), Post-Collisional tectonics and magmatism in the eastern Mediterranean region and Asia. *Geological Society of America Special Paper* 409, 463-474.

Bhattacharyya, B. K., Leu, L. K., 1977. Spectral analysis of gravity and magnetic anomalies due to rectangular prismatic bodies. *Geophysics* 41, 41-50.

Bozkurt, E., 2001. Neotectonics of Turkey-a synthesis. *Geodinamica Acta* 14, 3-30.

Büyüksaraç, A., 2007. Investigation into the regional wrench tectonics of inner East Anatolia (Turkey) using potential field data. *Physics of the Earth and Planetary Interiors* 160, 86-95.

Çoban, H., 2007. Basalt magma genesis and fractionation in collision- and extension-related provinces: a comparison between eastern, central and western Anatolia. *Earth-Science Reviews* 80, 219-238.

Dewey, J.F., Hempton, M.R., Kidd, W.S.F., Şaroğlu, F., Şengör, A.M.C., 1986. Shortening of continental lithosphere: the neotectonics of Eastern Anatolia-a young collision zone. In: Coward, M.P. and Ries, A.C (Eds.), *Collision Tectonics*. Geological Society, London, Special Publications 19, 1-36.

Dhont, D., Chorowicz, J., Luxey, P., 2006. Anatolian escape tectonics driven by Eocene crustal thickening and Neogene-Quaternary extensional collapse in the eastern Mediterranean region. In: Dilek, Y. and Pavlides, S. (Eds.), Post-Collisional tectonics and magmatism in the eastern Mediterranean region and Asia. *Geological Society of America Special Paper* 409, 441-462.

Dilek Y., Moores, E.M., 1990. Regional tectonics of the eastern Mediterranean ophiolites. In: Malpas. J., Moores, E.M., Panayiotou, A., Xenophontos C. (Eds.), Ophiolites, Oceanic Crustal Analogues, *Proceedings of the Symposium "Troodos 1987"*, Geological Survey Department, Nicosia, Cyprus, 295-309.

Dilek, Y., Whitney, D.L., 2000. Cenozoic crustal evolution in Central Anatolia: extension, magmatism and landscape development. In: Panayides, I., Xenophontos, C., Malpas, J. (Eds.), *Proceedings of the Third International Conference on the Geology of the Eastern Mediterranean*, September 1998, Nicosia, Cyprus, 183-192.

Elitok, Ö., Dolmaz, M.N., 2008. Mantle flow-induced crustal thinning in the area between the easternmost part of the Anatolian plate and the Arabian Foreland (E Turkey) deduced from the geological and geophysical data. *Gondwana Research* 13, 3, 302-318, doi: 10.1016/j.gr.2007.08.007.

Faccenna, C., Bellier, O., Martinod, J., Piromallo, C., Regard, V., 2006. Slab detachment beneath eastern Anatolia: A possible cause for the formation of the North Anatolian fault. *Earth and Planetary Science Letters* 242, 85-97.

Gautier, P., Brun, J.-P., Moriceau, R., Sokoutis, D., Martinod, J., Jolivet, L., 1999. Timing, kinematics and cause of Aegean extension: a scenario based on a comparsion with simple analogue experiments. *Tectonophysics* 315, 31-72.

Gök, R., Sandvol, E., Türkelli, N., Seber, D., Barazangi, M., 2003. Sn attenuation in the Anatolian and Iranian plateau and surrounding regions. *Geophysical Research Letters* 30, 8042, doi:10.1029/2003GL018912.

Gök, R., Pasyanos, M., Zor, E., 2007. Lithospheric structure of the continent–continent collision zone: eastern Turkey. *Geophysical Journal International* 169, 1079-1088.

Gürer, Ö.F., Gürer, A., 1999. Development of evaporites and the counterclockwise rotation of Anatolia, Turkey. *International Geology Review* 41, 607-622.

Hall, R., 1976. Ophiolite emplacement and the evolution of the Taurus suture zone, southeastern Turkey. *Geological Society of America Bulletin* 87, 1078-1088.

Hubert-Ferrari, A., King, G., Manighetti, I., Armijo, R., Meyer, B., Tapponnier, P., 2003. Long-term elasticity in the continental lithosphere; modelling the Aden Ridge propagation and the Anatolian extrusion process. *Geophysical Journal International* 153, 111-132.

Innocenti, F., Mazzuoli, R., Pasquare, G., Radicati Di Brozolo, F., Villari, L., 1982. Tertiary and Quaternary volcanism of the Erzurum-Kars area (eastern Turkey): Geochronological data and geodynamic evolution. *Journal of Volcanology and Geothermal Research* 13, 223-240.

Keskin, M., 2003. Magma generation by slab steepening and breakoff beneath a subduction-accretion complex: An alternative model for collision-related volcanism in Eastern Anatolia, *Turkey. Geophysical Research Letters* 30, 8046, doi:10.1029/2003GL018019.

Keskin, M., 2007. Eastern Anatolia: A hot spot in a collision zone without a mantle plume. In: Foulger, G.R. and Jurdy D. (Eds.), The Origins of Melting Anomalies: Plumes, Plates, and Planetary Processes (P4 Book). *Geological Society of America Special Paper* 430, 693-722.

Keskin, M., Pearce, J.A., Kempton, P.D., Greenwood, P., 2006. Magma-crust interactions and magma plumbing in a postcollisional setting: Geochemical evidence from the Erzuru-Kars volcanic plateau, eastern Turkey. In: Dilek, Y. and Pavlides, S. (Eds.), Post-Collisional tectonics and magmatism in the eastern Mediterranean region and Asia. *Geological Society of America Special Paper* 409, 475-505.

Koçyiğit, A., Yılmaz, A., Adamia, A., Kuloshvili, S., 2001. Neotectonics of East Anatolian Plateau (Turkey) and Lesser Caucasus: implication for transition from thrusting to strike-slip faulting. *Geodinamica Acta* 14, 177-195.

Lei, J., Zhao, D., 2007. Teleseismic evidence for a break-off subducting slab under Eastern Turkey. *Earth and Planetary Science Letters* 257, 14-28.

Lyberis, N., Yürür, T., Chorowicz, J., Kasapoğlu, E., Gündoğdu, N., 1992. The East Anatolian Fault: an oblique collision belt. *Tectonophysics* 204, 1-15.

McClusky, S., Balassanian, S., Barka, A., Demir, C., Ergintav, S., Georgiev, I., Gurkan, O., Hamburger, M., Hurst, K., Kahle, H., Kastens, K., Kekelidze, G., Ring, R., Kotzev, V., Lenk, O., Mahmoud, S., Mishin, A., Nadariya, M., Ouzounis, A., Paradissis, D., Peter, Y., Prilepin, M., Reilinger, R., Sanli, I., Seeger, H., Tealeb, A., Toksöz, M.N., Veis, G., 2000. Global positioning system constraints on plate kinematics and dynamics in the eastern Mediterranean and Caucasus. *Journal of Geophysical Research-Solid Earth* 105, 5695-5719.

Nagata, T., 1961. *Rock Magnetism*, 350 pp., Maruzen, Tokyo.

Okubo, Y., Graf, J.R., Hansen, R.O., Ogawa, K., Tsu, H., 1985. Curie Point Depths of the Island of Kyushu and Surrounding Areas, *Japan. Geophysics* 53, 481-494.

Özçep, F., Orbay, N., 2000. Paleomagnetic studies on the Anatolian (Turkish) plate and geodynamic implications: a review. *Terra Nostra* 10, From Secular Variation to Paleomagnetism-Workshop, Potsdam, pp 93-94.

Pearce, J.A., Bender, J.F., De Long, S.E., Kidd, W.S.F., Low, P.J., Güner, Y., Şaroğlu, F., Yılmaz, Y., Moorbath, S., Mitchell, J.G., 1990. Genesis of collision volcanism in Eastern Anatolia, Turkey. *Journal of Volcanology and Geothermal Research* 44, 189-229.

Platzman, E.S., Tapırdamaz, C., Sanver, M., 1998. Neogene anticlockwise rotation of central Anatolia (Turkey): preliminary paleomagnetic and geochronological results. *Tectonophysics* 299, 175-189.

Reilinger, R.E., McClusky, S.C., Oral, M.B., King, W., Toksöz, M.N., Barka, A.A., Kınık, I., Lenk, O., Sanlı, I., 1997, Global Positioning System measurements of present-day crustal movements in the Arabian-Africa-Eurasia plate collision zone. *Journal of Geophysical Research-Solid Earth* 102, 9983-9999.

Sandvol, E., Türkelli, N., Barazangi, M., 2003. The Eastern Turkey Seismic Experiment: The study of a young continent-continent collision. *Geophysical Research Letters* 30, 8038, doi:10.1029/2003GL018912.

Seyitoğlu, G., Scott, B.C., 1996. The cause of N-S extensional tectonics in western Turkey: tectonic escape vs back-arc spreading vs orogenic collapse. *Journal of Geodynamics* 22, 145-153.

Şengör, A.M.C., 1979. The North Anatolian transform fault: its age, offset and tectonic significance. *Journal of the Geological Society* 136, 269-282.

Şengör, A.M.C., Görür, N., Şaroğlu, F., 1985. Strike-slip faulting and related basin formation in zones of tectonic escape: Turkey as a case study. In: Biddle, K.T. and Christie-Blick, N. (Eds.), *Strike-slip Deformation, Basin Formation and Sedimentation*. Society of Economic Mineralogist and Paleontologists Special Publication 37, 227-264.

Şengör, A.M.C., Özeren, S., Genç, T., Zor, E., 2003. East Anatolian high plateau as a mantle-supported, North-south shortened domal structure. Geophysical Research Letters 30, 8045, doi:10.1029/2003GL017858.

Şengör, A.M.C., Yılmaz, Y., 1981. Tethyan evolution of Turkey: a plate tectonic approach. Tectonophysics 75, 181-241.

Tanaka, A., Okubo, Y., Matsubayashi, O., 1999. Curie Point Depth Based on Spectrum Analysis of the Magnetic Anomaly Data in East and Southeast Asia. *Tectonophysics* 306, 461–470.

Tanaka, A., Ishikawa, Y., 2005. Crustal thermal regime inferred from magnetic anomaly data and its relationship to seismogenic layer thickness: The Japanese islands case study. *Physics of the Earth and Planetary Interiors* 152, 257–266.

Tatar, O., Piper, J.D.A., Gürsoy, H., Temiz, H., 1996. Regional significance of Neotectonic counterclockwise rotation in Central Turkey. *International Geology Review* 38, 692-700.

Wilson, M., Bianchini, G., 1999. Tertiary-Quaternary magmatism within the Mediterranean and surrounding regions. In: Durand, B., Jolivet, L., Horvath, F., Seranne, M. (Eds.), *The Mediterranean Basins: Tertiary Extension within the Alpine Orogen*. Geological Society, London, Special Publications 156, 141-168.

Yazgan, E., 1981. Doğu Toroslarda etkin bir paleo-kıta kenarı etüdü (Üst Kretase-Orta Eosen) Malatya-Elazığ, Doğu Anadolu. *Journal of the Earth Sciences Application and Research Centre of Hacettepe University* 7, 83-104.

Yılmaz, Y., 1993. New evidence and model on the evolution of the southeast Anatolian orogen. *Geological Society of America Bulletin* 105, 251-271.

Yılmaz, Y., Yiğitbaş, E., Genç, Ş.C., 1993. Ophiolitic and metamorphic assemblages of southeast Anatolia and their significance in the geological evolution of the orogenic belt. *Tectonics* 12, 1280-1297.

Yılmaz Y., Güner, Y., Şaroğlu, F., 1998. Geology of the Quaternary volcanic centers of the east Anatolia. *Journal of Volcanology and Geothermal Research* 85, 173-210.

Yılmaz, Y., Yıldırım, M., 1996. Geology and evaluation of the Nap Region (the Metamorphic Massif) of the southeast Anatolian orogenic belt. *Turkish Journal of Earth Sciences* 5, 21-38.

Yiğitbaş., E., Genç, Ş.C., Yılmaz, Y., 1993. Güneydoğu Anadolu orojenik kuşağında Maden Grubunun tektonik konumu ve jeolojik önemi. A. *Suat Erk Sempozyumu Bildirileri*, 2-5 Eylül, A.Ü. Fen Fakültesi Jeoloji Mühendisliği Bölümü, Ankara, 251-264 (in Turkish with English abstract).

Yiğitbaş, E., Yılmaz, Y., 1996. New evidence and solution to Maden Complex controversy of the southeast Anatolian orogen (Turkey). *Geologische Rundschau* 85, 250-263.

Zor, E., Gürbüz, C., Türkelli, N., Sandvol, E., Seber, D. and Barazangi, M., 2003. The crustal structure of the East Anatolian Plateau from receiver functions. *Geophysical Research Letters* 30, 8044, doi:10.1029/2003GL018192.

12

Early-to-mid Pleistocene Tectonic Transition Across the Eastern Mediterranean Influences the Course of Human History

Uri Schattner
Dr. Moses Straus Dept. of Marine Geosciences, Charney School of Marine Sciences, University of Haifa, Israel

1. Introduction

1.1 Out of Africa through the Levantine corridor

The widely accepted 'Out-of-Africa' hypothesis considers northeastern Africa as the cradle of humankind, based on early archaeological evidence, (e.g., Bar-Yosef and Belfer-Cohen, 2001; Templeton, 2002). The earliest evidence for hominin activity was found in Kenya, Ethiopia and Tanzania. Amongst these findings were remnants of Oldowan and Acheulian stone tools, remnants of animals and remains of hominins in sites dated to the Pliocene (>1.8 Ma)(e.g., Semaw, 2000).

Hominin remains outside Africa are dated to the Pleistocene and Holocene periods, from ~1.8 Ma to present. Remains were found in sites spanning from northeast Africa to the Far East (Carto et al., 2009; Stringer, 2000). The Levantine corridor (Fig. 1), is a narrow land bridge connecting Africa with eastern Europe, central Asia, India and the Far East. It extends along the land area of the Sinai plate, between the Dead Sea fault and the Mediterranean continental margin. This corridor is considered to be one of the main pathways of hominin dispersal due to the discovery of some of the oldest prehistoric remains outside of Africa (Bar-Yosef and Belfer-Cohen, 2001).

Erk-el-Ahmar (1.96-1.78 Ma) is located in the central part of the Dead Sea Fault and reflects the first of three recognized dispersal pulses (Bar-Yosef and Belfer-Cohen, 2001). At this site, the earliest hominin related flint artifacts were found outside of Africa (Braun et al., 1991; Horowitz, 2001) (Fig. 2). The next pulse is indicated in Ubediya, ~1.4 Ma (Tchernov, 1987; Klein, 1989) where a rich complex of hominin and fauna remains was found. This complex represents numerous returns to the same location close to a lake. Acheulian artifacts of Ubediya are very similar to those found in contemporary assemblages of Upper Bed II of Olduvai Gorge (Bar-Yosef and Belfer-Cohen, 2001). The third pulse is represented by the mid-Acheulian site of Gesher Benot-Ya'aqov (Goren-Inbar et al., 2000) (0.78 Ma) and Ruhama (Laukhin et al., 2001) (0.99-0.85 Ma).

Dispersal routes of hominins to the rest of the world have been hypothesized based on archaeological evidence (Bar-Yosef and Belfer-Cohen, 2001)(Fig. 1). Reproduction of a species in a closed area limits the population growth and hence, given the right environment, this species will tend to quickly expand into the newly accessible area. Therefore, both physiographic modifications and their timing are crucial for the onset of dispersal.

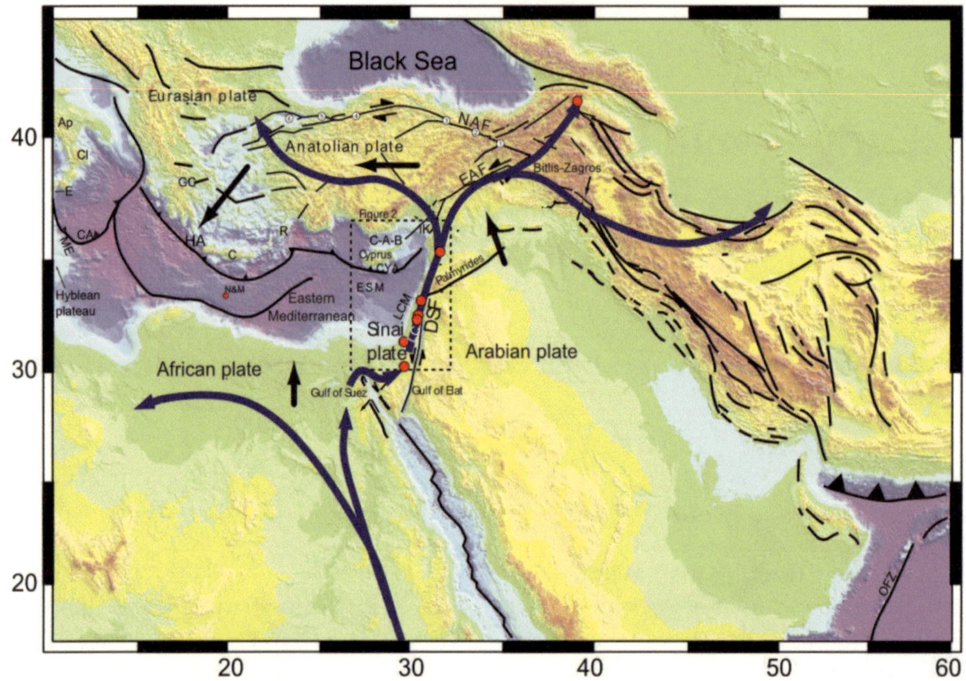

Fig. 1. Geodynamic settings of the eastern Mediterranean basin (after Reilinger et al., 2006) showing the differential convergence (subduction/collision) between Africa-Sinai-Arabia and Eurasia-Anatolia. Large black arrows show the direction of the general plate motion; small black arrows indicate relative plate motion; thin black lines show the major faults and plate boundaries. Black triangles symbol active subduction zones (overriding plate). Blue arrows indicate major pathways of hominin dispersal out of Africa; red dots denote the location of major Hominin sites along the Levantine Corridor (Bar-Yosef and Belfer-Cohen, 2001) (1.81-1.7 Ma). Ap – Apennine mountains; C – Crete; CA – Calabrian arc; C-A-B – Cilicia-Adana basins; Cl – Calabria; CYA – Cyprus Arc; DSF - Dead Sea fault; E – Etna volcano; EAF – East Anatolian fault; ESM – Eratosthenes Seamount; GC – Gulf of Corinth; HA – Hellenic arc; IK – Iskanderun bay; LC - Levantine Corridor; LCM - Levant continental margin; ME – Malta escarpment; N&M – Napoli and Milano mud volcanoes; NAF – North Anatolian fault; OFZ - Owen Fracture Zone; R – Rhode.Numbers along the NAF and DSF represent the following basins that developed as pull-aparts and were crossed by a diagonal through-going fault: (1) Erzincan, (2) Suşehri-Gölova, (3) Erbaa-Niksar, (4) Bolu-Yeniçağa, (5) Izmit-Sapanca, (6) Marmara Sea and (7) Hula. Inset: regional tectonic settings. Black rectangle shows the location of Fig. 2

This chapter highlights the temporal coincidence of the first out of Africa pulse with the initiation of major physiographic modifications of the mild topography of the Levantine corridor and the entire eastern Mediterranean basin. The chapter summarizes the works of Ben-Avraham et al. (2005), Schattner and Weinberger (2008), Schattner and Lazar (2009), Lazar and Schattner (2010), and Schattner (2010). It pieces together a wealth of accumulated knowledge published in the literature into a coherent tectonic reconstruction of the eastern Mediterranean convergence system during the beginning of the Pleistocene. The importance of prehistoric dates in this context is whether they are dated to pre, syn or post tectonic transition.

1.2 Eastern Mediterranean geodynamics

The eastern Mediterranean basin is a fossil remnant of the Meso and Neo-Tethys oceans (Fig. 1). The basin started to develop during the late Paleozoic when Eratosthenes block, amongst other platelets, drifted away from the Gondwana super-continent (Garfunkel, 1998; Schattner and Ben-Avraham, 2007). During the Paleogene–Neogene, fore-arc extension and subduction hinge rollback of the northern Afro-Arabian plate dominated the development of the eastern Mediterranean basin (Robertson, 1998). Towards the terminal stages of the Neo-Tethys, during the Oligocene to the present, the entire northern flank of the Afro-Arabian plate progressively subducted northward beneath Eurasia.

Until early–middle Eocene a single active subduction zone occupied the northern margin of the eastern Mediterranean basin, in southeastern Turkey, where arc volcanism continued into the early Neogene (Robertson, 1998). From the Neogene, the northwards subduction and hinge rollback continued with varying rates across the Alpine-Himalayan orogen (Ben-Avraham and Nur, 1976; Faccenna et al., 2006). While much of the subducted slab was consumed, some seafloor features were accreted onto the overriding plate (Şengör et al., 2003). During the Miocene the subduction front jumped to its present location south of Cyprus (Kempler and Ben-Avraham, 1987; Robertson, 1998).

At the forefront of subduction, slab break-off developed along the northern underthrusting edge of the Arabian plate (Faccenna et al., 2006). The break-off was suggested to propagate from the Owen fracture zone westward towards the northern edge of the Dead Sea fault, concurrent with increasing northwards indentation of Eurasia by Arabia (Fig. 1). The remaining of Arabian oceanic crust gradually collided and accreted to form the Bitlis–Zagros suture by the late Miocene to early Pliocene (Robertson, 1998; Faccenna et al., 2006).

In contrast, west of the Dead Sea fault, subduction of the eastern Mediterranean basin under the Cyprus Arc persisted without prominent interruptions until the early Pleistocene (Kempler, 1998; Robertson, 1998). Increased slab pull towards the north resulted in accelerated slab retreat and arc propagation southwards (Faccenna et al., 2006). Further to the south the continental margins of the Sinai and African plates remained passive throughout the Neogene.

2. Early-to-mid Pleistocene tectonic transition across the eastern Mediterranean

A major kinematic transition occurred across the eastern Mediterranean region during the early Pleistocene - the early-to-mid Pleistocene tectonic transition (Schattner, 2010). Approximately synchronous structural modifications were recorded along plate boundaries as well as within the plates are integrated below and Fig. 3.

2.1 Arabian plate and the Dead Sea fault

Eurasia indentation by the underthrusting northern Arabia resulted in the uplift of the Zagros belt (McQuarrie et al., 2003; Faccenna et al., 2006; Reilinger et al., 2006) and activation of the East Anatolian fault since the Pliocene (Fig. 1; Westaway and Arger, 2001). Along the River Euphrates in southeast Turkey, northern Syria and western Iraq uplift of terrace deposits increased in the late early Pleistocene (Demir et al., 2007). Further south, branches of the Dead Sea fault ruptured the Palmyrides and the stable part of the Arabian plate from the early Pleistocene (Rukieh et al., 2005). The regional stress field shifted from N to NW striking compression as evident by structural variations along the northern border of Arabia during the early Pleistocene (Zanchi et al., 2002).

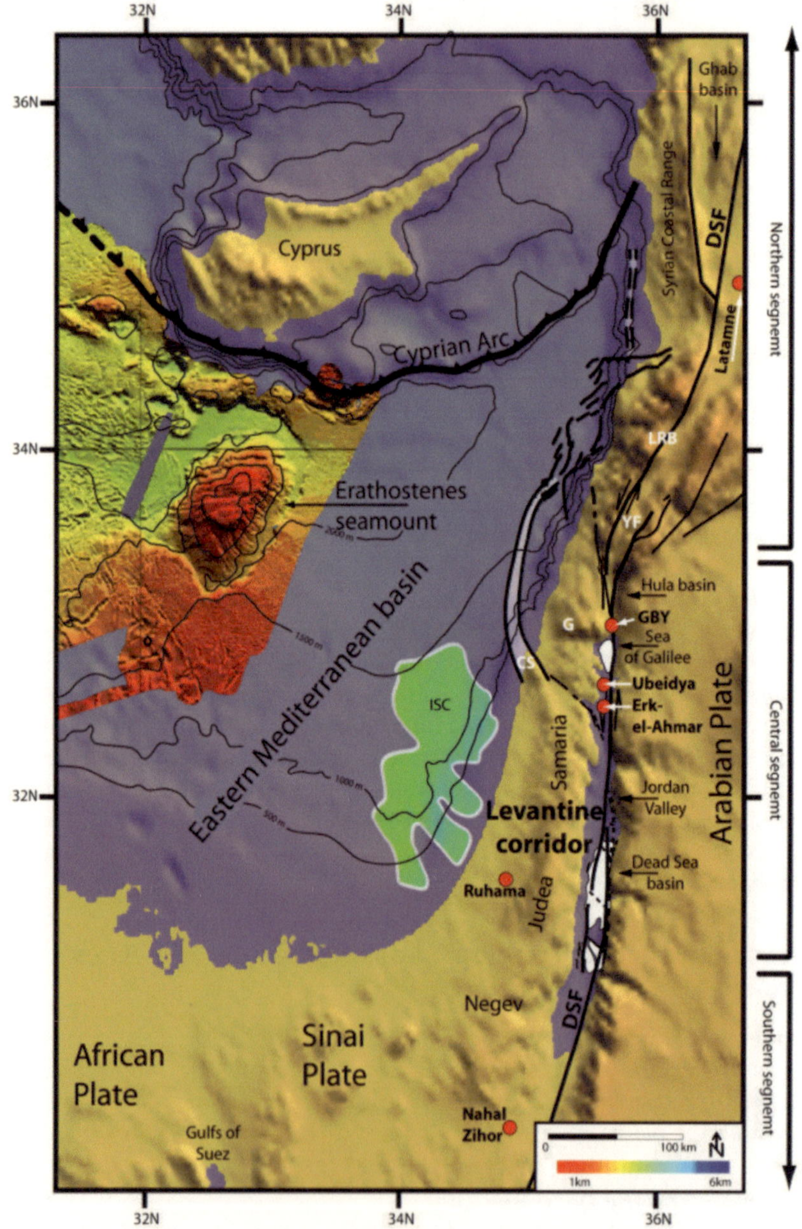

Fig. 2. Relief map of the eastern Mediterranean and its bathymetry (thin black lines) highlighting the bathymetry of Eratosthenes Seamount (Medimap Group et al., 2005; J.K. Hall, personal communication); main Plio-Pleistocene faulting along the Levant margins (Schattner et al., 2006 and references therein); and location of the Israel slump complex – ISC (Frey Martinez et al., 2005). Red dots denote the location of major Hominin sites along the Levantine Corridor mentioned in the text. DSF – Dead Sea fault; LRB – Lebanese restraining bend; YF – Yammunneh fault; KB – Korazim block; G – Galilee; C – Carmel fault; BTT – Beirut Tripoli thrust; Tr – Trodos massif; Kr – Kyrenia range

Increasing convergence between the Sinai and Arabian plates along the central and northern segments of the Dead Sea fault (Gomez et al., 2007) during the early Pleistocene, resulted in synchronous structural modifications along the Dead Sea fault axis. Along the northern Dead Sea fault, the Ghab basin subsided continuously during the Plio-Pleistocene (Brew, 2001). During the late early Pleistocene the Jisr ash Shugshur basalts (1.3–1.1 Ma; Sharkov et al., 1994) flowed to the northern part of the basin, covering a possible northern transverse faulting (Kopp et al., 1999). A notable phase of transpressive motion, folding and uplift deformed the Syrian coastal ranges (Gomez et al., 2006) and uplifted of the Nahr El-Kabir at the Mediterranean shore of NW Syria during the Pleistocene (Hardenberg and Robertson, 2007); uplifted the Lebanese restraining bend (Dubertret, 1955; Butler et al., 1998; Walley, 1998; Griffiths et al., 2000; Tapponnier et al., 2004; Elias, 2006); and branching developed along the Dead Sea fault Yammunneh main fault (Gomez et al., 2007). However, timing of initiation of the transpressive phase in Lebanon and Syria is often reported to the entire Pliocene - Pleistocene period.

Further south, basaltic units dated from the early Pleistocene onwards are found exclusively east of the Dead Sea fault axis (Weinstein et al., 2006). One of these units, the 1.5-0.5 Ma old Hazbani basalt (Sneh and Weinberger, 2003) flowed southward from its source in southern Lebanon along the Dead Sea fault axis towards the subsiding Hula basin (Heimann, 1990). Similar to the 1.3–1.1 Ma Jisr ash Shugshur basalts, the Hazbani basalt also covered the northern part of the basin, precluding a possible transverse fault from being identified at the surface (Sneh and Weinberger, 2006). At that time the Hula basin ceased to develop as a pull apart, when a diagonal through-going strike-slip fault propagated between its SE and NW corners. The trajectory of this fault was found to be parallel to the present-day motion along the Dead Sea fault (Schattner and Weinberger, 2008). In temporal and structural coincidence with the development of the Hula diagonal fault, block rotation initiated along the Korazim block (Heimann and Ron, 1993); rapid subsidence of the Sea of Galilee basin was accompanied by depocenter migration to the northeast (Hurwitz et al., 2002); two NE-striking anticlines developed immediately south of the Sea of Galilee (Rotstein et al., 1992; Zurieli, 2002); subsidence of two basins in the southern Jordan valley was accompanied by bending of overlaying monoclines (Lazar et al., 2006). At the same time (early Pleistocene) accelerated subsidence and extension prevailed across the southern and central Dead Sea fault segments (ten Brink and Ben-Avraham, 1989; Ben-Avraham et al., 2005). The Dead Sea basin subsided rapidly accumulating ~4 km of sediments that facilitated salt removal into prominent diapirs (Lisan and Sedom diapirs; Al-zoubi and ten Brink, 2001; Horowitz, 2001; Larsen et al., 2002; Ben-Avraham and Lazar, 2006). In addition, the depocenter of the southern Dead Sea basin migrated southeastwards while the largest vertical displacement occurred across its southern boundary. Further south a general tendency of fault localization towards the axis of the Dead Sea fault is recorded along the margins of the Gulf of Elat (Aqaba; Marco, 2007). This differential lateral motion along the entire Dead Sea fault, extensional in the south as opposed to contractional in the north, result from a change in relative plate motion which developed during the early Pleistocene (Schattner and Weinberger, 2008; Weinberger et al., 2009).

2.2 Sinai plate and the easternmost Mediterranean margins

The northward increase in contractional strike-slip motion across the Dead Sea fault (Gomez et al., 2007; Schattner and Weinberger, 2008) induced extensive deformation across the Sinai plate, between the Dead Sea fault and the eastern Mediterranean continental margin during

the early Pleistocene. Folding, faulting, uplift and tilting of the Syrian coastal ranges (Gomez et al., 2006), the Lebanese restraining bend (Walley, 1998), and the northern Galilee (Matmon et al., 2003), extended to the edge of the continental margin where shelf marginal wedges were tilted basinward (Ben-Avraham et al., 2006; Schattner et al., 2006). A 40-60 km wide N-trending arch developed during the early Pleistocene across the Sinai plate in the northern Galilee (Matmon et al., 1999, 2003), Samaria and Judea regions (Wdowinski and Zilberman, 1996, 1997) and the Negev, which as a result was uplifted and tilted (Ginat, 1997; Ginat et al. 1998; Avni, 1998; Avni et al., 2000).
As a consequence of the topographic uplift the catchment area of rivers from the Arabian plate to the Mediterranean Sea was severely shortened by a new water divide that developed close to the shore. At the same time the anomalously high subsidence rate of the Levant continental shelf which prevailed during the Pliocene diminished in the Pleistocene when the dominant landward aggradation shifted to basinward progradation (Ben-Gai et al., 2005). At the continental margin the Oligo-Miocene Jaffa basin became completely buried under a thick sedimentary cover (Gvirtzman et al., 2008). Marine channels crossing the continental shelf were gradually starved and buried by Nile derived sediments rather from onland. Seismic data reveals a succession of incised channels that drained land area during its tectonic uplift however today has no trace on the shelf seabed.
Offshore, several fault systems re-activated the northern Levant continental margin during the Pleistocene (Schattner and Ben-Avraham, 2007) from the marine extension of the Carmel fault (Schattner et al., 2006), along the Lebanese continental slope (Daëron et al., 2001, 2004; Carton et al., 2007, 2009; Elias et al., 2007), and were traced to a possible juvenile triple junction with the easternmost part of the Cyprus Arc (Butler et al., 1998; Schattner et al., 2006). These along-margin fault systems originate as westward branches of the Dead Sea fault, which dissect the uplifted folds of the Sinai plate. Over 15 slump bodies, 400 km^3 in volume, dated to the early Pleistocene, have been identified off the southern Levant margin (Frey Martinez et al., 2005).
Along the north African margin a well-documented NW to N shortening event occurred during the early Pleistocene (Guiraud, 1990). E-trending folds and reverse faults developed, associated with NW–SE dextral and NNE–SSW sinistral strike-slip faults. Major intra-plate fault zones were rejuvenated and regional uplifts intensified. However, with time, the intensity of shortening strongly decreased (Guiraud et al., 2005). During the Pleistocene the extensional axis direction across the Gulf of Suez rotated counterclockwise from NE to N (Bosworth and Taviani, 1996).

2.3 Cyprus and Eratosthenes Seamount

The progressive subduction of the Sinai plate beneath the Cyprus Arc was significantly disrupted during the late Pliocene–early Pleistocene with the initiation of the Eratosthenes Seamount-Cyprus Arc collision (Kempler, 1998). On the verge of subduction, tectonic subsidence of Eratosthenes accelerated and the Seamount reached its current vertical position (Robertson, 1998). Directly overlaying the top Messinian unconformity, early Pliocene succession reveals bathyal pelagic microfossils (Robertson, 1998). These deep-marine sediments uncovered on the northern flank of the Seamount (sites 965-967) were interpreted as indicators for a rapid subsidence of the Seamount (Ibid.). This rapid submergence at this stage could have also been greatly affected by the accelerated post Messinian sea level rise (e.g., Meijer and Krijgsman, 2005). However, the late Pliocene benthic foraminifera covered by Nile derived sediments indicate further tectonic deepening of the seamount (site 967, Robertson, 1998), this time without accompanying sea level rise.

Deformation pattern of the Seamount changes considerably from north southwards. While Pleistocene compressional folding deform the base of the northern lower slope (site 967), the central part of the Seamount's plateau experiences flexural loading and faulting (Limonov et al., 1994; Robertson et al., 1995). The Seamount and its basement were faulted by a low angle approximately E-W trending detachment zone which tilted its flanks to the north and south. Deformation of the Seamount resulted from crustal flexure, induced by southward overthrusting of the Cyprus active margin. Large scale catastrophic mass movements developed on the northern and southern slopes of the Seamount. Since identification of these slides is limited to the lines covered by seismic data it is assumed that mass movement did not occur exclusively to the north and south. The present-day bathymetry of the Seamount is faulted and scarred in the north relative to the much smoother southern slopes.

In a tomography study of the eastern Mediterranean Faccenna et al. (2006) image three different states of convergence. Beneath the Arabia-Eurasia collision zone tomography does not show any clear continuation of the Arabian slab northwards. Further to the west across the Cyprus Arc where incipient collision takes place, seismic velocities indicate a ~120 km gap in the descending Sinai slab. In their westernmost profile Faccenna et al. (2006) show that a continuous African slab is being consumed beneath the Hellenic Arc, where subduction still prevails. The relatively narrow gap in the northern Sinai slab suggests a younger initiation of the Seamount-Arc collision than the slab tear-off north of Arabia. This westwards tear propagation is in agreement with the nature of propagation described along other convergence margins of the Mediterranean region (e.g., Wortel and Spakman, 2000). While Tomography cannot infer directly to the state of convergence it strongly suggests a linkage between the continuity of the descending slab and the degree of collision at the upper lithosphere.

Convergence across the central Cyprus Arc shifted from subduction to collision (Salamon et al., 2003; Hall et al., 2010) while its eastern part exhibits strike-slip motion (Ben-Avraham et al., 1995). Uplift of the island of Cyprus which persisted since the late Miocene accelerated during the early Pleistocene when the island completely emerged from the sea (Kempler, 1998; Harrison et al., 2004). The Troodos Massif and the Kyrenia Range were uplifted in accelerated rates (Robertson and Woodcock, 1986; Robertson et al., 1995; Poole and Robertson, 1991, 1998) while NW to NE trending strike-slip faulting became dominant in northern Cyprus (Harrison et al., 2004). After the mid Pleistocene, uplift rates decreased (Poole and Robertson, 1992, 1998). While the deformations of the Cyprus region are straightforwardly associated with the collision, they consist a part of the synchronous structural deformations across the entire eastern Mediterranean region mentioned above - together comprising the early-to-mid Pleistocene tectonic transition (Fig. 3).

Deformations were also recorded in the marine surroundings of Cyprus, from Iskenderun through Adana-Cilicia to Antalya basins, and along the Larnaka and Latakia Ridges. In these regions, which were studied in detail in the last decade, the Pliocene-Quaternary is interpreted as a single seismic stratigraphic unit (Unit 1 in Aksu et al., 2005a; Aksu et al., 2005b; Burton-Ferguson et al., 2005; Calon et al., 2005a; Calon et al., 2005b; Hall et al., 2005a; Hall et al., 2005b; Isler et al., 2005; Aksu et al., 2009; Hall et al., 2009). In light of the findings brought here the young fill of these basins should be reexamined.

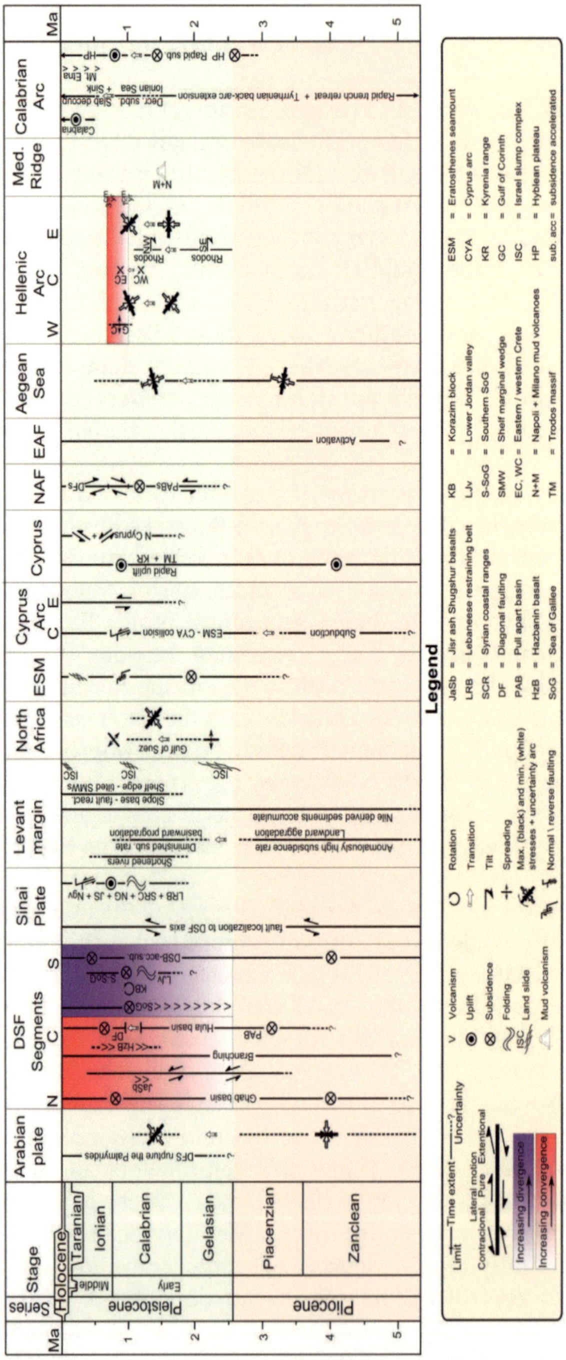

Fig. 3. Time–space diagram of the tectonic activity in the eastern Mediterranean region during the Pliocene to recent, modified after Schattner (2010). The time scale reflects the latest changes in the beginning of the Pleistocene (e.g., Mascarelli, 2009), however the data presented is based on works preceding this change

2.4 Anatolia and the Aegean sea

The early-to-mid Pleistocene tectonic transition was not limited to the easternmost part of the Mediterranean or the Levant but extended across the convergence front to the southern margin of Eurasia. Fault-kinematic analyses indicate a short-lived compressional stress regime across Anatolia and the North Anatolian fault around 0.7–1.0 Ma (mid Pleistocene; Over et al., 1997; Barka et al., 2000). As a result, a series of extensional basins located along the axis of the north Anatolian fault ceased to develop as pull-apart basins. The marginal faults of these basins (Erzincan, Suşehri–Gölova, Erbaa-Niksar, Bolu–Yeniçağa, Izmit-Sapanca and Marmara Sea basins, Fig. 1) became inactive when a through-going fault developed across each basin along the north Anatolian fault axis (Gürbüz and Gürer, 2009). This localization of motion to the axis of the main fault, in a contractional strike-slip motion, is similar to and synchronous, with the modifications along the Dead Sea fault (Marco, 2007). In particular, it is structurally similar to the through-going fault that developed at the same time diagonally across the Hula basin along the Dead Sea fault (Schattner and Weinberger, 2008). The short transpressional phase of the north Anatolian fault altered in the mid Pleistocene to transtension (Sorel et al., 1992; Bellier et al., 1997).

A marked change is observed during the same time frame in the direction of extension across the Aegean Sea (Figs. 1). NNE-SSW to N-S trending extensions which prevailed during the Pliocene shifted to NNW-SSE to N-S in the early-to-mid Pleistocene (Angelier et al., 1981; Mercier et al., 1989; Jolivet, 2001). Meanwhile the tectonic regime of the Hellenic arc changed drastically around 1 Ma (e.g., Duermeijer et al., 2000; Mantovani et al., 2002). In the western part of the arc, central and northern Greece, NE-SW tensional stresses shifted to NNW-SSE while the Gulf of Corinth underwent a rapid phase of rifting. An intense short-lived compressional phase separated the extensional periods (Sorel et al., 1992). Concurrently, at the eastern end of the Hellenic arc, N-S tension of western Turkey shifted toward NE-SW. The entire Hellenic arc was uplifted while several discontinuities developed in Crete-Rhodes region (Buttner and Kowalczyk, 1978; Armijo et al., 1992). Crustal stretching ceased in western Crete basin and initiated in the eastern Crete basin, with a roughly NW-SE extensional trend (Duermeijer et al., 2000). Rhodos, which was tilted southeastwards and submerged at depths of 500–600 m between 2.5 and 1.8 Ma, tilted to the northwest between 1.5 and 1.1 Ma. As a result some of its submerged relief re-emerged from the sea (van Hinsbergen et al., 2007).

2.5 Helenic and Calabrian arcs

At the outer circumference of the Hellenic arc, convergence rates increased from 1 to 3 cm/year (Rabaute and Chamot-Rooke, 2007). This shift probably occurred during the short but intense compressional phase of 1 to 0.7 Ma that prevailed between two extensional regimes (Sorel et al., 1992). The initiation of mud volcanism in Milano and Napoli volcanoes is dated to ~1.5 Ma (Kopf et al., 1998). The activity of both centers along the western and eastern branches of the Mediterranean Ridge was probably triggered by these changes in stress regime rather than from the on-going process of subduction. Rabaute and Chamot-Rooke (2007) suggest that most of the deformations took place along the Mediterranean ridge and the Hellenic backstop rather than along the north African margins.

Further west rapid trench retreat of the Calabrian arc (Figs. 1) and consequent Tyrrhenian back-arc extension were dominant since the Tortonian (Goes et al., 2004). However, towards the end of the early Pleistocene (~0.8 Ma), subduction rate of the Ionian oceanic lithosphere decreased considerably (Mattei et al., 2007). Back-arc opening in the southern Tyrrhenian

Sea and arc migration was suggested to die out when the arc arms ended their outward rotation (Sicily and Calabria - CW; Southern Apennines – CCW). The subducting slab began to decouple and sink into the mantle about 0.7-0.5 Ma, triggering the upward rebound of Calabria (Ibid.). This decoupling and associated uplift allowed an astanospheric sideflow which resulted in the initiation of volcanic activity of Mt. Etna at ~0.5 Ma (Gvirtzman and Nur, 1999). At the western end of the eastern Mediterranean, the late Pliocene-early Pleistocene rapid subsidence of the Hyblean Plateau (Branca et al., 2008) shifted to uplift at about 1 Ma while all its volcanic activity migrated northwards to the Catania Plain (Yellin-Dror et al., 1997).

3. Subduction to collision triggered the early-to-mid Pleistocene tectonic transition

The disruption in subduction of the Sinai plate is a relatively long-term process and not a singular moment in time. Unlike prehistoric developments, geodynamic effects on surface development often take several millions of years. However, in this case the effect was rapid (by geological terms) and somewhat catastrophic. The tectonic integration clearly shows that during the short period of the Early Pleistocene, the entire eastern Mediterranean was massively deformed and its mild regional topography was accentuated. Motion of tectonic plates are primarily driven by subducting slabs and their downwelling descent beneath overriding plates (Conrad and Lithgow-Bertelloni, 2002; Kincaid and Griffiths, 2003; Billen, 2008). When the Eratosthenes Seamount hindered the subduction of the Sinai plate a transition to collision commenced. The large dimensions and buoyant continental of the seamount (Ben-Avraham et al., 2002) (~25 km) induced a major regional disruption and uplift of the adjacent forearc. The incipient collision may have altered the lower crustal flow pattern (e.g., Westawey et al., 2009) which resulted in the regional change in topography. This major tectonic event could be considered as the immediate trigger for the synchronous structural modifications that occurred across the entire eastern Mediterranean region, constituting an Early Pleistocene tectonic transition.

Motion of tectonic plates is primarily driven by subducting slabs and their downwelling descent beneath overriding plates (Steiner and Conrad, 2007). The major driving force of this process, slab-pull, may vary depending on the stage of development of the overall subduction. Along the northern flanks of the African-Sinai-Arabian plates the stage of convergence plays a central role in the tectonic reorganization since the Neogene (Faccenna et al., 2006): Eurasia indentation by northward motion of Arabia, break-off of the northern Arabian slab beneath the Bitlis-Zagros suture, and consequent hinge rollback acceleration across the Hellenic trench. Tomography emphasizes the difference between a broken slab beneath Bitlis-Zagros (northern flank of Arabia) and a continuous Hellenic slab (northern flank of Africa; Ibid.). The transition between the two slabs was suggested to lie in the region between Rhodes and Cyprus where the slab extension to the north is not clear (Faccenna et al., 2006). These unique settings provoked the uplift of the Turkish-Iranian plateau, allowed the westward motion of the Anatolian plate and induced the formation of the north Anatolian fault. However, the occurrence of the sharp early-to-mid tectonic transition across the eastern Mediterranean still needs to be explained.

The Hellenic and Calabrian arcs propagated southwards mainly by hinge rollback of the African semi stationary plate which remained passive along its eastern Mediterranean margins (e.g., Rosenbaum and Lister, 2004; Doglioni et al., 2007). Similar relative motion could have prevailed also across the Cyprus Arc as part of the general closure of the eastern Mediterranean. However, since the late Pliocene–early Pleistocene, subduction across the Cyprus Arc was choked by the ~25 km thick and buoyant crust of the Eratosthenes Seamount. South of the Seamount, ~4 km of Nile derived sediments accumulated on the Mediterranean seafloor of the Sinai plate since the Pliocene (e.g., Segev et al., 2006). This thick load is exclusively linked with the vertical development of the Levant margin (Fig. 3 in Gvirtzman et al., 2008). Several works (e.g., Tibor et al., 1992) attributed both subsidence of the Levant margin and the uplift of Sinai plate (mainly in the Samaria and Judea) to the Nile sedimentary load. Others argue that the margin subsidence commenced prior to the Pliocene and therefore only part of it is attributed to the Nile sedimentary load (Gvirtzman et al., 2008). Either ways, Schattner (2010) assumes that this flexural load was not uni-directional but also affected regions north and west of the Nile cone. Instead, subsidence of the Seamount and its surroundings was recorded from the late Pliocene into the Pleistocene (e.g., Robertson, 1998) and no subduction-related flexural uplift was recorded in south of the Seamount. In this context the Nile load is likely to have inhibited the development of a flexural crustal fore-bulge. It is further suggested that the combined effect of Seamount buoyancy and fore-bulge depression resulted in lack of hinge roll-back and lower angle of subduction. These ultimately resulted in increased subduction friction, slab break-off and transition from subduction to collision.

The initial Eratosthenes Seamount collision with the Cyprus Arc fits into the center of the semi counterclockwise plate motion described above (black arrows on Fig. 1): Eurasia indentation by Arabia, westwards Anatolia escape and southwards Aegean propagation. However, in contrast to these long lasting motions the incipient Seamount-arc collision is constrained in time. With the transition to collision the relative convergent motion between Sinai and Anatolia increased abruptly, altering the stress regime. The immediate effect was recorded as the exhumation of Cyprus, however regional shortening was induced throughout the eastern Mediterranean, mainly radial to the collision. This convergence phase is exhibited by the folding, faulting and tilting that developed along the Levant margin (Sinai plate), and the diagonal faulting which ruptured basins along both the North Anatolian and the Dead Sea faults. On the other hand, extensional stresses were recorded in regions which are tangent to the collision, e.g., transtentional motion across the southern Dead Sea fault and the E to NE extensional event of northern Africa. In addition, cessation of subduction and slab break off also reflects reorganization of lower crustal flow (Westaway et al., 2009) and possibly the mantle lithosphere, which affect the regional isostasy distribution in the in the vicinity of the Cyprus Arc. Coupling between lower and upper crustal processes complexes the response pattern of local areas in the eastern Mediterranean, however, the identical timing points to a common trigger of the early-to-mid Pleistocene tectonic transition.

4. Conclusions

Subduction is a central geodynamic mechanism, which drives the entire plate-tectonic motion. Throughout Plio-Pleistocene time the progressive closure of the eastern Mediterranean basin continued across the Bitlis–Zagros collision zone and the Cyprus, Hellenic and Calabria arcs at variable rates. This convergence was severely interrupted during a relatively short period, late Pliocene–early Pleistocene, when subduction at the Cyprus Arc was transformed into collision due to the arrival of Eratosthenes Seamount from the south. Consequently, the entire eastern Mediterranean was massively deformed during the early-to-mid Pleistocene and its mild regional topography was accentuated. The Levantine corridor land-bridge became a more convenient gateway for hominin dispersal out of Africatowards Eurasia.

Co-occurrence of the structural modifications recorded across the entire region was synchronous rather than a cascade of events, with the exception of the somewhat younger deformations at the Calabrian arc. Timing of the deformations points to a temporal modification in the differential convergence across the boundary between Africa–Sinai–Arabia and Eurasia–Anatolia, which has been modified in the easternmost part of the basin by the Seamount–Arc collision and migrated to the western end. Timing of the transition is concurrent with the first pulse of hominin dispersal out of Africa. While the exact causes for human dispersal out of Africa are still highly debated, geodynamic forces set the stage for the first wave with the rest to follow.

5. References

Aksu, A. E., Calon, T. J., Hall, J., Mansfield, S., and Yasar, D., 2005a. The Cilicia-Adana basin complex, Eastern Mediterranean: Neogene evolution of an active fore-arc basin in an obliquely convergent margin. Marine Geology 221, 121-159.

Aksu, A. E., Calon, T. J., Hall, J., and Yasar, D., 2005b. Origin and evolution of the Neogene Iskenderun Basin, northeastern Mediterranean Sea. Marine Geology 221, 161-187.

Aksu, A. E., Hall, J., and Yaltlrak, C., 2009. Miocene-Recent evolution of Anaximander Mountains and Finike Basin at the junction of Hellenic and Cyprus Arcs, eastern Mediterranean. Marine Geology 258, 24-47.

Al-zoubi, A., and Ten Brink, U., 2001. Salt diapirs in the Dead Sea basin and their relationship with Quaternary extensional tectonics. Marine and Petroleum Geology 18, 779-797.

Angelier, J. et al., 1981. Analyses of fault mechanisms and expansion of southwestern Anatolia since the late Miocene. Tectonophysics, 75(3-4): T1-T9.

Armijo, R., Lyon-Caen, H., and Papanastassiou, D., 1992. East-west exstension and Holocene normal-fault scarps in the Hellenic arc. Geology 20, 491-494.

Avni, Y., 1998. Paleogeography and tectonics of the central Negev and the Dead Sea rift western margin during the Late Neogene and Quaternary, pp. 232. Geological Survey of Israel, Jerusalem.

Avni, Y., Bartov, Y., Garfunkel, Z., and Ginat, H., 2000. Evolution of the Paran drainage basin and its relation to the Plio-Pleistocene history of the Arava Rift western margin. Israel Journal of Earth Science 49, 215-238.

Barka, A., Akyuz, S., Cohen, H. A., and Watchorn, F., 2000. Tectonic evolution of the Niksar and Tasova–Erbaa pull-apart basins, North Anatolian Fault Zone: their significance for the motion of the Anatolian block. Tectonophysics 322, 243–264.

Bar-Yosef, O. and Belfer-Cohen, A., 2001. From Africa to Eurasia - early dispersals. Quaternary International, 75, 19-28.

Bellier, O., Over, S., Poisson, A., and Andreux, J., 1997. Recent temporal change in the stress state and modern stress field along the North Anatolian Fault Zone (Turkey). Geophysical Journal International 131, 60-86.

Ben-Avraham, Z., and Nur, A., 1976. Slip rates and morphology of continental collision belts. Geology 4, 661-664.

Ben-Avraham, Z., Ginzburg, A., Makris, J. and Eppelbaum, L., 2002. Crustal structure of the Levant Basin, eastern Mediterranean. Tectonophysics, 346, 23- 43.

Ben-Avraham, Z., Lazar, M., Schattner, U., and Marco, S., 2005. The Dead Sea Fault and its effect on civilization. In "Lecture Notes in Earth Sciences: Perspectives in Modern Seismology." (F. Wenzel, Ed.), pp. 147-170. Springer Verlag Heidelberg.

Ben-Avraham, Z., Schattner, U., Lazar, M., Hall, J. K., Ben-Gai, Y., Neev, D., and Reshef, M., 2006. Segmentation of the Levant continental margin, eastern Mediterranean. Tectonics 25, TC5002.

Ben-Avraham, Z., Tibor, G., Limonov, A. F., Leybov, M. B., Ivanov, M. K., Tokarev, M. Yu. and Woodside, J. M., 1995. Structure and tectonics of the eastern Cyprean arc. Marine and Petroleum Geology, 12(3), 263-271.

Ben-Gai, Y., Ben-Avraham, Z., Buchbinder, B., and Kendall, C., 2005. Post-Messinian evolution of the Levant basin based on two-dimensional stratigraphic modeling. Marine Geology 221, 359-379.

Billen, M.I., 2008. Modeling the Dynamics of Subducting Slabs. Annual Reviews of Earth and Planetary Sciences, 36, 325–356.

Bosworth, W., and Taviani, M., 1996. Late Quaternary reorientation of stress field and extension direction in the southern Gulf of Suez, Egypt: Evidence from uplifted coral terraces, mesoscopic fault arrays, and borehole breakouts. Tectonics 15, 791-802.

Branca, S., Coltelli, M., De Beni, E., and Wijbrans, J., 2008. Geological evolution of Mount Etna volcano (Italy) from earliest products until the first central volcanism (between 500 and 100 ka ago) inferred from geochronological and stratigraphic data. International Journal of Earth Sciences 97, 135-152.

Braun, D., Ron, H. and Marco, S., 1991. Magnetostratigraphy of the homonid tool-bearing Erk el Ahmar formation in the northern Dead Sea Rift. Israel Journal of Earth Sciences, 40, 191-197.

Brew, G., 2001. "Tectonic evolution of Syria interpreted from integrated geophysical and geological analysis." Unpublished Ph.D. thesis.

Burton-Ferguson, R., Aksu, A. E., Calon, T. J., and Hall, J., 2005. Seismic stratigraphy and structural evolution of the Adana Basin, eastern Mediterranean. Marine Geology 221, 189-222.

Butler, R. W. H., Spencer, S., and Griffiths, H., 1998. The structural response to evolving plate kinematics during transpression: evolution of the Lebanese restraining bend of the Dead Sea Transform. Spec. Publs. Geol. Soc. London 135, 81-106.

Buttner, D., and G., K., 1978. Late Cenozoic stratigraphy and paleogeography of Greece. A review. In: "Alps, Apennines, Hellenides." (H. Closs, D. Roeder, and K. Schmidt, Eds.), pp. 494-499. Schweizerbart'sche Verlagsbuchlandlung, Stuttgart.

Calon, T. J., Aksu, A. E., and Hall, J., 2005a. The Neogene evolution of the Outer Latakia Basin and its extension into the Eastern Mesaoria Basin (Cyprus), Eastern Mediterranean. Marine Geology b, 61-94.

Calon, T. J., Aksu, A. E., and Hall, J., 2005b. The Oligocene-Recent evolution of the Mesaoria Basin (Cyprus) and its western marine extension, Eastern Mediterranean. Marine Geology 221, 95-120.

Carton, H., Singh, S. C., Tapponier, P., Elias, A., Briais, A., Sursock, A., Jomma, R., King, G. C. P., Daëron, M., Jacques, E. and Barrier, L., 2009. Seismic evidence for Neogene and active shortening offshore of Lebanon (Shalimar cruise). Journal of Geophysical Research, 114: B07407, doi:10.1029/2007JB005391.

Carton, S., Singh, S. C., Tapponier, P., Elias, A., Briais, A., Sursock, A., Jomaa, R., King, G. C. P., Daëron, M., Jacques, E., and Barrier, E., 2007. Seismic Evidence for Neogene and Active Shortening Offshore Lebanon (SHALIMAR Cruise). In "American Geophysical Union, Fall Meeting 2007, abstract #T42B-07."

Carto, S.L., Weaver, A.J., Hetherington, R., Lam, Y., and Wiebe, E.C., 2009, Out of Africa and into an ice age: on the role of global climate change in the late Pleistocene migration of early modern humans out of Africa: Journal of Human Evolution, v. 56, p. 139-151.

Conrad, C.P. and Lithgow-Bertelloni, C., 2002. How mantle slabs drive plate tectonics. Science, 298, 207–209.

Daëron, M., Benedetti, L., Tapponnier, P., Sursock, A., and Finkel, R. C., 2004. Constraints on the post ~25-ka slip rate of the Yammouneh fault (Lebanon) using in situ cosmogenic 36Cl dating of offset limestone-clast fans. Earth and Planetary Science Letters 227, 105-119.

Daëron, M., Tapponnier, P., Jacques, E., Elias, A., King, G., Sursock, A., Geze, R., and Charbel, A., 2001. Evidence for Holocene Slip and Large Earthquakes on the Yammouneh Fault (Lebanon). In "Eos. Trans. AGU." pp. Abstract S52D-0666, Fall Meet. Suppl.

Demir, T., Westaway, R., Bridgland, D., Pringle, M., Yurtman, S., Beck, A., and Rowbotham, G., 2007. Ar-Ar dating of late Cenozoic basaltic volcanism in northern Syria: Implications for the history of incision by the River Euphrates and uplift of the northern Arabian Platform. Tectonics 26, TC3012, doi:10.1029/2006TC001959.

Doglioni, C., Carminati, E., Cuffaro, M. and Scrocca, D., 2007. Subduction kinematics and dynamic constraints. Earth-Science Reviews, 83, 125-175.

Dubertret, L., 1955. Carte géologique du Liban au 1/200 000. République libanaise, Ministère des Travaux publiques, Institut géographique nationnal.

Duermeijer, C. E., Nyst, M., Meijer, P. T., Langereis, C. G., and Spakman, W., 2000. Neogene evolution of the Aegean arc: paleomagnetic and geodetic evidence for a rapid and young rotation phase. Earth and Planetary Science Letters 176, 509-525.

Elias, A., 2006. "Le chevauchement de Tripoli-Saida: Croissance du Mont-Liban et risque sismique." Unpublished PhD thesis, Institut de Physique du Globe de Paris.

Elias, A., Tapponier, P., Singh, S. C., King, G. C. P., Briais, A., Daëron, M., Carton, S., Sursock, A., Jacques, E., Jomaa, R., and Klinger, Y., 2007. Active thrusting offshore Mount Lebanon: Source of the tsunamigenic A.D. 551 Beirut-Tripoli earthquake. Geology 35, 755-758.

Faccenna, C., Bellier, O., Martinod, J., Piromallo, C., and Regard, V., 2006. Slab detachment beneath eastern Anatolia: A possible cause for the formation of the North Anatolian fault. Earth and Planetary Science Letters 242, 85-97.

Frey Martinez, J., Cartwright, J., and Hall, B., 2005. 3D seismic interpretation of slump complexes: examples from the continental margin of Israel. Basin Research 17, 83–108.

Garfunkel, Z., 1998. Constrains on the origin and history of the Eastern Mediterranean basin. Tectonophysics, 298, 5-35.

Ginat, H., 1997. "The geology and landscape evolution of the Hiyyon and Zihor valleys." Unpublished PhD thesis, (in Hebrew, English abstract).

Ginat, H., Zilberman, E., and Amit, R., 2002. Red sedimentary units as indicators of Early Pleistocene tectonic activity in the southern Negev desert, Israel. Geomorphology 45, 127-146.

Goes, S., Giardini, D., Jennya, S., Hollensteinb, C., Kahleb, H.-G., and Geigerb, A., 2004. A recent tectonic reorganization in the south-central Mediterranean. Earth and Planetary Science Letters 226, 335-345.

Gomez, F., Karam, G., Khawlie, M., McClusky, S., Vernant, P., Reilinger, R., Jaafar, R., Tabet, C., Khair, K., and Barazangi, M., 2007. Global Positioning System measurements of strain accumulation and slip transfer through the restraining bend along the Dead Sea fault system in Lebanon. Geophysical Journal International 168, 1021-1028.

Gomez, F., Khawlie,M., Tabet, C., Darkal, A. N., Khair, K. and Barazangi, M., 2006. Late Cenozoic uplift along the northern Dead Sea transform in Lebanon and Syria. Earth and Planetary Science Letters, 241, 913-931.

Goren-Inbar, N., Feibel, C.S., Verosub, K.L., Melamed, Y., Kislev, M.E., Tchernov, E. and Saragusti, I., 2000. Pleistocene milestones on the Out-of-Africa Corridor at Gesher Benot Ya'aqov, Israel. Science, 289(5481), 944-947.

Griffiths, H. M., Clark, R. A., Thorp, K. M., and Spencer, S., 2000. Strain accommodation at the lateral margin of an active transpressive zone: geological and seismological evidence from the Lebanese restraining bend. Geological Society, London, Special Publications 157, 289-302.

Guiraud, R., 1990. Guiraud, R., 1990. Evolution post-triasique de l'avant-pays de la chaîne alpine en Algérie, d'après l'étude du bassin du Hodna et des régions voisines. Service Géologique de l'Algérie Mémoire 3, 271

Guiraud, R., Bosworth, W., Thierry, J., and Delplanque, A., 2005. Phanerozoic geological evolution of Northern and Central Africa: an overview. Journal of African Earth Sciences 43, 83-143.

Gürbüz, A., and Gürer, O. F., 2009. Middle Pleistocene extinction process of pull-apart basins along the North Anatolian Fault Zone. Physics of the Earth and Planetary Interiors 173, 177-180.

Gvirtzman, Z., and Nur, A., 1999. The formation of Mount Etna as the consequence of slab rollback. Nature 401, 782-785.

Gvirtzman, Z., Zilberman, E., and Folkman, Y., 2008. Reactivation of the Levant passive margin during the late Tertiary and formation of the Jaffa Basin offshore central Israel. Journal of the Geological Society 165, 563-578.

Hall, J., Aksu, A. E., Calon, T. J., and Yasar, D., 2005a. Varying tectonic control on basin development at an active microplate margin: Latakia Basin, Eastern Mediterranean. Marine Geology 221, 15-60.

Hall, J., Calon, T. J., Aksu, A. E., and Meade, S. R., 2005b. Structural evolution of the Latakia Ridge and Cyprus Basin at the front of the Cyprus Arc, Eastern Mediterranean Sea. Marine Geology 221, 261-297.

Hall, J., Aksu, A. E., Yaltlrak, C., and Winsor, J. D., 2009. Structural architecture of the Rhodes Basin: A deep depocentre that evolved since the Pliocene at the junction of Hellenic and Cyprus Arcs, eastern Mediterranean. Marine Geology 258, 1-23.

Hall, J., C. Hübscher, et al. (2010). From subduction to collision: the role of the Eratosthenes Seamount in the eastern Mediterranean. Tectonic Crossroads: Evolving Orogens of Eurasia-Africa-Arabia, Ankara, Turkey, Geological Society of America (GSA).

Hardenberg, M. F., and Robertson, A. H. F., 2007. Sedimentology of the NW margin of the Arabian plate and the SW-NE trending Nahr El-Kabir half-graben in northern Syria during the latest Cretaceous and Cenozoic. Sedimentary Geology 201, 231-266.

Harrison, R.W., Newell, W.L., Batihanli, H., Panayides, I., McGeehin, J.P., Mahan, S.A., Ozhur, A., Tsiolakis, E. and Necdet, M., 2004. Tectonic framework and Late Cenozoic tectonic history of the northern part of Cyprus: implications for earthquake hazards and regional tectonics. Journal of Asian Earth Sciences, 23: 191-210.

Heimann, A., 1990. The development of the Dead Sea rift and its margins in the northern Israel during the Pliocene and the Pleistocene. Golan Research Institute and Geological Survey of Israel.

Heimann, A., and Ron, H., 1993. Geometrical changes of plate boundaries along part of the northern Dead Sea transform: Geochronologic and paleomagnetic evidence. Tectonics 12, 477-491.

Horowitz, A., 2001. The Jordan Rift Valley. A. A. Balkema, 748 pp.

Hurwitz, S., Garfunkel, Z., Ben-Gai, Y., Reznikov, M., Rotstein, Y., and Gvirtzman, H., 2002. The tectonic framework of a complex pull-apart basin: seismic reflection observations in the Sea of Galilee, Dead Sea transform. Tectonophysics 359, 289–306.

Isler, F. I., Aksu, A. E., Hall, J., Calon, T. J., and Yasar, D., 2005. Neogene development of the Antalya Basin, Eastern Mediterranean: An active forearc basin adjacent to an arc junction. Marine Geology 221, 299-330.

Jolivet, L., 2001. A comparison of geodetic and finite strain pattern in the Aegean, geodynamic implications. Earth and Planetary Science Letters 187, 95-104.

Kempler, D., 1998. Eratosthenes Seamount: the possible spearhead of incipient continental collision in the eastern Mediterranean. In "Proceedings of the Ocean Drilling Program, Scientific Results, Vol. 160." (A. H. F. Robertson, K.-C. Emeis, C. Richter, and A. Camerlenghi, Eds.), pp. 709-721.

Kempler, D., and Ben-Avraham, Z., 1987. The tectonic evolution of the Cyprean arc. Annales Tectonicae 1, 58-71.

Kincaid, C. and Griffiths, R.W., 2003. Laboratory models of the thermal evolution of the mantle during rollback subduction. Nature, 425, 58-62.

Klein R. G. 1989. The Human Career: Human Biological and Cultural Origins. University of Chicago Press, Chicago, 544 pp.

Kopf, A., Robertson, A. H. F., Clennell, M. B., and Flecker, R., 1998. Mechanisms of mud extrusion on the Mediterranean Ridge Accretionary Complex. Geo-Marine Letters 18, 97-114.

Kopp, M. P., Adzhamyan, Zh., Il'yas, K., Fakiani, F., and Khafez, A., 1999. Mechanism of Formation of the El Ghab Wrench Graben (Syria). Geotectonics, 33, 408–422.

Larsen, B. D., Ben-Avraham, Z., and Shulman, H., 2002. Fault and salt tectonics in the southern Dead Sea basin. Tectonophysics 346, 71-90.

Laukhin, S.A., Ronen, A., Pospelova, G.A., Sharonova, Z.V., Ranov, V.A., Burdukiewicz, J.M., Volgina, V.A. and Tsatskin, A., 2001. New data on the geology and geochronology of the Lower Paleolithic site Bizat Ruhama in the Southern Levant. Paléorient 27(1), 69-80.

Lazar, M., Ben-Avraham, Z., and Schattner, U., 2006. Formation of sequential basins along a strike-slip fault-Geophysical observations from the Dead Sea basin. Tectonophysics 421, 53-69.

Limonov, A.F., Woodside, J.M., Ivanov, M.K. (Eds.), 1994. Mud Volcanism in the Mediterranean and Black Seas and Shallow Structure of the Eratosthenes Seamount. Initial Results of the Geological and Geophysical Investigations During the Third 'Training Through Research' Cruise of the R/V 'Glendzhik' (June–July, 1993): UNESCO Reports in Marine Sciences, 64. 173 pp.

Mantovani, E., Albarello, D., Babbucci, D., Tamburelli, C., and Viti, M., 2002. Trench-Arc-BackArc Systems in the Mediterranean area: Examples of Extrusion Tectonics. Journal of the Virtual Explorer 8, 131-147.

Marco, S., 2007. Temporal variation in the geometry of a strike-slip fault zone: Examples from the Dead Sea Transform. Tectonophysics, 445: 186-199.

Mascarelli, A.L., 2009. Quaternary geologistswin timescale vote. Nature, 459: 624.

Matmon, A., Enzel, Y., Zilberman, E., and Heimann, A., 1999. Late Pliocene and Pleistocene reversal of drainage systems in northern Israel: tectonic implications. Geomorphology 28, 43-59.

Matmon, A., Wdowinski, S. and Hall, J.K., 2003. Morphological and structural relations in the Galilee extensional domain, northern Israel. Tectonophysics, 371, 223-241.

Mattei, M., Cifelli, F., and D'Agostino, N., 2007. The evolution of the Calabrian Arc: Evidence from paleomagnetic and GPS observations. Earth and Planetary Science Letters 263, 259-274.

McQuarrie, N., Stock, J.M., Verdel, C. and Wernicke, B.P., 2003. Cenozoic evolution of Neotethys and implications for the causes of plate motions. Geophysical Research Letters, 30, doi:10.1029/2003GL017992.

Medimap Group, Loubrieu B., Mascle J., and al., E., 2005. Morpho-bathymetry of the Mediterranean Sea, (C. I. edition, Ed.). CIESM/Ifremer edition, 2 maps at 1/2000000.

Meijer, P.T. and Krijgsman, W., 2005. A quantitative analysis of the desiccation and re-filling of the mediterranean during the Messinian salinity crisis. Earth and Planetary Science Letters, 240 510-520.

Mercier, J.L., Sorel, D., Vergeley, P. and Simeakis, C., 1989, 1989. Extensional tectonic regimes in the Aegean basins during the Cenozoic. Basin Research, 2, 49-71.

Over, S., Bellier, O., Poisson, A., and Andrieux, J., 1997. Late Cenozoic stress state changes along the central North Anatolian Fault Zone (Turkey). Annales Tectonicae XI 1/2.

Poole, A. J., and Robertson, A. H. F., 1991. Quaternary uplift and sea-level change at an active plate boundary, Cyprus. Journal of the Geological Society, London 148, 909-921.

Poole, A.J. and Robertson, A.H.F., 1991. Quaternary uplift and sea-level change at an active plate boundary, Cyprus. Journal of the Geological Society, London, 148, 909-921.

Rabaute, A., and Chamot-Rooke, N., 2007. Quantitative mapping of mud volcanism at the western Mediterranean Ridge - backstop contact. Marine Geophysical Research 28, 271–295.

Reilinger, R., McClusky, S., Vernant, P., Lawrence, S., Ergintav, S., Cakmak, R., Ozener, H., Kadirov, F., Guliev, I., Stepanyan, R., Nadariya, M., Hahubia, G., Mahmoud, S., Sakr, K., ArRajehi, A., Paradissis, D., Al-Aydrus, A., Prilepin, M., Guseva, T., Evren, E., Dmitrotsa, A., Filikov, S. V., Gomez, F., Al-Ghazzi, R., and Karam, G., 2006. GPS constraints on continental deformation in the Africa-Arabia-Eurasia continental collision zone and implications for the dynamics of plate interactions. Journal of Geophysical Research 111, B05411.

Robertson, A.H.F., Clift, P.D., and Jones, G., 1991, Paleogeographic and paloetectonic evolution of the Eastern Mediterranean Neotethys: Palaeogeography, Palaeoclimatology, Palaeoecology, v. 87, p. 289-343.

Robertson, A. H. F., 1998. Tectonic significance of the Eratosthenes Seamount: a continental fragment in the process of collision with a subduction zone in the eastern Mediterranean (Ocean Ddilling Program Leg 160. Tectonophysics 298, 63-82.

Robertson, A. H. F., and Woodcock, N. H., 1986. The role of the Kyrenia Range Lineament, Cyprus, in the geological evolution of the eastern Mediterranean area. Philosophical Transactions of the Royal Society of London A317, 141–177.

Robertson, A. H. F., Kidd, R. B., Ivanov, M. K., Limonov, A. F., Woodside, J. M., Galindo-Zaldivar, J., and Nieto, L., 1995. Eratosthenes Seamount: collisional processes in the easternmost Mediterranean in relation to the Plio-Quaternary uplift of southern Cyprus. Terra Nova 7, 254-264.

Rosenbaum, G. and Lister, G.S., 2004. Neogene and Quaternary rollback evolution of the Tyrrhenian Sea, the Apennines, and the Sicilian Maghrebides. Tectonics, 23: TC1013, doi:10.1029/2003TC001518.

Rotstein, Y., Bartov, Y., and Frieslander, U., 1992. Evidence for local shifting of the main fault and changes in the structural setting, Kinarot basin, Dead Sea transform. Geology 20, 251-254.

Rukieh, M., Trifonov, V. G., Dodonov, A. E., Minini, H., Ammar, O., Ivanova, T. P., Zaza, T., Yusef, A., Al-Shara, M., and Jobaili, Y., 2005. Neotectonic map of Syria and some aspects of Late Cenozoic evolution of the northwestern boundary zone of the Arabian plate. Journal of Geodynamics 40, 235-256.

Salamon, A., Hofstetter, A., Garfunkel, Z., and Ron, H., 2003. Seismotectonics of the Sinai subplate - the eastern Mediterranean region. Geophysical Journal International 155, 149-173.

Schattner, U. and Weinberger, R., 2008. A mid-Pleistocene deformation transition in the Hula basin, northern Israel: Implications for the tectonic evolution of the Dead Sea Fault. Geochemistry Geophysics Geosystems, 9, Q07009.

Schattner, U., and Ben-Avraham, Z., 2007. Transform margin of the northern Levant, eastern Mediterranean: From formation to reactivation. Tectonics 26, 1-17.

Schattner, U., and Lazar, M., 2009, Subduction, collision and initiation of hominin dispersal: Quaternary Science Reviews, v. 28, p. 1820-1824.

Schattner, U., 2010, What triggered the early-to-mid Pleistocene tectonic transition across the entire eastern Mediterranean?: Earth and Planetary Science Letters, v. 289, p. 539-548.

Schattner, U., Ben-Avraham, Z., Lazar, M., and Hübscher, C., 2006. Tectonic isolation of the Levant basin offshore Galilee-Lebanon - Effects of the Dead Sea fault plate boundary on the Levant continental margin, eastern Mediterranean. Journal of Structural Geology 28, 2049-2066.

Segev, A., Rybakov, M., Lyakhovsky, V., Hofstetter, A., Tibor, G., Goldshmidt, V., and Ben Avraham, Z., 2006. The structure, isostasy and gravity field of the Levant continental margin and the southeast Mediterranean area. Tectonophysics 425, 137-157.

Semaw, S., 2000. The World's Oldest Stone Artefacts from Gona, Ethiopia: Their Implications for Understanding Stone Technology and Patterns of Human Evolution Between 2·6-1·5 Million Years Ago. Journal of Archaeological Science, 27(12): 1197-1214.

Şengör, A., Özeren, S., Genç, T. and Zor, E., 2003. East Anatolian high plateau as a mantle supported, north-south shortened domal structure. Geophysical Research Letters, 30.

Sharkov, E. V., Chernyshev, I. V., Devyatkin, E. V., and et al., 1994. Geochronology of Late Cenozoic Basalts in Western Syria, Petrologiya. Petrologiya 2, 439–448.

Sneh, A., and Weinberger, G., 2003. Geological Map of Israel. Sheet 2-II Metulla. Geological Survey of Israel.

Sneh, A., and Weinberger, G., 2006. Geological Map of Israel. Sheet 2-IV Rosh Pinna, scale 1:50,000. Geological Survey of Israel.

Sorel, D., Bizon, G., Aliaj, S., and Hasani, L., 1992. Calage stratigraphique de l'gge et de la duree des phases compressives des Hellenides externes (Grece Nord-occidentale et Albanie), du Miocene a 1'Actuel. Bull. Soc. GLol. France 163.

Stringer, C., 2000, Coasting out of Africa: Nature, v. 405, p. 24-27.

Steiner, S. A., and Conrad, C. P., 2007. Does active mantle upwelling help drive plate motions? Physics of the Earth and Planetary Interiors 161, 103-114.

Tapponnier, P. E., Daëron, M., Sursock, A., Jomaa, R., Briais, A., Carton, H., Singh, S., Elias, A., King, G. C., and E., J., 2004. Passive-active margin inversion along the Levant plate-boundary: Subduction birth and growth of Mt Lebanon. In "AGU fall meeting." pp. Abstract T52B-05, San Francisco.

Tchernov, E., 1987. The age of the 'Ubediya Formation, an Early Pleistocene hominid site in the Jordan Valley Israel, Israel Journal of Earth Sciences 36: 3–30.

Templeton, A., 2002. Out of Africa again and again. Nature, 416(6876): 45-51.

ten Brink, U. and Ben-Avraham, Z., 1989. The anatomy of a pull-apart basin: seismic reflection observations of the Dead Sea basin. Tectonics, 8: 333-350.

Tibor, G., Ben Avraham, Z., Steckler, M. and Fligelman, H., 1992. Late Tertiary subsidence history of the southern Levant margin, Eastern Mediterranean Sea, and its implications to the understanding of the Messinian event. Journal of Geophysical Research, 97, 17593–17614.

van Hinsbergen, D. J. J., Krijgsman, W., Langereis, C. G., Cornee, J.-J., Duermeijer, C. E., and van Vugt, N., 2007. Discrete Plio-Pleistocene phases of tilting and counterclockwise rotation in the southeastern Aegean arc (Rhodos, Greece): early Pliocene formation of the south Aegean left-lateral strike-slip system. Journal of the Geological Society 164, 1133-1144.

Walley, C.D., 1998. Some outstanding issues in the geology of Lebanon and their importance in the tectonic evolution of the Levantine region. Tectonophysics, 298, 37-62.

Wdowinski, S., and Zilberman, E., 1996. Kinematic modeling of large-scale structural asymmetry across the Dead Sea Rift. Tectonophysics 266, 187-201.

Wdowinski, S., and Zilberman, E., 1997. Systematic analyses of the large-scale topography and structures across the Dead Sea Rift. Tectonics 16, 409-424.

Weinberger, G., Gross, M.R., and Sneh, A., 2009, Evolving deformation along a transform plate boundary: Example from the Dead Sea Fault in northern Israel: Tectonics, 28, TC5005, doi:10.1029/2008TC002316.

Weinstein, Y., Navon, O., Altherr, R., and Stein, M., 2006. The Role of Lithospheric Mantle Heterogeneity in the Generation of Plio-Pleistocene Alkali Basaltic Suites from NW Harrat Ash Shaam (Israel). Journal of Petrology 47, 1017-1050.

Westaway, R. Guillou, H., Seyrek, A., Demir, T., Bridgland, D., Scaillet, S. and Beck, A., 2009. Late Cenozoic surface uplift, basaltic volcanism, and incision by the River Tigris around Diyarbakır, SE Turkey. International Journal of Earth Sciences, 98(3): 601-625.

Westaway, R., and Arger, J., 2001. Kinematics of the Malatya–Ovacik Fault Zone. Geodynamica Acta 14, 103-131.

Wortel, M. J. R., and Spakman, W., 2000. Subduction and Slab Detachment in the Mediterranean-Carpathian Region. Science 290, 1910-1917.

Yellin-Dror, A., Grasso, M., Ben-Avraham, Z., and Tibor, G., 1997. The subsidence history of the northern Hyblean Plateau margin, southeastern Sicily. Tectonophysics 282, 277-289.

Zanchi, A., Crosta, G. B., and Darkal, A. N., 2002. Paleostress analyses in NW Syria: constraints on the Cenozoic evolution of the northwestern margin of the Arabian plate. Tectonophysics 357, 255-278.

Zurieli, A., 2002. Structure and neotectonics in Kinarot Valley based on high-resolution seismic reflection. Tel-Aviv University.

Part 4

The Persian Gulf

13

The Qatar–South Fars Arch Development (Arabian Platform, Persian Gulf): Insights from Seismic Interpretation and Analogue Modelling

C.R. Perotti[1], S. Carruba[2], M. Rinaldi[1], G. Bertozzi[2], L. Feltre[2] and M. Rahimi[3]
[1]*Dipartimento di Scienze della Terra, Università di Pavia, Pavia,*
[2]*Edison S.p.A., Milano,*
[3]*NIOC Exploration, 1st Dead End, Tehran*
[1,2]*Italy*
[3]*Iran*

1. Introduction

The Qatar-South Fars Arch is a major regional anticline that runs through the central Persian Gulf (figure 1), warping the sedimentary cover of the Arabian Platform. The structure is detectable from offshore seismic data in the Iranian sector of the Persian Gulf, where it has a northeast-southwest direction, and extends southwards into the Qatar peninsula. Offshore, the arch hosts the biggest gas and condensate field in the world (South Pars–North Field), which straddles Iranian and Qatari waters. The Qatar–South Fars Arch represents a first-order structure and separates the Persian Gulf basin into two areas characterized by significant Proterozoic Hormuz salt diapirism (Northern and Southern Gulf Salt Basins, figure 1). The reported absence of salt-related phenomena on its crest led the authors to consider it as cored by an Infracambrian basement horst block, initiated during the Infracambrian Najd rifting, and repeatedly reactivated during subsequent geological time (Al-Husseini, 2000; Konert et al., 2001; Edgell, 1996; Talbot & Alavi, 1996). Although basement tectonics in response to geodynamic events is considered the main cause of the deformations associated with basement-cored structures in the Arabian Platform, the development history of the Qatar–South Fars Arch may not fit neatly into this scenario. Its continuous growth and extension, which is about one order of magnitude greater than the other north-south trending "Arabian structures", the absence of major faults bounding the structure that actually encompasses several salt diapirs, and its north-northeast trend poorly matching with possible compressional and extensional reactivations that acted in a northeast-southwest direction point to a more complex origin for this intracratonic regional deformation.

In this paper, the interpretation of a grid of 2D seismic lines crossing the Qatar–South Fars Arch axis in the Iranian sector of the Central Persian Gulf allows the study of a regional geological section perpendicular to the arch and the reconstruction of a structural map of the area, with the aim of defining the deformation history of this structure. Considering the widespread and long-lasting Hormuz salt tectonics in the neighbourhood of the Qatar-

South Fars Arch, we built scaled physical models in order to investigate the possible role of regional basal salt movements in the warping of the sedimentary succession over the structure, and to propose an alternative possible model of deformation of this intracratonic arch.

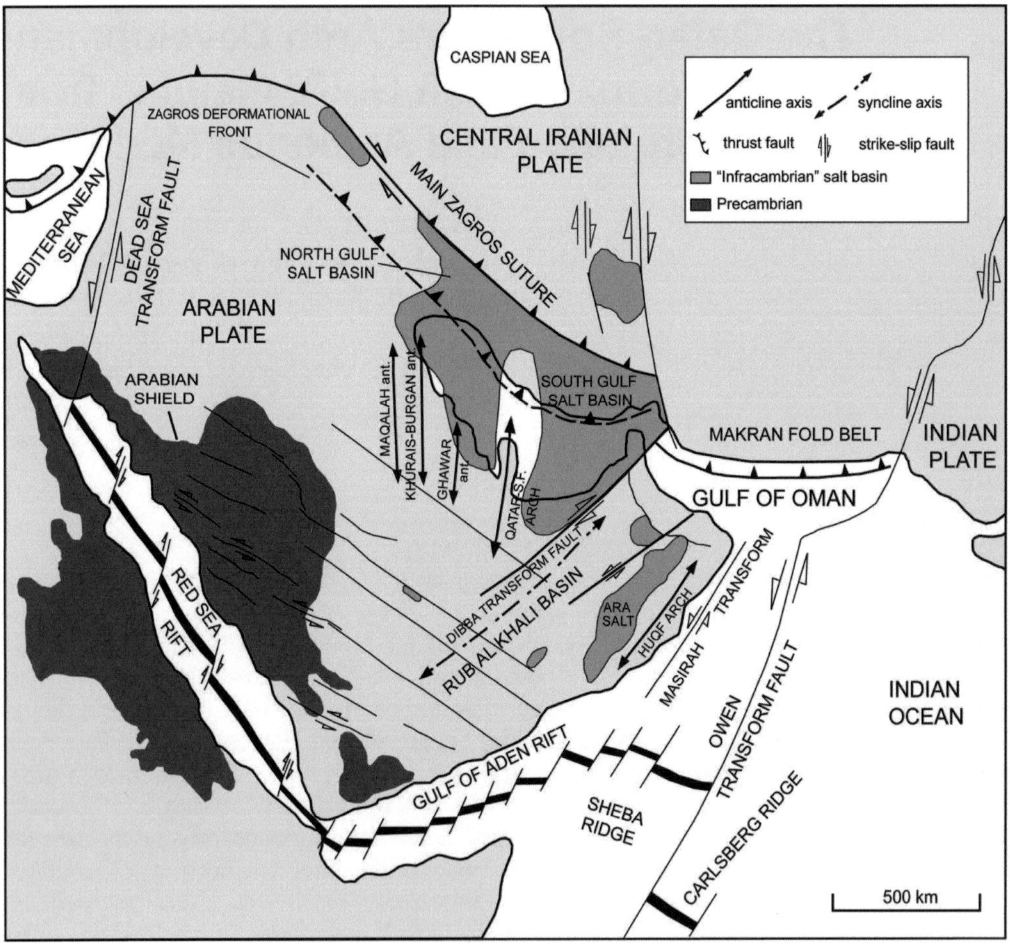

Fig. 1. Tectonic sketch map of the Persian Gulf and Arabian Peninsula, modified from Al-Husseini (2000), Ziegler (2001) and Pollastro (2003)

2. Geological setting and evolution of the Arabian Platform

The Arabian Platform is a vast sedimentary basin where a very thick, virtually continuous sequence of sediments (figure 2) accumulated from the Late Proterozoic to Holocene on the northeastern margin of the Arabian sector of Gondwana (Alsharan & Nairn, 1997; Beydoun, 1991; Edgell, 1996; Konert et al., 2001; Sharland et al,. 2001; Ziegler 2001). In a southwest-northeast section (figure 3), the basin is wedge-shaped with sediments gradually thickening from a few metres around the Arabian Shield (southwest), to about 18 km (Edgell, 1996) in the Zagros (northeast), which corresponds to the sector of the Arabian Platform deformed as a result of the Arabian-Iranian plate convergence in the Neogene. The present-day Zagros deformational front runs approximately along the Iranian coastline.

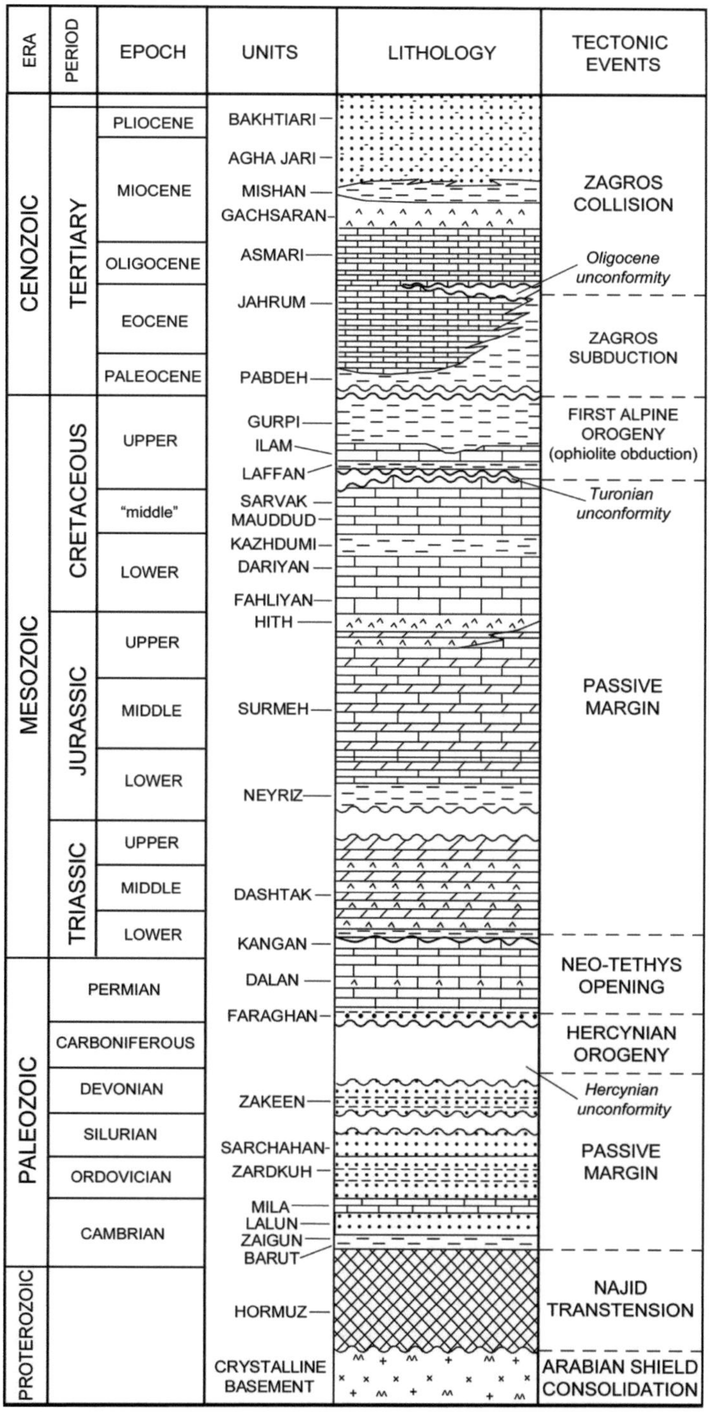

Fig. 2. Stratigraphic section, litho-stratigraphic units and major tectonic events of the central Persian Gulf, modified from Sharland et al. (2001) and Al-Husseini (2008)

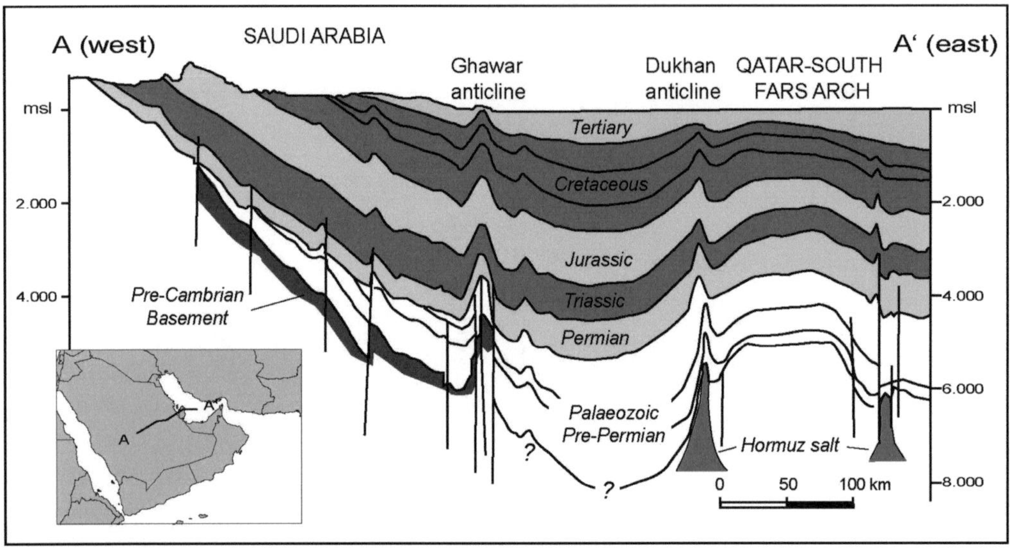

Fig. 3. Simplified geologic cross section across Saudi Arabia and Qatar from Arabian Shield to Arabian-Persian Gulf, modified from Alsharhan and Nairn (1997), Konert et al. (2001) and Pollastro (2003)

The geological history of the Arabian Platform (Alsharan & Nairn, 1997; Konert et al., 2001; Sharland et al., 2001; Ziegler, 2001) can be summarized as follows (figure 4). After the consolidation of the Arabian Shield, a Late Proterozoic extensional phase (Najd rifting: Husseini, 1988; Al-Husseini, 2000) created a number of basins that are currently located in the Persian Gulf sector and in Oman, in which the thick (up to 2.5 km: Edgell, 1996) evaporites, predominantly composed of halite, of the Hormuz Formation (and the equivalent Ara formation in Oman) were deposited. The subsequent geological history of the area was characterized, during most of the Palaeozoic, by uniform clastic continental and shallow-marine sedimentation on a stable passive margin–interior margin setting on northeastern Gondwana. The Hercynian events of the Carboniferous affected the area, creating regional uplift, widespread erosion and basement tectonism along the inherited, mechanically weak Late Proterozoic trends (Konert et al., 2001). From the Permian to the Palaeogene the area was a broad, stable platform on the newly-formed passive margin at the northeastern border of the African plate, where the deposition of mainly shallow-water carbonates with minor anhydrites and shales occurred. The Neothethys Ocean to the northeast separated this margin from the Laurasia continental assemblage. In the Upper Cretaceous and then since the Oligocene, the northeastern part of the basin has been undergoing shortening as a consequence of collision of the Arabian Plate with Laurasia (Zagros orogeny). The resulting flexure of the Arabian Plate underneath the Zagros fold-and-thrust belt created a wedge-shaped, low-angle (less than 2°) foreland basin, that has been the site of mixed evaporitic, carbonate and clastic sedimentation. Several regional unconformities and hiatuses punctuate the sedimentary succession as a consequence of eustatic sea-level changes and epeirogenic movements, most of them having occurred during plate-wide events (Sharland et al., 2001).

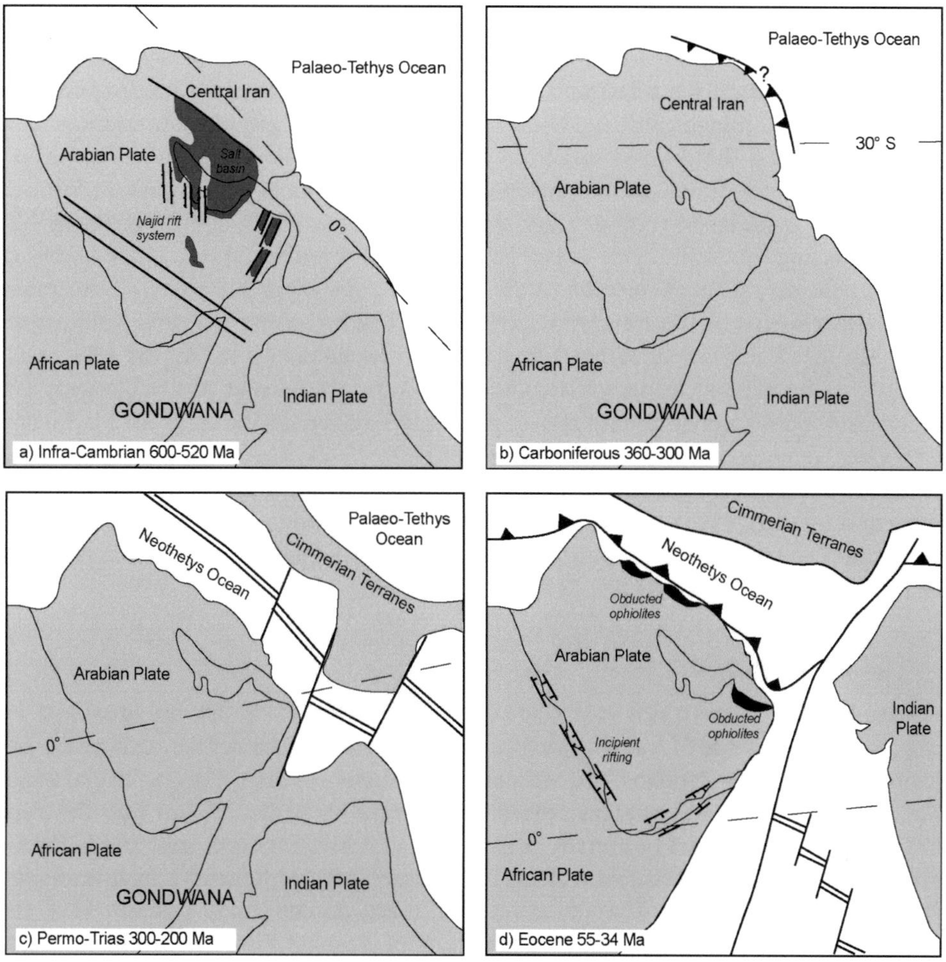

Fig. 4. Simplified plate tectonic evolution of the Persian Gulf and Arabian Peninsula, modified from Loosveld et al. (1996), Sharland et al. (2001), Stampfli and Borel (2004)

It is widely accepted that the tectonic history of the Arabian Platform has been influenced by the inheritance of the Precambrian structural highs forming an unstable basement for the basin. According to most authors, these deep structures have been repeatedly reactivated during the Phanerozoic, triggering the uplift of salt diapirs (Edgell, 1996) and basement-cored structures (Wender et al., 1998), and controlling the geometry and deposition of the overlying sedimentary cover (Edgell, 1992; Konert, 2001; Pollastro, 2003). The deformation structures in the Gulf area are essentially caused by the movements of the Hormuz evaporites of late Proterozoic to earliest Cambrian age (often referred to informally as 'Infracambrian') at the base of the sedimentary succession (Al-Husseini, 2008), which have created salt domes, anticlines and diapirs characterized by continuous growth since at least the Jurassic (Edgell, 1996; Sugden, 1962). A steady diapiric rise during the Early Palaeozoic has also been described, as well as several main phases of salt movement in the Permo-Triassic, middle Cretaceous, Eocene–Oligocene and Neogene (Carruba et al., 2007). The time-correspondence between the phases of diapiric uplift and the regional tectonic events suggests that the latter could have triggered salt movements. The Arabian coastal sector and its interior are mainly affected by very large (hundreds of kilometres long by tens of kilometres wide), roughly north–south trending anticlines, with gently dipping flanks and sometimes a subtle topographic expression. These structures are of paramount economic importance, forming the oil and gas traps of the most prolific hydrocarbon basin in the world. Most of these anticlines are interpreted (e.g. Ghawar and Khurais) as formed by draping of the sedimentary cover over systems of major north-south basement blocks (originated during the Najd rifting episodes) delimited by faults (Edgell, 1992). Horst uplift associated with strong crestal erosion in the Carboniferous is particularly well documented in the Ghawar structure (Wender et al., 1998). The folding of the overlying succession in Ghawar has been attributed to horst reactivations in response to Zagros rifting (Early Triassic), the First Alpine Orogeny (Late Cretaceous) and the Second Alpine Orogeny (Mid–Late Tertiary). Some elongated anticlines, generally smaller, in the Arabian Platform and Persian Gulf, have cores of deep-seated salt (Edgell, 1992). The basement at the core of structures is revealed by geophysical (gravity and magnetic) investigations (Edgell, 1992; Johnson and Stewart, 1995), and local subsurface data (Wender et al., 1998), while seismic evidence has been gained so far only in some cases in the internal parts of the platform in Arabia (Edgell, 1992; Sharland et al., 2001).

3. The Qatar–South Fars Arch

The Qatar-South Fars Arch is a very large (over 100 km wide and 300 km long) and long-lasting (several hundreds of Ma) warping of the Phanerozoic sedimentary succession of the Arabian Platform in the Persian Gulf. Its onshore expression is the Qatar peninsula, which is marked by a low-relief topographic feature that protrudes some 150 km into the Persian Gulf. The peninsula consists of outcropping Eocene and Mio-Pliocene units (Alsharhan and Nairn, 1994; Dill et al., 2003; Nasir et al., 2008), very gently deformed in a north–south regional anticline (figure 5). The Qatar regional anticline continues offshore in a north-northeast to northeast direction where the arching of Arabian Platform sedimentary cover underneath the Quaternary sediments of the Persian Gulf is known from geophysical data. Talbot and Alavi (1996) considered the continuation of the Qatar–South Fars Arch farther north into the Zagros belt, where a north–south trending zone, up to 100 km wide, is aligned with the north–south lineaments of the Arabian Platform that have been repeatedly reactivated. This zone constitutes the persistently high area (as revealed by stratigraphically thinned units) of the Fars Platform, and is characterized by the absence of salt diapirs, which occur estensively to the west and east of it.

The Qatar–South Fars Arch has been described as a comparatively reduced Phanerozoic sedimentary cover draping a repeatedly rejuvenated Precambrian basement horst block. Indications of that include:

1. the distribution of salt diapirs of Late Proterozoic Hormuz formations in the area, which are present only southeast and northwest of the Arch, and could delineate a north-northeast to northeast basement feature aligned with the other basement structures of the "Arabian trend" that affect the Arabian coastal sector and its interior (Al-Husseini, 2000; Edgell, 1992);
2. the relative thinning of the Mesozoic sedimentary section over the Qatar–South Fars Arch (Alsharan and Nairn, 1997);

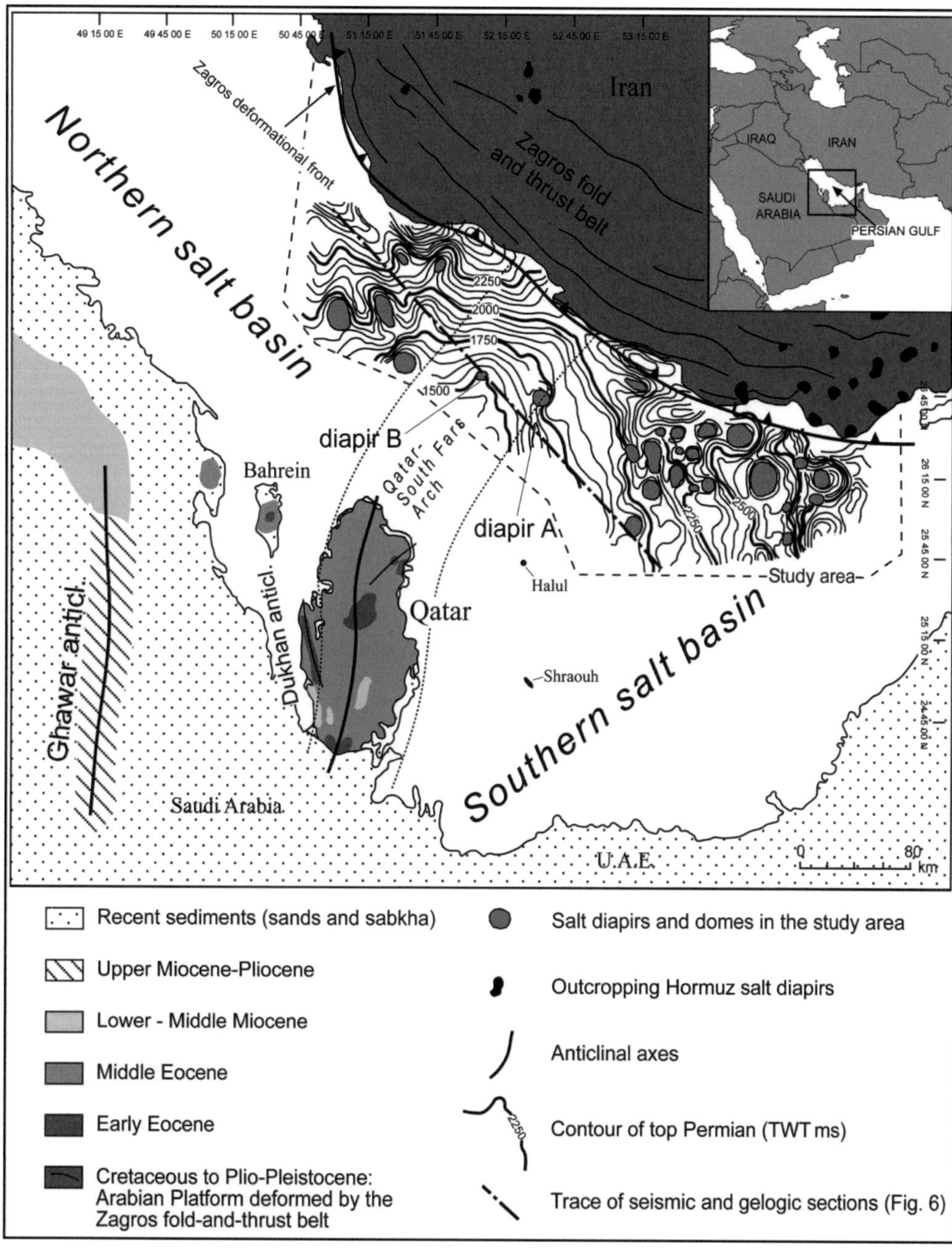

Fig. 5. Tectonic map of the central Persian Gulf showing the seismic contour map in time (twt ms) of near top Permian in the study area

3. the growth of the Qatar–South Fars Arch during several periods of time: Alsharan and Nairn (1994) report emergence at the end of the Triassic to Early Jurassic, and during the Turonian; minor uplifts in the Early–Middle Eocene and at the end of the Middle Eocene. Then, during the Late Eocene to Oligocene a main phase of Qatar–South Fars

Arch uplift occurred (concurrently with the development of a major unconformity). Another uplift phase started in the Early Miocene and increased in intensity during Late Miocene–Pliocene;

4. the Qatar peninsula anticlinorium with outcropping Tertiary units, strongly contrasting with the surrounding Quaternary cover (figure 5).

So far, there is no published subsurface evidence (well or seismic data) of the basement high below the Qatar–South Fars Arch and adjacent areas in the Persian Gulf. As far as its depth is concerned, the top of basement high below the Qatar peninsula, as shown by the tentative basement map drawn by Konert et al. (2001), is placed at a depth of 4–5 km. Depth to basement rapidly increases to 8–9 km in the Persian Gulf west, north and east of Qatar. The thickness of the sedimentary cover in the Persian Gulf has been generally quoted by other authors as being more than 10 km (Alsharhan and Nairn, 1994), 12 km (Pollastro, 2003) or 13 km (Edgell, 1992, 1996).

4. Seismic interpretation

A 2D seismic coverage with a 2 x 2 km grid has been interpreted in the Iranian sector of the Central Persian Gulf (figure 5). Acquisition was performed in 1999 with a 4600 m streamer and 7 s of record length; processing consisted of a routine including deconvolution, radon multiple removal, common offset DMO (Dip Move Out), stack and migration.

Following well-to-seismic calibration we interpreted and correlated 17 key horizons on a 340 km long composite seismic profile stretching in a northwest–southeast direction perpendicular to the Qatar–South Fars Arch (figure 5). Well calibration was available down to the top of the Permian (1500–2200 ms TWT); from this level downwards the seismic quality and continuity of reflections tend to rapidly degrade. Seismic evidence of basement is lacking below the culmination of the arch. On the flanks of the Qatar–South Fars Arch, the Hormuz salt is buried at a depth greater than 7 s TWT, beyond the lower limit of seismic data, as indicated by concave reflections bordering relatively seismic-transparent zones at these levels (figure 6).

The time interpretation was converted to depth using interval velocities measured from wells. A constant velocity of 4000 m/s has been used for the interval from top-Permian to the base of the section, giving a profile as deep as 14 km (figure 7). The depth conversion did not significantly alter the geometric features visible in the time section. The section across the Qatar–South Fars Arch (figure 7) is characterized by an overall convex-upward arrangement of reflectors, featuring a 250–300 km wide regional warping with extremely low-dipping limbs (0.4–0.7° at top-Permian) and an overall increase in structural dip downwards. Several thinning phenomena highlighted by the interpretation of the Meso-Cenozoic horizons are visible on both limbs of the arch in several intervals. The line drawing below the top of the Permian shows seismic reflections with progressively increasing dips in the whole pre-Permian succession (figure 8). No discontinuities in seismic features and reflector geometries are visible in the deepest parts of the Arch culmination, where the reflections appear to be warped conformably with the overlying units. The sequence above the Oligocene unconformity does not appear to be folded, but tilted to the northwest with a series of onlap phenomena occurring in different phases. The main erosional unconformities (Turonian, Oligocene) are underlain by truncations whose geometries testify to structural growth before the erosional period (figure 9).

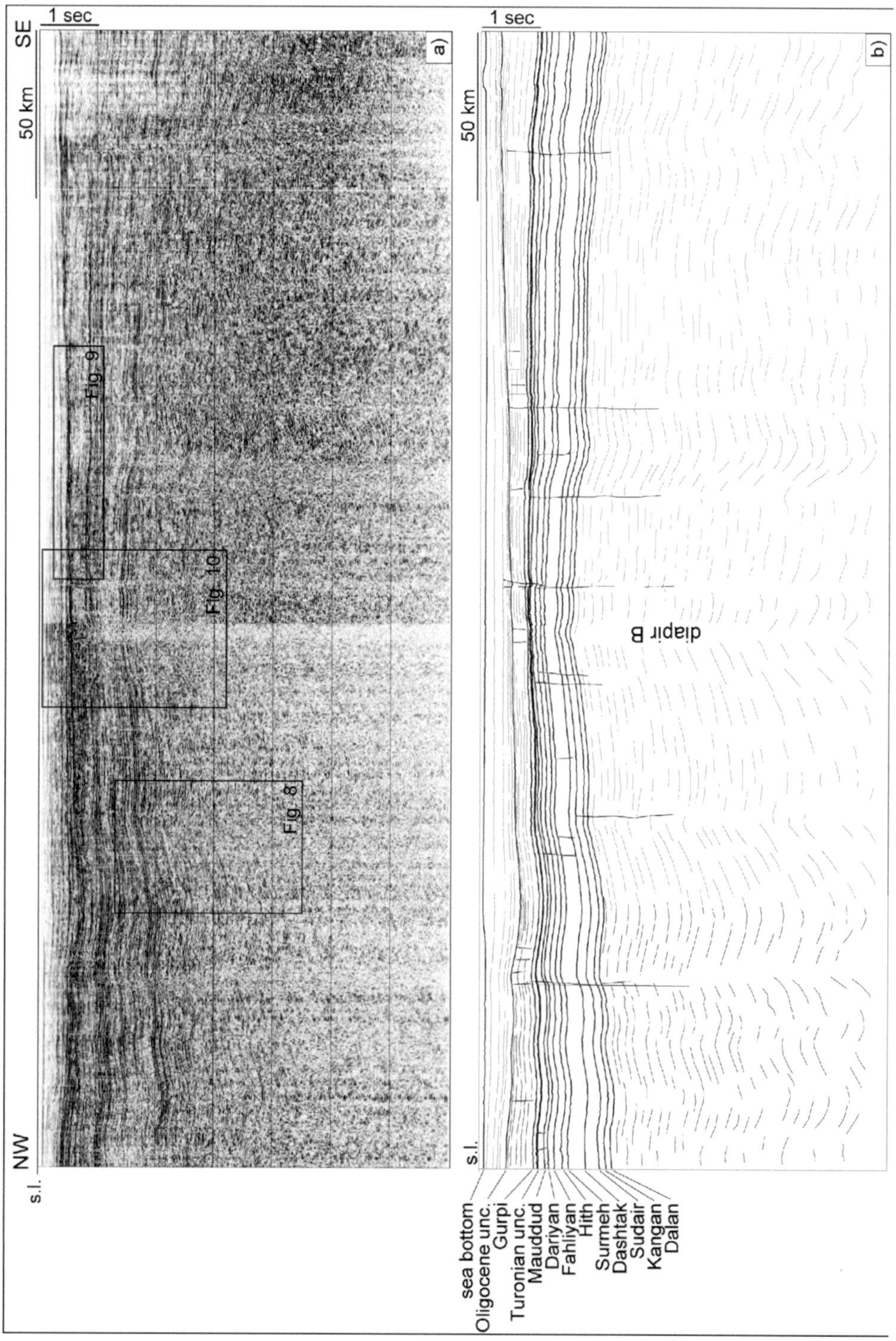

Fig. 6. Migrated seismic section across the Qatar-South Fars Arch (a) and interpreted line-drawing (b) (for location see Fig. 5)

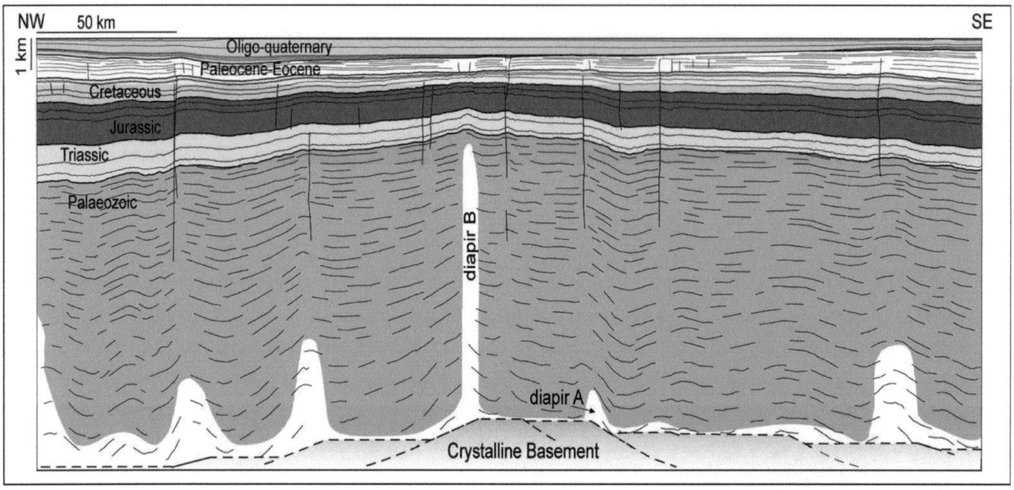

Fig. 7. Geologic section across the Qatar-South Fars Arch derived from depth conversion of the seismic line of Fig. 6 (for location see Fig. 5). Interpreted salt diapirs in white

Relatively abrupt changes in reflection dips (flexures) are visible along subvertical lineaments, particularly at the sides of the central 100-km-wide zone that can be considered the culmination of the Arch. These lineaments correspond to very small displacement faults in the Mesozoic section; there is no appreciable displacement of reflections in the deeper part. A seismic transparent zone, about 4 km in width, crosses the section from its base up to the Oligocene unconformity few kilometres west of the crest of the arch, and it is surrounded by concave-upwards reflections (figure 10). At its top, seismic reflections of the Oligocene unconformity and the underlying section appear to be arranged in a collapse-like structure 250–300 ms TWT thick. Other similar features are present on the crest of the Arch in our study area, but they appear as transparent zones beneath collapse structures below the Turonian and Oligocene unconformities with limited thickness (100–200 ms) and with an absence of lateral concave-upwards reflections.

The interpretation and correlation of the seismic horizons over the area covered by the seismic grid led to the construction of a time structural map of the top-Permian (figure 5). Because of the general parallelism of the overlying horizons (except the Oligocene unconformity and Neogene reflectors), the maps of the Mesozoic and Lower Tertiary reflectors will tend to mimic that of the top Permian (with reduced dips). The map of figure 5 highlights the distribution of salt structures indicated by the circular to elliptical, 5 to 20 km wide, folds of the structural surface on both sides of the Qatar–South Fars Arch. In particular, the map shows that the extent of the regional warping reaches a wavelength as great as 200 km and covers a large part of the area with diapirs: the folded area of the Qatar–South Fars Arch actually encompasses areas with original salt (sub)basins or their borders. In figure 11 a seismic section across a diapir in the Northern Gulf Salt Basin shows the typical features of these kinds of structures: diapirs with an average diameter of 10 km (5-15 km) piercing the stratigraphic sequence at different levels. The salt bodies appear to have vertical or sometimes slightly overhanging flanks, although the actual shape of steep diapirs flanks is not properly imaged by the conventional seismic acquisition and processing. The folded overburden is characterized by rim synclines surrounding the salt diapirs and downward increasing of the dip of the layers.

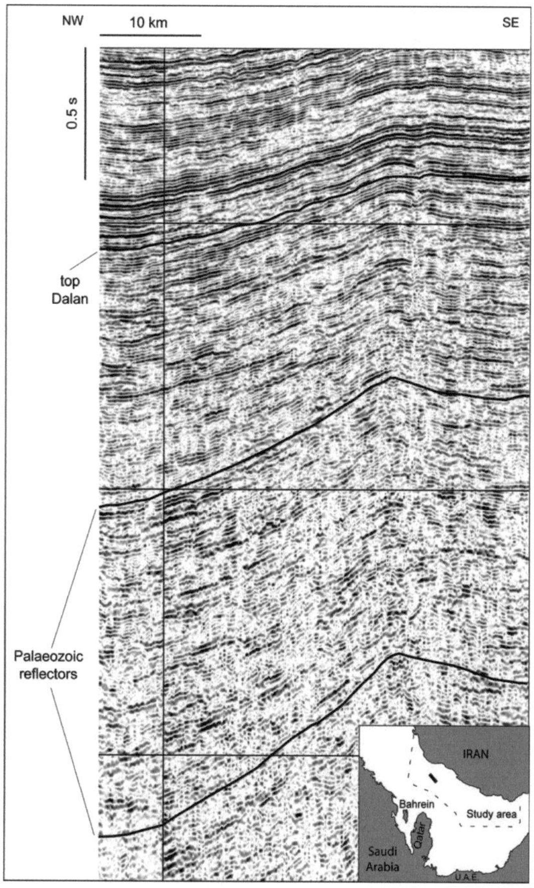

Fig. 8. Detailed view of part of the seismic section of figure 6 showing the progressive increase of the dip of the reflectors with depth caused by a continuous growth of the Qatar-South Fars structure. Inset map: location of the seismic section

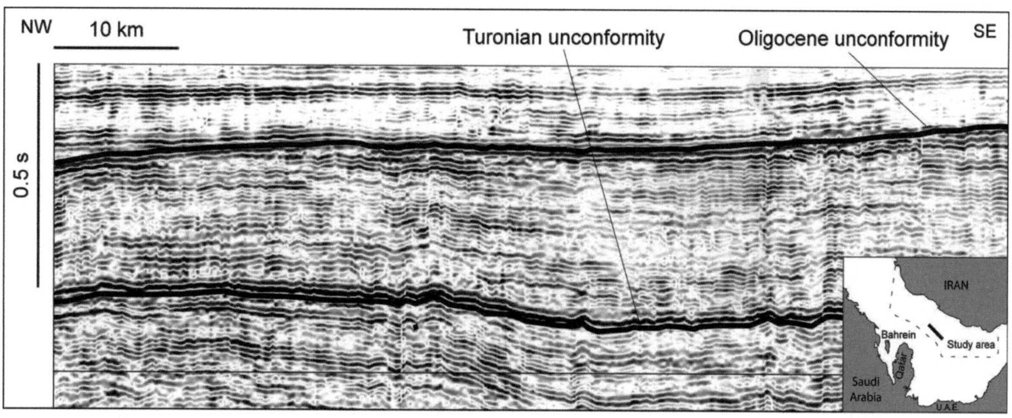

Fig. 9. Detailed view of part of the seismic section of figure 6 showing the Turonian and Oligocene unconformities. Inset map: location of the seismic section

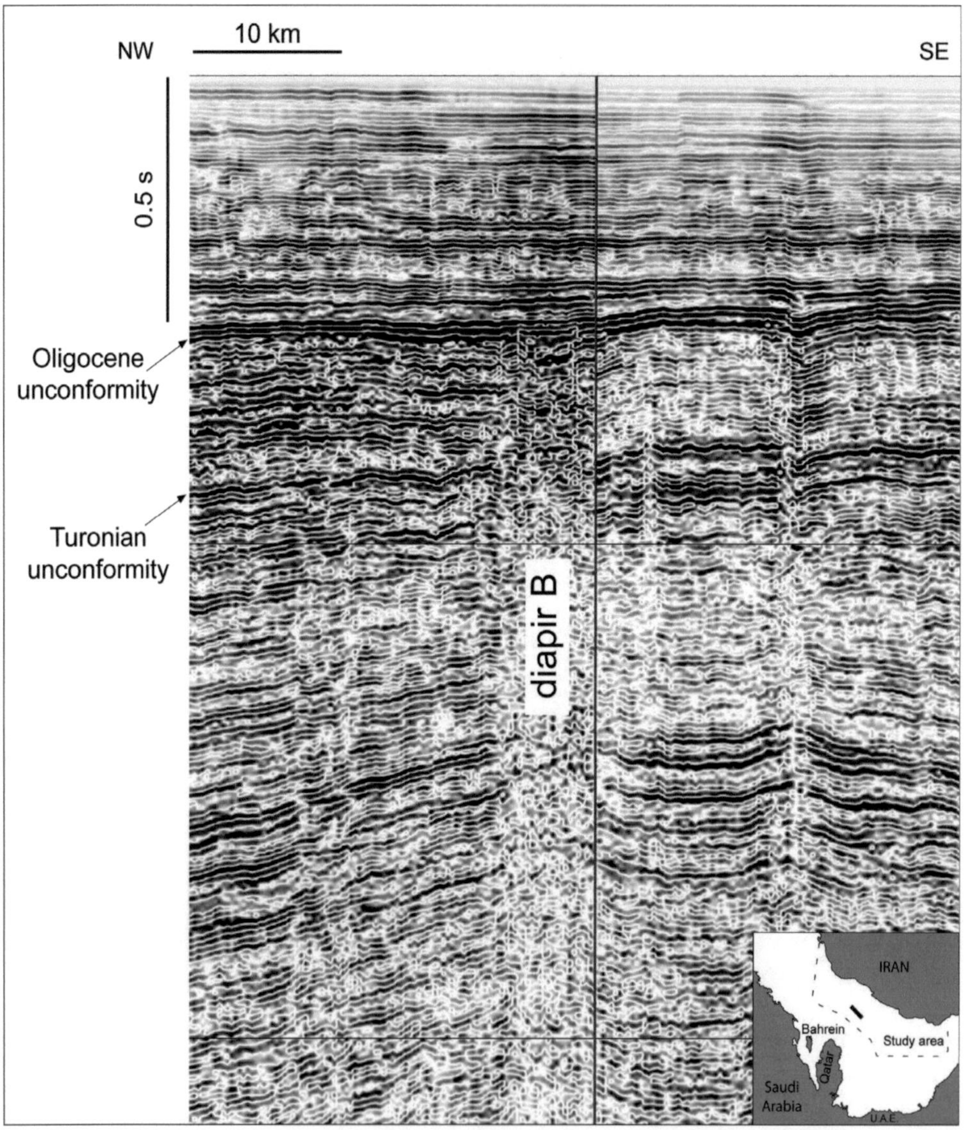

Fig. 10. Detailed view of part of the seismic section of figure 6 showing the diapir B. On the right (SE) of the section a minor fault zone cutting the Cretaceous-Lower Tertiary succession is visible. Inset map: location of the seismic section

The progressive flattening (restoration to a horizontal datum) of the horizons in the regional geological section of figure 7 shows the development of the Qatar-South Fars Arch with time. This process has been carried out under the assumption that the flattened horizons were the tops of originally flat rock units, which we consider a fairly good approximation in a 2D regional section within a basically undeformed intracratonic, passive margin setting. Compaction and erosional effects have not been considered due to lack of data; however, we think that flattening of the main unconformities has not introduced significant geometric errors at this scale because of their regional nature. In any case, flattening at unconformities aimed to show the relative geometric relationships between the under- and overlying strata in order to illustrate tectonic movements of the underlying section, regardless of the initial sub-horizontal or uniformly tilted geometry of the unconformity surface.

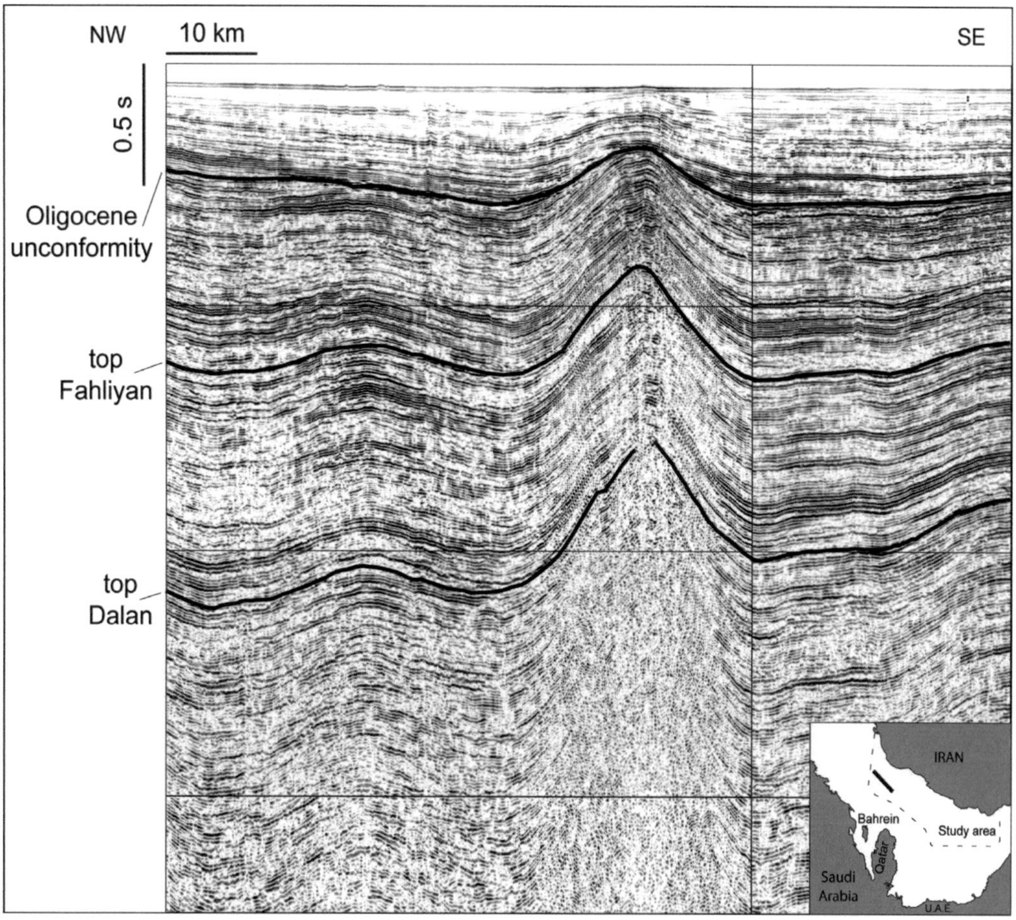

Fig. 11. Seismic section running across a diapir in Northern Gulf Salt Basin. Inset map: location of the seismic section

5. Geological interpretation

The regional geological section of figure 7 shows all of the main structural features occurring in the central Persian Gulf: the Qatar–South Fars Arch, salt domes and diapirs (although salt cores are not directly crossed by the section), flexuring and minor faulting of the sedimentary cover. The first-order structure is the Qatar–South Fars Arch regional anticline, several tens of kilometres wide. The Arch is a very gentle, broad warping of the sedimentary cover, an order of magnitude larger than the diapiric structures. The contrast between the wavelengths of the Arch and the salt diapirs and pillow deformations is remarkable. The whole, very thick sedimentary sequence (at least 14 km) is draped across the basement block to form the arch without apparent major faulting. The deformation related to the Qatar–South Fars Arch involves large parts of the Southern and Northern Gulf Salt basins. According to the currently accepted models of the Arabian Plate basement structure, the crystalline basement is assumed to be split by faults into several blocks with different structural elevations. A basement high is thus inferred in the core of the Qatar–South Fars Arch. This basement high is not visible on the seismic sections and could be located at a depth of at least 13–14 km (figure 7). In plan view, following the traditional approach, we can set the limits of the basement high beneath the Qatar–South Fars Arch by mapping the areas without salt-related phenomena. The salt structures closest to the crest of the Arch are the Dukhan anticline (elongated structure assumed to be a deep-seated salt anticline: Qatar General Petroleum Corporation and Amoco Qatar Petroleum Company 1991) to the west, and the circular diapir A to the east (figure 5). The diapir A is located in an area of rapidly changing reflector dips, suggesting the presence of a deep-seated sharp discontinuity at its base (figure 7). A basement fault would also account for the geometry of the Dukhan anticline (Edgell, 1996). The circular feature on the crest of the arch (figure 5) and corresponding to the seismic transparent zone bordered by concave reflectors visible in the regional section (figures 6, 7 and 10) may be interpreted as a small diapir (diapir B) with salt dissolution phenomena at its top (collapse-like structures).

The general progressive increase in structural relief of the Qatar–South Fars Arch with depth indicates that this structure acted as a long-lasting area of positive relief in the Phanerozoic, with continuous phases of uplift interspersed with short periods of more uniform subsidence. Seismic data reveal its continuous activity during the Palaeozoic, and well log correlations confirm its persistence as a structural high at least from the Jurassic until the mid-Cretaceous, with a reduction of the Cretaceous to Middle Jurassic succession at the top of the Arch of about 30% (Edison unpublished data).

At the present time the salt structures appear to be different on the two sides of the Qatar–South Fars Arch: they are more numerous and generally smaller (5–10 km to 20 km in few cases) in the Southern Gulf Salt Basin and fewer and larger (15–20 km) in the Northern Gulf Salt Basin. Moreover, in the Southern Basin most of them have a circular shape in plan view, while in the Northern Basin they mostly have an elliptical shape and are connected by elongated ridges interpreted as possible deep-seated salt walls (figures 5, 6). In some cases, there is no seismic evidence of salt below these domes, but their circular shape in plan view and growth history (inferred from the continuous downward increasing dips of the limbs) support the conclusion of a component of salt diapirism in the creation of these structures. In other cases, salt appears to pierce most of the sedimentary succession and push up the most recent sediments into positive bulges of the seabed. The dips of the limbs of the salt domes are about 3–4° (at top-Permian), but this could represent a minimum, approximate value because the processing of seismic data was not addressed with regard to imaging of the diapir limbs. An attempt of a quantitative or semi-quantitative evaluation of the amount of salt in the diapirs is hampered by their incomplete imaging, both in their lateral and bottom limits. No data on the amount of salt dissolution is available.

The abrupt changes in reflection dips along subvertical lineaments in the deeper part of the section are mostly lined up with overlying, small displacement faults in the Mesozoic section. This minor faulting can be interpreted as the effect of basement faults on the sedimentary cover, in the presence of an interlayered ductile level that tends to mechanically decouple the deformations (e.g. Higgins and Harris, 1997; Koyi and Petersen, 1993; Maurin, 1995; Nalpas and Brun, 1993; Vendeville et al., 1995; Withjack et al. 1989, 1990), the degree of decoupling being related to the thickness of the ductile layer (Withjack and Callaway, 2000). The most accentuated flexures/faults are localized in correspondence of diapir A (figures 5, 7) and symmetrically to the west of the crest of the Arch, suggesting deep-seated major faulting along these trends and therefore the location of the main high separating the northern and southern Infracambrian salt basins. Notwithstanding the presence of a number of these faults, it can be seen that the overall warping of the sedimentary sequence to form the Qatar-South Fars Arch has been accommodated by folding or draping without apparent major faulting.

In figure 12 the interpreted regional geological section is represented together with five flattenings related to the subsequent stratigraphic levels: Oligocene unconformity (circa 28–30 Ma), Turonian unconformity (circa 90 Ma), top Hith (Tithonian, circa145 Ma), top Dashtak (Norian, circa 210 Ma) and Dalan (Permian, circa 250 Ma). The analysis of each stage of the retro-deformation allows us to reconstruct the history of the Qatar-South Fars Arch during the following time intervals.

1. Palaeozoic (figure 12f): a continuous and symmetric growth of the structure can be inferred by the progressive increase of the dip of the reflectors with depth (figure 8), particularly noticeable toward the lower part of the section (Early? Palaeozoic).
2. Early Triassic-Norian (figure 12e): a considerable thickening of the sedimentary succession is visible, especially in the northwestern sector of the arch, indicating an asymmetric development of the anticline. This phenomenon can be explained by the reactivation during this period of the north-south trending structures located near the Qatar Peninsula, such as the Ghawar fault system (Wender et al. 1998), and possibly by a subsequent stronger halokinetic activity in the Northern Gulf Salt Basin.
3. Norian-Jurassic (figure 12d): a thickening of the sedimentary succession in the northwestern part of the arch is detectable, probably for the same reasons as already indicated for the early Triassic.
4. Early Cretaceous-Turonian (figure 12c): a significant growth of the Qatar-South Fars Arch is indicated by the thickening of the sedimentary succession symmetrically along the two limbs of the structure. The movements were practically continuous and uninterrupted, with only modest changes of velocity.
5. Turonian-Oligocene (figure 12b): during the Late Cretaceous and Paleogene the deformation activity decreases, as indicated by a minor thickening of the succession along the two limbs of the arch. The Coniacian probably represents an interval of stasis. At the top of this interval, several erosional truncations of the reflectors are visible (figure 9). A pre-Oligocene period of growth is shown by discordant convex reflectors truncated by the Oligocene unconformity.
6. Oligocene-Present (figures 7, 12a): the Qatar-South Fars Arch is not growing in the study area. The Zagros foredeep development is the only main deformational event, registered in the section by the thickening of the succession in the northwestern sector of the Arch, where the Zagros external deformational front is closer.

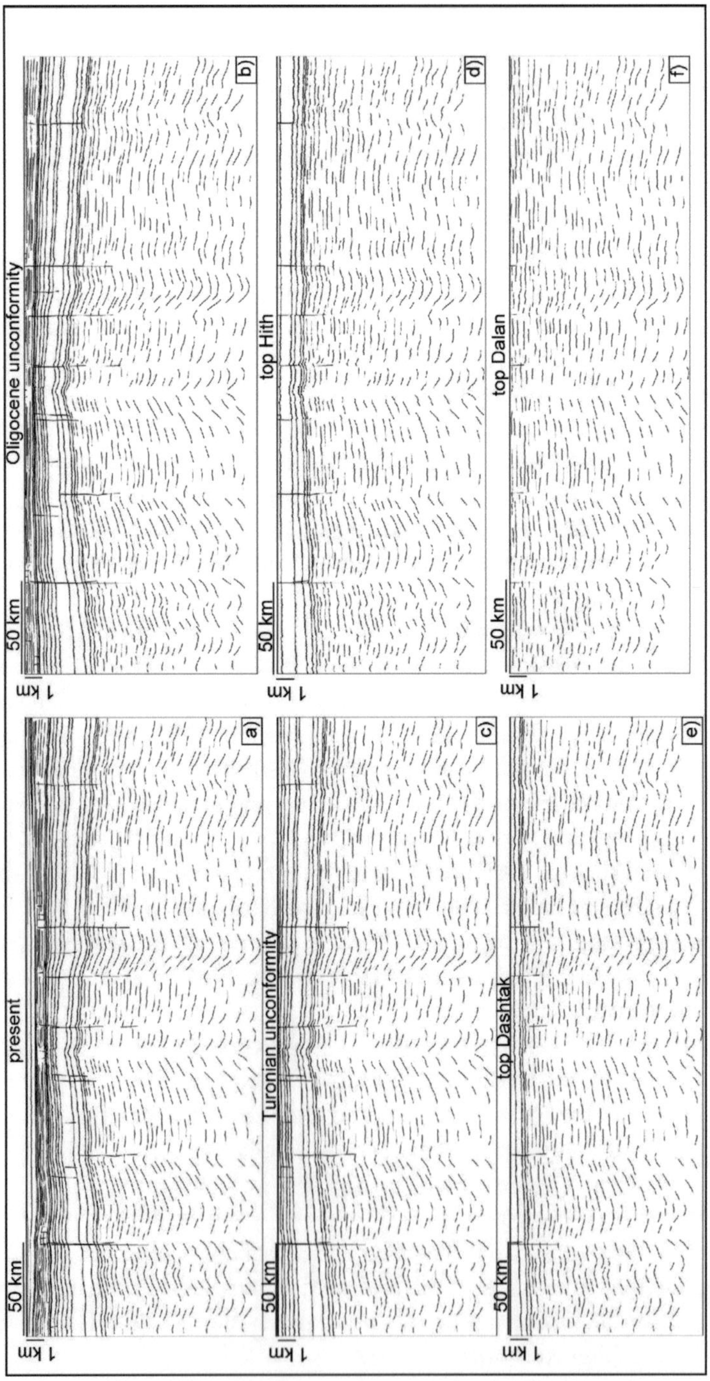

Fig. 12. Undeformed regional geologic section (a) across the Qatar-South Fars Arch (see figure 7). The same section flattened on the Oligocene unconformity (b), Turonian unconformity (c), top Hith (Tithonian) (d), top Dashtak (e) (top Norian) and top Dalan (f) (near top Permian), illustrating the progressive development of the structure

5.1 Possible mechanisms of development

As far as the origin and dynamics of this type of intracratonic deformation is concerned, compressive intraplate deformations due to far-field stresses acting at the plate boundaries and warping sectors of the plate interior (e.g. Nikishin et al. ,1993) or, alternatively, east-west (trans)tensional stresses reactivating the supposed basement high coring the Qatar–South Fars Arch, do not adequately explain the continuing activity of the arch over several hundreds of millions of years. These two mechanisms contrast with the accepted geodynamic reconstructions, which consider major episodes of divergence and convergence in a northeast-southwest direction, along the northwest–southeast margin of the Arabian Plate; these tectonic trends are almost perpendicular to the Qatar–South Fars Arch anticline (Sharland et al., 2001), which has a north-south to northeast–southwest axis. Moreover, the dimensions of the arch are about one order of magnitude greater that the other north-south trending "Arabian Structures" and, in the seismic interpretation, no major faults cutting the arch have been detected. The existence of widespread and long-lasting salt tectonics in the Persian Gulf suggests an alternative mechanism to basement block reactivation that can account for the peculiar structural features of the Qatar–South Fars Arch and its evolution and persistence through time. The following elements are at the heart of our model:

1. salt-related structures over the Qatar–South Fars Arch are basically absent or very limited in number compared with adjacent areas (Northern and Southern Gulf Salt basins): substantially, the small-diameter diapir B (the less than 4 km wide) is the only diapir interpreted on the top of the arch. This implies the presence of a general reduced thickness of the mobile Hormuz salt layer over a fault-controlled basement high, possibly intersected by small salt basins (the presence of a large belt of non-evaporitic facies is also possible). In fact, a strong positive relationship between salt thickness and diapir wavelength was found, for example, in the North Sea by Hughes and Davison (1993);
2. The abundance of diapirs in the Southern and Northern Gulf Salt basins. This means that widespread salt diapirism or rise (and possibly subsequent dissolution) could have removed an unquantified but large amount of the Hormuz evaporites from the base of the sedimentary succession;
3. There are good time correlations between the deformation of the arch and the diapiric phenomena. Specifically, Carruba et al. (2007) indicated the continuous rise during the Palaeozoic of a salt structure immediately to the north of the Qatar–South Fars anticline that is in good agreement with the continuous flexure observed during the same period in the regional section after the flattening of the top-Permian (figure 12f);
4. Approximately, the height difference between the Qatar–South Fars Arch and the adjoining regions (corresponding in our interpretation to the height difference between the crest and trough of the Qatar–South Fars mega- anticline) appears compatible with the salt Hormuz Formation thickness (probably more than 2500 m).

These conditions lead us to think that the basal salt withdrawal on a regional scale in the Infracambrian sub-basins could have contributed to the regional warping of the overburden over the main basement block system of the Qatar–South Fars Arch.

6. Analogue modelling

In order to verify if the Qatar–South Fars mega-anticline could have been caused by the Hormuz salt withdrawal at the sides of a basement structural high where the salt is lacking or very thin, a set of analogue experiments have been carried out in the laboratory, using a silicone putty layer in order to simulate the salt behaviour of the Hormuz Formation, and loose quartz sand to simulate the overlying competent sedimentary succession,. In our models we have tried to reproduce the regional geological section depicted in figure 7, which crosses the Qatar–South Fars Arch and has been reconstructed by seismic interpretation. The models do not aim to simulate the actual three-dimensional distribution of salt and basement structures, which is unknown in details, but are built to be appropriately scaled in two dimensions to give insights on the possibility of the mechanism proposed.

Models with silicone and sand have been developed by several groups to simulate the processes of gravity-induced deformation of a sedimentary pile above salt, with or without synchronous sedimentation (Brun and Fort, 2004; Brun and Mauduit, 2008; Cobbold et al., 1989; Cobbold and Szatmari, 1991; Fort et al., 2004a, 2004b; Ge et al., 1997; Gaullier et al., 1993; Mauduit and Brun, 1998; Mauduit et al., 1997a, 1997b; McClay et al., 1998; Vendeville et al., 1987; Vendeville and Jackson, 1992a, 1992b). Most of these analogue experiments simulating gravity-driven deformation use models in which brittle sediments are represented by sand, and salt deposits by silicone putty. These experiments have been extremely useful in understanding salt tectonics and associated structures as diapirs and folds. Detailed descriptions of the equipment, rheology of materials and analysis of models have already been presented in a number of previous studies (Faugère and Brun, 1984; Gaullier et al., 1993; Vendeville and Cobbold, 1987), which discuss scaling with regard to nature.

6.1 Materials and scaling

Our physical models use dry quartz sand to simulate brittle sedimentary rocks. The sand shows near-perfect Navier-Coulomb behaviour, with a grain size of 100–300 μm. It has negligible cohesion, an internal friction angle of about 34° ($\Phi i = 0.67$) and a bulk density of 1.3 g/cm^3. The dry oxides used to colour the sand did not significantly modify its rheological characteristics. Sand is the best analogue for most sedimentary rocks in the upper continental crust, which obey Mohr-Coulomb behaviour (Byerlee, 1978; Weijermars et al., 1993). Silicone putty has been commonly used as a Newtonian material with a strain-rate-dependent viscosity for modelling the ductile flow of incompetent décollement rocks such as evaporites (e.g. Cotton and Koyi, 2000; Weijermars et al., 1993). The Newtonian silicone polymer used in our experiments is polydimethyl-siloxane (PDMS, Dow Corning, SGM36) with a density of 965 kg/m^3 and a viscosity of 5×10^4 Pa s. For detailed rheology of this material and suitability as model analogue, see Weijermars (1986) and Weijermars et al. (1993). For a small-scale model to be representative of a natural example (a prototype), a dynamic similarity in terms of distribution of stresses, rheologies and densities between the model and the prototype is required (Hubbert, 1937; Ramberg, 1981). However, in gravity-driven physical experiments, where no external forces or displacements are applied, the modelling of ductile salt flow below a brittle substratum, with a mechanism of passive diapirism or gravity spreading, scaling of geometry, densities and rheology is sufficient for achieving dynamic similarity between the models and natural prototypes (Weijermars et al., 1993). In our experiments the scale ratios between models and prototype are $L^* = 2.9 \times 10^{-6}$ (1 cm in models represents approximately 3.5 km in nature), $g^* = 1$ (models and geological prototype both deformed in same gravity field), ρ^*_b (density ratio of the brittle overburdens)

= 0.5 (considering about 2600 kg m^{-3} as the average density of the sedimentary succession), ρ^*_d (density ratio of the basal ductile layers) = 0.4 (considering about 2200 kg m^{-3} as the average density of the salt). The difference between these last two ratios is considered acceptable as the density ratio is not the primary parameter affecting the rise of diapirs under normal gravity (Weijermars et al., 1993). Modelling the flow and extrusion of salt diapirs due to lithostatic pressure requires scaling the viscosity. Viscosity (μ) may be scaled according to the expression:

$$\mu^* = (\rho^*_b \times L^*) / \varepsilon^* \quad (1)$$

where μ^* is the viscosity ratio and ε^* is the strain-rate ratio between the model and prototype (respectively μ_m/μ_p and $\varepsilon_m/\varepsilon_p$).
Considering a displacement rate of the Hormuz salt of at least 1mm/yr (Bruthans et al., 2006) and a thickness of 3000 m, we can evaluate a strain rate of about 10^{-14} s^{-1}. Analysing the experiments, we can calculate an approximate strain rate of the silicone of about 10^{-5} s^{-1}. Considering that we used in the experiments silicone putty with a Newtonian viscosity of 5 x 10^4 Pa s, the viscosity of the salt is in the range of the right order of magnitude (~3 x 10^{19} Pa s). In fact, the effective viscosity of the salt in nature ranges from 10^{17} Pa s for small grain size and high temperature, up to 10^{20} Pa s for large grain size and low temperature, and the viscosity is strongly dependent upon grain size and moderately dependent upon temperature (van Keken et al., 1993). The ratio of the time ($t^* = t_m/t_p$) is not particularly meaningful because the diapiric rise of salt in the study area is a discontinuous process, with long periods of inactivity. In the experiments with sedimentation during the diapirism, the sedimentation rate is then adapted to the rate of diapiric rise. Table 1 shows the values of the main physical parameters for both the natural prototype and analogue materials, and the corresponding scaling factors, according to the approach of Weijermars et al. (1993).

Parameter	Model	Prototype	Scale factor
Lenght l (m)	1 x 10^{-2}	3.5 x 10^3	2.9 x 10^{-6}
Density overburden ρ_b (kg m^{-3})	1300	~2600	0.5
Density ductile layer ρ_d (kg m^{-3})	965	2200	0.4
Gravity g (m s^2)	9.81	9.81	1
Strain rate ε (s^{-1})	~10^{-5}	~10^{-14}	~10^9
Viscosity μ (Pa s)	5 x 10^4	~3x10^{19}	~10^{-15}

Table 1. Nature and model parameters

6.2 Experimental procedures

The experimental apparatus used in the models was a glass-sided sand-box, 90 cm long, 15 cm wide and 8 cm high, with a rigid structural rise (figure 13) placed in the middle of the glass basal plate. Friction along the sidewalls was reduced by coating them with graphite powder. A set of three experiments, with the same initial geometry, stratigraphy and boundary development conditions, were performed to verify the repeatability of the obtained results.

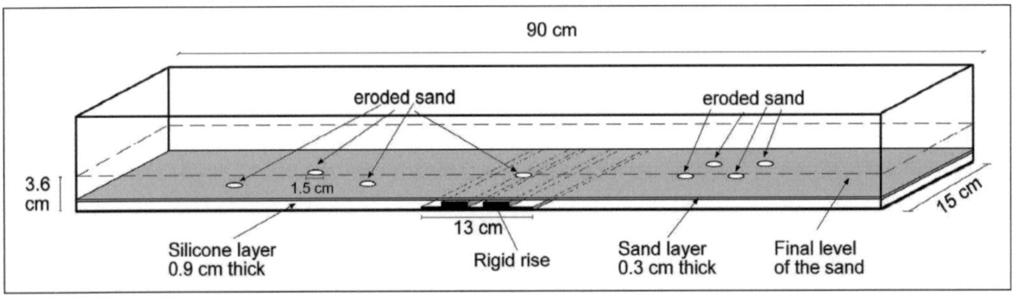

Fig. 13. Experimental apparatus and model settings used in the set of experiments to simulate the flow and withdrawal of a basal salt layer around a structural high, and the deformation of an overlying sedimentary succession

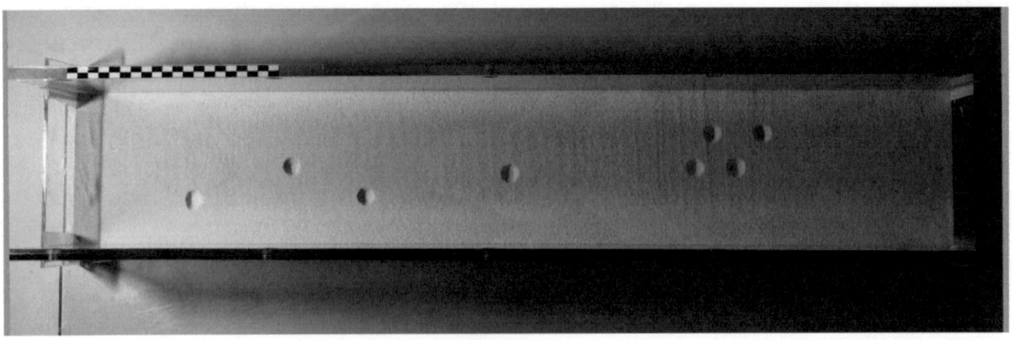

Fig. 14. Top view of the initial stage of one of the experiments of the second set. After the deposition of 0.3 cm of flat sand, the formation of 8 diapirs was induced by locally eroding the sand over 8 circular areas of 1.5 cm diameter

The models were prepared with a 0.9-cm-thick basal layer of silicone that covered with a thin layer (less than 1 mm) the central structural rise infilling the small graben between the two horsts (figure 13). After the initial deposition of 0.3 cm of flat sand, the formation of 8 diapirs was induced by locally eroding the sand, over 8 circular areas of 1.5 cm diameter (figures 13, 14). Subsequently, a total thickness of 2.7 cm of sand simulating the overburden was progressively deposited with a sedimentation rate adapted to the rate of diapiric rise. In particular, when the top of at least one of the eight diapirs reached the surface of the model, the next thin horizontal layer of 0.2 cm of sand was added to cover the diapir(s) arriving at the surface and all the others that had not completely pierced the overburden. In this way a passive diapiric mechanism was simulated in the experiments. At the end of the experiments (the total duration was about 72 hours), internal sections were cut away after the additional coverage of sand on the top and saturation with water. The final geometries of deformation and the mechanism of development of the deformation observed in all the three experiments were very similar.

6.3 Experimental results

The results of this set of experiments can be summarized as follows:

a. the nucleation of several smaller diapiric bodies and the contemporary sedimentation of the overburden produced the formation of a wide symmetric anticline with a subvertical axial plane and a hinge zone located over the rigid structural rise of the basement plate, where the basal silicone layer was very thin (figure 15);
b. the flow and withdrawal of the silicone affected the whole basal horizon and not just the areas surrounding the diapirs (rim synclines): the overall flow induced the differential lowering of the entire upper sand layer and the progressive formation of a passive anticline that was much wider than the underlying rigid block;
c. The growth of the anticline was progressive and clearly connected with the extrusion of the silicone putty in the diapirs;
d. at the end of the experiments (figure 15), the average dip of the two limbs of the accommodation passive anticline caused by the generalized flow of silicone towards the diapirs ranged from 1.9° to 0.5°, progressively decreasing from the base to the top of the sand layers. Moreover, the two limbs of the fold were affected by weak undulations due to the presence of the diapirs;
e. During the experiments, at the surface near some diapiric bodies (figure 16), slight rim synclines were observed, showing a ring-shaped lowering of the surface of less of 1 mm;
f. The average rate of diapir rise was about 0.04 cm/hr ($4x10^{-7}$ m s^{-1}), but not all the diapiric bodies reached the same vertical development (figure 16), probably due to the different initial conditions (the initial erosion of the first sand layer over 8 circular areas was not exactly the same);
g. Over the graben located in the middle of the rigid structural high, the initial surface erosion induced the growth of a small diapir, which, however, did not reach the surface at the end of the experiments.

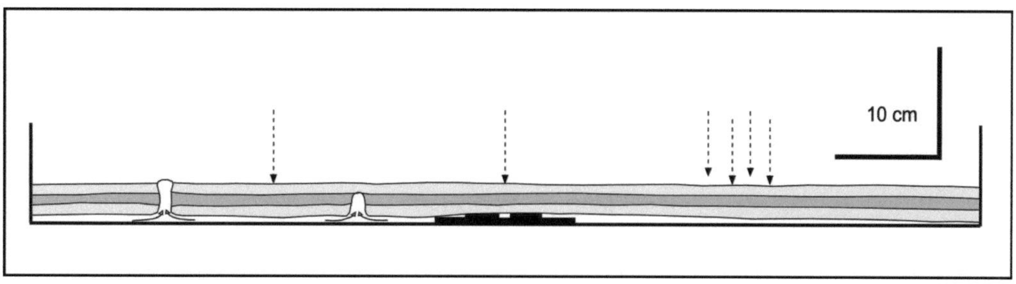

Fig. 15. Sketch of an internal section (5 cm from the removed glass wall) of the experiment. Two diapirs are visible at the left side of the model, while the dotted arrows indicate the position of the other diapirs not visible along this section. The extrusion of the silicone putty in the diapirs caused the formation of the overall passive accommodation anticline, whose limbs have an average dip ranging from 1.9° to 0.5°

Fig. 16. (a) Top view of two diapirs formed at the final stage of one of the experiments of the second set. A weak lowering of the surface around some diapir (rim synclines), not noticeable in the photograph, was observed during the experiments. (b) Detailed view of a section of a silicone diapir

6.4 Discussion

The experiments were focused on the possible regional consequences that a diffuse diapiric rise from a deep discontinuous ductile layer can induce in the overlying sedimentary succession (overburden). In the sand-box models the diapirs pierced the overburden with passive diapirism or downbuilding mechanisms caused by erosion. In fact, they rose continually with respect to surrounding strata and remained exposed while sediments accumulated around them and partially on their roofs. The small diapir formed over the graben located in the middle of the rigid structural high, which did not reach the surface at the end of the experiments suggests that smaller and thinner salt tectonic basins intersecting the basement high under the Qatar–South Fars Arch could have generated small-diameter diapirs (see diapir B in figures 7, 10). In synthesis, the models show that the extrusion of diapiric bodies from a deep salt layer around a structural high characterized by the absence or extremely reduced thickness of salt could induce the formation of a passive accommodation mega-anticline, whose geometry (extension, dip of the limbs, tightness) is comparable to the Qatar-South Fars warping structure (figure 17). In particular, the experiments performed in a sand-box 90 cm long (corresponding to more than 300 km in nature) indicate that the flow and withdrawal of the salt can operate over very great distances, inducing deformation (lowering) of the overburden. From this point of view, taking into account the different volumes involved and the peculiar geometries of the diapirism, the bulk flow of the salt connected to spreading phenomena and salt extrusion (both gravity-induced and with no external tectonic forces applied) is substantially the same: the salt migrates laterally for long distances and accumulates into the rising diapirs. The consequent withdrawal of basal salt causes the lowering of the overburden with respect to the area where salt is absent (the Qatar–South Fars basement high). The very low amount of these differential vertical movements along with the fact that no horizontal stresses are applied (e.g., basement extension, gravitational sliding of the overburden) is probably the reason for the absence of significant deformations in the thick sedimentary succession above the salt.

It should be noted that our model (figure 17) does not exclude the repeated basement reactivation of the supposed Qatar–South Fars Arch fault-bounded basement high (basement reactivation has been shown in other structures in the region not surrounded by Hormuz salt; see e.g. Wender et al., 1998) but, rather, points out the possible role of basal salt withdrawal as an important control factor in the warping of the sedimentary overburden above pronounced basement structures. With real data (2D seismic sections coverage) it was not possible to compare quantitatively the amount of salt removed from the base of the sedimentary succession with that accumulated into diapirs (and possibly dissolved), that is, to verify salt mass conservation. A good match between these two quantities would point to basal salt withdrawal as a main cause of overburden arching, while a non-correspondence would indicate other processes for overburden deformation. Despite the lack of quantitative assessment of mass conservation in the study area, however, the analogue experiments suggest that the proposed mechanism for the development of the Qatar–South Fars Arch is viable (figure 17).

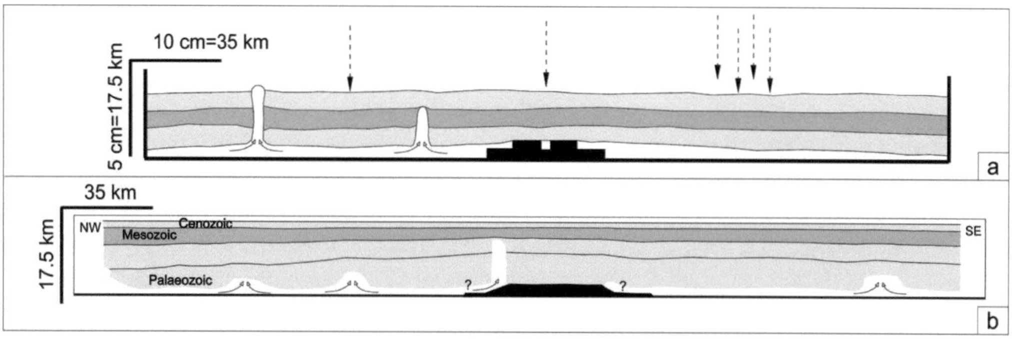

Fig. 17. Comparison between the physical model of figure 17 (a) and the interpreted geologic section (figure 7) across the Qatar-South Fars Arch (b) (vertical exaggeration x2). Both sections show mega-accommodation passive anticlines formed as a result of silicone-salt withdrawal at the sides of a basal structural high. The two anticlines have limbs with a similar average dip and the same extent with respect to the basement high and salt diapirs

7. Conclusions

The Qatar–South Fars Arch is the largest structure in the Persian Gulf province. It is a very gentle mega-anticline (more than 100 km wide and 300 km long), with a north-northeast to northeast orientation, that developed over a long time-span stretching from the Lower (?) Palaeozoic to the Oligocene. Although the origin of the Qatar–South Fars Arch long-lasting deformation has been explained so far (Edgell, 1992; Konert et al. 2001; Wender et al., 1998) by the repeated tectonic reactivation of Proterozoic faults bordering the basement high at the core of the Arch, no main faults bounding the overlying sedimentary succession of the Qatar–South Fars Arch are detectable in the available seismic sections. In addition to that, the following elements have been highlighted and must be considered.

1. One of the most important features of the Persian Gulf is the presence of several diapiric bodies that locally deform the sea bed or form islands (Edgell, 1996; Nasir et al., 2008). Salt diapirism created numerous elongate or circular (especially in the Southern Gulf Salt basin) salt domes. However, diapiric phenomena are practically absent over the top of the Qatar–South Fars Arch, possibly due to the lack or reduced (diapirism directly adjacent to the horst) thickness of the Hormuz Formation on an emerged

basement horst block during the deposition of the evaporites (late Proterozoic–early Palaeozoic).

2. The height difference between the crest and trough of the Qatar–South Fars anticline seems approximately compatible with the Hormuz Formation thickness (probably more than 2500 m).
3. A tectonic reactivation of the basement high coring the Qatar–South Fars Arch is hardly compatible with a north to northeast direction of the arch. In fact, the major episodes of compression and extension had roughly a northeast- southwest axis, which was perpendicular to the northwest-southeast margin of the Arabian plate and therefore not congruent with the development of the Qatar–South Fars Arch anticline.

All these elements and the widespread salt mobilization in the Gulf led us to think that the arching of the sedimentary cover over a basement high during several phases in the Phanerozoic might have been linked to the lateral withdrawal of the deep Infracambrian Hormuz salt, continuously feeding the several salt plugs and diapirs occurring to the northwest and southeast of the arch.

Drape folding of the succession by salt withdrawal into vertically rising diapirs has been investigated with physical models. The experimental results suggest that the mobilization of a discontinuous basal Hormuz salt layer may have induced the passive folding of the overlying sedimentary succession over a "rigid" Qatar–South Fars Arch core (a basement high or a non-evaporitic rock unit). In other words, the Qatar–South Fars Arch could be a mega-accommodation anticline due to salt withdrawal of an unevenly distributed salt layer.

The proposed salt-controlled evolution of the Qatar–South Fars Arch could account for its continuous, long-lasting evolution in a basically stable tectonic setting, where tectonic movements have been limited to periodic salt diapir growth and reactivation (Carruba et al., 2007).

8. Acknowledgments

The authors are grateful to NIOC and EDISON managements for granting to publish this study. The ideas expressed in this article are those of the authors and may not reflect any official position of the two companies. Many thanks also to the reviewers, who substantially helped improve the original manuscript.

9. References

Al Husseini, M.I. (2000). Origin of the Arabian plate structures: Amar collision and Najd rift. *GeoArabia*, Vol. 5, No. 4, pp. 527-542

Al Husseini, M.I. (2008). Middle East Geological Time Scale 2008. *GeoArabia*, Vol. 13, No. 4

Ala, M.A. (1974). Salt diapirism in Southern Iran. *American Association of Petroleum Geologists Bulletin*, Vol. 58, No. 9, (September 1974), pp. 1758-1770

Alsharhan, A.S., & Nairn, A.E.M. (1994). Geology and hydrocarbon habitat in the Arabian Basin: the Mesozoic of the State of Qatar. *Geologie en Mijnbouw*, Vol. 72, pp. 265-294

Alsharhan, A.S., & Nairn, A.E.M. (1997). *Sedimentary basins and petroleum geology of the Middle East*, Elsevier, Amsterdam

Beydoun, Z.R. (1991). Arabian Plate Hydrocarbon Geology and Potential – A Plate Tectonic Approach. *American Association of Petroleum Geologists Studies in Geology*, Vol. 33

Brun, J.P., & Fort, X. (2004). Compressional salt tectonics (Angolan Margin). *Tectonophysics*, Vol. 382, No. 3-4, (April 2004), pp. 129-150

Brun, J.P., & Mauduit, T. (2008). Rollovers in salt tectonics: the inadequacy of the listric fault model. *Tectonophysics*, Vol. 457, No. 1-2, (September 2008), pp. 1–11

Bruthans, J., Filippi, M., Geršl, M., Zare, M., Melková, J., Pazdur, A., & Bosák, P. (2006). Holocene marine terraces on two salt diapirs in Persian Gulf (Iran): age, depositional history and uplift rates. *Journal of Quaternary Science*, Vol.21, No. 8, (December 2006), pp. 843–857

Byerlee, J.D. (1978). Friction of Rocks, In: *Pure and Applied Geophysics. Contribution to Current Research in Geophysics,* J.D. Byerlee, & M. Wyss, (Eds), pp. 615–626, Birkhäuser Verlag, Basel

Carruba, S., Bertozzi, G., Perotti, C.R., & Rinaldi, M. (2007). Alcuni aspetti del diapirismo salino nel Golfo Persico. *Rendiconti della Società Geologica Italiana*, Vol. 4, pp. 188-190

Cobbold, P.R., Rossello, E., & Vendeville, B. (1989). Some experiments on interacting sedimentation and deformation above salt horizons. *Bulletin de la Société Géologique de France*, Vol. 3, pp. 453–460

Cobbold, P.R., & Szatmari, P. (1991). Radial gravitational gliding on passive margins. *Tectonophysics*, Vol. 188, No. 3-4, (March 1991), pp. 249-289

Cotton, J., & Koyi, H. (2000). Modeling of thrust front above ductile and frictional detachments: application to structures in the Salt Range and Potwar Plateau, Pakistan. *Geological Society of America Bulletin*, Vol. 112, No. 3, (March 2000), pp. 351-363

Dill, H.G., Nasir, S., & Al-Saad, H. (2003). Lithological and structural evolution of the northern sector of the Dukhan Anticline, Qatar, during the early Tertiary: with special reference to bounding surfaces of sequence stratigraphical relevance. *Geoarabia*, Vol. 82, pp. 201-226

Edgell, H.S. (1992). Basement tectonics of Saudi Arabia as related to oil field structures, In: *Basement tectonics 9,* M.H. Rickard, & al., (Eds), pp. 169-193, Kluwer Academic Publishers, Dordrecht

Edgell, H.S. (1996). Salt tectonism in the Persian Gulf, In: *Salt tectonics*, G.I. Alsop, D.J. Blundell, & I. Davison, (Eds), pp. 129-151, Geological Society of London Special Publication No. 100, The Geological Society Publishing House, Bath

Faugère, E., & Brun, J.P. (1984). Modélisation expèrimentale de la distension continentale. *Comptes Rendus Académie des Sciences Paris*, Vol. 299, pp. 365-370

Fort, X., Brun, J.P., & Chauvel, F. (2004a). Salt tectonics on the Angolan margin, synsedimentary deformation processes. *American Association of Petroleum Geologists Bulletin*, Vol. 88, No. 11, (November 2004), pp. 1523–1544

Fort, X., Brun, J.P., & Chauvel, F. (2004b). Contraction induced by block rotation above salt. *Marine and Petroleum Geology*, Vol. 21, No. 10, (December 2004), pp. 1281-1294

Gaullier, V., Brun, J.P., Guerin, G., & Lecanu, H. (1993). Raft tectonics: the effects of residual topography below a salt décollement. *Tectonophysics*, Vol. 228, No. 3-4, (December 1993), pp. 363-381

Ge H., Jackson M.P.A., & Vendeville, B. (1997). Kinematics and dynamics of salt tectonics driven by progradation. *American Association of Petroleum Geologists Bulletin*, Vol. 81, No. 3, (March 1997), pp. 398-423

Higgins, R.I., & Harris, L.B. (1997). The effects of cover composition on extensional faulting above re-activated basement faults: results from analogue modelling. *Journal of Structural Geology*, Vol. 19, No. 1 (January 1997), pp. 89-98

Hubbert, M.K. (1937). Theory of scale models as applied to the study of geologic structures. *Geological Society of America Bulletin*, Vol. 48, No. 10, (October 1937), pp.1459-1520

Hughes, M., & Davison, I. (1993). Geometry and growth kinematics of salt pillows in the southern North Sea. *Tectonophysics*, Vol. 228, No. 3-4, (December 1993), pp. 239-254

Husseini, M.I. (1988). The Arabian Infracambrian extensional system. *Tectonophysics*, Vol. 148, No. 1-2, (April 1988), pp. 93-103

Johnson, P.R., & Stewart, I.C.F. (1995). Magnetically inferred basement structure in central Saudi Arabia. *Tectonophysics*, Vol. 245, No. 1-2, (May 1995), pp. 37-52

Kent, P.E. (1958). Recent studies of South Persian salt plugs. *American Association of Petroleum Geologists Bulletin*, Vol. 42, No. 12, (December 1958), pp. 2951-2972

Konert, G., Afifi, A.M., Al-Hajri, S.A., & Droste, H.J. (2001). Paleozoic stratigraphy and hydrocarbon habitat of the Arabian Plate. *GeoArabia*, Vol. 6, No. 3, pp. 407-442

Koyi, H., & Petersen, K. (1993). Influence of basement faults on the development of salt structures in the Danish basin. *Marine and Petroleum Geology*, Vol. 10, No. 2 (April 1993), pp. 82-94

Mauduit, T., Gaullier. V., Guérin, G., & Brun, J.P. (1997a). On the asymmetry of turtle back growth anticlines. *Marine and Petroleum Geology*, Vol. 14, No. 7-8, (November-December 1997), pp. 763-771

Mauduit, T., Guérin, G., Brun, J.P., & Lecanu, H. (1997b). Raft tectonics, the effects of basal slope value and sedimentation rate on progressive deformation. *Journal of Structural Geology*, Vol. 19, No. 9, (September 1997), pp. 1219–1230

Mauduit, T., & Brun, J.P. (1998). Development of growth fault/rollover systems. *Journal of Geophysical Research*, Vol. 103, No. B8, pp. 18119-18130

Maurin, J.C. (1995). Drapage et décollement des series jurassiques sur la faille de détachement majeure du rift rhénan sud: implications sur la géométrie des depots syn-rifts oligocenes. *Comptes Rendus Académie des Sciences Paris*, Vol. 321, pp. 1025-1032

McClay, K.R., Dooley, T., & Lewis, G. (1998). Analog modeling of progradational delta systems. *Geology*, Vol. 26, No. 9, (September 1998), pp. 771-774

Nalpas, T., & Brun, J.P. (1993). Salt flow and diapirism related to extension at crustal scale. *Tectonophysics*, Vol. 228, No. 3-4, (December 1993), pp. 349-362

Nasir, S., Al-Saad, H., Alsayigh, A., & Weidlich, O. (2008). Geology and petrology of the Hormuz dolomite, Infra-Cambrian: Implications for the formation of the salt-cored Halul and Shraouh islands, Offshore, State of Qatar. *Journal of Asian Earth Sciences*, Vol. 33, No. 5-6, (August 2008), pp. 353-365

Nikishin, A.M., Cloetingh, S., Lobkovsky, L.I., Burov, E.B., Lankreijer, A.C. (1993). Continental lithosphere folding in Central Asia (part I): constraints from geological observations. *Tectonophysics*, Vol. 226, No. 1-4, (November 1993), pp. 59-72

Pollastro, R.M. (2003). Total petroleum systems of the Paleozoic and Jurassic, Greater Ghawar Uplift and adjoining provinces of Central Saudi Arabia and Northern Arabia-Persian Gulf. *United States Geological Survey Bulletin*, Vol. 2202-H, Available from http://pubs.usgs.gov/bul/b2202-h/

Qatar General Petroleum Corporation (QGPC), Amoco Qatar Petroleum Company (AQPC) (1991). Dukhan Field - Arabian Platform, In: *Structural traps V, Treatise of Petroleum Geology, Atlas of Oil and Gas Fields,* N.H. Foster, & E.A. Beaumont (Eds), pp. 103-120, American Association of Petroleum Geologists, Tulsa

Ramberg, H. (1981). *Gravity, Deformation and Earth's Crust,* (second edition), Elsevier, New York

Sharland, P.R., Archer, R., Casey, D.M., Davies, R.B., Hall, S.H., Heward, A.P., Horbury, A.D., Simmons, M.D. (2001). *Arabian Plate sequence stratigraphy.* GeoArabia Special Publication No.2, Gulf PetroLink, Manama

Stampfli, G.M., & Borel, G.D. (2004). The TRANSMED Transects in Space and Time: Constraints on the Paleotectonic Evolution of the Mediterranean Domain, In: *The TRANSMED Atlas: the Mediterranean Region from Crust to Mantle,* W. Cavazza, F. Roure, W. Spakman, G.M. Stampfli, P. Ziegler (Eds.), pp. 53-80, Springer-Verlag, Berlin Heidelberg

Sugden, W. (1962). Structural analysis and geometrical prediction for change of form with depth of some Arabian plains-type folds. *American Association of Petroleum Geologists Bulletin,* Vol. 46, No. 12 (December 1962), pp. 2213-2228

Talbot, C.J., & Alavi, M. (1996). The past of a future syntaxis across the Zagros. In: *Salt Tectonics,* G.I. Alsop, D.J. Blundell, & I. Davison (Eds.), pp. 89-109, Geological Society of London Special Publication No. 100, The Geological Society Publishing House, Bath

Talbot, C., Aftabi, P., Chemia, Z. (2009). Potash in a salt mushroom at Hormoz Island, Hormoz Strait, Iran. *Ore Geology Reviews,* Vol. 35, No. 3-4, (June 2009), pp. 317-332

van Keken, P.E., Spiers, C.J., Van den Berg, A.P., Muyzert, E.J. (1993). The effective viscosity of rocksalt: implementation of steady-state creep laws in numerical models of salt diapirism. *Tectonophysics,* Vol. 225, No. 4, (October 1993), pp. 457-476

Vendeville, B., & Cobbold, P.R. (1987). Glissements gravitaires synsédimentaires et failles normales listriques: modèles expérimentaux. *Comptes Rendus Académie des Sciences Paris,* Vol. 305, pp. 1313-1319

Vendeville, B., Cobbold, P.R., Davy, P., Brun, J.P., & Choukroune, P. (1987). Physical models of extensional tectonics at various scales. In: *Continental Extensional Tectonics,* M. Coward, J.F. Dewey, & P.L. Hancock (Eds.), pp. 95-107, Geological Society of London Special Publication No. 28, Blackwell Scientific Publication, Oxford

Vendeville, B., & Jackson, M.P.A. (1992a). The rise of diapirs during thin skinned extension. *Marine and Petroleum Geology,* Vol 9, No. 4, (August 1992), pp. 331–353

Vendeville, B., & Jackson, M.P.A. (1992b). The fall of diapirs during thin skinned extension. *Marine and Petroleum Geology,* Vol. 9, No. 4, (August 1992), pp. 354-371

Vendeville, B., Hongxing, G., & Jackson, M.P.A. (1995). Scale models of salt tectonics during basement-involved extension. *Petroleum Geoscience,* Vol. 1, No. 2, (May 1995), pp. 179-183

Weijermars, R. (1986). Flow behavior and physical chemistry of bouncing putties and related polymers in view of tectonic laboratory applications. *Tectonophysics,* Vol. 124, No. 3-4, (April 1986), pp. 325-358

Weijermars, R., Jackson, M.P.A., & Vendeville, B. (1993). Rheological and tectonic modeling of salt provinces. *Tectonophysics,* Vol. 217, No. 1-2, (January 1993), pp. 143-174

Wender, L.E., Bryant, J.W., Dickens, M.F., Neville, A.S., & Al-Moqbel, A.M. (1998). Paleozoic (Pre-Khuff) Hydrocarbon Geology of the Ghawar Area, Eastern Saudi Arabia. *GeoArabia*, Vol. 3, pp. 273-302

Withjack, M.O., Meisling, K., & Russel, L. (1989). Forced folding and basement-detached normal faulting in the Haltenbanken area, offshore Norway. In: *Extensional tectonics and stratigraphy of the North Atlantic margins*, pp. 567-575, A.J. Tankard, & H.R. Balkwill (Eds), American Association of Petroleum Geologists Memoir No.46, Tulsa

Withjack, M.O., Olson, J., & Peterson, E. (1990). Experimental models of extensional forced folds. *American Association of Petroleum Geologists Bulletin*, Vol. 74, No. 7, (July 1990), pp. 1038-1054

Withjack, M.O., & Callaway, S. (2000). Active Normal Faulting Beneath a Salt Layer: An Experimental Study of Deformation Patterns in the Cover Sequence. *American Association of Petroleum Geologists Bulletin*, Vol. 84, No. 5, (May 2000), pp. 627-651

Ziegler, M.A. (2001). Late Permian to Holocene Paleofacies Evolution of the Arabian Plate and its Hydrocarbon Occurrences. *GeoArabia*, Vol. 6, pp. 445-504